Conversion Factors to SI …

English	*SI*	*SI symbo*		
		Power, Heat …		
horsepower	kilowatt	kW	0.7457	1.341
foot-pound/sec	watt	W	1.356	0.7376
Btu/hour	watt	W	0.2929	3.414
		Pressure		
pound/square inch	kilopascal	kPa	6.895	0.1450
pound/square foot	kilopascal	kPa	0.04788	20.89
feet of H_2O	kilopascal	kPa	2.983	0.3352
inches of Hg	kilopascal	kPa	3.374	0.2964
		Temperature		
Fahrenheit	Celcius	°C	$5/9(°F - 32)$	$9/5 \times °C + 32$
Fahrenheit	kelvin	K	$5/9(°F + 460)$	$9/5 \times K - 460$
		Velocity		
foot/second	meter/second	m/s	0.3048	3.281
mile/hour	meter/second	m/s	0.4470	2.237
mile/hour	kilometer/hour	km/h	1.609	0.6215
		Acceleration		
foot/second squared	meter/second squared	m/s^2	0.3048	3.281
		Torque		
pound-foot	newton-meter	$N \cdot m$	1.356	0.7376
pound-inch	newton-meter	$N \cdot m$	0.1130	8.85
		Viscosity, Kinematic Viscosity		
pound-sec/square foot	newton-sec/square meter	$N \cdot s/m^2$	47.88	0.02089
square foot/second	square meter/second	m^2/s	0.09290	10.76
		Flow Rate		
cubic foot/second	cubic meter/second	m^3/s	0.02832	35.32
cubic foot/second	liter/second	L/s	28.32	0.03532

Mechanics and Durability of Solids, Volume I

MIT–PRENTICE HALL SERIES ON CIVIL, ENVIRONMENTAL, AND SYSTEMS ENGINEERING

Titles in the Series

Feniosky Peña-Mora, Carlos E. Sosa, and Sean D. McCone
Introduction to Construction Dispute Resolution
0-13-047089-9

Franz-Josef Ulm and Olivier Coussy,
Mechanics and Durability of Solids, Volume I
0-13-047957-8

Jerome J. Connor
Introduction to Structural Motion Control
0-13-009138-3

Mechanics and Durability of Solids, Volume I

Solid Mechanics

Franz-Josef Ulm
Massachusetts Institute of Technology

Olivier Coussy
Laboratoire Central des Ponts et Chaussèes

Pearson Education, Inc., *Upper Saddle River, New Jersey 07458*

Library of Congress Data Available

Vice President and Editorial Director, ECS: *Marcia J. Horton*
Acquisitions Editor: *Laura Fischer*
Editorial Assistant: *Erin Katchmar*
Vice President and Director of Production and Manufacturing, ESM: *David W. Riccardi*
Executive Managing Editor: *Vince O'Brien*
Managing Editor: *David A. George*
Production Editor: *Rose Kernan*
Director of Creative Services: *Paul Belfanti*
Creative Director: *Carole Anson*
Art Director: *Jayne Coate*
Cover Designer: *Bruce Kenselaar*
Art Editor: *Greg Dulles*
Manufacturing Manager: *Trudy Pisciotti*
Manufacturing Buyer: *Lisa McDowell*
Senior Marketing Manager: *Holly Stark*

Upper Saddle River, New Jersey 07458

Printed in the United States of America
10 9 8 7 6 5 4 3 2 1

ISBN 0-13-0479578

Pearson Education Ltd., *London*
Pearson Education Australia Pty. Ltd., *Sydney*
Pearson Education Singapore, Pte. Ltd.
Pearson Education North Asia Ltd., *Hong Kong*
Pearson Education Canada, Inc., *Toronto*
Pearson Educacion de Mexico, S.A. de C.V.
Pearson Education—Japan, *Tokyo*
Pearson Education Malaysia, Pte. Ltd.
Pearson Education, Inc., Upper Saddle River, New Jersey

Contents

Preface, xiii

I Deformation and Strain 1

1 Description of Finite Deformation 2
1.1 The Continuum Model 2
1.2 Deformation Gradient 3
1.3 Transport Formulas 6
1.3.1 Volume Transport 6
1.3.2 Transport of an Oriented Material Surface 8
1.4 Finite Strain Measures 9
1.4.1 Cauchy Dilatation Tensor 10
1.4.2 Green–Lagrange Strain Tensor 12
1.4.3 Linear Dilatation and Distortion 14
1.5 Strain-Displacement Relation 16
1.6 Training Set: Single Shear 17
1.6.1 Deformation Gradient 17
1.6.2 Transport Formulas for Volume and Oriented Surface 18
1.6.3 Finite Strain Measures 19
1.6.4 Linear Dilatation and Distortion 20
1.7 Appendix Chapter 1: Tensor Notation 20
1.7.1 Scalar: Zero-Order Tensor 20
1.7.2 Vector: First-Order Tensor 20
1.7.3 Second-Order Tensor and Matrix Representation 21

1.8 Problem Set: Finite Bending Deformation of a Beam 22
1.8.1 Finite Deformation 24
1.8.2 Infinitesimal Deformation 26

2 Infinitesimal Deformation 30
2.1 The Hypothesis of Infinitesimal Deformation 30
2.1.1 The Hypothesis 30
2.1.2 Linearization of Transport Formulas 32
2.2 The Linearized Strain Tensor 33
2.2.1 Geometric Interpretation: Relative Length and Angle Variation 33
2.2.2 Polar Decomposition in the Linear Theory (Optional) 35
2.3 The Mohr Strain Plane 39
2.4 Extensometer Measurements: Strain Gage Rosette 42
2.5 Training Set: Torsion of a Cylinder 44
2.5.1 Kinematically Admissible Displacement Field 45
2.5.2 Linearized Strain Tensor 46
2.5.3 Strain Rosette Measurements and Mohr Circle 47
2.5.4 Torsion and Extension 48
2.6 Appendix Chapter 2: Differential Operators 50
2.6.1 Orthonormal Cartesian Coordinates 50
2.6.2 Cylinder Coordinates 51
2.6.3 Spherical Coordinates 52

II Momentum Balance, Stress, and Stress States 55

3 Momentum Balance and Stress 56
3.1 The Hypothesis of Local Contact Forces 56
3.1.1 Body Forces 57
3.1.2 Surface Forces 57
3.2 Momentum Balance 57
3.3 The Cauchy Stress Tensor 59
3.3.1 Action–Reaction Law 59
3.3.2 The Tetrahedron Lemma 60
3.3.3 Relation with Force and Moment Resultants: Reduction Formulas 63
3.4 Equation of Motion 64
3.4.1 The Dynamic Resultant Theorem and the Local Equilibrium Equation 64
3.4.2 The Dynamic Moment Theorem and the Symmetry of the Stress Tensor 67
3.4.3 Principal Stresses and Principal Stress Directions 68
3.5 Training Set: Pressure Vessel Formula 68
3.5.1 Cylinder Stress Components and Boundary Conditions 68
3.5.2 Momentum Balance in Cylinder Coordinates 70
3.5.3 One Statically Admissible Stress Field 71

4 Stress and Stress States **74**
4.1 Statically Admissible Stress Field 74
4.2 Practical Stress Plane and Stress Space Quantities 75
4.2.1 Surface Tension and Shear Stress 75
4.2.2 Principal Stresses 76
4.2.3 The Hydrostatic Pressure and Other Stress Invariants 76
4.2.4 Stress Deviator 77
4.3 The Mohr Stress Plane 79
4.4 Admissible Stress States and Strength Criterion 84
4.4.1 The Tresca Criterion 85
4.4.2 The Mohr–Coulomb Criterion 87
4.5 Training Set: Excavation Pit 90
4.5.1 Statically Admissible Stress Fields 90
4.5.2 A Critical Excavation Height 92
4.6 Problem Set: Triaxial Test and Limit Load of a Circular Foundation 94
4.6.1 Part I: Triaxial Test 95
4.6.2 Part II: Limit Load of a Circular Foundation 98
4.7 Problem Set: Statically Admissible Stress Fields of a Beam 100
4.7.1 Pure Bending 101
4.7.2 Combined Bending and Axial Load 103
4.8 Problem Set: Why Sandcastles Fall 105
4.8.1 Dry Sandpile 106
4.8.2 Humid Sandpile 109

III Elasticity and Elasticity Bounds **111**

5 Thermoelasticity **112**
5.1 The Necessity of Material Laws 112
5.2 The Notion of Elasticity and Elasticity Potential 113
5.2.1 1D Think Model 113
5.2.2 1D Thermodynamics of Thermoelasticity 115
5.2.3 Linear 3D Thermoelasticity 118
5.3 Isotropic Linear Thermoelastic Material Properties 120
5.3.1 Bulk Modulus 121
5.3.2 Shear Modulus 122
5.3.3 Thermal Dilatation Coefficient 123
5.3.4 Young's Modulus and Poisson Ratio 124
5.3.5 An Instructive Exercise: Dimensional Analysis of the Boussinesq Problem 125
5.4 Direct Solving Methods of Elastic Problems 127
5.4.1 Small Perturbation 127
5.4.2 Governing Equations 128
5.4.3 Theorem of Superposition 129
5.4.4 Displacement Method 129
5.4.5 Stress Method 131
5.5 Training Set: From the Cylinder Tube to Deep Tunneling 131

5.5.1 Elastic Equilibrium of a Cylinder Tube 132
5.5.2 Case Study: The Pressure Vessel Revisited. Elastic Yield Limit 136
5.5.3 Case Study: Deep Tunneling 137
5.6 Problem Set: The Boussinesq Problem—Refined Analysis 140
5.6.1 Dimensional Analysis 140
5.6.2 Displacement Method in Spherical Coordinates 142
5.6.3 The Uniform Pressure Disk Problem 145
5.7 Problem Set: Elastic Bending of a Beam 146
5.7.1 Direct Solving Method: Stress Approach 147
5.7.2 Stress-Based Variational Method 149

6 The Theorem of Virtual Work and Variational Methods in Elasticity 151
6.1 Theorem of Virtual Work 151
6.1.1 The Theorem 151
6.1.2 Application to Heterogeneous Material Systems: The Hill Lemma 153
6.1.3 Potential Energy and Complementary Energy 156
6.1.4 Convexity 157
6.2 Variational Method I: Theorem of Minimum Potential Energy 158
6.2.1 1D Think Model 158
6.2.2 3D Displacement-Based Variational Formulation 160
6.2.3 Displacement Method: Application to Linear Isotropic Elastic Material Systems 162
6.2.4 Relation with Displacement-Based Finite Element Method . . 165
6.3 Variational Method II: Theorem of Minimum Complementary Energy 165
6.3.1 1D Think Model 165
6.3.2 3D Stress-Based Variational Formulation 168
6.3.3 Stress Method: Application to Linear Isotropic Elastic Material Systems 169
6.3.4 Stress Method in Linear Elastic Structural Mechanics 171
6.4 Upper and Lower Bounds: Clapeyron's Formula 171
6.5 Training Set: Effective Modulus of a Heterogeneous Material System 173
6.5.1 Clapeyron's Formula 174
6.5.2 Upper Bound: Minimum Potential Energy 175
6.5.3 Lower Bound: Minimum Complementary Energy 177
6.5.4 The Voigt–Reuss Elasticity Bounds 179
6.6 Appendix Chapter 6: A Thermodynamic Argument 180
6.7 Problem Set: Water Filling of a Gravity Dam 182
6.7.1 First Approximation: Linear Triangle Element 183
6.7.2 Second Approximation: Higher-Order Polynomials 187
6.7.3 Finite Element Approximation 189
6.8 Problem Set: Elastic Settlement Bounds below a Circular Foundation 189
6.8.1 Statically Admissible Stress Field 191
6.8.2 Lower Energy Bound–Upper Displacement Bound 191

6.8.3 Upper Energy Bound–Lower Displacement Bound 194
6.9 Problem Set: Torsion of an Elastic Heterogeneous Cylinder 194
6.9.1 Equivalent Homogeneous Sample. 196
6.9.2 Heterogeneous Sample 197
6.9.3 Composite Cylinder Model 200
6.10 Problem Set: Elasticity Bounds of Microflexural Structures (MEMS Type) 202
6.10.1 The Equivalent Homogeneous Sample 204
6.10.2 Heterogeneous Sample 204
6.10.3 Application to Two-Layer Beam 207

IV Plasticity and Yield Design 209

7 1D Plasticity: An Energy Approach 210
7.1 Ideal Plasticity 210
7.1.1 Friction Element 210
7.1.2 1D Think Model of Ideal Elastoplasticity 212
7.1.3 Energy Considerations: The Principle of Maximum Plastic Work 213
7.1.4 1D Thermodynamics of Ideal Plasticity 215
7.2 Hardening Plasticity 218
7.2.1 1D Think Model of Hardening Plasticity 218
7.2.2 An Instructive Exercise: An Alternative 1D Think Model of Hardening Plasticity 221
7.2.3 Energy Considerations: The Frozen Energy 224
7.2.4 1D Thermodynamics of Hardening Plasticity 226
7.2.5 Kinematic versus Isotropic Hardening 228
7.3 Viscoplasticity 230
7.3.1 1D Viscoplasticity Model 231
7.3.2 Apparent Rate Effects 234
7.3.3 1D Thermodynamics 234
7.3.4 Normal Dissipative Mechanism 235
7.4 Training Set: 1D Cyclic Plasticity 236
7.4.1 The Stéfani Model 237
7.4.2 Energy Approach: The Stéfani Model 239
7.4.3 Model Response to 1D Cyclic Loading 242
7.5 Problem Set: The Three-Truss Analogy 245
7.5.1 Part I: Loading 248
7.5.2 Part II: Unloading 252
7.5.3 Equivalent Macroscopic Model 255
7.6 Problem Set: Creep Hesitancy 258
7.6.1 Elasticity Domain 260
7.6.2 Constitutive Equations 261
7.6.3 Instantaneous and Time-Dependent Deformation 262

8 **Plasticity Models** 265
8.1 Elements of 3D Ideal Plasticity Models . . . 265
8.1.1 Ideal Plasticity Criterion . . . 266
8.1.2 Flow Rule of Ideal Plasticity . . . 267
8.1.3 Von–Mises Plasticity . . . 267
8.1.4 Drucker–Prager Plasticity . . . 271
8.1.5 Energy Considerations: Plastic Dissipation and Principle of Maximum Plastic Work . . . 276
8.2 Thermodynamics of 3D Hardening Plasticity . . . 277
8.2.1 Elements of 3D Thermodynamics of Irreversible Processes . . 278
8.2.2 Linear and Isotropic Thermoelastoplasticity . . . 279
8.2.3 Inversion of State Equations . . . 279
8.2.4 Complementary Evolution Laws: Flow Rule and Hardening Rule . . . 281
8.3 Hardening Plasticity Models . . . 283
8.3.1 Loading Function and Consistency Condition . . . 283
8.3.2 Isotropic and Kinematic Hardening . . . 284
8.3.3 Hardening Variables . . . 285
8.3.4 Example: Drucker–Prager Criterion with Isotropic and Kinematic Hardening . . . 286
8.3.5 Incremental State Equation: Tangent Modulus . . . 287
8.4 Training Set: The Cambridge (or Cam–Clay) Model . . . 289
8.4.1 Cam–Clay Type of Loading Surface . . . 289
8.4.2 Cam–Clay Type of Flow Rule . . . 291
8.4.3 Cam–Clay Type of Hardening Rule . . . 292
8.4.4 Thermodynamic Consistency . . . 294
8.4.5 Cam–Clay Type of Viscoplasticity Model . . . 295
8.5 Problem Set: Strength Estimates by Microindentation (Microhardness) . . . 297
8.5.1 Triaxial Stress State and Von–Mises Strength Criterion . . . 298
8.5.2 Maximum Force . . . 299
8.5.3 Summary: Model versus Experiment . . . 302
8.6 Problem Set: Thin-Walled Cylinder Subjected to Tension and Torsion 302
8.6.1 Loading Surface: Von–Mises Plasticity . . . 304
8.6.2 Plastic Hardening . . . 304
8.6.3 Elastic Ideal Plastic Loading . . . 306
8.7 Problem Set: Champagne Method . . . 310
8.7.1 Deformation and Strain . . . 312
8.7.2 Elasticity . . . 313
8.7.3 Elastic Strength Limit . . . 314
8.7.4 Plasticity . . . 315

9 **Limit Analysis and Yield Design** 321
9.1 Elements of Limit Analysis . . . 321
9.1.1 Plastic Collapse Load . . . 322
9.1.2 Plastic Collapse Kinematics . . . 324

9.2 Lower Limit Theorem 327
9.2.1 The Theorem 328
9.2.2 Static Approach from Inside: Constructing the Domain of Safe Loads 332
9.3 Upper Limit Theorem 334
9.3.1 The Theorem 335
9.3.2 Kinematic Fields Involving Surfaces of Discontinuity 339
9.3.3 Kinematic Approach from Outside versus Static Approach from Inside 340
9.3.4 A Classical Exercise: The Excavation Pit Revisited from Outside 340
9.4 Application to Structural Elements 348
9.4.1 Static Approach from Inside Applied to Structural Elements 348
9.4.2 Kinematic Approach from Outside: Plastic Hinge Design . . 350
9.5 Training Set: Strength Domain of Fiber-Reinforced Composite Materials 354
9.5.1 Lower and Upper Bounds 354
9.5.2 Voigt–Reuss Strength Bounds 355
9.5.3 An Improved Lower Bound: Enriched Statically Admissible Stress Field 358
9.5.4 Refinement with Interface Strength Criterion 364
9.6 Appendix Chapter 9: Dissipation Functions 366
9.6.1 Tresca Criterion 367
9.6.2 Von–Mises Criterion 367
9.6.3 Mohr–Coulomb Criterion 367
9.6.4 Drucker–Prager Criterion 368
9.7 Problem Set: Section Strength for Combined Bending and Axial Force 368
9.7.1 Improved Lower Bound 369
9.7.2 Upper Bound 370
9.8 Problem Set: Design Thickness of a Pressure Vessel 372
9.8.1 Lower Bound 374
9.8.2 Upper Bound 375

Appendix: Further Reading Volume I **379**

Index **382**

Preface

This textbook is the first of two volumes dealing with Mechanics and Durability of Solids. It provides an introduction to continuum mechanics and material modeling of engineering materials based on first energy principles. The second volume extends the approach to fracture and durability mechanics of solids. The overall theme of both volumes is a unified 'mechanistic' approach that uses energy concepts for modeling a large range of engineering material behavior, while generating the basis of a common language with other core disciplines in engineering sciences.

The first volume is composed of four parts: (I) Deformation and Strain; (II) Momentum Balance, Stress and Stress States; (III) Elasticity and Elasticity Bounds; (IV) Plasticity and Yield Design. Parts I and II introduce the two pillars of continuum mechanics and focus on geometrical and physical interpretation of strain and stresses, starting with the finite deformation theory, which is consistently linearized. Part III is dedicated to non-dissipative material behavior, with a focus on thermoelasticity and variational methods in elasticity and its application to heterogeneous material systems. Part IV starts with 1D plasticity, introducing ideal plasticity, hardening plasticity, and associated energy transformations. It is within the energy approach that the 1D Think models are extended to three dimensions, introducing the notion of associated and non-associated plasticity. Finally, the plastic collapse is introduced, leading to the development of the upper and lower bound theorem of limit analysis as bounds of the maximum admissible dissipation at plastic collapse of material systems and structures.

From the onset, our approach to writing this textbook was nourished by the multicultural flavor of our educational backgrounds: the pragmatism of the traditional German Engineering Mechanics education and the modern mathematical eloquence of "La Mécanique Rationelle." In such an endeavor, the need for a common language is critical. We developed this language over the years with our students on blackboards through 1D Think Models in France, Germany, Brazil, and finally at M.I.T. The outcome of this cultural adventure is this textbook; it is situated at the interface of Applied Mechanics and Engineering Mechanics.

The first ideas about writing this textbook go back to France, where we taught Continuum Mechanics together to undergraduate students in a joint program of L'École Normale Supérieure de Cachan and Université de Marne-La-Vallée. But it was M.I.T. that gave us the occasion to develop a spoken language into lecture notes for undergraduate and graduate students. Still, some of the Problem Sets in this textbook have a much longer history, rooted in the teaching of "La Mécanique Rationelle" by the most gifted educators, who instilled in us the beauty of Mechanics: Jean Mandel, Paul Germain, Jean Salençon, Yves Bamberger, Bernard Halphen; and with our colleagues and friends: Patrick de Buhan, Luc Dormieux, and many more. By recycling some of the Problem Sets from our drawers into this textbook, we trust that we remain true to our roots.

We wish to thank Professor Stein Sture of the University of Colorado at Boulder and Professor John Rudnicki of Nortwestern University for their assistance in reviewing the textbook.

We trust that this textbook will be a source of imagination.

FRANZ-JOSEF ULM
Cambridge, Massachusetts

OLIVIER COUSSY
Paris, France

PART ONE

DEFORMATION AND STRAIN

CHAPTER 1

Description of Finite Deformation

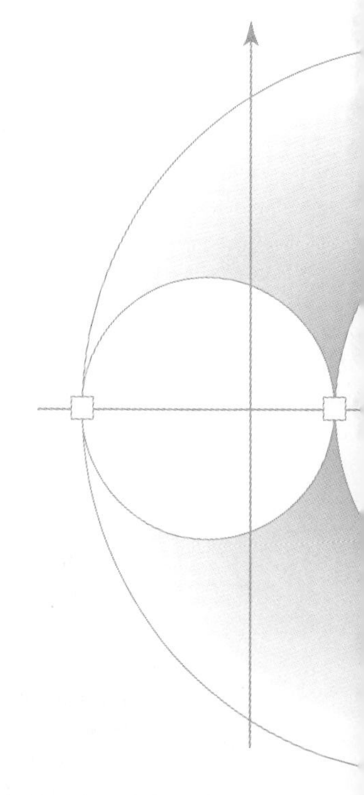

This first chapter is devoted to the description of finite deformation of continuous material systems. By this we mean material systems, of which the behavior is described by means of continuum mechanics. This first chapter develops the essential mathematical ingredients for the description of the deformation without restriction on the order of magnitude of the deformation. These are the deformation gradient and the transport formulas of a material vector, of an elementary volume, and of an oriented surface. In addition, based on the analysis of the transport of the scalar product of two material vectors in deformation, the appropriate strain measures are derived: the Cauchy dilatation tensor and the Green–Lagrange strain tensor. They account for length and angle variations due to deformation and are invariant with respect to rigid body motion. Finally, the link between strain tensors and displacement is derived.

1.1 THE CONTINUUM MODEL

Continuum mechanics is concerned with the continuous description of the transformation of a material point within a system. A system is the part of the world in which we have a special interest; we denote this domain Ω. In engineering, the system is the structure under consideration (e.g., in civil engineering a tunnel, a bridge, or a foundation) and the interest that we have in it as engineers is the analysis of its deformation, stresses, and so on when subjected to loading. Every structural system is characterized by a length scale, defining the structural dimension (substratum height H, foundation width B, tunnel radius R, etc.), as sketched

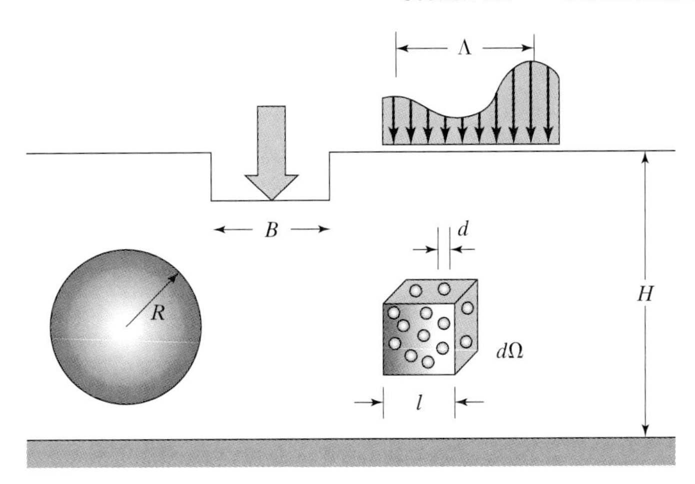

FIGURE 1.1: Modeling scales in mechanics.

in Figure 1.1. Furthermore, on this structural level, we may have to deal with load cases, of which the space variation is characterized by a wave length Λ. At a level below, we consider a material point of elementary volume $d\Omega$ composing the domain, which includes sufficient matter for it to be representative of the studied macroscopic phenomena. It is therefore also called the representative elementary volume (REV). We associate a macroscopic scale with it characterized by a length l (typically the laboratory test specimen size). Below this scale, the matter is heterogeneous, and this material system is characterized by the scale of its components at the microscopic level (for instance, the grain diameter or aggregate size d). To give a continuous description of a medium that is heterogeneous at the microscopic level thus implies the adoption of a macroscopic scale. Its characteristic length l must be much larger than the dimension d of the components it contains, and much smaller than the structural dimensions H, B, and R and the wave length Λ. Formally, we can define the macroscopic scale of continuum mechanics by

$$d \ll l \ll (H, B, R, \Lambda) \tag{1.1}$$

The hypothesis of continuity stipulates that the physical properties vary continuously from one material point to another. This confirms the existence of a scale that defines the dimension of the elementary volume of an intended application. It does not prove it. More precisely, the characteristic length l is not intrinsic to the matter (in contrast to the microscopic dimension d). It is the application that fixes it.

1.2 DEFORMATION GRADIENT

We consider now the material point of volume $d\Omega$ in a reference configuration. In this configuration, the material point is located by the position vector $\mathbf{X}$ of components X_α ($\alpha = 1, 2, 3$) in a Cartesian coordinate frame of orthonormal (Cartesian) basis $\mathbf{e}_\alpha$ ($\alpha = 1, 2, 3$); thus (see Figure 1.2):

$$\mathbf{X} = X_1\mathbf{e}_1 + X_2\mathbf{e}_2 + X_3\mathbf{e}_3 = X_\alpha\mathbf{e}_\alpha \tag{1.2}$$

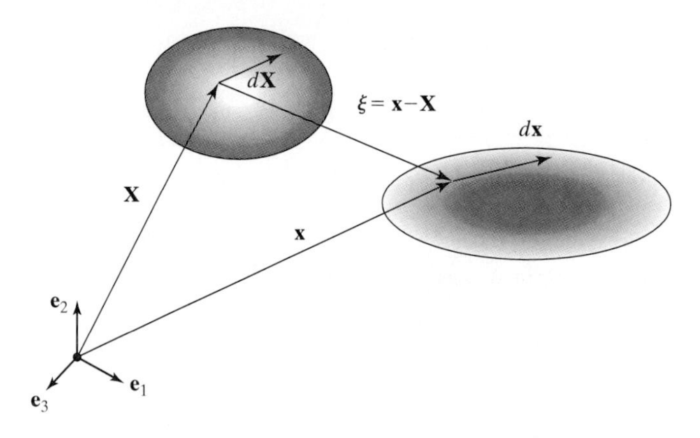

FIGURE 1.2: Transport of a material vector.

Repeated subscript indicates summation. Voluntarily, we employed here a Greek letter to indicate that the quantity is expressed with regard to the initial configuration. At time t, after deformation, the material point is in a new configuration, called the current configuration. In this configuration, it is located by the position vector $\mathbf{x}$ of components x_i:

$$\mathbf{x} = x_i \mathbf{e}_i \tag{1.3}$$

To indicate that the quantity refers to the current configuration, we employ a Latin subscript. The deformation that occurred depends on the initial position defined by position vector $\mathbf{X}$ and on time t:

$$\mathbf{x} = \mathbf{x}(\mathbf{X}, t); \qquad x_i = x_i(X_\alpha, t) \tag{1.4}$$

From (1.4) it is readily understood that the description of the deformation in continuum mechanics aims at describing the geometrical transformation of the material point with respect to the initial configuration. It seeks for appropriate operators to pass from one configuration to the other. In other words, time t refers here only to the instant of observation.

Consider now an infinitesimal vector $d\mathbf{X} = dX_\alpha \mathbf{e}_\alpha$ attached to the material point—similar to a scotch tape attached to the material point with which it transforms. Such an infinitesimal vector attached to the matter is called a material vector. After deformation the infinitesimal material vector (*cf.* Figure 1.2) is $d\mathbf{x} = dx_i \mathbf{e}_i$. From expression (1.4), it follows:

$$d\mathbf{x} = \frac{\partial \mathbf{x}}{\partial \mathbf{X}} \cdot d\mathbf{X}; \quad dx_i = \frac{\partial x_i}{\partial X_\alpha} dX_\alpha \tag{1.5}$$

or, in matrix form,

$$\begin{pmatrix} dx_1 \\ dx_2 \\ dx_3 \end{pmatrix} = \begin{bmatrix} \frac{\partial x_1}{\partial X_1} & \frac{\partial x_1}{\partial X_2} & \frac{\partial x_1}{\partial X_3} \\ \frac{\partial x_2}{\partial X_1} & \frac{\partial x_2}{\partial X_2} & \frac{\partial x_2}{\partial X_3} \\ \frac{\partial x_3}{\partial X_1} & \frac{\partial x_3}{\partial X_2} & \frac{\partial x_3}{\partial X_3} \end{bmatrix} \begin{pmatrix} dX_1 \\ dX_2 \\ dX_3 \end{pmatrix} \tag{1.6}$$

This partial derivative, that is,

$$\mathbf{F} = \frac{\partial \mathbf{x}}{\partial \mathbf{X}} = \mathrm{Grad}\,\mathbf{x} = F_{i\alpha}\mathbf{e}_i \otimes \mathbf{e}_\alpha; \quad F_{i\alpha} = \frac{\partial x_i}{\partial X_\alpha} \tag{1.7}$$

is called the deformation gradient. It transports and transforms a material vector from the initial configuration into the current one (i.e., $d\mathbf{X} \to d\mathbf{x}$). Analogously, its inverse $\mathbf{F}^{-1}$ transports and transforms the material vector from the current into the initial configuration (i.e., $d\mathbf{x} \to d\mathbf{X}$):

$$d\mathbf{X} = \mathbf{F}^{-1} \cdot d\mathbf{x}; \quad dX_\alpha = F_{\alpha i}^{-1} dx_i \tag{1.8}$$

with

$$\mathbf{F}^{-1} = \frac{\partial \mathbf{X}}{\partial \mathbf{x}} = \mathrm{grad}\,\mathbf{X} = F_{\alpha i}^{-1}\mathbf{e}_\alpha \otimes \mathbf{e}_i; \quad F_{\alpha i}^{-1} = \frac{\partial X_\alpha}{\partial x_i} \tag{1.9}$$

The capital *G*rad and, respectively, the small *g*rad of the gradient operator in (1.7) and (1.9) indicate that the spatial derivations are carried out with regard to, respectively, the initial configuration and the current configuration. Finally, Grad and grad are related by

$$\mathrm{Grad}\,(.) = \mathrm{grad}\,(.) \cdot \mathbf{F} \tag{1.10}$$

Exercise 1. Consider the uniaxial extension of a truss shown in Figure 1.3. After deformation, the position of any point of the beam is given by

$$x_1 = X_1(1+\alpha); \quad x_2 = X_2(1-\beta); \quad x_3 = X_3(1-\beta)$$

Determine the deformation gradient $\mathbf{F}$, together with its transpose ${}^t\mathbf{F}$ and its inverse $\mathbf{F}^{-1}$.

The deformation gradient follows from (1.5) and (1.6):

$$\begin{pmatrix} dx_1 \\ dx_2 \\ dx_3 \end{pmatrix} = \begin{bmatrix} 1+\alpha & 0 & 0 \\ 0 & 1-\beta & 0 \\ 0 & 0 & 1-\beta \end{bmatrix} \begin{pmatrix} dX_1 \\ dX_2 \\ dX_3 \end{pmatrix}$$

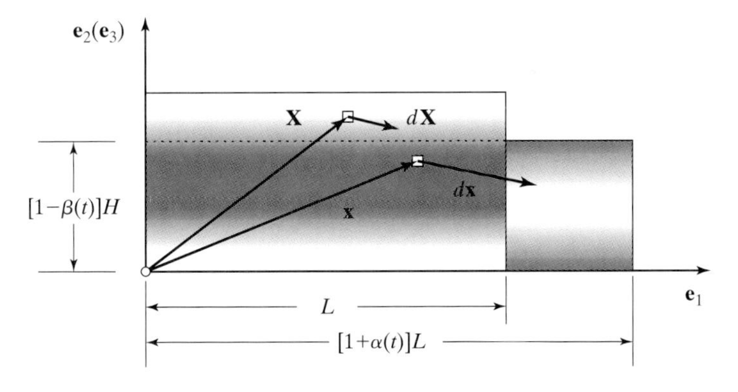

FIGURE 1.3: Uniaxial extension: position vectors and material vectors.

or, equivalently, in tensor notation

$$\mathbf{F} = (1+\alpha)\mathbf{e}_1 \otimes \mathbf{e}_1 + (1-\beta)[\mathbf{e}_2 \otimes \mathbf{e}_2 + \mathbf{e}_3 \otimes \mathbf{e}_3]$$

Due to the symmetry of $\mathbf{F}$, it is

$$\mathbf{F} =^t \mathbf{F}; \quad F_{i\alpha} = F_{\alpha i}$$

Due to the diagonal structure of $\mathbf{F}$, the inverse reads

$$(F_{i\alpha}^{-1}) = \begin{bmatrix} \frac{1}{1+\alpha} & 0 & 0 \\ 0 & \frac{1}{1-\beta} & 0 \\ 0 & 0 & \frac{1}{1-\beta} \end{bmatrix}$$

or, equivalently, in tensor notation

$$\mathbf{F}^{-1} = (1+\alpha)^{-1}\mathbf{e}_1 \otimes \mathbf{e}_1 + (1-\beta)^{-1}[\mathbf{e}_2 \otimes \mathbf{e}_2 + \mathbf{e}_3 \otimes \mathbf{e}_3]$$ ■

1.3 TRANSPORT FORMULAS

The deformation gradient $\mathbf{F}$ gives a local description of the deformation, transporting and transforming a material vector from the initial configuration into the current one. We now see some further application of the deformation gradient.

1.3.1 Volume Transport

Through deformation, the elementary volume $d\Omega = dX_1 dX_2 dX_3$ becomes the volume $d\Omega_t = dx_1 dx_2 dx_3$. This volume transport is described by the Jacobian J of the deformation:

$$d\Omega_t = J d\Omega; \quad J = \det \mathbf{F} \tag{1.11}$$

Exercise 2. Derive volume transport formula (1.11) from pure geometrical considerations.

We express the volumes, $d\Omega$ and $d\Omega_t$, before and after deformation in form of the scalar triple products:

$$d\Omega = d\mathbf{X}_1 \cdot (d\mathbf{X}_2 \times d\mathbf{X}_3); \quad d\Omega_t = d\mathbf{x}_1 \cdot (d\mathbf{x}_2 \times d\mathbf{x}_3)$$

Indeed, the absolute value of a scalar triple product is equal to the volume of the parallelepiped[1] with $d\mathbf{X}_\alpha$ (respectively, $d\mathbf{x}_i$) as adjacent edges. This is shown in Figure 1.4. The vectors $d\mathbf{X}_\alpha$ and $d\mathbf{x}_i$ are the projection of the material vectors on the axis, that is,

$$d\mathbf{X}_\alpha = (d\mathbf{X} \cdot \mathbf{e}_\alpha)\mathbf{e}_\alpha; \quad d\mathbf{x}_i = (d\mathbf{x} \cdot \mathbf{e}_i)\mathbf{e}_i$$

We now apply to each vector $d\mathbf{x}_i$ in the scalar triple product, the transformation formula (1.5):

$$d\mathbf{x}_1 = \mathbf{F} \cdot d\mathbf{X}_1; \quad d\mathbf{x}_2 = \mathbf{F} \cdot d\mathbf{X}_2; \quad d\mathbf{x}_3 = \mathbf{F} \cdot d\mathbf{X}_3$$

[1] A parallelepiped is a solid with three pairs of parallel sides. If the sides are perpendicular, it is a box; and if, moreover, the edges all have the same length, it is a cube.

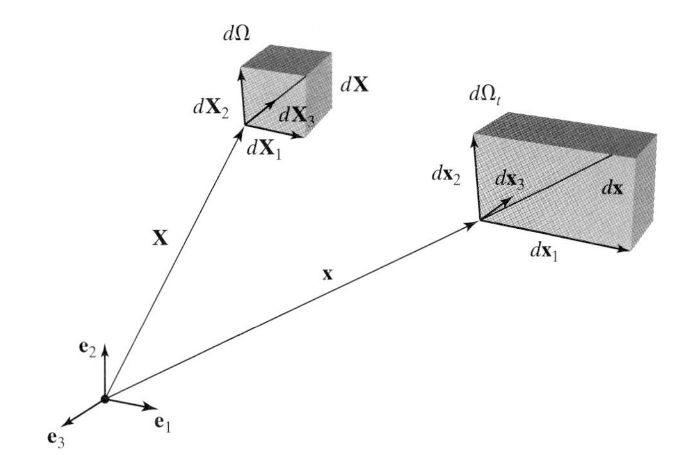

FIGURE 1.4: Geometrical interpretation of volume transport.

This yields

$$\begin{aligned} d\Omega_t &= \mathbf{F}\cdot d\mathbf{X}_1\cdot(\mathbf{F}\cdot d\mathbf{X}_2\times\mathbf{F}\cdot d\mathbf{X}_3) \\ &= \det\mathbf{F}\left[\,d\mathbf{X}_1\cdot(d\mathbf{X}_2\times d\mathbf{X}_3)\right] = Jd\Omega \end{aligned}$$

which is (1.11). ■

From this geometrical interpretation of transport formula (1.11), it follows that the Jacobian J ($=\det\mathbf{F}$) must remain strictly positive to ensure that an elementary material volume never becomes zero. Mathematically, this ensures that the tangent linear application $\mathbf{F}$ can always be inverted.

Exercise 3. Apply the volume transport formula (1.11) to the uniaxial extension.

In this case,

$$J = \det\mathbf{F} = \begin{vmatrix} 1+\alpha & 0 & 0 \\ 0 & 1-\beta & 0 \\ 0 & 0 & 1-\beta \end{vmatrix} = (1+\alpha)(1-\beta)^2$$

Hence,

$$d\Omega_t = (1+\alpha)(1-\beta)^2 d\Omega$$

By way of example, let us also determine $d\Omega_t$ from the geometrical interpretation given before. In the uniaxial extension case, it is

$$d\mathbf{x} = (1+\alpha)dX_1\mathbf{e}_1 + (1-\beta)[dX_2\mathbf{e}_2 + dX_3\mathbf{e}_3]$$

or, equivalently,

$$d\mathbf{x} = F_{11}dX_1\mathbf{e}_1 + F_{22}dX_2\mathbf{e}_2 + F_{33}dX_3\mathbf{e}_3$$

The three vectors defining the three edges of the parallelepiped are given by

$$d\mathbf{x}_1 = (d\mathbf{x} \cdot \mathbf{e}_1)\mathbf{e}_1 = F_{11} dX_1 \mathbf{e}_1; \quad d\mathbf{x}_2 = F_{22} dX_2 \mathbf{e}_2; \quad d\mathbf{x}_3 = F_{33} dX_3 \mathbf{e}_3$$

Use of these vectors in the triple scalar product yields

$$\begin{aligned} d\Omega_t &= d\mathbf{x}_1 \cdot (d\mathbf{x}_2 \times d\mathbf{x}_3) = F_{11} dX_1 \mathbf{e}_1 \cdot (F_{22} dX_2 \mathbf{e}_2 \times F_{33} dX_3 \mathbf{e}_3) \\ &= F_{11} F_{22} F_{33}\, \mathbf{e}_1 \cdot (\mathbf{e}_2 \times \mathbf{e}_3) dX_1 dX_2 dX_3 = F_{11} F_{22} F_{33} d\Omega \\ &= J d\Omega \end{aligned}$$

with $\mathbf{e}_2 \times \mathbf{e}_3 = \mathbf{e}_1$, and $\mathbf{e}_1 \cdot \mathbf{e}_1 = 1$. ■

1.3.2 Transport of an Oriented Material Surface

Next, we consider a material surface $d\mathbf{A}$ of infinitesimal (nominal) area $dA = |d\mathbf{A}|$, which is oriented by unit normal $\mathbf{N}$ (i.e., $d\mathbf{A} = dA\mathbf{N}$). During deformation, both the nominal area dA and the unit normal $\mathbf{N}$ transform:

$$d\mathbf{A} = dA\,\mathbf{N} \rightarrow d\mathbf{a} = da\,\mathbf{n} \tag{1.12}$$

where $da = |d\mathbf{a}|$ is the nominal area of the deformed material surface, oriented by unit normal $\mathbf{n}$ in the current configuration. The description of the transport of an oriented material surface, therefore, consists in describing simultaneously the transformation of the nominal area $dA \rightarrow da$, and that of the unit normal $\mathbf{N} \rightarrow \mathbf{n}$. We may now wrongly believe that transport formula (1.5) can be applied to $\mathbf{N} \rightarrow \mathbf{n}$. However, we should note that (1.5) applies only to material vectors (i.e., to vectors attached to the material point), while unit normal vectors are not attached to the matter. In addition, since convective transport transforms the norm of vectors (i.e., change in length $|d\mathbf{X}| \rightarrow |d\mathbf{x}|$), unit vector $\mathbf{n}$ is not the convective transportee of unit vector $\mathbf{N}$; thus—except for rigid body motion, for which $\mathbf{F} = \mathbf{1}$,

$$|\mathbf{N}| = |\mathbf{n}| = 1 \Leftrightarrow \mathbf{n} \neq \mathbf{F} \cdot \mathbf{N} \tag{1.13}$$

In order to derive the transport formula of an oriented surface, consider a cylinder generated by $d\mathbf{A}$ in a translation of a vector $\mathbf{U}$ attached to the material

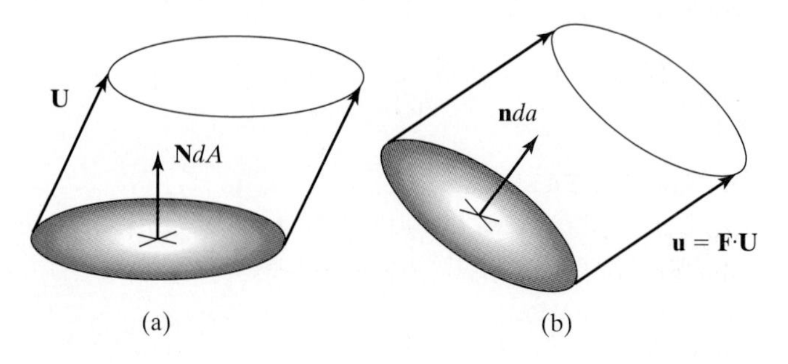

FIGURE 1.5: Transport of an oriented material surface: (a) before deformation, (b) after deformation.

point. This is sketched in Figure 1.5. The initial volume of this cylinder is

$$d\Omega = \mathbf{N} \cdot \mathbf{U} dA \tag{1.14}$$

After deformation, the cylinder has transformed, being now generated by $d\mathbf{a}$ in a translation of the material vector $\mathbf{u}$, to which transport formula (1.5) applies (i.e., $\mathbf{u} = \mathbf{F} \cdot \mathbf{U}$). The volume after deformation, therefore, is

$$d\Omega_t = \mathbf{n} \cdot \mathbf{u} da = \mathbf{n} \cdot \mathbf{F} \cdot \mathbf{U} da \tag{1.15}$$

Let us now apply volume transport formula (1.11) to (1.14) and (1.15); that is

$$d\Omega_t = {}^t\mathbf{F} \cdot \mathbf{n} \cdot \mathbf{U} da = J\mathbf{N} \cdot \mathbf{U} dA \tag{1.16}$$

where ${}^t\mathbf{F}$ is the transpose of the deformation gradient. It is readily understood that the demonstration is independent of the particular choice of material vector $\mathbf{U}$; and relation (1.16) yields the transport formula of an oriented surface[2]:

$$\forall \mathbf{U} : \quad \mathbf{n} da = J {}^t\mathbf{F}^{-1} \cdot \mathbf{N} dA \tag{1.17}$$

Exercise 4. Apply transport formula (1.17) to the uniaxial extension.

Let us first consider $\mathbf{N} = \mathbf{e}_1$. In matrix notation, (1.17) reads

$$\begin{aligned}
\begin{pmatrix} n_1 \\ n_2 \\ n_3 \end{pmatrix} da &= (1+\alpha)(1-\beta)^2 \begin{bmatrix} \frac{1}{1+\alpha} & 0 & 0 \\ 0 & \frac{1}{1-\beta} & 0 \\ 0 & 0 & \frac{1}{1-\beta} \end{bmatrix} \begin{pmatrix} 1 \\ 0 \\ 0 \end{pmatrix} dA \\
&= (1-\beta)^2 \begin{pmatrix} 1 \\ 0 \\ 0 \end{pmatrix} dA
\end{aligned}$$

or, in tensor notation,

$$\mathbf{N} = \mathbf{e}_1 \rightarrow \mathbf{n} da = (1-\beta)^2 \mathbf{e}_1 dA; \quad da = (1-\beta)^2 dA; \quad \mathbf{n} = \mathbf{e}_1$$

Analogously,

$$\mathbf{N} = \mathbf{e}_2 \ (= \mathbf{e}_3) \rightarrow da = (1+\alpha)(1-\beta) dA; \quad \mathbf{n} = \mathbf{e}_2 \ (= \mathbf{e}_3)$$ ■

1.4 FINITE STRAIN MEASURES

With the transport formulas in hand, we can proceed to further applications to determine appropriate strain measurements for the description of the deformation. In fact, the deformation gradient $\mathbf{F}$ alone is not sufficient to describe the straining of the material. For instance, let us consider a rigid body motion. A rigid body motion is an orthogonal transformation that preserves the length of any vector

[2]Note that $({}^t\mathbf{F})^{-1} = {}^t(\mathbf{F}^{-1}) = {}^t\mathbf{F}^{-1}$.

and the angle between any two vectors. The deformation gradient reduces to the orthogonal operator $\mathbf{R}$, which satisfies the orthogonality properties:

$$\mathbf{F} = \mathbf{R}; \quad {}^t\mathbf{R} = \mathbf{R}^{-1}; \quad J = \det \mathbf{R} = 1 \tag{1.18}$$

Hence, the deformation is a rotation (possibly combined with a translation), and the deformation gradient $\mathbf{F} = \mathbf{R}$ changes its values in an orthogonal transformation (rigid body motion), for which the strains are obviously zero. Indeed, a rigid body motion has an important physical property: It has a zero energy mode (i.e., there is no transformation of [internal] energy).

Exercise 5. Consider a plane rotation of the position vector $\mathbf{X}$ around $\mathbf{e}_3$ by an angle θ. Determine the deformation gradient.

The initial position vector is given by

$$\mathbf{X} = X_1\mathbf{e}_1 + X_2\mathbf{e}_2 + X_3\mathbf{e}_3$$

The axis after the rotation is oriented by the unit vectors:

$$\begin{aligned} \mathbf{e}'_1 &= \cos\theta\mathbf{e}_1 + \sin\theta\mathbf{e}_2 \\ \mathbf{e}'_2 &= -\sin\theta\mathbf{e}_1 + \cos\theta\mathbf{e}_2 \\ \mathbf{e}'_3 &= \mathbf{e}_3 \end{aligned}$$

The position vector after the rotation is given by

$$\begin{aligned} \mathbf{x} &= X_1\mathbf{e}'_1 + X_2\mathbf{e}'_2 + X_3\mathbf{e}'_3 \\ &= X_1(\cos\theta\mathbf{e}_1 + \sin\theta\mathbf{e}_2) + X_2(-\sin\theta\mathbf{e}_1 + \cos\theta\mathbf{e}_2) + X_3\mathbf{e}_3 \\ &= (X_1\cos\theta - X_2\sin\theta)\mathbf{e}_1 + (X_1\sin\theta + X_2\cos\theta)\mathbf{e}_2 + X_3\mathbf{e}_3 \end{aligned}$$

Hence, the transport and transformation of a material vector in this rotation reads according to (1.5):

$$\begin{pmatrix} dx_1 \\ dx_2 \\ dx_3 \end{pmatrix} = \begin{bmatrix} \cos\theta & -\sin\theta & 0 \\ \sin\theta & \cos\theta & 0 \\ 0 & 0 & 1 \end{bmatrix} \begin{pmatrix} dX_1 \\ dX_2 \\ dX_3 \end{pmatrix}$$

It is straightforward to verify the orthogonality properties (1.18) of the deformation gradient. ■

1.4.1 Cauchy Dilatation Tensor

We will study here the transport of a scalar product of two material vectors. In fact, a scalar product has an interesting property: During deformation, it accounts for both length- and angle variation. Let $d\mathbf{X}$ and $d\mathbf{Y}$ be two infinitesimal material vectors attached to the same material point. This is sketched in Figure 1.6. After deformation, these material vectors become

$$d\mathbf{x} = \mathbf{F} \cdot d\mathbf{X}; \quad d\mathbf{y} = \mathbf{F} \cdot d\mathbf{Y} \tag{1.19}$$

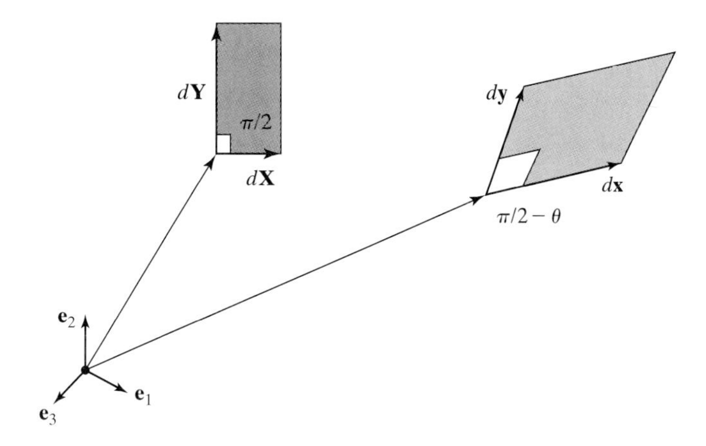

FIGURE 1.6: Transport of a scalar product of two material vectors.

and the scalar product of these transformed material vectors reads

$$d\mathbf{x} \cdot d\mathbf{y} = \mathbf{F} \cdot d\mathbf{X} \cdot \mathbf{F} \cdot d\mathbf{Y} = d\mathbf{X} \cdot {}^t\mathbf{F} \cdot \mathbf{F} \cdot d\mathbf{Y} \tag{1.20}$$

With regard to the very definition of the scalar product, the length and the angle variation in cause of deformation are included in

$$\mathbf{C} = {}^t\mathbf{F} \cdot \mathbf{F}; \quad C_{\alpha\beta} = F_{\alpha\gamma} F_{\gamma\beta}; \quad \mathbf{C} = C_{\alpha\beta} \mathbf{e}_\alpha \otimes \mathbf{e}_\beta \tag{1.21}$$

Tensor $\mathbf{C}$ is known as Cauchy dilatation tensor. In contrast to deformation gradient $\mathbf{F}$, $\mathbf{C}$ is symmetric, since[3]

$${}^t\mathbf{C} = {}^t\left({}^t\mathbf{F} \cdot \mathbf{F}\right) = {}^t\mathbf{F} \cdot \mathbf{F} = \mathbf{C} \tag{1.22}$$

Furthermore, since symmetric tensor $\mathbf{C}$ is associated with a scalar product (i.e., with a real inner product space), the Cauchy dilatation tensor is positive definite. Hence, tensor $\mathbf{C}$ has three real positive eigenvalues C_J $(J = I, II, III)$ associated to the orthonormal set of eigenvectors $\mathbf{u}_J$ $(J = I, II, III)$:

$$\mathbf{C} \cdot \mathbf{u}_J = C_J \mathbf{u}_J \tag{1.23}$$

In addition, in the deformation the orthogonality of the eigenvectors is preserved, that is,

$$I \neq J : \mathbf{u}_I \cdot \mathbf{C} \cdot \mathbf{u}_J = 0 \leftrightarrow \left(\mathbf{u}_I \cdot {}^t\mathbf{F}\right) \cdot \left(\mathbf{F} \cdot \mathbf{u}_J\right) = 0 \tag{1.24}$$

This is a remarkable property of tensor $\mathbf{C}$. In fact, the orthonormal set of eigenvectors forms the principal axis of deformation.

Finally, in a rigid body motion with $\mathbf{F} = \mathbf{R}$ according to (1.18), tensor $\mathbf{C}$ remains constant:

$$\mathbf{F} = \mathbf{R} \leftrightarrow \mathbf{C} = {}^t\mathbf{R} \cdot \mathbf{R} = \mathbf{R}^{-1} \cdot \mathbf{R} = \mathbf{1}; \quad C_{\alpha\beta} = \delta_{\alpha\beta} \tag{1.25}$$

where $\delta_{\alpha\beta}$ denotes the Kronecker delta ($\delta_{\alpha\beta} = 1$ for $\alpha = \beta$; $\delta_{\alpha\beta} = 0$ for $\alpha \neq \beta$). Hence, in contrast to the deformation gradient $\mathbf{F}$, the dilatation tensor is invariant with respect to rigid body motion.

[3] Recall that ${}^t(\mathbf{a} \cdot \mathbf{b}) = {}^t\mathbf{b} \cdot {}^t\mathbf{a}$.

1.4.2 Green–Lagrange Strain Tensor

The Cauchy dilatation tensor has one drawback: It has a nonzero value under rigid body motion. To overcome this drawback, we consider the difference of the scalar products of the material vectors before and after deformation:

$$d\mathbf{x} \cdot d\mathbf{y} - d\mathbf{X} \cdot d\mathbf{Y} = d\mathbf{X} \cdot (\mathbf{C} - 1) \cdot d\mathbf{Y} \tag{1.26}$$

This leads to the definition of the Green–Lagrange strain tensor:

$$2\mathbf{E} = \mathbf{C} - 1 = {}^t\mathbf{F} \cdot \mathbf{F} - 1 \tag{1.27}$$

The Green–Lagrange strain tensor preserves the properties of the Cauchy dilatation tensor: symmetry, realness of eigenvalues, and invariance with regard to rigid body motion, for which it takes zero values. In addition, tensors $\mathbf{C}$ and $\mathbf{E}$ have the same eigenvectors $\mathbf{u}_J$ $(J = I, II, III)$. The eigenvalues E_J $(J = I, II, III)$ of $\mathbf{E}$ are, therefore, obtained from (1.23) and (1.27):

$$2E_J = C_J - 1 \tag{1.28}$$

The eigenvalues E_J are the principal strains in the principal directions $\mathbf{u}_J$ of the deformation.

Exercise 6. Determine the Cauchy dilatation tensor and the Green–Lagrange strain tensor for the uniaxial extension experiment.

Carrying out the tensor contraction (1.21) in matrix notation reads

$$\begin{aligned}(C_{\alpha\beta}) &= \begin{bmatrix} 1+\alpha & 0 & 0 \\ 0 & 1-\beta & 0 \\ 0 & 0 & 1-\beta \end{bmatrix} \begin{bmatrix} 1+\alpha & 0 & 0 \\ 0 & 1-\beta & 0 \\ 0 & 0 & 1-\beta \end{bmatrix} \\ &= \begin{bmatrix} (1+\alpha)^2 & 0 & 0 \\ 0 & (1-\beta)^2 & 0 \\ 0 & 0 & (1-\beta)^2 \end{bmatrix}\end{aligned}$$

or, in tensor notation,

$$\mathbf{C} = (1+\alpha)^2 \mathbf{e}_1 \otimes \mathbf{e}_1 + (1-\beta)^2 [\mathbf{e}_2 \otimes \mathbf{e}_2 + \mathbf{e}_3 \otimes \mathbf{e}_3]$$

The diagonal form of the tensor shows that the principal deformation axis is the coordinate axis $\mathbf{u}_J = \mathbf{e}_i$, and the eigenvalues are

$$C_I = (1+\alpha)^2; \quad C_{II} = C_{III} = (1-\beta)^2$$

With these eigenvalues in hand, the principal (Green–Lagrange) strains are obtained from (1.28) in the form (recall that the eigenvectors of $\mathbf{C}$ coincide with those of $\mathbf{E}$)

$$E_I = \frac{1}{2}(C_I - 1) = \alpha + \frac{\alpha^2}{2}; \quad E_{II} = E_{III} = -\beta + \frac{\beta^2}{2}$$

The Green–Lagrange strain tensor then reads

$$\mathbf{E} = \sum_{J=I}^{J=III} E_J \mathbf{u}_J \otimes \mathbf{u}_J = \left(\alpha + \frac{\alpha^2}{2}\right) \mathbf{e}_1 \otimes \mathbf{e}_1 + \left(-\beta + \frac{\beta^2}{2}\right) [\mathbf{e}_2 \otimes \mathbf{e}_2 + \mathbf{e}_3 \otimes \mathbf{e}_3] \quad \blacksquare$$

Exercise 7. Polar Decomposition (Optional): Consider a two-step deformation composed of a pure stretching defined by deformation gradient $\mathbf{D}$, which is followed by a rigid body rotation of deformation gradient $\mathbf{R}$. Show that the Green–Lagrange strain tensor is invariant with respect to rigid body motion.

We apply in a step-to-step fashion the transport formula (1.5) of a material vector to the pure deformation state:

$$d\mathbf{x}' = \mathbf{D} \cdot d\mathbf{X}$$

and then to the rigid body rotation:

$$d\mathbf{x} = \mathbf{R} \cdot d\mathbf{x}' = \mathbf{R} \cdot \mathbf{D} \cdot d\mathbf{X}$$

Hence, the overall deformation gradient reads

$$\mathbf{F} = \mathbf{R} \cdot \mathbf{D}$$

This decomposition is known as polar decomposition. The first operation $d\mathbf{X} \rightarrow d\mathbf{x}'$ corresponds to a pure stretching in the eigenvector basis $\mathbf{u}_{J=I,II,III}$, while the second operation, $d\mathbf{x}' \rightarrow d\mathbf{x}$, characterized by the rotation tensor $\mathbf{R}$, carries the principal axis of deformation into the final configuration. The Cauchy dilatation tensor is obtained from (1.21) in the form

$$\mathbf{C} = {}^t(\mathbf{R} \cdot \mathbf{D}) \cdot \mathbf{R} \cdot \mathbf{D} = {}^t\mathbf{D} \cdot {}^t\mathbf{R} \cdot \mathbf{R} \cdot \mathbf{D} = {}^t\mathbf{D} \cdot \mathbf{D}$$

Here we made use of the orthogonality condition (1.18), according to which ${}^t\mathbf{R} \cdot \mathbf{R} = \mathbf{R}^{-1} \cdot \mathbf{R} = 1$. Hence, the Cauchy dilatation tensor is independent of the rigid body motion, as is the Green–Lagrange strain tensor, which reads

$$\mathbf{E} = \frac{1}{2}({}^t\mathbf{D} \cdot \mathbf{D} - 1)$$

Tensor $\mathbf{D}$ has the same principal axis as dilatation tensor $\mathbf{C}$ and strain tensor $\mathbf{E}$. Indeed,

$$\mathbf{D} \cdot \mathbf{e}_J = \mathbf{R}^{-1} \cdot \mathbf{F} \cdot \mathbf{e}_J = (1 + \lambda_J)\mathbf{e}_J$$

where $(1+\lambda_J)$ are the eigenvalues of tensor $\mathbf{D}$. This implies the symmetry of tensor $\mathbf{D} ={}^t \mathbf{D}$, since it has the same eigenvectors (i.e., the same vectorial base) as $\mathbf{C}$ and $\mathbf{E}$, and thus

$$\mathbf{E} = \frac{1}{2}(\mathbf{D}^2 - 1)$$

For instance, the considered uniaxial extension is a pure deformation state, and any rigid body motion will not alter the Cauchy dilatation and the Green–Lagrange strain tensor. ■

Remark The same end result (i.e., deformation gradient $\mathbf{F}$ can be obtained by inverting the polar decomposition, starting with the rigid body rotation, set forth by $\mathbf{R}'$, followed by a pure deformation, characterized by a stretch tensor $\mathbf{D}'$, such that

$$\mathbf{F} = \mathbf{D}' \cdot \mathbf{R}'$$

It is left to the reader to show that $\mathbf{D} \neq \mathbf{D}'$ and $\mathbf{R} \neq \mathbf{R}'$.

1.4.3 Linear Dilatation and Distortion

So far, we have used the scalar product of two material vectors to derive appropriate strain measurements. We will now apply the Green–Lagrange strain tensor to express the relative length variation and the relative angle variation between two materials vectors (i.e., the linear dilatation and the distortion).

Length Variation of a Material Vector: Linear Dilatation.

The length of the material vector $|d\mathbf{X}| = \sqrt{d\mathbf{X} \cdot d\mathbf{X}}$ becomes after deformation the length $|d\mathbf{x}| = \sqrt{d\mathbf{x} \cdot d\mathbf{x}}$. The length variation due to deformation can be expressed in terms of the linear dilatation λ of material vector $d\mathbf{X}$:

$$|d\mathbf{x}| = (1 + \lambda(d\mathbf{X}))\,|d\mathbf{X}| \tag{1.29}$$

To determine $\lambda(d\mathbf{X})$ we take the previous equation in square:

$$|d\mathbf{x}|^2 = d\mathbf{x} \cdot d\mathbf{x} = (1 + \lambda(d\mathbf{X}))^2 d\mathbf{X} \cdot d\mathbf{X} \tag{1.30}$$

We encounter again the scalar product of the material vectors, which we have used to derive the Green–Lagrange strain tensor. Use of transport formula (1.5) and (1.27) on the left-hand side of (1.30) yields

$$|d\mathbf{x}|^2 = d\mathbf{X} \cdot {}^t\mathbf{F} \cdot \mathbf{F} \cdot d\mathbf{X} = 2d\mathbf{X} \cdot \mathbf{E} \cdot d\mathbf{X} + |d\mathbf{X}|^2 \tag{1.31}$$

The two previous equations lead to determine the relative length change $\lambda(d\mathbf{X})$ in the form

$$\lambda(d\mathbf{X}) = \sqrt{1 + \frac{2d\mathbf{X} \cdot \mathbf{E} \cdot d\mathbf{X}}{|d\mathbf{X}|^2}} - 1 \tag{1.32}$$

Equation (1.32) can be rewritten in a more convenient form if we note that $d\mathbf{X} \cdot \mathbf{E} \cdot d\mathbf{X} = E_{\alpha\alpha} |d\mathbf{X}|^2$, where $E_{\alpha\alpha} = \mathbf{e}_\alpha \cdot \mathbf{E} \cdot \mathbf{e}_\alpha$ is the diagonal strain component in the direction $\mathbf{e}_\alpha$. Thus, the linear dilatation in the direction $\mathbf{e}_\alpha$ reads

$$\lambda(\mathbf{e}_\alpha) = \sqrt{1 + 2E_{\alpha\alpha}} - 1 \tag{1.33}$$

With (1.33) in hand, we can determine the relative length change in the direction of the principal strain $\mathbf{u}_J$:

$$\lambda(\mathbf{u}_J) = \sqrt{1 + 2E_J} - 1 = \sqrt{C_J} - 1 \tag{1.34}$$

Here, E_J and C_J are the eigenvalues (principal strain and principal dilatation) of the Green–Lagrange strain tensor $\mathbf{E}$ and the Cauchy Dilatation tensor $\mathbf{C}$, respectively.

Exercise 8. Show that the linear dilatations $\lambda(\mathbf{u}_J)$ of the uniaxial extension are α and $-\beta$, respectively. Determine the relative length change $\lambda(\mathbf{e}_\alpha)$ in the direction $\mathbf{e}_\alpha = \cos\theta \mathbf{e}_1 + \sin\theta \mathbf{e}_2$.

Use of the previously determined principal dilatations in (1.34) yields

$$\lambda(\mathbf{u}_I = \mathbf{e}_1) = \sqrt{(1+\alpha)^2} - 1 = \alpha$$
$$\lambda(\mathbf{u}_{II} = \mathbf{e}_2) = \lambda(\mathbf{u}_{III} = \mathbf{e}_3) = \sqrt{(1-\beta)^2} - 1 = -\beta$$

We first determine $E_{\alpha\alpha} = \mathbf{e}_\alpha \cdot \mathbf{E} \cdot \mathbf{e}_\alpha$, in matrix form:

$$\begin{aligned} E_{\alpha\alpha} &= \begin{pmatrix} \cos\theta & \sin\theta & 0 \end{pmatrix} \begin{bmatrix} \alpha + \frac{\alpha^2}{2} & 0 & 0 \\ & -\beta + \frac{\beta^2}{2} & 0 \\ & & -\beta + \frac{\beta^2}{2} \end{bmatrix} \begin{pmatrix} \cos\theta \\ \sin\theta \\ 0 \end{pmatrix} \\ &= \left(\alpha + \frac{\alpha^2}{2}\right)\cos^2\theta + \left(-\beta + \frac{\beta^2}{2}\right)\sin^2\theta \end{aligned}$$

Use of $E_{\alpha\alpha}$ in (1.33) yields the linear dilatation in the direction $\mathbf{e}_\alpha$:

$$\lambda(\mathbf{e}_\alpha) = \sqrt{1 + 2\left[\left(\alpha + \frac{\alpha^2}{2}\right)\cos^2\theta + \left(-\beta + \frac{\beta^2}{2}\right)\sin^2\theta\right]} - 1$$

■

Angle Variation of Two Material Vectors: Distortion.

In a similar fashion, we can determine the angle variation between two material vectors, which occurs during deformation. Since we are just interested in the relative angle variation before and after deformation, we consider two material vectors, $d\mathbf{X}$ and $d\mathbf{Y}$, which are initially perpendicular (i.e., $d\mathbf{X} \cdot d\mathbf{Y} = 0$). Due to deformation they transform into $d\mathbf{x}$ and $d\mathbf{y}$, with an angle between them of $(d\mathbf{x}, d\mathbf{y}) = \pi/2 - \theta$, where θ is the sought angle variation due to deformation, called distortion. The geometrical situation is sketched in Figure 1.6. The angle between the material vectors, $d\mathbf{x}$ and $d\mathbf{y}$, is obtained from the scalar product:

$$d\mathbf{x} \cdot d\mathbf{y} = |d\mathbf{x}|\,|d\mathbf{y}|\cos(\pi/2 - \theta) = |d\mathbf{x}|\,|d\mathbf{y}|\sin\theta \tag{1.35}$$

Now, we can use the length variation formula just derived for

$$|d\mathbf{x}| = (1 + \lambda(d\mathbf{X}))\,|d\mathbf{X}|\,; \quad |d\mathbf{y}| = (1 + \lambda(d\mathbf{Y}))\,|d\mathbf{Y}| \tag{1.36}$$

In addition, we can replace the left-hand side of (1.35) by (1.26), that is,

$$d\mathbf{x} \cdot d\mathbf{y} = 2d\mathbf{X} \cdot \mathbf{E} \cdot d\mathbf{Y} + d\mathbf{X} \cdot d\mathbf{Y} = 2d\mathbf{X} \cdot \mathbf{E} \cdot d\mathbf{Y} \tag{1.37}$$

where we made use of the fact that $d\mathbf{X}$ and $d\mathbf{Y}$ are initially perpendicular. The previous equations lead to

$$\sin\theta = \frac{2d\mathbf{X} \cdot \mathbf{E} \cdot d\mathbf{Y}}{(1 + \lambda(d\mathbf{X}))(1 + \lambda(d\mathbf{Y}))\,|d\mathbf{X}|\,|d\mathbf{Y}|} \tag{1.38}$$

Equation (1.38) expresses the distortion between two initially perpendicular material vectors, $d\mathbf{X}$ and $d\mathbf{Y}$. Furthermore, if we note that $d\mathbf{X} \cdot \mathbf{E} \cdot d\mathbf{Y} = E_{\alpha\beta}\,|d\mathbf{X}|\,|d\mathbf{Y}|$ with $E_{\alpha\beta} = \mathbf{e}_\alpha \cdot \mathbf{E} \cdot \mathbf{e}_\beta$, and $\mathbf{e}_\alpha \cdot \mathbf{e}_\beta = 0$ (perpendicular), the distortion can be rewritten in the more convenient form

$$\sin\theta(\mathbf{e}_\alpha, \mathbf{e}_\beta) = \frac{2E_{\alpha\beta}}{(1 + \lambda(\mathbf{e}_\alpha))(1 + \lambda(\mathbf{e}_\beta))} = \frac{2E_{\alpha\beta}}{\sqrt{(1 + 2E_{\alpha\alpha})(1 + 2E_{\beta\beta})}} \tag{1.39}$$

(no summation on repeated subscript). Equation (1.39) expresses the angle variation (or distortion) between the two initially perpendicular directions $\mathbf{e}_\alpha$ and $\mathbf{e}_\beta$.

From a comparison of (1.33) and (1.39), we note that the diagonal terms $E_{\alpha\alpha} = \mathbf{e}_\alpha \cdot \mathbf{E} \cdot \mathbf{e}_\alpha$ of the Green–Lagrange strain tensor define relative length variations, while the nondiagonal terms $E_{\alpha\beta} = \mathbf{e}_\alpha \cdot \mathbf{E} \cdot \mathbf{e}_\beta$ (with $\alpha \neq \beta$) give rise to distortions.

1.5 STRAIN-DISPLACEMENT RELATION

We still miss a relation that links strains and displacements. The displacement is the difference between the position vector before and after deformation (*cf.* Figure 1.2):

$$\boldsymbol{\xi} = \mathbf{x} - \mathbf{X} \tag{1.40}$$

The deformation gradient (1.7), therefore, can be rewritten in the form

$$\mathbf{F} = \frac{\partial(\mathbf{X} + \boldsymbol{\xi})}{\partial \mathbf{X}} = 1 + \operatorname{Grad} \boldsymbol{\xi}; \quad F_{i\alpha} = \delta_{i\alpha} + \frac{\partial \xi_i}{\partial X_\alpha} \tag{1.41}$$

where $\delta_{i\alpha}$ is the Kronecker delta.[4] Use of (1.41) in (1.27) leads to the sought strain-displacement relation:

$$\mathbf{E} = \frac{1}{2}\left(\operatorname{Grad} \boldsymbol{\xi} + {}^t\operatorname{Grad} \boldsymbol{\xi} + {}^t\operatorname{Grad} \boldsymbol{\xi} \cdot \operatorname{Grad} \boldsymbol{\xi}\right) \tag{1.42}$$

or in components (with summation on repeated subscript)

$$E_{\alpha\beta} = \frac{1}{2}\left(\frac{\partial \xi_\alpha}{\partial X_\beta} + \frac{\partial \xi_\beta}{\partial X_\alpha} + \frac{\partial \xi_\gamma}{\partial X_\alpha}\frac{\partial \xi_\gamma}{\partial X_\beta}\right) \tag{1.43}$$

We note the nonlinearity of the strain-displacement relation in the general theory of finite deformations. In the next chapter, we will see how the relations here derived without any restriction on the magnitude of the deformation linearize within the hypothesis of infinitesimal deformation.

Exercise 9. For the uniaxial extension test, determine the displacement vector and the components of the strain tensor.

The displacement vector reads

$$\boldsymbol{\xi} = \mathbf{x} - \mathbf{X} = \xi_1 \mathbf{e}_1 + \xi_2 \mathbf{e}_2 + \xi_3 \mathbf{e}_3$$

and its components in uniaxial extension are

$$\xi_1 = x_1 - X_1 = \alpha X_1; \quad \xi_2 = -\beta X_2; \quad \xi_3 = -\beta X_3$$

The components of $\mathbf{E}$ are obtained from (1.43) in the form

$$E_{11} = \frac{1}{2}\left[\frac{\partial \xi_1}{\partial X_1} + \frac{\partial \xi_1}{\partial X_1} + \frac{\partial \xi_1}{\partial X_1}\frac{\partial \xi_1}{\partial X_1} + \frac{\partial \xi_2}{\partial X_1}\frac{\partial \xi_2}{\partial X_1} + \frac{\partial \xi_3}{\partial X_1}\frac{\partial \xi_3}{\partial X_1}\right] = \alpha + \alpha^2/2$$

Analogously,

$$E_{22} = E_{33} = -\beta + \beta^2/2; \quad E_{ij} = 0 \text{ for } i \neq j$$

■

[4] $\delta_{i\alpha} = 1$ if $i = \alpha$; $\delta_{i\alpha} = 0$ if $i \neq \alpha$.

1.6 TRAINING SET: SINGLE SHEAR

As a challenging training set, reviewing the developed concepts of finite deformation, let us consider the single shear deformation sketched in Figure 1.7. The displacement vector $\boldsymbol{\xi}$ has only one component in the $\mathbf{e}_1$-direction and reads

$$\boldsymbol{\xi} = \xi_1 \mathbf{e}_1 = 2\gamma X_2 \mathbf{e}_1$$

Determine

1. The deformation gradient
2. Transport of volume and surface
3. Finite strain measures
4. Linear dilatation and distortion

1.6.1 Deformation Gradient

For the problem shown in Figure 1.7a, the components dX_2 and dX_3 of the material vector preserve their length in the deformation, and the only nonzero length variation is $dx_1 \neq dX_1$. This can be expressed from the very definition of the deformation gradient (1.6) in the form

$$dx_1 = F_{11} dX_1 + F_{12} dX_2 + F_{13} dX_3$$

Using (1.41), the components of the deformation gradients are

$$F_{11} = 1 + \frac{\partial \xi_1}{\partial X_1} = 1; \quad F_{12} = \frac{\partial \xi_1}{\partial X_2} = 2\gamma; \quad F_{13} = \frac{\partial \xi_1}{\partial X_3} = 0$$

The other nonzero components of $\mathbf{F}$ are

$$F_{22} = F_{33} = 1$$

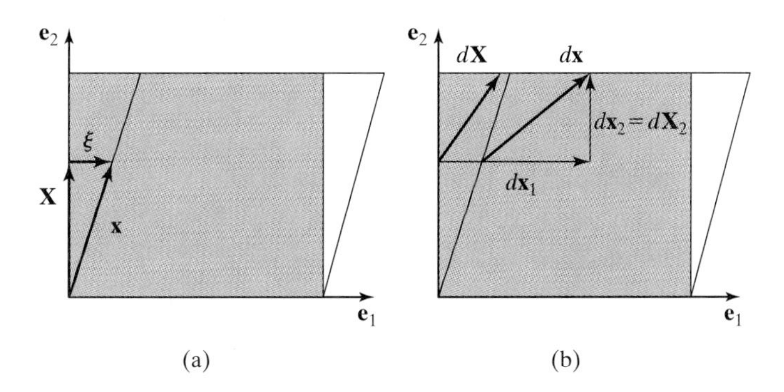

FIGURE 1.7: Simple shear deformation: (a) displacement vector; (b) material vectors.

Thus

$$(F_{i\alpha}) = \begin{bmatrix} 1 & 2\gamma & 0 \\ 0 & 1 & 0 \\ 0 & 0 & 1 \end{bmatrix}$$

or, in tensor notation,

$$\mathbf{F} = \mathbf{1} + 2\gamma \mathbf{e}_1 \otimes \mathbf{e}_2$$

Let us note that the deformation gradient is nonsymmetric. The transpose and the inverse of $\mathbf{F}$ read

$$^t\mathbf{F} = \mathbf{1} + 2\gamma \mathbf{e}_2 \otimes \mathbf{e}_1; \quad \mathbf{F}^{-1} = \mathbf{1} - 2\gamma \mathbf{e}_1 \otimes \mathbf{e}_2$$

or, in matrix form,

$$^t(F_{i\alpha}) = \begin{bmatrix} 1 & 0 & 0 \\ 2\gamma & 1 & 0 \\ 0 & 0 & 1 \end{bmatrix}; \quad (F_{i\alpha}^{-1}) = \begin{bmatrix} 1 & -2\gamma & 0 \\ 0 & 1 & 0 \\ 0 & 0 & 1 \end{bmatrix}$$

1.6.2 Transport Formulas for Volume and Oriented Surface

Since $J = \det \mathbf{F} = 1$, there is volume conservation in the considered simple shear:

$$J = \det \mathbf{F} = 1 \Leftrightarrow d\Omega_t = d\Omega$$

In return, from Figure 1.7 it clearly appears that the surface does not preserve neither its nominal surface, nor its orientation. Indeed, let us consider $\mathbf{N} = \mathbf{e}_1$. Writing (1.17) in matrix form yields

$$\begin{pmatrix} n_1 \\ n_2 \\ n_3 \end{pmatrix} da = 1 \begin{bmatrix} 1 & 0 & 0 \\ -2\gamma & 1 & 0 \\ 0 & 0 & 1 \end{bmatrix} \begin{pmatrix} 1 \\ 0 \\ 0 \end{pmatrix} dA = \begin{pmatrix} 1 \\ -2\gamma \\ 0 \end{pmatrix} dA$$

or, in tensor notation,

$$\mathbf{n}da = (1\mathbf{e}_1 - 2\gamma \mathbf{e}_2)\, dA$$

Thus, $\mathbf{N} \neq \mathbf{n}$ in the single shear deformation. Recalling that $|\mathbf{n}| = 1$, we determine the nominal surface from

$$da = |\mathbf{n}da| = |1\mathbf{e}_1 - 2\gamma \mathbf{e}_2|\, dA = \sqrt{1 + 4\gamma^2} dA$$

Substitution in the surface transport formula leads to

$$\begin{pmatrix} n_1 \\ n_2 \\ n_3 \end{pmatrix} \sqrt{1 + 4\gamma^2} dA = \begin{pmatrix} 1 \\ -2\gamma \\ 0 \end{pmatrix} dA$$

This allows for the determination of the components n_i of the unit vector

$$n_1 = \frac{1}{\sqrt{1+4\gamma^2}}; \quad n_2 = \frac{-2\gamma}{\sqrt{1+4\gamma^2}}; \quad n_3 = 0$$

and

$$\mathbf{n} = \frac{1}{\sqrt{1+4\gamma^2}} \mathbf{e}_1 - \frac{2\gamma}{\sqrt{1+4\gamma^2}} \mathbf{e}_2$$

1.6.3 Finite Strain Measures

Next, with the deformation gradient $\mathbf{F}$ in hand, the Cauchy dilatation tensor $\mathbf{C} = {}^t\mathbf{F} \cdot \mathbf{F}$ is obtained from (1.21); in matrix notation

$$(C_{\alpha\beta}) = \begin{bmatrix} 1 & 0 & 0 \\ 2\gamma & 1 & 0 \\ 0 & 0 & 1 \end{bmatrix} \begin{bmatrix} 1 & 2\gamma & 0 \\ 0 & 1 & 0 \\ 0 & 0 & 1 \end{bmatrix} = \begin{bmatrix} 1 & 2\gamma & 0 \\ 2\gamma & 1+4\gamma^2 & 0 \\ 0 & 0 & 1 \end{bmatrix}$$

We note that $\mathbf{C}$ is symmetric (but not diagonal). Finally, use of $\mathbf{C}$ in (1.27) yields the Green–Lagrange strain tensor:

$$(E_{\alpha\beta}) = \frac{1}{2}\left\{\begin{bmatrix} 1 & 2\gamma & 0 \\ 2\gamma & 1+4\gamma^2 & 0 \\ 0 & 0 & 1 \end{bmatrix} - \begin{bmatrix} 1 & 0 & 0 \\ 0 & 1 & 0 \\ 0 & 0 & 1 \end{bmatrix}\right\} = \begin{bmatrix} 0 & \gamma & 0 \\ \gamma & 2\gamma^2 & 0 \\ 0 & 0 & 0 \end{bmatrix}$$

or, in tensor notation,

$$\mathbf{E} = \gamma\left[\mathbf{e}_1 \otimes \mathbf{e}_2 + \mathbf{e}_2 \otimes \mathbf{e}_1\right] + 2\gamma^2 \mathbf{e}_2 \otimes \mathbf{e}_2$$

From the nondiagonal structure of both $\mathbf{C}$ and $\mathbf{E}$, it follows that $\mathbf{e}_1$,$\mathbf{e}_2$ are not the principal axis of deformation.

The eigenvalues E_J of $\mathbf{E}$ can be obtained from

$$\det(\mathbf{E} - E_J\mathbf{1}) = 0$$

that is,

$$\begin{vmatrix} -E_J & \gamma & 0 \\ \gamma & 2\gamma^2 - E_J & 0 \\ 0 & 0 & E_J \end{vmatrix} = E_J^2(2\gamma^2 - E_J) + \gamma^2 E_J = 0$$

One eigenvalue is obviously $E_{III} = 0$ associated with the eigenvector $\mathbf{u}_{III} = \mathbf{e}_3$. The other follows from the characteristic polynomial:

$$E_J^2 - 2\gamma^2 E_J - \gamma^2 = 0 \leftrightarrow E_{I,II} = \gamma^2 \pm \sqrt{\gamma^4 + \gamma^2}$$

The associated principal directions (eigenvectors) to these eigenvectors can be determined analogously to (1.23):

$$\mathbf{E} \cdot \mathbf{u}_J = E_J \mathbf{u}_J$$

Thus, here, for $\mathbf{u}_J = \mathbf{u}_I$,

$$\begin{bmatrix} 0 & \gamma \\ \gamma & 2\gamma^2 \end{bmatrix}\begin{pmatrix} u_{I1} \\ u_{I2} \end{pmatrix} = E_I \begin{pmatrix} u_{I1} \\ u_{I2} \end{pmatrix}$$

where u_{Ii} denote the components of the (unit) eigenvector (and we considered $u_{I3} = 0$), which satisfies

$$|\mathbf{u}_I| = 1 \leftrightarrow u_{I1}^2 + u_{I2}^2 + u_{I3}^2 = 1$$

The two previous equations define four equations for three unknowns. One expression satisfying these relations is

$$u_{I1} = \frac{\gamma}{\sqrt{E_I^2 + \gamma^2}}; \quad u_{I2} = \frac{E_I}{\sqrt{E_I^2 + \gamma^2}}; \quad u_{I3} = 0$$

and the eigenvector reads in tensor notation

$$\mathbf{u}_I = \frac{\gamma \mathbf{e}_1 + E_I \mathbf{e}_2}{\sqrt{E_I^2 + \gamma^2}}$$

Finally, the third eigenvector $\mathbf{u}_{II}$ can be obtained from the orthonormality of the eigenvectors, that is,

$$\mathbf{u}_{II} = \mathbf{u}_{III} \times \mathbf{u}_I = -u_{I2}\mathbf{e}_1 + u_{I1}\mathbf{e}_2$$

1.6.4 Linear Dilatation and Distortion

With the eigenvalues of $\mathbf{E}$ in hand, the principal dilatations are straightforwardly determined using (1.34):

$$\lambda(\mathbf{u}_I) = \sqrt{1 + 2\gamma^2 + 2\sqrt{\gamma^4 + \gamma^2}} - 1$$
$$\lambda(\mathbf{u}_{II}) = \sqrt{1 + 2\gamma^2 - 2\sqrt{\gamma^4 + \gamma^2}} - 1$$
$$\lambda(\mathbf{u}_{III}) = 0$$

In turn, the distortion in the $(\mathbf{e}_1 \times \mathbf{e}_2)-$ plane reads

$$\sin\theta_{12} = \frac{2E_{12}}{\sqrt{(1 + 2E_{11})(1 + 2E_{22})}} = \frac{2\gamma}{\sqrt{1 + 4\gamma^2}}$$

1.7 APPENDIX CHAPTER 1: TENSOR NOTATION

Any tensor is composed of coordinates and basis, which define its order.

1.7.1 Scalar: Zero-Order Tensor

A zero-order tensor is a scalar that has no orientation but that has the same value at a given point in all directions, that is, in Cartesian coordinates x, y, and z:

$$a(\mathbf{x}, t) = a(x, y, z, t) \tag{1.44}$$

1.7.2 Vector: First-Order Tensor

A vector, say $\mathbf{b}$, is a first-order tensor. It is composed of coordinates $b_i = b_i(\mathbf{x}, t)$ and basis $\mathbf{e}_i$:

$$\mathbf{b} = b_i(\mathbf{x}, t)\mathbf{e}_i = b_x(x, y, z, t)\mathbf{e}_x + b_y(x, y, z, t)\mathbf{e}_y + b_z(x, y, z, t)\mathbf{e}_z \tag{1.45}$$

It is useful to recall the following vector operations:

1. A scalar product (symbol "$\cdot$") of two vectors gives a scalar (summation on repeated subscript):

$$\mathbf{b} \cdot \mathbf{b}' = a; \quad b_i b_i' = a \tag{1.46}$$

or, in matrix notation,

$$\begin{pmatrix} b_1 & b_2 & b_3 \end{pmatrix} \begin{pmatrix} b_1' \\ b_2' \\ b_3' \end{pmatrix} = b_1 b_1' + b_2 b_2' + b_3 b_3' = a$$

For instance, the length of a vector is given by

$$|\mathbf{b}| = \sqrt{\mathbf{b} \cdot \mathbf{b}} \tag{1.47}$$

2. A vector product (symbol "$\times$") of two vectors gives a vector, which is normal to both vectors:

$$\mathbf{b} \times \mathbf{b}' = \mathbf{b}''; \quad \mathbf{b} \cdot \mathbf{b}'' = 0; \quad \mathbf{b}' \cdot \mathbf{b}'' = 0 \tag{1.48}$$

or, in matrix form,

$$\begin{pmatrix} b_1 \\ b_2 \\ b_3 \end{pmatrix} \times \begin{pmatrix} b_1' \\ b_2' \\ b_3' \end{pmatrix} = \begin{pmatrix} b_1'' \\ b_2'' \\ b_3'' \end{pmatrix} = \begin{pmatrix} b_2 b_3' - b_2' b_3 \\ -(b_1 b_3' - b_1' b_3) \\ b_1 b_2' - b_1' b_2 \end{pmatrix}$$

3. A tensor product (symbol "$\otimes$") of two vectors gives a second-order tensor:

$$\mathbf{b} \otimes \mathbf{b}' = \mathbf{c}; \quad b_i b_j' = c_{ij} \tag{1.49}$$

or, in matrix form,

$$\begin{pmatrix} b_1 \\ b_2 \\ b_3 \end{pmatrix} \begin{pmatrix} b_1' & b_2' & b_3' \end{pmatrix} = \begin{bmatrix} c_{11} = b_1 b_1' & c_{12} = b_1 b_2' & c_{13} = b_1 b_3' \\ c_{21} = b_2 b_1' & c_{22} = b_2 b_2' & c_{23} = b_2 b_3' \\ c_{31} = b_3 b_1' & c_{32} = b_3 b_2' & c_{33} = b_3 b_3' \end{bmatrix}$$

As an application, show that the following operation holds:

$$(\mathbf{a} \otimes \mathbf{b}) \cdot \mathbf{c} = (\mathbf{b} \cdot \mathbf{c})\mathbf{a} \tag{1.50}$$

1.7.3 Second-Order Tensor and Matrix Representation

A second-order tensor, say $\mathbf{c} = \mathbf{c}(\mathbf{x}, t)$, is composed of coordinates $c_{ij}(\mathbf{x}, t)$ and basis $\mathbf{e}_i \otimes \mathbf{e}_j$:

$$\mathbf{c}(\mathbf{x}, t) = c_{ij}(\mathbf{x}, t)\mathbf{e}_i \otimes \mathbf{e}_j \tag{1.51}$$

The second-order basis $\mathbf{e}_i \otimes \mathbf{e}_j$ is obtained by the tensor product [see Eq. (1.49)] of the two base vectors $\mathbf{e}_i$ and $\mathbf{e}_j$. A second-order tensor can be represented in matrix form composed of nine components:

$$(\mathbf{c}(\mathbf{x}, t)) = \begin{bmatrix} c_{11} & c_{12} & c_{13} \\ c_{21} & c_{22} & c_{32} \\ c_{31} & c_{23} & c_{33} \end{bmatrix}$$

The specific component c_{ij} is obtained by a double scalar contraction of tensor $\mathbf{c}(\mathbf{x},t)$:

$$c_{ij} = \mathbf{e}_i \cdot \mathbf{c}(\mathbf{x},t) \cdot \mathbf{e}_j \tag{1.52}$$

For instance, the component c_{13} is obtained from

$$\begin{aligned} c_{13} &= \mathbf{e}_1 \cdot \mathbf{c}(\mathbf{x},t) \cdot \mathbf{e}_3 \\ &= \begin{pmatrix} 1 & 0 & 0 \end{pmatrix} \begin{bmatrix} c_{11} & c_{12} & c_{13} \\ c_{21} & c_{22} & c_{32} \\ c_{31} & c_{23} & c_{33} \end{bmatrix} \begin{pmatrix} 0 \\ 0 \\ 1 \end{pmatrix} \\ &= \begin{pmatrix} c_{11} & c_{12} & c_{13} \end{pmatrix} \begin{pmatrix} 0 \\ 0 \\ 1 \end{pmatrix} = c_{13} \end{aligned}$$

The same could have been obtained by the following double tensor contraction (symbol ":") of two second-order tensors:

$$c_{ij} = \mathbf{c}(\mathbf{x},t) : [\mathbf{e}_i \otimes \mathbf{e}_j] \tag{1.53}$$

The tensor contraction in (1.52) can be carried out as the sum of products of each element of matrix $(\mathbf{c}(\mathbf{x},t))$ with the elements of matrix $(\mathbf{e}_i \otimes \mathbf{e}_j)$.

1.8 PROBLEM SET: FINITE BENDING DEFORMATION OF A BEAM

We want to study the deformation of the beam shown in Figure 1.8. After deformation, the initially horizontal beam fibers form a circle of radius $R' = R - Y$, where R is the distance between the center of the circle and the "neutral" axis of the deformed beam ($Y = 0$), and Y denotes the coordinate of the position vector in the $\mathbf{e}_y$-direction in the initial configuration:

$$\mathbf{X} = X\mathbf{e}_x + Y\mathbf{e}_y + Z\mathbf{e}_z$$

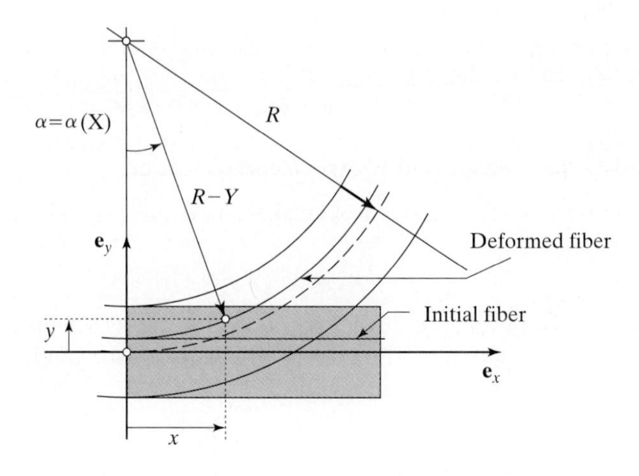

FIGURE 1.8: Problem set: bending deformation of a beam.

Height and width of the beam are assumed to remain constant throughout the deformation.

1. **Finite Deformation**: We are interested in describing the beam's deformation without restriction of the order of deformation (i.e., finite theory).

   (a) In the Cartesian coordinate setting, the position of any point after deformation reads

   $$\mathbf{x} = x\mathbf{e}_x + y\mathbf{e}_y + z\mathbf{e}_z$$

   Show that the position vector $\mathbf{x}$ depends only on Y and $\alpha = \alpha(X)$. Determine the components x, y, z of $\mathbf{x}$.

   (b) By using the previously determined position vector $\mathbf{x}$, determine the deformation gradient $\mathbf{F} = \text{Grad}\ \mathbf{x}$. Show that this overall deformation can be decomposed in a pure deformation characterized by a symmetric deformation gradient $\mathbf{D} = {}^t\mathbf{D}$ and an orthogonal rigid body rotation characterized by the skew-symmetric deformation gradient $\mathbf{R}$, such that

   $$\mathbf{F} = \mathbf{R} \cdot \mathbf{D}$$

   Give an interpretation of this polar decomposition for the beam deformation, determine $\mathbf{D}$ and $\mathbf{R}$, and proof the orthogonality properties of $\mathbf{R}$. Determine the volume variation and the oriented surface perpendicular to the beam's reference axis $(Y = 0)$.

   (c) Determine the Cauchy dilatation tensor $\mathbf{C}$ and the Green–Lagrange strain tensor $\mathbf{E}$. Determine the linear dilatations. By using the linear dilatation $\lambda(\mathbf{e}_x)$, determine the condition for which $\alpha = X/R$. Give a geometrical explanation.

2. **Infinitesimal Deformation**: [after study of Chapter 2] We now turn to the linearized theory of infinitesimal deformation of the beam. In what follows, we will assume that $\alpha = X/R$.

   (a) Specify the hypothesis of infinitesimal deformation for the considered beam problem. (*Hint*: There are 2.)

   (b) Show that, in the linear theory, the deformation gradient of the beam deformation can be developed in the form

   $$\mathbf{F} = \mathbf{1} + \boldsymbol{\varepsilon} + \boldsymbol{\omega}$$

   where $\boldsymbol{\varepsilon}$ is the linearized (symmetric) strain tensor, and $\boldsymbol{\omega}$ the skew-symmetric rotation tensor:

   $$\boldsymbol{\varepsilon} = \frac{1}{2}\left(\text{grad}\,\boldsymbol{\xi} + {}^t\text{grad}\,\boldsymbol{\xi}\right); \quad \boldsymbol{\omega} = \frac{1}{2}\left(\text{grad}\,\boldsymbol{\xi} - {}^t\text{grad}\,\boldsymbol{\xi}\right)$$

   Determine $\boldsymbol{\varepsilon}$ and $\boldsymbol{\omega}$. Compare the results with those obtained in the finite deformation theory.

(c) Beam theories are generally based on the Navier–Bernoulli hypothesis: An initially straight section remains straight throughout the deformation and normal to the middle axis of the beam. How many strain rosettes are required to check this assumption for the given infinitesimal 2D-beam deformation? Justify your answer in the material plane and Mohr strain plane.

1.8.1 Finite Deformation

a. Position Vector x

$$\begin{aligned} x &= (R-Y)\sin\alpha(X) \\ y &= R-(R-Y)\cos\alpha(X) \\ z &= Z \end{aligned}$$

b-1. Deformation Gradient F

$$\mathbf{F} = \operatorname{Grad}\mathbf{x}$$

$$(\mathbf{F}) = \begin{bmatrix} \frac{\partial x}{\partial X} & \frac{\partial x}{\partial Y} & \frac{\partial x}{\partial Z} \\ \frac{\partial y}{\partial X} & \frac{\partial y}{\partial Y} & \frac{\partial y}{\partial Z} \\ \frac{\partial z}{\partial X} & \frac{\partial z}{\partial Y} & \frac{\partial z}{\partial Z} \end{bmatrix} = \begin{bmatrix} (R-Y)\cos\alpha(X)\frac{\partial\alpha}{\partial X} & -\sin\alpha(X) & 0 \\ (R-Y)\sin\alpha(X)\frac{\partial\alpha}{\partial X} & \cos\alpha(X) & 0 \\ 0 & 0 & 1 \end{bmatrix}$$

b-2. Polar Decomposition

$$d\mathbf{x}' = \mathbf{D}\cdot d\mathbf{X} \rightarrow d\mathbf{x} = \mathbf{R}\cdot d\mathbf{x}' = (\mathbf{R}\cdot\mathbf{D})\cdot d\mathbf{X} = \mathbf{F}\cdot d\mathbf{X}$$

Components of material vector $d\mathbf{x}$:

$$dx = (R-Y)\cos\alpha(X)\frac{\partial\alpha}{\partial X}dX - \sin\alpha(X)\,dY$$

$$dy = (R-Y)\sin\alpha(X)\frac{\partial\alpha}{\partial X} + \cos\alpha(X)\,dY$$

$$dz = dZ$$

or, equivalently,

$$dx = \cos\alpha\,dx' - \sin\alpha\,dy'$$

$$dy = \sin\alpha\,dx' + \cos\alpha\,dy'$$

$$dz = dz'$$

with dx', dy', dz' the components of the material vector

$$dx' = (R-Y)\frac{\partial\alpha}{\partial X}dX; \quad dy' = dY; \quad dz' = dZ$$

Thus

$$(\mathbf{D}) = \begin{bmatrix} (R-Y)\frac{\partial\alpha}{\partial X} & 0 & 0 \\ & 1 & 0 \\ \text{sym} & & 1 \end{bmatrix}; \quad (\mathbf{R}) = \begin{bmatrix} \cos\alpha & -\sin\alpha & 0 \\ \sin\alpha & \cos\alpha & 0 \\ 0 & 0 & 1 \end{bmatrix}$$

Orthogonality Properties:

$$\det \mathbf{R} = \cos^2 \alpha + \sin^2 \alpha = 1; \quad {}^t\mathbf{R} = \mathbf{R}^{-1}$$

Note that the pure stretch by $\mathbf{D}$ occurs in the $\mathbf{e}_x$-direction (i.e., along the fiber orientation of the beam).

b-3. Volume Variation

$$d\Omega_t = J d\Omega; \quad J = \det \mathbf{D} = (R - Y)\frac{\partial \alpha}{\partial X}$$

b-4. Oriented Surface in the $\mathbf{e}_y$-Direction (Perpendicular to Beam's Axis) The infinitesimal surface dA oriented by $\mathbf{N} = \mathbf{e}_y$ is situated in the $(\mathbf{e}_x \times \mathbf{e}_z)$-plane. According to the expression of $\mathbf{D}$, the only nonzero stretch is in the $\mathbf{e}_x$-direction. Since the width of the beam is constant, the change in nominal surface is

$$da = (R - Y)\frac{\partial \alpha}{\partial X} dA$$

In turn, the change in orientation of the surface $\mathbf{N} \to \mathbf{n}$ results from the change of orientation of the fiber during deformation (i.e., from the rigid body rotation $\mathbf{R}$); thus

$$\mathbf{n} = \mathbf{R} \cdot \mathbf{e}_y; \quad \begin{pmatrix} n_x \\ n_y \\ n_z \end{pmatrix} = \begin{bmatrix} \cos\alpha & -\sin\alpha & 0 \\ \sin\alpha & \cos\alpha & 0 \\ 0 & 0 & 1 \end{bmatrix} \begin{pmatrix} 0 \\ 1 \\ 0 \end{pmatrix} = \begin{pmatrix} -\sin\alpha \\ \cos\alpha \\ 0 \end{pmatrix}$$

We can obtain the same result from application of the transport formula of an oriented surface:

$$da\mathbf{n} = J\,{}^t\mathbf{F}^{-1} \cdot \mathbf{N} dA$$

$$({}^t\mathbf{F}^{-1}) = \begin{bmatrix} \frac{\cos\alpha}{(R-Y)\frac{\partial\alpha}{\partial X}} & -\sin\alpha & 0 \\ \frac{\sin\alpha}{(R-Y)\frac{\partial\alpha}{\partial X}} & \cos\alpha & 0 \\ 0 & 0 & 1 \end{bmatrix}$$

Hence, for $\mathbf{N} = \mathbf{e}_y$,

$$\begin{aligned} da \begin{pmatrix} n_x \\ n_y \\ n_z \end{pmatrix} &= (R - Y)\frac{\partial \alpha}{\partial X} \begin{bmatrix} \frac{\cos\alpha}{(R-Y)\frac{\partial\alpha}{\partial X}} & -\sin\alpha & 0 \\ \frac{\sin\alpha}{(R-Y)\frac{\partial\alpha}{\partial X}} & \cos\alpha & 0 \\ 0 & 0 & 1 \end{bmatrix} \begin{pmatrix} 0 \\ 1 \\ 0 \end{pmatrix} dA \\ &= (R - Y)\frac{\partial \alpha}{\partial X} \begin{pmatrix} -\sin\alpha \\ \cos\alpha \\ 0 \end{pmatrix} dA \end{aligned}$$

Thus,

$$da = (R - Y)\frac{\partial \alpha}{\partial X} dA; \quad \mathbf{n} = -\sin\alpha \mathbf{e}_x + \cos\alpha \mathbf{e}_y$$

c-1. Cauchy Dilatation Tensor

$$\mathbf{C} = \mathbf{D}^2; \quad (\mathbf{C}) = \begin{pmatrix} (R-Y)^2\left(\frac{\partial\alpha}{\partial X}\right)^2 & 0 & 0 \\ & 1 & 0 \\ \text{sym} & & 1 \end{pmatrix}$$

Note the diagonal structure of $\mathbf{C}$, which means that the base vectors $(\mathbf{e}_x, \mathbf{e}_y, \mathbf{e}_z)$ are the principal directions of deformation $\mathbf{u}_J$ $(J = 1, 2, 3)$.

c-2. Green–Lagrange Strain Tensor

$$2\mathbf{E} = \mathbf{C} - \mathbf{1}$$

Since eigenvectors of $\mathbf{C}$ and $\mathbf{E}$ are the same, it follows that

$$\mathbf{E} = \frac{1}{2}\left[(R-Y)^2\left(\frac{\partial\alpha}{\partial X}\right)^2 - 1\right]\mathbf{e}_x \otimes \mathbf{e}_x$$

c-3. Linear Dilatations The only length variation is in the $\mathbf{e}_x$-direction; therefore,

$$\lambda(\mathbf{e}_y) = \lambda(\mathbf{e}_z) = 0 \text{ (i.e., no change in height and width)}$$

Relative length variation of any fiber $\lambda(\mathbf{e}_x)$ is the length ratio after and before deformation. Noting the fiber length before dX, and after deformation $(R-Y)\,d\alpha$, it follows that

$$\lambda(\mathbf{e}_x) = \frac{(R-Y)\,d\alpha - dX}{dX}$$

The same result could have been obtained from the formula

$$\begin{aligned} \lambda(\mathbf{e}_x &= \mathbf{u}_I) = \sqrt{1+2E_I} - 1 = \sqrt{1 + \left[(R-Y)^2\left(\frac{\partial\alpha}{\partial X}\right)^2 - 1\right]} - 1 \\ &= (R-Y)\left(\frac{\partial\alpha}{\partial X}\right) - 1 \end{aligned}$$

c-4. Condition for $\alpha = X/R$ For $\alpha = X/R$,

$$\lambda(\mathbf{e}_x = \mathbf{u}_I) = \frac{R-Y}{R} - 1 = -\frac{Y}{R}$$

Thus for $Y = 0$, it is $\lambda(\mathbf{e}_x, Y = 0) = 0$. This means that the length of the neutral axis remains constant during deformation.

1.8.2 Infinitesimal Deformation

a. Hypothesis of Infinitesimal Deformation

$$\|\text{Grad}\,\boldsymbol{\xi}\| \ll 1$$

The hypothesis of infinitesimal deformation must apply to both the pure stretch and the rigid body rotation:

$$d\mathbf{x}' = \mathbf{D} \cdot d\mathbf{X}; \quad \mathbf{D} = \mathbf{1} + \text{Grad}\ \boldsymbol{\xi}'$$

$$d\mathbf{x} = \mathbf{R} \cdot d\mathbf{x}'; \quad \mathbf{R} = \mathbf{1} + \text{Grad}\ \boldsymbol{\xi}''$$

where

$$(\text{Grad}\,\boldsymbol{\xi}') = \begin{bmatrix} (R-Y)\frac{\partial \alpha}{\partial X} - 1 & 0 & 0 \\ & 0 & 0 \\ \text{sym} & & 0 \end{bmatrix}$$

$$(\text{Grad}\ \boldsymbol{\xi}'') = \begin{bmatrix} \cos\alpha - 1 & -\sin\alpha & 0 \\ \sin\alpha & \cos\alpha - 1 & 0 \\ 0 & 0 & 0 \end{bmatrix}$$

Therefore,

$$\left\|\text{Grad}\ \boldsymbol{\xi}'\right\| = \left((R-Y)\frac{\partial \alpha}{\partial X} - 1\right)^2 \ll 1$$

$$\left\|\text{Grad}\ \boldsymbol{\xi}''\right\| = 4(1-\cos\alpha) \ll 1$$

For $\alpha = X/R$, the assumption of infinitesimal deformation reads

$$\left|\frac{Y}{R}\right| \ll 1; \quad |\alpha| = \left|\frac{X}{R}\right| \ll 1$$

Remark: The first assumption implies the infinitesimal order of the strains, the second the infinitesimal order of the rotation.

b. Polar Decomposition in the Infinitesimal Theory

$$\begin{aligned} \mathbf{F} &= \mathbf{R} \cdot \mathbf{D} = \left(\mathbf{1} + \text{Grad}\ \boldsymbol{\xi}''\right) \cdot \left(\mathbf{1} + \text{Grad}\ \boldsymbol{\xi}'\right) \\ &= \mathbf{1} + \text{Grad}\ \boldsymbol{\xi}' + \text{Grad}\ \boldsymbol{\xi}'' + \text{Grad}\ \boldsymbol{\xi}'' \cdot \text{Grad}\,\boldsymbol{\xi}' \end{aligned}$$

In the linear theory (merging initial and current configuration),

$$\text{Grad}\ \boldsymbol{\xi}' \simeq \text{grad}\ \boldsymbol{\xi}' = -\frac{Y}{R}\mathbf{e}_x \otimes \mathbf{e}_x$$

The only nonzero component of this displacement gradient corresponds to the relative length variation $\lambda(\mathbf{e}_x) = -Y/R$. In addition, since Grad $\boldsymbol{\xi}'$ is the displacement gradient, which provokes strains, all other strain elements are zero, and therefore

$$\boldsymbol{\varepsilon} = \frac{1}{2}\left(\text{grad}\,\boldsymbol{\xi} +^t \text{grad}\,\boldsymbol{\xi}\right) = \text{grad}\,\boldsymbol{\xi}' = \lambda(\mathbf{e}_x)\mathbf{e}_x \otimes \mathbf{e}_x$$

Analogously, Grad $\boldsymbol{\xi}''$ = grad $\boldsymbol{\xi}'' = \boldsymbol{\omega}$; hence for $|\alpha| \ll 1 \leftrightarrow \cos\alpha = 1; \sin\alpha = 0$,

$$\boldsymbol{\omega} = \frac{1}{2}\left(\text{grad}\,\boldsymbol{\xi} - {}^t\text{grad}\,\boldsymbol{\xi}\right) = \begin{bmatrix} 0 & -\alpha & 0 \\ \alpha & 0 & 0 \\ 0 & 0 & 0 \end{bmatrix}$$

Finally,

$$\begin{aligned}\operatorname{Grad}\boldsymbol{\xi}'' \cdot \operatorname{Grad}\boldsymbol{\xi}' &= \lambda(\mathbf{e}_x)[(\cos\alpha - 1)\mathbf{e}_x \otimes \mathbf{e}_x + \sin\alpha\mathbf{e}_y \otimes \mathbf{e}_x] \\ &\simeq \lambda\alpha\mathbf{e}_y \otimes \mathbf{e}_x\end{aligned}$$

With $\lambda \ll 1$ and $\alpha \ll 1$, the term $\lambda\alpha$ is an order of magnitude smaller than λ and α, and therefore negligible in the infinitesimal theory of deformation.

c. Navier–Bernoulli Assumption for 2D-Beam Problems In the $(\mathbf{e}_x \times \mathbf{e}_y)$-plane, let $d\mathbf{X}$ and $d\mathbf{Y}$ be two perpendicular material vectors in the initial configuration, which transform into $d\mathbf{x}$ and $d\mathbf{y}$ during deformation. The Navier–Bernoulli assumption states

$$d\mathbf{X} \cdot d\mathbf{Y} = 0; \quad d\mathbf{x} \cdot d\mathbf{y} = 0$$

Thus,

$$d\mathbf{x} \cdot d\mathbf{y} = |d\mathbf{x}|\,|d\mathbf{y}| \cos(\pi/2 - \theta) = 0 \rightarrow \theta = 0$$

The Navier–Bernoulli hypothesis, therefore, implies that the distortion in the $(\mathbf{e}_x \times \mathbf{e}_y)$-plane is zero; that is, in infinitesimal deformation,

$$\theta = 2\gamma(\mathbf{e}_x, \mathbf{e}_y) = 2\varepsilon_{xy} = 0$$

This also implies that $\mathbf{e}_x$ and $\mathbf{e}_y$ are principal directions of deformation.

Consider one strain gage rosette on the beam's surface in the $(\mathbf{e}_x \times \mathbf{e}_y)$-plane, inclined with regard to the beam's axis by 45°, such that the gage in between the two orthonormal gages coincides with the beam's axis $\mathbf{n}_b = \mathbf{e}_x$. This is sketched in Figure 1.9a. This strain gage will measure the fiber strain $\varepsilon_{xx} = \lambda(\mathbf{n}_b = \mathbf{e}_x)$. If the shear stress $\varepsilon_{xy} = 0$ was zero, the relative length increase recorded by the

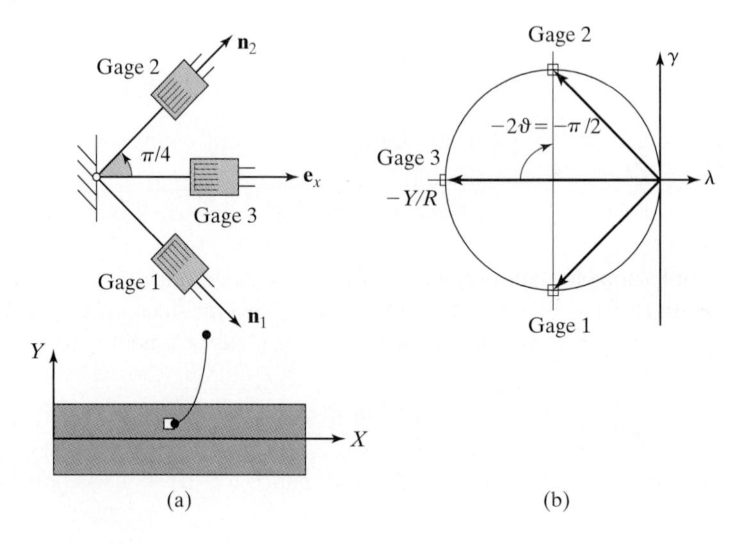

FIGURE 1.9: Experimental verification of Navier–Bernoulli's hypothesis in (a) material plane, (b) Mohr strain plane.

other strain gages, oriented respectively by $\mathbf{n}_1 = \frac{\sqrt{2}}{2}[\mathbf{e}_x - \mathbf{e}_y]$ and $\mathbf{n}_2 = \frac{\sqrt{2}}{2}[\mathbf{e}_x + \mathbf{e}_y]$, should be the same, and—provided the beam's height constant—equal to half the longitudinal strain, $\lambda(\mathbf{n}_1) = \lambda(\mathbf{n}_2) = \lambda(\mathbf{n}_b)/2$. This is sketched in the Mohr strain plane in Figure 1.9b. If, in addition, the longitudinal strain is of the form $\lambda(\mathbf{n}_b) = -Y/R$, then a second strain gage in the same section, but situated on another height, would allow verification of the linear strain distribution suggested by the Navier–Bernoulli hypothesis.

CHAPTER 2

Infinitesimal Deformation

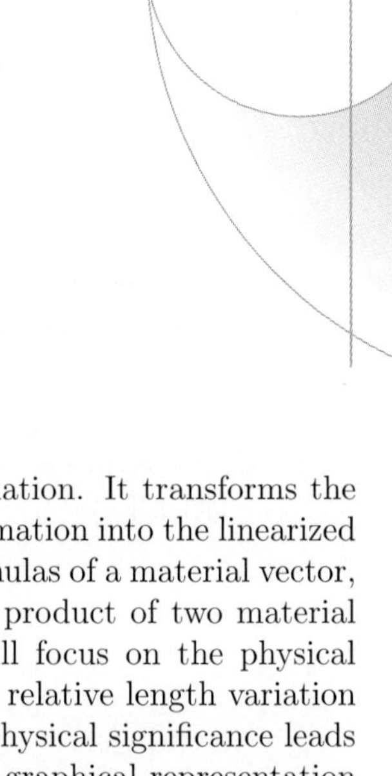

Chapter 2 is devoted to the linear theory of deformation. It transforms the mathematical tools developed in Chapter 1 for finite deformation into the linearized framework of infinitesimal deformation: the transport formulas of a material vector, an elementary volume, an oriented surface, and a scalar product of two material vectors associated with the strain tensor. Attention will focus on the physical significance and geometric interpretation of the strains as relative length variation (linear dilatation) and angle variation (distortion). This physical significance leads us straightforward to the Mohr circle in strain space, the graphical representation of the symmetric linearized second-order strain tensor. A direct application of this geometric interpretation and its Mohr plane representation is the strain rosette (i.e., the extensometer measurement of infinitesimal strains using strain rosettes).

2.1 THE HYPOTHESIS OF INFINITESIMAL DEFORMATION

In many continuous engineering material problems, a linear approximation of the finite theory can be carried out, provided that the deformation is infinitesimal.

2.1.1 The Hypothesis

Infinitesimal deformation means an infinitesimal order of magnitude of the deformation, expressed by a norm (e.g., $||...||$). In addition, we have already seen that the deformation is characterized by the gradient of a vector—for instance, by that of the displacement vector $\boldsymbol{\xi} = \boldsymbol{\xi}(\mathbf{X})$. The order of magnitude of deformation needs to be characterized by a nondimensional norm of the displacement gradient[1]:

$$\|\mathrm{Grad}\,\boldsymbol{\xi}\| = \mathrm{tr}\,(\mathrm{Grad}\,\boldsymbol{\xi} \cdot {}^t\mathrm{Grad}\,\boldsymbol{\xi}) \tag{2.1}$$

where tr stands for the trace. The hypothesis of infinitesimal deformation is expressed in the following way:

$$\forall \mathbf{X}: \quad \|\mathrm{Grad}\,\boldsymbol{\xi}\| \ll 1 \tag{2.2}$$

This hypothesis necessarily implies that the strains are infinitesimal:

$$\|\mathrm{Grad}\,\boldsymbol{\xi}\| \ll 1 \Rightarrow \|\mathbf{E}\| \ll 1 \tag{2.3}$$

However, the inverse is not necessarily true: The strains can be infinitesimal, while the deformation is not. For instance, the strains in a rigid body motion are zero, while the deformation has a finite value with $\mathbf{F} = \mathbf{R}$, and the displacement gradient $\mathrm{Grad}\,\boldsymbol{\xi} = \mathbf{R} - 1$ may take any value [see Eqs. (1.18) and (1.41)].

Exercise 10. Proof this statement for the pure rotation deformation of the position vector $\mathbf{X}$ around $\mathbf{e}_3$ by an angle θ.

The deformation gradient of this rigid body motion $\mathbf{F} = \mathbf{R}$ was determined in the Exercise in Section 1.4, Eq. (1.18). From these results, $\mathrm{Grad}\,\boldsymbol{\xi} = \mathbf{R} - 1$ reads in matrix form:

$$\left[\frac{\partial \xi_i}{\partial X_\alpha}\right] = (R_{i\alpha}) - (\delta_{i\alpha}) = \begin{bmatrix} \cos\theta - 1 & -\sin\theta & 0 \\ \sin\theta & \cos\theta - 1 & 0 \\ 0 & 0 & 0 \end{bmatrix}$$

Substitution in (2.1) yields

$$\|\mathrm{Grad}\,\boldsymbol{\xi}\| = 2[(\cos\theta - 1)^2 + \sin^2\theta] = 4(1 - \cos\theta)$$

Here, the norm of the displacement gradient can take any value in $\|\mathrm{Grad}\,\boldsymbol{\xi}\,(\theta)\| \in [0, 8]$, while the strain is zero. For instance, use of (1.43) gives

$$\begin{aligned} E_{11} &= \frac{1}{2}\left[\frac{\partial \xi_1}{\partial X_1} + \frac{\partial \xi_1}{\partial X_1} + \frac{\partial \xi_1}{\partial X_1}\frac{\partial \xi_1}{\partial X_1} + \frac{\partial \xi_2}{\partial X_1}\frac{\partial \xi_2}{\partial X_1} + \frac{\partial \xi_3}{\partial X_1}\frac{\partial \xi_3}{\partial X_1}\right] \\ &= \frac{1}{2}\left[2(\cos\theta - 1) + (\cos\theta - 1)^2 + \sin^2\theta\right] = 0 \end{aligned}$$

■

[1] Note, however, that the displacement gradient $\|\mathrm{Grad}\,\boldsymbol{\xi}\|$ belongs to a vector space of finite dimensions with respect to which all the norms are equivalent; and hence the norm $\|...\|$ does not need to be specified.

2.1.2 Linearization of Transport Formulas

We have extensively used the gradient operator (spatial derivatives) in the previous chapter to describe the deformation. Let us now see how the gradient operator is affected by the hypothesis of infinitesimal deformation. Using (1.41) in (1.10), the Gradient operator (spatial derivative with respect to the initial configuration) can be developed in the form

$$\text{Grad}\,(.) = \text{grad}\,(.) \cdot (1 + \text{Grad}\,\boldsymbol{\xi}) = \text{grad}\,(.) + \text{grad}\,(.) \cdot \text{Grad}\,\boldsymbol{\xi} \tag{2.4}$$

Within the hypothesis of infinitesimal deformation, the second-order term on the right-hand side of (2.4) [i.e., the term $\text{grad}\,(.) \cdot \text{Grad}\,\boldsymbol{\xi}$] can be neglected with respect to the linear term:

$$\|\text{Grad}\,\boldsymbol{\xi}\| \ll 1 \Rightarrow \text{Grad}\,(.) \simeq \text{grad}\,(.); \quad \frac{\partial(.)}{\partial \mathbf{X}} \simeq \frac{\partial(.)}{\partial \mathbf{x}} \tag{2.5}$$

In other words, concerning the spatial derivatives (and only them), we can merge initial and current configuration.

Exercise 11. Linearize the volume transport formula (1.11) and the transport formula of an oriented surface (1.17).

Starting with (1.11), the Jacobian J describing the volume transport $d\Omega_t = J d\Omega$ can be developed in the form

$$J = \det(1 + \text{grad}\,\boldsymbol{\xi}) \simeq 1 + \frac{\partial \xi_1}{\partial x_1} + \frac{\partial \xi_2}{\partial x_2} + \frac{\partial \xi_3}{\partial x_3} + ... = 1 + \text{div}\,\boldsymbol{\xi}$$

where "$\text{div}\,(.) = \nabla \cdot (.)$" is the divergence operator. It is useful to recall that the divergence represents the trace of a gradient; for instance, here

$$\nabla \cdot \boldsymbol{\xi} = \text{div}\,\boldsymbol{\xi} = \text{tr}(\text{grad}\,\boldsymbol{\xi}) = \frac{\partial \boldsymbol{\xi}}{\partial \mathbf{x}} : 1 = \frac{\partial \xi_i}{\partial x_j} \delta_{ij}$$

where δ_{ij} is the Kronecker delta.

The linearization of the transport formulas (1.17) of an oriented surface requires the linearized expression of ${}^t\mathbf{F}^{-1}$. It can be obtained from (1.5), (1.41), and (2.5):

$$d\mathbf{x} = (1 + \text{grad}\,\boldsymbol{\xi}) \cdot d\mathbf{X}$$

which inverts into

$$d\mathbf{X} = d\mathbf{x} - \text{grad}\,\boldsymbol{\xi} \cdot d\mathbf{X} \simeq (1 - \text{grad}\,\boldsymbol{\xi}) \cdot d\mathbf{x}$$

This delivers the linearized form of $\mathbf{F}^{-1}$:

$$\mathbf{F}^{-1} \simeq 1 - \text{grad}\,\boldsymbol{\xi}$$

With the linearized expression of J and $\mathbf{F}^{-1}$ in hand, the transport formula (1.17) becomes

$$\mathbf{n}da = (1 + \text{div}\,\boldsymbol{\xi})(1 - {}^t\text{grad}\,\boldsymbol{\xi}) \cdot \mathbf{N}dA$$

■

2.2 THE LINEARIZED STRAIN TENSOR

With (2.3), the Green–Lagrange strain tensor (1.42) linearizes to

$$\mathbf{E} = \frac{1}{2}\left(\mathrm{Grad}\,\boldsymbol{\xi} + {}^t\mathrm{Grad}\,\boldsymbol{\xi}\right) \tag{2.6}$$

In addition, using (2.5), the linearized Green–Lagrange strain tensor can equally be written in the form[2]

$$\mathbf{E} \simeq \boldsymbol{\varepsilon}; \quad E_{\alpha\beta} \simeq E_{ij} = \varepsilon_{ij} \tag{2.7}$$

Here, $\boldsymbol{\varepsilon}$ is the linearized strain tensor defined by

$$\boldsymbol{\varepsilon} = \frac{1}{2}\left(\mathrm{grad}\,\boldsymbol{\xi} + {}^t\mathrm{grad}\,\boldsymbol{\xi}\right); \quad \varepsilon_{ij} = \frac{1}{2}\left(\frac{\partial \xi_i}{\partial x_j} + \frac{\partial \xi_j}{\partial x_i}\right); \quad \boldsymbol{\varepsilon} = \varepsilon_{ij}\mathbf{e}_i \otimes \mathbf{e}_j \tag{2.8}$$

The three real eigenvalues ε_J associated with the three principal directions of deformation $\mathbf{u}_J$ read

$$\boldsymbol{\varepsilon} \cdot \mathbf{u}_J = \varepsilon_J \mathbf{u}_J \tag{2.9}$$

2.2.1 Geometric Interpretation: Relative Length and Angle Variation

We seek a geometric interpretation of the components of the linearized strain tensor $\boldsymbol{\varepsilon}$. Consider an infinitesimal material facet oriented by unit normal $\mathbf{n}$. Using (1.33) while linearizing it,[3] following assumption (2.3), the linear dilatation in the direction $\mathbf{n}$ reads

$$\lambda(\mathbf{n}) = \mathbf{n} \cdot \boldsymbol{\varepsilon} \cdot \mathbf{n} = \varepsilon_{nn} \tag{2.10}$$

Hence, in the linear theory, the diagonal terms of the strain tensor coincide with the linear dilatations (i.e., with the relative length variations). For instance, if we consider $\mathbf{n} = \mathbf{e}_i$, the diagonal term ε_{ii} (no summation on repeated subscript) of tensor $\boldsymbol{\varepsilon}$ represents the linear dilatation $\lambda(\mathbf{e}_i)$ in this physical direction $\mathbf{e}_i$:

$$i = j : \lambda(\mathbf{e}_i) = \mathbf{e}_i \cdot \boldsymbol{\varepsilon} \cdot \mathbf{e}_i = \varepsilon_{ii} \tag{2.11}$$

From this geometric significance it naturally appears that the relative volume variation in the linear theory can be written in the form

$$\frac{d\Omega_t - d\Omega}{d\Omega} = \lambda(\mathbf{e}_1) + \lambda(\mathbf{e}_2) + \lambda(\mathbf{e}_3) = \mathrm{tr}\,\boldsymbol{\varepsilon} = \boldsymbol{\varepsilon} : 1 = \varepsilon_{ij}\delta_{ij} \tag{2.12}$$

Furthermore, for $\mathbf{e}_i = \mathbf{u}_J$, we obtain from (2.11) the principal strains (or principal linear dilatations):

$$\varepsilon_J = \lambda(\mathbf{u}_J) = \mathbf{u}_J \cdot \boldsymbol{\varepsilon} \cdot \mathbf{u}_J \tag{2.13}$$

With this geometrical significance, it appears that the vector defined by (2.9) is the strain vector built on a material surface oriented by unit vector $\mathbf{u}_J$, which has as its only component the principal strain ε_J in the $\mathbf{u}_J$-direction. This is sketched in Figure 2.1a.

---

[2]Note that we can drop the difference between Greek and Latin subscripts, introduced in Chapter 1 to distinguish initial and current configuration in the finite deformation theory.

[3]Consider this a challenging exercise! *Hint*: Develop the square root in (1.33) in a series truncated after the first-order term.

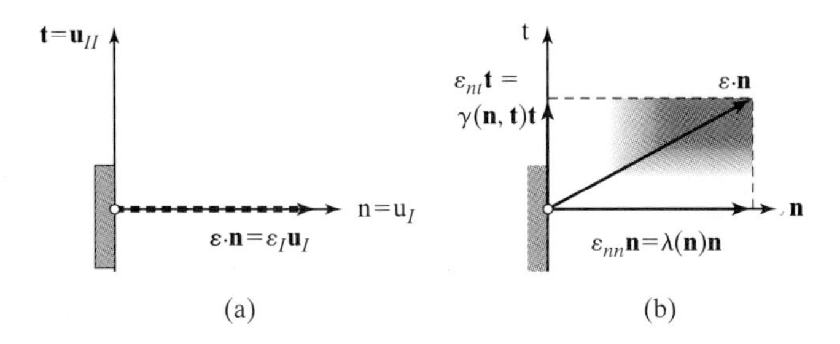

FIGURE 2.1: Strain vector built on a material plane oriented by unit vector $\mathbf{n}$: (a) in principal strain plane $(\mathbf{u}_I \times \mathbf{u}_{II})$; (b) in the $(\mathbf{n} \times \mathbf{t})$-plane.

In a similar fashion, we can linearize the expression of the angle variation (1.39) between two orthonormal vectors $\mathbf{t}$ and $\mathbf{n}$:

$$\sin\theta \simeq \theta(\mathbf{t}, \mathbf{n}) = \mathbf{t} \cdot 2\boldsymbol{\varepsilon} \cdot \mathbf{n} \tag{2.14}$$

Hence, the nondiagonal terms ε_{ij} $(i \neq j)$ of the linearized strain tensor represent the change in the angle (i.e., the distortion), which occurs through deformation between the two orthonormal vectors $\mathbf{e}_i$ and $\mathbf{e}_j$ of the material basis locally attached to the material point:

$$i \neq j : \theta(\mathbf{e}_i, \mathbf{e}_j) = \mathbf{e}_i \cdot 2\boldsymbol{\varepsilon} \cdot \mathbf{e}_j = 2\varepsilon_{ij} = 2\gamma(\mathbf{e}_i, \mathbf{e}_j) \tag{2.15}$$

In other words, the linearized strain tensor can be represented in the following way:

$$(\varepsilon_{ij}) = \begin{bmatrix} \lambda(\mathbf{e}_1) & \gamma(\mathbf{e}_1, \mathbf{e}_2) & \gamma(\mathbf{e}_1, \mathbf{e}_3) \\ & \lambda(\mathbf{e}_2) & \gamma(\mathbf{e}_2, \mathbf{e}_3) \\ \text{sym} & & \lambda(\mathbf{e}_3) \end{bmatrix} \tag{2.16}$$

with $\gamma(\mathbf{e}_i, \mathbf{e}_j)$ the half-distortion in the $(\mathbf{e}_i \times \mathbf{e}_j)$-material plane. In a material plane, any strain vector $\boldsymbol{\varepsilon} \cdot \mathbf{n}$ built on a material plane oriented by $\mathbf{n}$ has three components: the linear dilatation $\lambda(\mathbf{n})$, and the half-distortions $\gamma(\mathbf{n}, \mathbf{t})$ and $\gamma(\mathbf{n}, \mathbf{b})$. This is illustrated in Figure 2.1b.

Finally, we should note that the demonstration here is independent of any particular choice of the set of orthonormal basis. In other words, this geometric interpretation applies equally to non-Cartesian coordinate systems.

Exercise 12. Reconsider the uniaxial extension (*cf.* Figure 1.3) in the linear theory. Specify the hypothesis of infinitesimal deformation, and determine the linearized strain tensor, the volume variation, and the transport of the oriented surfaces.

The displacement vector in uniaxial extension was given by

$$\boldsymbol{\xi} = \mathbf{x} - \mathbf{X} = \alpha X_1 \mathbf{e}_1 - \beta[X_2 \mathbf{e}_2 + X_3 \mathbf{e}_3]$$

The displacement gradient reads

$$\operatorname{Grad} \boldsymbol{\xi} = \alpha \mathbf{e}_1 \otimes \mathbf{e}_1 - \beta[\mathbf{e}_2 \otimes \mathbf{e}_2 + \mathbf{e}_3 \otimes \mathbf{e}_3]$$

Use of the norm (2.1) allows us to specify the hypothesis of infinitesimal deformation:

$$\mathrm{tr}\,(\mathrm{Grad}\,\boldsymbol{\xi}\cdot{}^t\mathrm{Grad}\,\boldsymbol{\xi}) = \alpha^2 + 2\beta^2 \ll 1$$

The linearized strain tensor follows from (2.8) and the diagonal form of the displacement gradient in uniaxial extension:

$$\boldsymbol{\varepsilon} = \mathrm{Grad}\,\boldsymbol{\xi} = \mathrm{grad}\,\boldsymbol{\xi} = {}^t\mathrm{grad}\,\boldsymbol{\xi}$$

Hence, in comparison to the finite theory developed in Chapter 1, the quadratic terms in the strain tensor skip out; and the diagonal terms of the linearized strain tensor $\boldsymbol{\varepsilon}$ coincide with the linear dilatations of the finite theory:

$$\lambda(\mathbf{e}_1) = \alpha; \quad \lambda(\mathbf{e}_2) = \lambda(\mathbf{e}_3) = -\beta$$

Furthermore, according to (2.12), the relative volume variation becomes here

$$J - 1 \simeq \mathrm{tr}\,\boldsymbol{\varepsilon} = \alpha - 2\beta > 0$$

Finally, the transport of the oriented surface in the linear theory reads (see Exercise 11)

$$\mathbf{n}da = (1 + \mathrm{tr}\,\boldsymbol{\varepsilon})(1 - {}^t\mathrm{grad}\,\boldsymbol{\xi}) \cdot \mathbf{N}dA$$

For the uniaxial extension, we have in matrix form

$$\begin{pmatrix} n_1 \\ n_2 \\ n_3 \end{pmatrix} da = (1 + \alpha - 2\beta) \begin{bmatrix} 1-\alpha & & \\ & 1+\beta & \\ & & 1+\beta \end{bmatrix} \begin{pmatrix} N_1 \\ N_1 \\ N_2 \end{pmatrix} dA$$

Application for $\mathbf{N} = \mathbf{e}_1$ yields

$$\begin{aligned} \mathbf{n}da &= (1 + \alpha - 2\beta)(1 - \alpha)\mathbf{e}_1 dA \\ &= (1 - \alpha^2 - 2\beta(1 - \alpha)\mathbf{e}_1 dA \\ &\simeq (1 - 2\beta)\mathbf{e}_1 dA \end{aligned}$$

Analogously, for $\mathbf{N} = \mathbf{e}_2$ (or $\mathbf{N} = \mathbf{e}_3$),

$$\begin{aligned} \mathbf{n}da &= (1 + \alpha - 2\beta)(1 + \beta)\mathbf{e}_2 dA \\ &\simeq (1 + \alpha - \beta)\mathbf{e}_2 dA \end{aligned}$$

Compare these results with those obtained for the finite theory. ■

2.2.2 Polar Decomposition in the Linear Theory (Optional)

The uniaxial extension test has a particular feature: The displacement gradient is symmetric, which means that the deformation is a pure stretch (i.e., $\mathbf{F} = \mathbf{D}$; see Chapter 1). This is not the case in some applications, in particular when rotational terms enter the deformation. This suggests a decomposition of the displacement gradient in a symmetric and a skew-symmetric (or asymmetric) part:

$$\operatorname{grad} \boldsymbol{\xi} = \boldsymbol{\varepsilon} + \boldsymbol{\omega} \tag{2.17}$$

The symmetric part is the linearized strain tensor $\boldsymbol{\varepsilon} = {}^t\boldsymbol{\varepsilon}$ (of components: $\varepsilon_{ij} = \varepsilon_{ji}$), while $\boldsymbol{\omega} = -{}^t\boldsymbol{\omega}$ (of components: $\omega_{ij} = -\omega_{ji}$):

$$\boldsymbol{\varepsilon} = \frac{1}{2}\left(\operatorname{grad} \boldsymbol{\xi} + {}^t\operatorname{grad} \boldsymbol{\xi}\right); \quad \boldsymbol{\omega} = \frac{1}{2}\left(\operatorname{grad} \boldsymbol{\xi} - {}^t\operatorname{grad} \boldsymbol{\xi}\right) \tag{2.18}$$

This decomposition lends itself readily to the polar decomposition of the deformation gradient. Indeed, we have seen in Chapter 1 that the successive application of a pure stretch followed by a rigid body motion, that is,

$$d\mathbf{X} \rightarrow d\mathbf{x}' = \mathbf{D} \cdot d\mathbf{X} \rightarrow d\mathbf{x} = \mathbf{R} \cdot d\mathbf{x}'$$

is expressed by

$$d\mathbf{x} = (\mathbf{R} \cdot \mathbf{D}) \cdot d\mathbf{X} \tag{2.19}$$

The pure stretch deformation is expressed by $\mathbf{D} = \mathbf{1} + \boldsymbol{\varepsilon}$, and the rigid body motion by $\mathbf{R} = \mathbf{1} + \boldsymbol{\omega}$:

$$\mathbf{F} = \mathbf{R} \cdot \mathbf{D} = (\mathbf{1} + \boldsymbol{\omega}) \cdot (\mathbf{1} + \boldsymbol{\varepsilon}) = \mathbf{1} + \boldsymbol{\varepsilon} + \boldsymbol{\omega} \cdot (\mathbf{1} + \boldsymbol{\varepsilon}) \tag{2.20}$$

Hence, it appears that the term $\boldsymbol{\omega} \cdot (\mathbf{1} + \boldsymbol{\varepsilon}) \cdot d\mathbf{X} = \boldsymbol{\omega} \cdot d\mathbf{x}'$ corresponds to a rigid body rotation of the material vector $d\mathbf{x}' = (\mathbf{1} + \boldsymbol{\varepsilon}) \cdot d\mathbf{X}$ attached to the already deformed matter. Within the hypothesis of infinitesimal deformation, the second-order term $\boldsymbol{\omega} \cdot \boldsymbol{\varepsilon}$ in (2.20) can be neglected. This reads formally

$$|\boldsymbol{\omega} \cdot \boldsymbol{\varepsilon} \cdot d\mathbf{X}| \ll |d\mathbf{X}| \tag{2.21}$$

The remaining linear term $\boldsymbol{\omega} \cdot d\mathbf{X}$ corresponds to the rigid body rotation of the material vector $d\mathbf{X}$ defined by the rotation vector $\boldsymbol{\Omega}$:

$$\boldsymbol{\omega} \cdot d\mathbf{X} = \boldsymbol{\Omega} \times d\mathbf{X}; \quad \boldsymbol{\Omega} = \frac{1}{2}\operatorname{curl} \boldsymbol{\xi} \tag{2.22}$$

where "curl" stands for the curl operator.[4] With the help of (2.22), we can express the transport of a material vector in the linear theory in the form

$$d\mathbf{x} = d\mathbf{X} + \operatorname{grad} \boldsymbol{\xi} \cdot d\mathbf{X} = d\mathbf{X} + \boldsymbol{\varepsilon} \cdot d\mathbf{X} + \boldsymbol{\Omega} \times d\mathbf{X} \tag{2.23}$$

This difference in polar decomposition between the finite and the infinitesimal theory of deformation is shown in Figure 2.2.

[4]In a rigid body motion, the curl of the linearized displacement field motion has the direction of the axis of rotation, and its magnitude equals twice the rotation angle θ. We should also note that the displacement vector in this rigid body motion in infinitesimal deformation can be expressed by

$$\boldsymbol{\xi} = \boldsymbol{\Omega} \times \mathbf{X} - \mathbf{X}$$

with $\boldsymbol{\Omega}$ defined by (2.22). For purpose of refreshment of mathematical basis, show this for the pure rotation deformation of the position vector $\mathbf{X}$ around $\mathbf{e}_3$ by an angle θ.

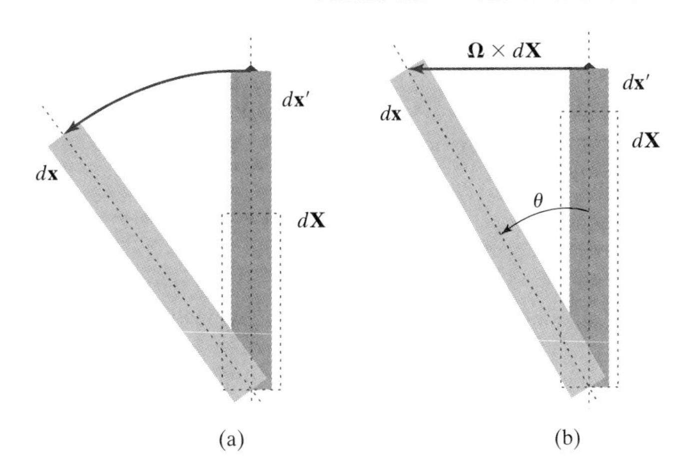

FIGURE 2.2: Polar decomposition: (a) finite deformation; (b) infinitesimal theory.

Exercise 13. Single Shear and Double Shear—The Rotation Makes the Difference. We reconsider the single shear in the linear theory defined by the displacement vector:

$$\xi = 2\gamma X_2 \mathbf{e}_1$$

Specify the hypothesis of infinitesimal deformation and determine the linearized strain tensor, the polar decomposition, and the associated rotation vector. Subsequently, consider the double shear defined by the displacement vector:

$$\xi = \gamma X_2 \mathbf{e}_1 + \gamma X_1 \mathbf{e}_2$$

Show that the double shear is a pure deformation and that the single shear is nothing but a pure deformation followed by a rotation.

For the single shear the displacement gradient reads

$$\operatorname{grad} \boldsymbol{\xi} = 2\gamma \mathbf{e}_1 \otimes \mathbf{e}_2$$

Application of (2.1) allows us to specify the hypothesis of infinitesimal deformation:

$$\gamma^2 << 1$$

Application of (2.18) delivers the linearized strain tensor $\boldsymbol{\varepsilon}$ corresponding to the symmetric part of the displacement gradient:

$$\boldsymbol{\varepsilon} = \frac{1}{2}\left(\operatorname{grad} \boldsymbol{\xi} + {}^t\operatorname{grad} \boldsymbol{\xi}\right) = \gamma \mathbf{e}_1 \otimes \mathbf{e}_2 + \gamma \mathbf{e}_2 \otimes \mathbf{e}_1$$

The asymmetric part $\boldsymbol{\omega}$ of the displacement gradient is given by

$$\boldsymbol{\omega} = \frac{1}{2}\left(\operatorname{grad} \boldsymbol{\xi} - {}^t\operatorname{grad} \boldsymbol{\xi}\right) = \gamma \mathbf{e}_1 \otimes \mathbf{e}_2 - \gamma \mathbf{e}_2 \otimes \mathbf{e}_1$$

The associated rotation vector $\boldsymbol{\Omega}$ has to satisfy (2.22) whatever vector $d\mathbf{X}$ considered:

$$\forall d\mathbf{X};\quad \boldsymbol{\Omega} \times d\mathbf{X} = \boldsymbol{\omega} \cdot d\mathbf{X}$$

The left-hand side reads

$$\begin{pmatrix} \Omega_1 \\ \Omega_2 \\ \Omega_3 \end{pmatrix} \times \begin{pmatrix} dX_1 \\ dX_2 \\ dX_3 \end{pmatrix} = \begin{pmatrix} \Omega_2 dX_3 - \Omega_3 dX_2 \\ \Omega_3 dX_1 - \Omega_1 dX_3 \\ \Omega_1 dX_2 - \Omega_2 dX_1 \end{pmatrix}$$

The right-hand side yields:

$$\begin{pmatrix} 0 & \gamma & 0 \\ -\gamma & 0 & 0 \\ 0 & 0 & 0 \end{pmatrix} \cdot \begin{pmatrix} dX_1 \\ dX_2 \\ dX_3 \end{pmatrix} = \begin{pmatrix} \gamma dX_2 \\ -\gamma dX_1 \\ 0 \end{pmatrix}$$

A comparison of the two previous equations yields

$$\boldsymbol{\Omega} = -\gamma \mathbf{e}_3$$

This result determined from $(2.22)_1$ could have been directly obtained by applying the curl operator to the displacement vector $\boldsymbol{\xi}$:

$$\boldsymbol{\Omega} = \frac{1}{2}\operatorname{curl}\boldsymbol{\xi} = \frac{1}{2}\nabla \times \boldsymbol{\xi} = \frac{1}{2}\begin{pmatrix} \frac{\partial}{\partial X_1} \\ \frac{\partial}{\partial X_2} \\ \frac{\partial}{\partial X_3} \end{pmatrix} \times \begin{pmatrix} 2\gamma X_2 \\ 0 \\ 0 \end{pmatrix} = \begin{pmatrix} 0 \\ 0 \\ -\gamma \end{pmatrix}$$

In the case of double shear, things are somehow different. The displacement gradient associated with the double shear displacement $\boldsymbol{\xi} = \gamma X_2 \mathbf{e}_1 + \gamma X_1 \mathbf{e}_2$ is symmetric:

$$\boldsymbol{\varepsilon} = \operatorname{grad}\boldsymbol{\xi} = {}^t\operatorname{grad}\boldsymbol{\xi} = \gamma \mathbf{e}_1 \otimes \mathbf{e}_2 + \gamma \mathbf{e}_2 \otimes \mathbf{e}_1$$

This symmetry implies $\boldsymbol{\omega} = 0$: The double shear is a pure deformation, and its associated strain tensor coincides with that of the single shear. It follows that the single shear deformation of intensity 2γ can be decomposed into a double shear deformation of intensity γ—that is, a pure deformation—followed by a rotation of angle γ around the $\mathbf{e}_3$-axis. This is sketched in Figure 2.3.

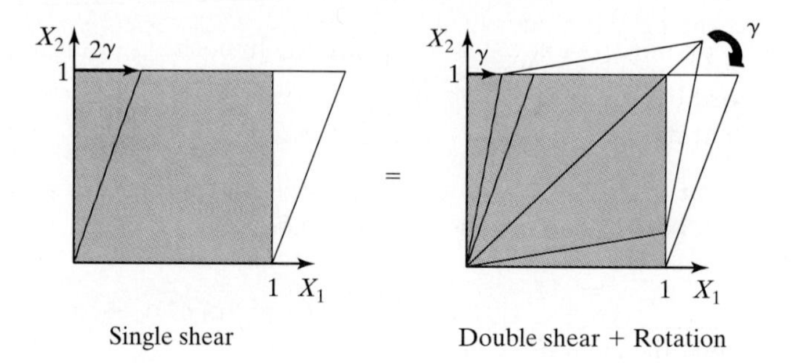

FIGURE 2.3: Polar decomposition: Single shear = Double shear + Rotation.

■

2.3 THE MOHR STRAIN PLANE

We have seen that the linearized strain tensor (2.16) represents both relative length variations $\varepsilon_{ii} = \lambda_i = \lambda(\mathbf{e}_i)$ and half-distortions $\varepsilon_{ij} = \theta(\mathbf{e}_i, \mathbf{e}_j)/2 = \gamma(\mathbf{e}_i, \mathbf{e}_j)$ (for $i \neq j$). This physical significance allows for an (almost) direct experimental determination of the strain components. These measurements are normally carried out in a plane characterized by two orthonormal unit vectors in this plane, denoted $\mathbf{n}$ and $\mathbf{t}$. In this plane measurement situation, the deformation is characterized by two independent linear dilatations, $\lambda(\mathbf{n})$ and $\lambda(\mathbf{t})$, and the half-distortion $\gamma(\mathbf{n}, \mathbf{t}) = \theta(\mathbf{n}, \mathbf{t})/2$ defining the angular variation between the vectors $\mathbf{n}$ and $\mathbf{t}$. Hence, from the overall symmetric second-order strain tensor $\boldsymbol{\varepsilon}$ we extract these three components:

$$\bar{\boldsymbol{\varepsilon}} = \varepsilon_{nn}\mathbf{n} \otimes \mathbf{n} + \varepsilon_{tt}\mathbf{t} \otimes \mathbf{t} + \varepsilon_{nt}(\mathbf{n} \otimes \mathbf{t} + \mathbf{t} \otimes \mathbf{n}) \tag{2.24}$$

or, equivalently, in matrix form,

$$(\bar{\boldsymbol{\varepsilon}}) = \begin{bmatrix} \varepsilon_{nn} & \varepsilon_{nt} \\ \varepsilon_{tn} & \varepsilon_{tt} \end{bmatrix} = \begin{bmatrix} \lambda(\mathbf{n}) & \gamma(\mathbf{n}, \mathbf{t}) \\ \gamma(\mathbf{n}, \mathbf{t}) & \lambda(\mathbf{t}) \end{bmatrix} \tag{2.25}$$

Projection of this reduced measurement strain tensor $\bar{\boldsymbol{\varepsilon}}$ on $\mathbf{n}$ leads to the strain vector:

$$\bar{\mathbf{E}}(\mathbf{n}) = \bar{\boldsymbol{\varepsilon}} \cdot \mathbf{n} = \lambda(\mathbf{n})\mathbf{n} + \gamma(\mathbf{n}, \mathbf{t})\mathbf{t} \tag{2.26}$$

Let us clearly note that $\bar{\mathbf{E}}(\mathbf{n})$ is the reduced strain vector of in-plane strain components $\varepsilon_{nn} = \lambda(\mathbf{n})$ and $\varepsilon_{nt} = \gamma(\mathbf{n}, \mathbf{t})$ built on a material surface oriented by unit normal $\mathbf{n}$, and that it does not include the out-of-plane distortions $\varepsilon_{nb} = \gamma(\mathbf{n}, \mathbf{b})$, which are not accessible by measurements in the $\mathbf{n} \times \mathbf{t}$-plane.[5]

In the same (physical) material plane, let us consider the material basis $(\mathbf{e}_1, \mathbf{e}_2)$, to which the measurement basis $(\mathbf{n}, \mathbf{t})$ is related by

$$\mathbf{n} = \cos\vartheta \mathbf{e}_1 + \sin\vartheta \mathbf{e}_2; \quad \mathbf{t} = -\sin\vartheta \mathbf{e}_1 + \cos\vartheta \mathbf{e}_2 \tag{2.27}$$

Here, $\vartheta = \vartheta(\mathbf{e}_1, \mathbf{n})$ is the angle between $\mathbf{e}_1$ and $\mathbf{n}$. More particularly, let us consider that $\mathbf{e}_1$ and $\mathbf{e}_2$ represent principal directions of deformation (i.e., $\mathbf{e}_1 \to \mathbf{u}_I$ and $\mathbf{e}_2 \to \mathbf{u}_{II}$). The angle $\vartheta = \vartheta(\mathbf{u}_I, \mathbf{n}) = \vartheta(\mathbf{u}_{II}, \mathbf{t})$ in (2.27), therefore, is the angle between the principal axis of $\bar{\boldsymbol{\varepsilon}}$ (i.e., the eigenvectors $\mathbf{u}_I, \mathbf{u}_{II}$) and the measurement basis $(\mathbf{n}, \mathbf{t})$. This is sketched in Figure 2.4a. Letting $\mathbf{e}_1 = \mathbf{u}_I$ and $\mathbf{e}_2 = \mathbf{u}_{II}$ in (2.27), use of (2.27) in (2.10) and (2.14) yields a relation that links the measured linear dilatation $\lambda(\mathbf{n})$ and the half-distortion $\gamma(\mathbf{n}, \mathbf{t})$ to the principal strains of $\bar{\boldsymbol{\varepsilon}}$:

$$\begin{aligned} \lambda(\mathbf{n}) &= \mathbf{n} \cdot \bar{\boldsymbol{\varepsilon}} \cdot \mathbf{n} = \begin{pmatrix} \cos\vartheta & \sin\vartheta \end{pmatrix} \begin{bmatrix} \bar{\varepsilon}_I & 0 \\ 0 & \bar{\varepsilon}_{II} \end{bmatrix} \begin{pmatrix} \cos\vartheta \\ \sin\vartheta \end{pmatrix} \\ &= \bar{\varepsilon}_I \cos^2\vartheta + \bar{\varepsilon}_{II} \sin^2\vartheta \end{aligned} \tag{2.28}$$

[5]Note that the complete strain vector would read

$$\boldsymbol{\varepsilon} \cdot \mathbf{n} = \varepsilon_{nn}\mathbf{n} + \varepsilon_{nt}\mathbf{t} + \varepsilon_{nb}\mathbf{b}$$

with $\boldsymbol{\varepsilon}$ the total strain tensor. The strain component $\varepsilon_{nb} = \gamma(\mathbf{n}, \mathbf{b})$ requires measurements in the $(\mathbf{n} \times \mathbf{b})$-plane.

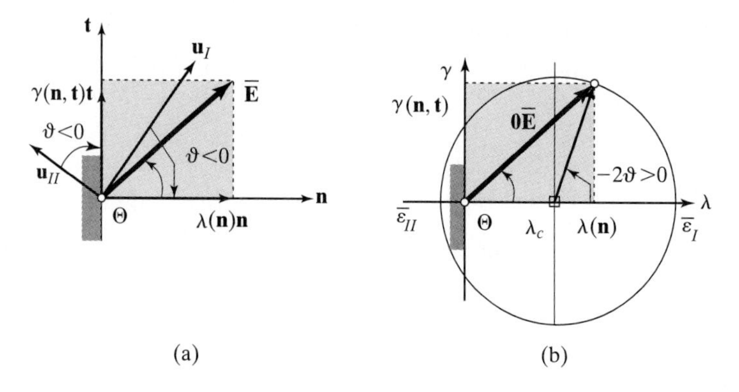

FIGURE 2.4: Representation of the strain vector: (a) material or measurement plane $(\mathbf{n} \times \mathbf{t})$; (b) Mohr strain plane $(\lambda \times \gamma)$.

$$\begin{aligned}\gamma(\mathbf{n},\mathbf{t}) &= \mathbf{t}\cdot\bar{\boldsymbol{\varepsilon}}\cdot\mathbf{n} = \begin{pmatrix} -\sin\vartheta & \cos\vartheta \end{pmatrix}\begin{bmatrix} \bar{\varepsilon}_I & 0 \\ 0 & \bar{\varepsilon}_{II} \end{bmatrix}\begin{pmatrix} \cos\vartheta \\ \sin\vartheta \end{pmatrix} \\ &= -(\bar{\varepsilon}_I - \bar{\varepsilon}_{II})\sin\vartheta\cos\vartheta \end{aligned} \tag{2.29}$$

or, in parameter form,

$$\lambda(\mathbf{n}) = \frac{\bar{\varepsilon}_I + \bar{\varepsilon}_{II}}{2} + \frac{\bar{\varepsilon}_I - \bar{\varepsilon}_{II}}{2}\cos 2\vartheta \tag{2.30}$$

$$\gamma(\mathbf{n},\mathbf{t}) = -\frac{\bar{\varepsilon}_I - \bar{\varepsilon}_{II}}{2}\sin 2\vartheta \tag{2.31}$$

where $\bar{\varepsilon}_I$ and $\bar{\varepsilon}_{II}$ are the eigenvalues of the reduced strain tensor $\bar{\boldsymbol{\varepsilon}}$ associated with eigenvalues $\mathbf{u}_I, \mathbf{u}_{II}$.[6] Equations (2.30) and (2.31) express nothing but the coordinates of a circle in the $\lambda \times \gamma$-plane, with center

$$\lambda_c = \frac{\bar{\varepsilon}_I + \bar{\varepsilon}_{II}}{2}; \quad \gamma_c = 0 \tag{2.32}$$

and radius

$$R = \frac{\bar{\varepsilon}_I - \bar{\varepsilon}_{II}}{2} \tag{2.33}$$

This circle is known as Mohr circle in strain space, and the corresponding $(\lambda \times \gamma)$-plane as the Mohr strain plane (i.e., length variation *versus* distortion). In this space, (2.30) and (2.31) are the vector coordinates of the strain vector $\mathbf{0\bar{E}} = (\lambda(\mathbf{n})\ \gamma(\mathbf{n},\mathbf{t}))$. Vector $\mathbf{0\bar{E}}$ represents in the Mohr strain plane the physical strain vector $\bar{\mathbf{E}}$ defined by (2.26). This is shown in Figure 2.4b. Finally, with regard to the sign of $-\sin 2\vartheta$ in (2.31), we note that any positive angle, $+\vartheta$, in the material plane

[6] From (2.25), it clearly appears that the eigenvalues $\bar{\varepsilon}_I, \bar{\varepsilon}_{II}$ and the associated eigenvectors $\mathbf{u}_I, \mathbf{u}_{II}$ of the reduced strain tensor $\bar{\boldsymbol{\varepsilon}}$ defined in the measurement plane do not *a priori* coincide with the eigenvalues of the complete strain tensor $\boldsymbol{\varepsilon}$ defined in the three-dimensional space. Indeed, the reduced strain vector $\bar{\mathbf{E}}$ is the projection of the three-dimensional strain vector $\boldsymbol{\varepsilon}\cdot\mathbf{n} = \lambda(\mathbf{n})\mathbf{n} + \gamma(\mathbf{n},\mathbf{t})\mathbf{t} + \gamma(\mathbf{n},\mathbf{b})\mathbf{b}$ in the $(\mathbf{n},\mathbf{t})$-plane, where the measurements take place. In other words, the eigenvalues $\bar{\varepsilon}_I, \bar{\varepsilon}_{II}$ determined in this plane are only the eigenvalues of the total tensor if $\mathbf{b} = \mathbf{u}_{III}$ is one principal strain direction, such that $\gamma(\mathbf{n},\mathbf{b}) = 0$.

corresponds in the Mohr plane to the angle -2ϑ (positive in the Mohr plane is the angle going from the linear dilatation axis λ to the distortion axis γ; *cf.* Figure 2.4b).

Finally, it is instructive to note that the Mohr plane in general is a graphical representation of any symmetric second-order tensor. We will see later further applications of this graphical representation of second-order tensors. Here, we applied it to the reduced linearized strain tensor.

Exercise 14. For the uniaxial extension, determine the direction of the maximum distortion. Specify the corresponding strain vector in the material plane and the Mohr plane.

We have already seen that the linear dilatations are the principal strains:

$$\lambda(\mathbf{e}_1 = \mathbf{u}_I) = \alpha; \quad \lambda(\mathbf{e}_2 = \mathbf{u}_{II}) = \lambda(\mathbf{e}_3 = \mathbf{u}_{III}) = -\beta$$

The corresponding material planes and reduced strain tensors are

$$(\mathbf{u}_I \times \mathbf{u}_{II}) \text{ and } (\mathbf{u}_I \times \mathbf{u}_{III}) : (\bar{\varepsilon}) = \begin{bmatrix} \alpha & 0 \\ 0 & -\beta \end{bmatrix}$$

$$(\mathbf{u}_{II} \times \mathbf{u}_{III}) : (\bar{\varepsilon}) = \begin{bmatrix} -\beta & 0 \\ 0 & -\beta \end{bmatrix}$$

In the Mohr plane, these three measurement basis are represented by three circles with centers and radii:

$$(\mathbf{u}_I \times \mathbf{u}_{II}) \text{ and } (\mathbf{u}_I \times \mathbf{u}_{III}) : \lambda_c = \frac{(\alpha - \beta)}{2}; \quad R = \frac{(\alpha + \beta)}{2}$$

$$(\mathbf{u}_{II} \times \mathbf{u}_{III}) : \lambda_c = -\beta; \quad R = 0$$

The Mohr representations are given in Figure 2.5b. The maximum distortion corresponds to the maximum value of γ of the Mohr circles (i.e., from Figure 2.5b):

$$\max \gamma = \frac{(\alpha + \beta)}{2}; \quad \lambda = \frac{(\alpha - \beta)}{2}$$

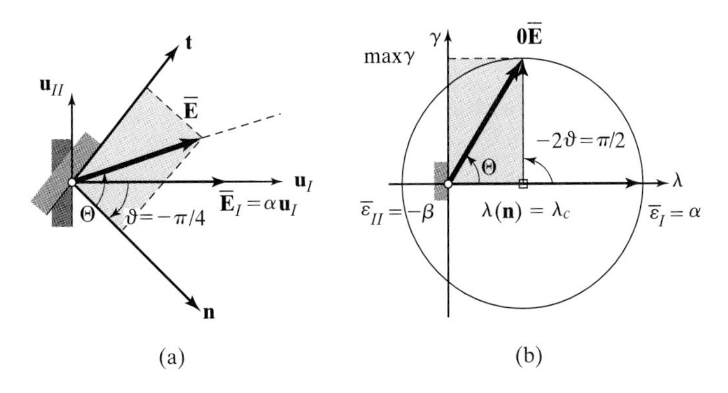

FIGURE 2.5: The uniaxial extension: (a) material plane; (b) Mohr strain plane.

In the Mohr plane, it is located with regard to the principal dilatation axis $\lambda(\mathbf{u}_I)$ by an angle $+\pi/2$; thus in the material plane

$$-2\vartheta(\mathbf{u}_I, \mathbf{e}_{\max\gamma}) = +\pi/2 \rightarrow \vartheta(\mathbf{u}_I, \mathbf{e}_{\max\gamma}) = -\pi/4$$

(i.e., in the opposite direction of the coordinate orientation $\mathbf{u}_I \rightarrow \mathbf{u}_{II}$ or $\mathbf{u}_I \rightarrow \mathbf{u}_{III}$). This is shown in Figure 2.5a. For instance, in the $(\mathbf{u}_I \times \mathbf{u}_{II})$-material plane, the maximum distortion occurs—according to (2.27)—in the material basis:

$$\mathbf{n} = \frac{\sqrt{2}}{2}[\mathbf{u}_I - \mathbf{u}_{II}]; \quad \mathbf{t} = \frac{\sqrt{2}}{2}[\mathbf{u}_I + \mathbf{u}_{II}]$$

Finally, the corresponding strain vector in the Mohr plane is given by (*cf.* Figure 2.5b)

$$\mathbf{0\bar{E}} = \begin{pmatrix} \frac{1}{2}(\alpha - \beta) \\ \frac{1}{2}(\alpha + \beta) \end{pmatrix}$$

and, using (2.26), in the material plane (*cf.* Figure 2.5a):

$$\bar{\mathbf{E}} = \frac{\sqrt{2}}{2}[\alpha \mathbf{u}_I + \beta \mathbf{u}_{II}]$$

■

2.4 EXTENSOMETER MEASUREMENTS: STRAIN GAGE ROSETTE

Extensometer measurements are concerned with the measurement of strains. Representing relative length variations, the diagonal terms $\lambda(\mathbf{e}_i)$ of the strain tensor $\boldsymbol{\varepsilon}$ can be determined by length measurements. In return, the direct access to distortions $\gamma(\mathbf{e}_i, \mathbf{e}_j)$ by means of experiment is more difficult and requires costly instruments of great accuracy—given the "infinitesimal" order of magnitude of the angle variations. In this light, extensometer measurements circumvents this problem: The nondiagonal terms of the strain tensor are accessed by pure length measurements.[7]

Consider a strain rosette attached to the matter in the plane defined by the measurement base vectors $\mathbf{e}_1$ and $\mathbf{e}_2$. The rosette is composed of three strain gages, measuring relative length variations in the directions of $\mathbf{e}_1$ and $\mathbf{e}_2$, and in a third direction $\mathbf{n}_b$ inclined to $\mathbf{e}_1$ and $\mathbf{e}_2$ by an angle $\alpha = \pm\pi/4$:

$$\mathbf{n}_b = \frac{\sqrt{2}}{2}[\mathbf{e}_1 + \mathbf{e}_2] \tag{2.34}$$

This measurement set-up is sketched in Figure 2.6a. The corresponding (directly) measured relative length variations are noted as $\lambda(\mathbf{e}_1)$, $\lambda(\mathbf{e}_2)$, and $\lambda(\mathbf{n}_b)$. From (2.25), $\lambda(\mathbf{e}_1)$ and $\lambda(\mathbf{e}_2)$ are the diagonal strain components ε_{11} and ε_{22} in the measurement basis $(\mathbf{e}_1 \times \mathbf{e}_2)$:

$$\varepsilon_{11} = \lambda(\mathbf{e}_1); \quad \varepsilon_{22} = \lambda(\mathbf{e}_2); \quad (\bar{\varepsilon}) = \begin{bmatrix} \lambda(\mathbf{e}_1) & \varepsilon_{12} \\ \varepsilon_{21} & \lambda(\mathbf{e}_2) \end{bmatrix} \tag{2.35}$$

[7]We should note that strain-measuring transducers such as electrical resistance strain gages measure the average normal strain under the sensing wire or foil element of the gage (along some gage length) and not the strain at a point. As long as the gage length is kept small, errors associated with such measurements can be kept within acceptable limits.

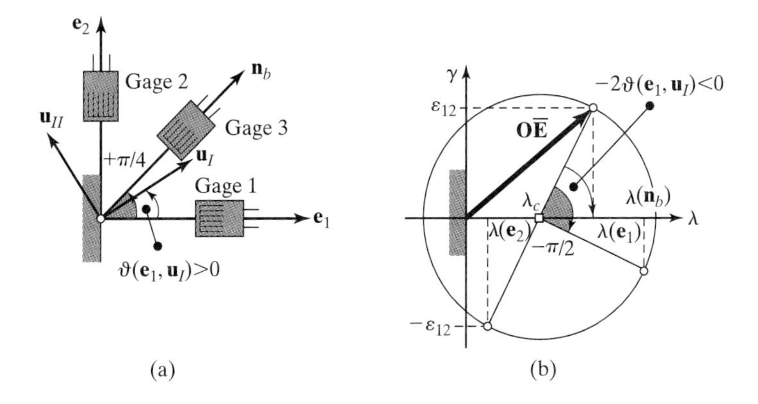

FIGURE 2.6: Strain rosette: (a) measurement set-up in the material plane; (b) representation in the Mohr strain plane.

Since no angle variations are measured in the experiment, the nondiagonal distortion term $\varepsilon_{12} = \varepsilon_{21} = \gamma(\mathbf{e}_1, \mathbf{e}_2)$ is still not known. However, we can determine it by using the third length variation measurement $\lambda(\mathbf{n}_b)$ in the $\mathbf{n}_b$-direction. Indeed, application of (2.11) to the reduced strain tensor defined by (2.35) yields

$$\lambda(\mathbf{n}_b) = \mathbf{n}_b \cdot \bar{\varepsilon} \cdot \mathbf{n}_b = \frac{1}{2}\left(\lambda(\mathbf{e}_1) + 2\varepsilon_{12} + \lambda(\mathbf{e}_2)\right) \tag{2.36}$$

which gives access to the sought half-distortion:

$$\varepsilon_{12} = \gamma(\mathbf{e}_1, \mathbf{e}_2) = \lambda(\mathbf{n}_b) - \frac{1}{2}\left(\lambda(\mathbf{e}_1) + \lambda(\mathbf{e}_2)\right) \tag{2.37}$$

Hence, the three length measurements give access to the components of ε in the measurement basis (i.e., $\bar{\varepsilon}$).

Exercise 15. By means of strain gage rosettes, verify that the strain tensor in uniaxial extension reads

$$\varepsilon = \alpha \mathbf{e}_1 \otimes \mathbf{e}_1 - \beta[\mathbf{e}_2 \otimes \mathbf{e}_2 + \mathbf{e}_3 \otimes \mathbf{e}_3]$$

We first consider a strain gage rosette in the $(\mathbf{e}_1 \times \mathbf{e}_2)$-measurement plane. With

$$(\bar{\varepsilon}) = \begin{bmatrix} \alpha & 0 \\ 0 & -\beta \end{bmatrix}$$

we need to verify $\varepsilon_{12} = 0$. From (2.37), we need to measure the following relative length variations:

$$\varepsilon_{12} = 0 \rightarrow \lambda(\mathbf{e}_1) = \alpha; \quad \lambda(\mathbf{e}_2) = -\beta; \quad \lambda(\mathbf{n}_b) = \frac{1}{2}(\alpha - \beta)$$

Analogously, in the $(\mathbf{e}_2 \times \mathbf{e}_3)$-measurement plane, we will verify

$$\lambda(\mathbf{e}_2) = \lambda(\mathbf{e}_3) = \lambda(\mathbf{n}_b) = -\beta$$

■

With the components of $\bar{\varepsilon}$ in hand, we can proceed to determine the principal strains and the principal strain directions in the $(\mathbf{e}_1 \times \mathbf{e}_2)$-measurement plane. We can determine them analytically or graphically using the Mohr strain plane. The graphical determination is sketched in Figure 2.6b. The first step consists of determining center and radius of the Mohr circle in the measurement plane. The center on the λ-axis in the Mohr plane is given by

$$\lambda_c = \frac{1}{2}(\varepsilon_{11} + \varepsilon_{22}) = \frac{1}{2}(\lambda(\mathbf{e}_1) + \lambda(\mathbf{e}_2)); \quad \gamma_c = 0 \tag{2.38}$$

In turn, one point on the Mohr circle is given by the components of the strain vector $\bar{\mathbf{E}}$; for instance,

$$\bar{\mathbf{E}} = \lambda(\mathbf{e}_1)\mathbf{e}_1 + \varepsilon_{12}\mathbf{e}_2 \tag{2.39}$$

Hence, the Mohr coordinates of this point are

$$\mathbf{0}\bar{\mathbf{E}} = \begin{pmatrix} \varepsilon_{11} \\ \varepsilon_{12} \end{pmatrix} = \begin{pmatrix} \lambda(\mathbf{e}_1) \\ \gamma(\mathbf{e}_1, \mathbf{e}_2) \end{pmatrix} \tag{2.40}$$

with ε_{12} from (2.37). The radius of the Mohr circle is therefore given by

$$R = \left|\mathbf{0}\bar{\mathbf{E}} - \mathbf{0}\bar{\mathbf{E}}_c\right| = \sqrt{(\varepsilon_{11} - \lambda_c)^2 + \varepsilon_{12}^2} = \sqrt{\frac{1}{2}(\varepsilon_{11} - \varepsilon_{22})^2 + \varepsilon_{12}^2} \tag{2.41}$$

with $\mathbf{0}\bar{\mathbf{E}}_c = (\ \lambda_c \quad 0\ )$. From Figure 2.6b, the principal strains (for which $\gamma = 0$) are

$$\bar{\varepsilon}_{I,II} = \lambda_c \pm R = \frac{1}{2}(\varepsilon_{11} + \varepsilon_{22}) \pm \sqrt{\frac{1}{2}(\varepsilon_{11} - \varepsilon_{22})^2 + \varepsilon_{12}^2} \tag{2.42}$$

and the angle $\vartheta(\mathbf{e}_1, \mathbf{u}_I) = -\vartheta(\mathbf{u}_I, \mathbf{e}_1)$ reads

$$\tan(2\vartheta) = \frac{\varepsilon_{12}}{\varepsilon_{11} - \lambda_c} = \frac{2\varepsilon_{12}}{\varepsilon_{11} - \varepsilon_{22}} = \frac{\theta(\mathbf{e}_1, \mathbf{e}_2)}{\lambda(\mathbf{e}_1) - \lambda(\mathbf{e}_2)} \tag{2.43}$$

Here, we focused on the graphical determination of the principal strains and directions in the measurement plane, using the Mohr strain plane. The analytical solution is left to the reader.

2.5 TRAINING SET: TORSION OF A CYLINDER

This training set is devoted to the deformation induced by torsion of a cylinder of height h and circular section of radius R. It reviews the developed concepts of infinitesimal deformation. The vertical fibers of the cylinder are initially parallel to the Oz-axis. The lower surface $z = 0$ is fixed throughout the deformation. The upper surface $z = h$ is attached to a rigid plateau, which is subjected to an infinitesimal rotation $\omega \ll 1$. The displacement of any point $\mathbf{A}$ of this plateau rotating around the $\mathbf{OM}$-axis is given by

$$\boldsymbol{\xi}^d = \omega \mathbf{e}_z \times \mathbf{MA}$$

where $\mathbf{MA} = r\mathbf{e}_r$ is the position vector locating point A in the rigid plateau, and $\mathbf{e}_r$ is the radial base vector. This torsion set-up is given in Figure 2.7.

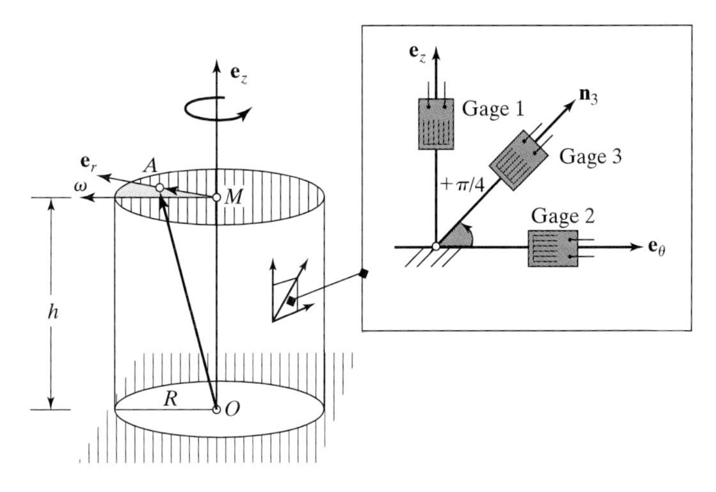

FIGURE 2.7: Torsion set-up of a cylinder sample with a strain rosette at the surface.

2.5.1 Kinematically Admissible Displacement Field

We first need to find a displacement field $\boldsymbol{\xi}$ of the cylinder that is compatible with the displacements $\boldsymbol{\xi}^d = \boldsymbol{\xi}(z = h)$ and $\boldsymbol{\xi}^d = \mathbf{0}(z = 0)$ prescribed at the boundaries. The superscript $(.)^d$ stands for given *d*ata. A displacement field that satisfies the displacement boundary conditions is called *kinematically admissible.* We keep this notion in mind for later applications.

Exercise 16. Consider as displacement field

$$\boldsymbol{\xi} = \alpha \frac{z}{h} \mathbf{e}_z \times \mathbf{OA}'$$

where A' denotes any point in the cylinder. Determine the condition for which this displacement field is kinematically compatible with the boundary conditions.

With $\mathbf{OA}' = z\mathbf{e}_z + r\mathbf{e}_r$, the chosen displacement field can be rewritten in the form

$$\boldsymbol{\xi} = \alpha \frac{z}{h} \mathbf{e}_z \times (z\mathbf{e}_z + r\mathbf{e}_r) = \alpha \frac{zr}{h} \mathbf{e}_\theta$$

(recalling that $\mathbf{e}_z \times \mathbf{e}_z = \mathbf{0}$, $\mathbf{e}_z \times \mathbf{e}_r = \mathbf{e}_\theta$). This displacement field satisfies the boundary condition

$$z = 0 : \boldsymbol{\xi} = \mathbf{0}$$

In addition, the boundary condition at the rotating surface implies

$$z = h : \boldsymbol{\xi} = \alpha r \mathbf{e}_\theta = \boldsymbol{\xi}^d = \omega r \mathbf{e}_\theta$$

Hence, the chosen displacement field is kinematically admissible provided that $\alpha = \omega$. This displacement field is of the orthoradial form

$$\boldsymbol{\xi} = \xi_\theta(z, r)\mathbf{e}_\theta; \quad \xi_\theta = \omega \frac{zr}{h}$$ ■

2.5.2 Linearized Strain Tensor

Throughout this chapter, we were only concerned with Cartesian coordinate systems. For the torsion problem at hand, we now have to deal with cylinder coordinates of orthonormal basis $(\mathbf{e}_r, \mathbf{e}_\theta, \mathbf{e}_z)$. The basis $\mathbf{e}_r, \mathbf{e}_\theta$ transforms when spatial derivatives are applied, as required when determining, for example, the strain tensor ε. However, we can circumvent these specific derivation rules for curvilinear coordinate systems by making use of the geometric interpretation of the strain tensor components.

Exercise 17. From geometric considerations, determine the components of the linearized strain tensor ε. *Hint*: Show that any vertical cylinder fiber (generatrices) subjected to the displacement field $\boldsymbol{\xi} = \xi_\theta(z,r)\mathbf{e}_\theta$ deforms by an angle $\theta = \omega(r/h)$ with respect to the cylinder axis Oz.

Figures 2.8a and 2.8b show a vertical and a horizontal section of the cylinder. In the horizontal section view, the orthoradial displacement is given by $\xi_\theta = \omega rz/h$, and in the vertical section view $\xi_\theta \simeq z\gamma$ (recalling that for $\omega \ll 1 \rightarrow \tan\omega \simeq \omega$). Hence, after deformation any vertical cylinder fiber is inclined by an angle

$$\theta(\mathbf{e}_z, \mathbf{e}_\theta) = \omega\frac{r}{h}$$

This angle represents the distortion between the vertical basis $\mathbf{e}_z$ and the orthoradial basis $\mathbf{e}_\theta$:

$$\theta(\mathbf{e}_z, \mathbf{e}_\theta) = 2\varepsilon_{\theta z} = 2\varepsilon_{z\theta}$$

With respect to the infinitesimal order of magnitude of the rotation, all other strain components are zero. The linearized strain tensor, therefore, is readily determined:

$$\varepsilon = \frac{\omega r}{2h}(\mathbf{e}_z \otimes \mathbf{e}_\theta + \mathbf{e}_\theta \otimes \mathbf{e}_z)$$

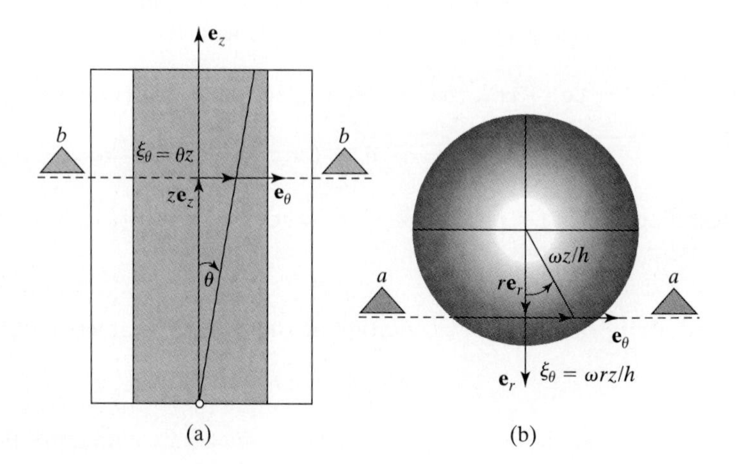

FIGURE 2.8: Vertical and horizontal section view of torsional deformation of one cylinder fiber. ∎

2.5.3 Strain Rosette Measurements and Mohr Circle

The strain measurements by means of a strain rosette also apply to non-Cartesian geometric situations, provided the infinitesimal order of magnitude of the deformation.

Exercise 18. Consider a strain rosette on the cylinder surface $(r = R, z)$, as shown in Figure 2.7. Strain gage 1 is oriented in the direction of the cylinder, and gage 2 in the direction of the orthoradial direction. Gage 3 is inclined with respect to the other two by an angle of 45°. Determine the relative length variations λ_i $(i = 1, 2, 3)$ of each strain gage, and give a graphical representation in the Mohr strain plane. In both the material plane and the Mohr plane, determine the principal strains and their directions.

The measurement plane is the $(\mathbf{e}_\theta \times \mathbf{e}_z)$-plane, and the corresponding reduced strain tensor is

$$(\mathbf{e}_\theta \times \mathbf{e}_z) : (\bar{\varepsilon}) = \begin{bmatrix} 0 & \gamma \\ \gamma & 0 \end{bmatrix}; \quad \gamma = \frac{\omega R}{2h} = \theta/2$$

The Mohr representation of $\bar{\varepsilon}$ is given in Figure 2.9b: a circle around the origin $(\lambda_c = 0)$ with radius $R = \omega R/2h$. The points representing the strain gage measurements are also given in this figure. Let us note that gage 1 has as normal and tangential vectors

$$\mathbf{n}_1 = \mathbf{e}_z; \quad \mathbf{t}_1 = -\mathbf{e}_\theta$$

From (2.26) it follows that

$$\begin{aligned}\bar{\mathbf{E}}_1 &= \bar{\varepsilon} \cdot \mathbf{n}_1 = \frac{\omega R}{2h}(\mathbf{e}_z \otimes \mathbf{e}_\theta + \mathbf{e}_\theta \otimes \mathbf{e}_z) \cdot \mathbf{e}_z \\ &= \frac{\omega R}{2h}[(\mathbf{e}_\theta \cdot \mathbf{e}_z)\mathbf{e}_z + (\mathbf{e}_z \cdot \mathbf{e}_z)\mathbf{e}_\theta] \\ &= \frac{\omega R}{2h}\mathbf{e}_\theta = -\frac{\omega R}{2h}\mathbf{t}_1\end{aligned}$$

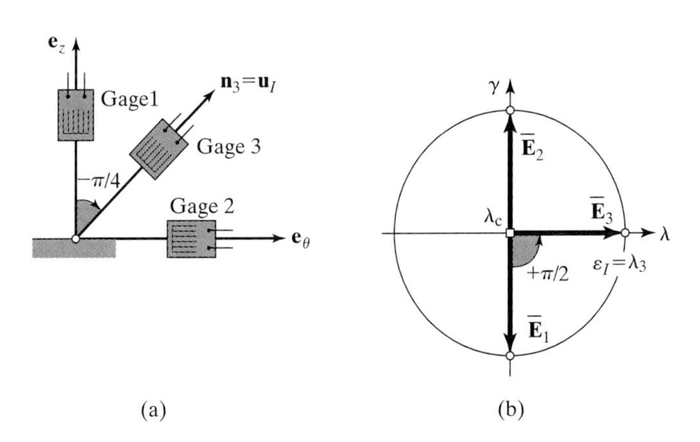

FIGURE 2.9: Pure torsional deformation at the cylinder surface $(r = R)$: (a) material plane; (b) Mohr plane.

Hence, in the Mohr plane, strain gage 1 corresponds to

$$\mathbf{0}\bar{\mathbf{E}}_1 = \left(0; -\frac{\omega R}{2h}\right)$$

With one point in hand, we can determine the other points corresponding to the strain gages in the Mohr plane shown in Figure 2.9b. For instance, the angle between $\mathbf{n}_1 = \mathbf{e}_z$ and $\mathbf{n}_3$ defined by (2.34) is $\vartheta = -45°$. This corresponds in the Mohr plane to an angle of $-2\vartheta = +90°$ starting from gage 1. We obtain

$$\mathbf{0}\bar{\mathbf{E}}_3 = \left(\frac{\omega R}{2h}; 0\right); \quad \mathbf{0}\bar{\mathbf{E}}_2 = \left(0; +\frac{\omega R}{2h}\right)$$

The measured relative length variations are

$$\lambda_1 = \lambda_2 = 0; \quad \lambda_3 = \frac{\omega R}{2h}$$

In addition, λ_3 corresponds to one principal strain. Therefore, the principal strains and their directions are given by

$$\bar{\varepsilon}_{I,II} = \pm\frac{\omega R}{2h}; \quad \mathbf{u}_{I,II} = \frac{\sqrt{2}}{2}\left[\mathbf{e}_z \pm \mathbf{e}_\theta\right] \qquad \blacksquare$$

2.5.4 Torsion and Extension

The linear theory allows the superposition of different deformations. For instance, in addition to the torsion, the application of an uniaxial extension $\xi_z(z)$ in the cylinder axis direction reads, in terms of displacements,

$$\boldsymbol{\xi} = \xi_\theta(z, r)\mathbf{e}_\theta + \xi_z(z)\mathbf{e}_z$$

Recalling that the gradient operator is a linear tangent application, the displacement gradient can be expressed by the superposition of the gradient operations:

$$\operatorname{grad}\left(\xi_\theta(z,r)\mathbf{e}_\theta + \xi_z(z)\mathbf{e}_z\right) = \operatorname{grad}\left(\xi_\theta(z,r)\mathbf{e}_\theta\right) + \operatorname{grad}\left(\xi_z(z)\mathbf{e}_z\right))$$

Hence, in the linear theory, in which spatial gradients intervene only in linear form, the related strain measures of torsion and extension can be superposed.

Exercise 19. In addition to the previously determined torsion, consider an axial displacement $\xi_z(z) = Wz/h$, with W a positive constant. Determine the linearized strain field and the new relative length variations Λ_i of the strain gages of the rosette on the cylinder surface $r = R$. Show the modified Mohr circle, and determine graphically the new principal strains and associated principal directions.

The linearized strain tensor is given by

$$\boldsymbol{\varepsilon} = \boldsymbol{\varepsilon}(\xi_\theta) + \boldsymbol{\varepsilon}(\xi_z) = \frac{\omega r}{2h}(\mathbf{e}_z \otimes \mathbf{e}_\theta + \mathbf{e}_\theta \otimes \mathbf{e}_z) + \frac{W}{h}\mathbf{e}_z \otimes \mathbf{e}_z$$

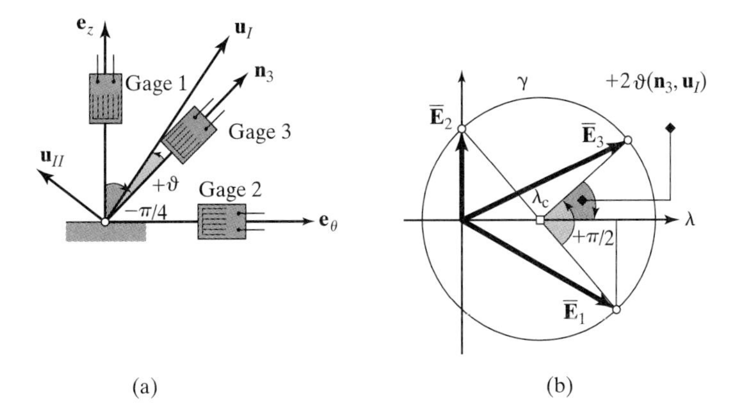

FIGURE 2.10: Combined torsional and extension deformation: (a) material plane; (b) Mohr Strain Plane.

In the $(\mathbf{e}_z \times \mathbf{e}_\theta)$-measurement plane, the reduced strain tensor now reads

$$(\mathbf{e}_\theta, \mathbf{e}_z) : (\bar{\varepsilon}) = \begin{bmatrix} 0 & \gamma \\ \gamma & \lambda_z \end{bmatrix}; \quad \gamma = \frac{\omega R}{2h}; \quad \lambda_z = \frac{W}{h}$$

Proceeding as before, the strain vectors read (here determined analytically)[8]

$$(\mathbf{n}_1 = \mathbf{e}_z, \mathbf{t}_1 = -\mathbf{e}_\theta) : \bar{\mathbf{E}}_1 = \bar{\varepsilon} \cdot \mathbf{e}_z = \lambda_z \mathbf{e}_z + \gamma \mathbf{e}_\theta = \lambda_z \mathbf{n}_1 - \gamma \mathbf{t}_1$$

$$(\mathbf{n}_2 = \mathbf{e}_\theta, \mathbf{t}_2 = \mathbf{e}_z) : \bar{\mathbf{E}}_2 = \bar{\varepsilon} \cdot \mathbf{e}_\theta = \gamma \mathbf{e}_z = \gamma \mathbf{t}_2$$

$$\left(\mathbf{n}_3 = \frac{\sqrt{2}}{2}\left[\mathbf{e}_z + \mathbf{e}_\theta\right], \mathbf{t}_3 = \frac{\sqrt{2}}{2}\left[\mathbf{e}_z - \mathbf{e}_\theta\right]\right) : \bar{\mathbf{E}}_3 = \bar{\varepsilon} \cdot \mathbf{n}_3 = (\lambda_z/2 + \gamma)\mathbf{n}_3 + \lambda_z/2\mathbf{t}_3$$

The relative length variations of the three strain gages are

$$\Lambda_1 = \lambda_z; \quad \Lambda_2 = 0; \quad \Lambda_3 = \lambda_z/2 + \gamma$$

The corresponding Mohr circle is given in Figure 2.10b. The following modifications with respect to the pure torsion are noted:

1. Translation of the Mohr circle center by

$$\lambda_c = \lambda_z/2 = \frac{W}{2h}$$

[8] In detail,

$$\begin{aligned} \bar{\mathbf{E}}_3 &= \bar{\varepsilon} \cdot \mathbf{n}_3 = \frac{\sqrt{2}}{2}(\bar{\mathbf{E}}_1 + \bar{\mathbf{E}}_2) \\ &= \frac{\sqrt{2}}{2}[\lambda_z \mathbf{e}_z + \gamma(\mathbf{e}_\theta + \mathbf{e}_z)] = \frac{1}{2}\lambda_z(\mathbf{n}_3 + \mathbf{t}_3) + \gamma \mathbf{n}_3 \end{aligned}$$

where we made use of $\mathbf{e}_\theta + \mathbf{e}_z = \frac{\sqrt{2}}{2}\mathbf{n}_3$ and $\mathbf{e}_z = \frac{\sqrt{2}}{2}(\mathbf{n}_3 + \mathbf{t}_3)$.

2. Increase of the radius by a factor

$$\rho = \sqrt{\left(\frac{W}{\omega R}\right)^2 + 1}$$

3. Change of principal strains

$$\bar{\varepsilon}_{I,II} = \frac{W}{2h} \pm \frac{\omega R}{2h}\rho$$

4. Change of associated principal strain directions

$$\tan[2\vartheta(\mathbf{n}_3, \mathbf{u}_I)] = \tan[-2\vartheta(\mathbf{u}_I, \mathbf{n}_3)] = \frac{\gamma_3}{\Lambda_3 - \lambda_c} = \frac{\lambda_z}{2\gamma} = \frac{W}{\omega R}$$

As a concluding exercise, compare the results of the torsion training set with the linearized simple shear training set, of which the finite deformation was described in Section 1.6. ■

2.6 APPENDIX CHAPTER 2: DIFFERENTIAL OPERATORS

2.6.1 Orthonormal Cartesian Coordinates

A point M has coordinates x_i $(i = 1, 2, 3)$, respectively x, y, z, in a vector base $\mathbf{e}_i$ $(i = 1, 2, 3)$, respectively $(\mathbf{e}_x, \mathbf{e}_y, \mathbf{e}_z)$. The position vector $\mathbf{OM}$ reads

$$\mathbf{OM} = x_i \mathbf{e}_i = x\mathbf{e}_x + y\mathbf{e}_y + z\mathbf{e}_z$$

Scalar field $a(\mathbf{x}, t)$

$$\operatorname{grad} a = \nabla a = \frac{\partial a}{\partial x_i}\mathbf{e}_i$$

Vector field $\mathbf{b}(\mathbf{x}, t)$

$$\mathbf{b}(\mathbf{x}, t) = b_i(\mathbf{x}, t)\mathbf{e}_i$$

$$\operatorname{div} \mathbf{b} = \nabla \cdot \mathbf{b} = \frac{\partial b_i}{\partial x_i}$$

$$\begin{aligned} \operatorname{curl} \mathbf{b} &= \nabla \times \mathbf{b} = \begin{vmatrix} \mathbf{e}_x & \mathbf{e}_y & \mathbf{e}_z \\ \frac{\partial}{\partial x} & \frac{\partial}{\partial y} & \frac{\partial}{\partial z} \\ b_x & b_y & b_z \end{vmatrix} \\ &= \left(\frac{\partial b_z}{\partial x} - \frac{\partial b_y}{\partial z}\right)\mathbf{e}_x + \left(\frac{\partial b_x}{\partial z} - \frac{\partial b_z}{\partial x}\right)\mathbf{e}_y + \left(\frac{\partial b_y}{\partial x} - \frac{\partial b_x}{\partial y}\right)\mathbf{e}_z \end{aligned}$$

$$\operatorname{grad} \mathbf{b} = \nabla \mathbf{b} = \frac{\partial b_i}{\partial x_j}\mathbf{e}_i \otimes \mathbf{e}_j$$

Second-order tensor field $\mathbf{c}(\mathbf{x}, t)$

$$\mathbf{c}(\mathbf{x}, t) = c_{ij}\mathbf{e}_i \otimes \mathbf{e}_j$$

$$\operatorname{div}\mathbf{c} = \nabla \cdot \mathbf{c} = \frac{\partial c_{ij}}{\partial x_j}\mathbf{e}_i$$

2.6.2 Cylinder Coordinates

A point M is defined by the cylinder coordinates r, θ, z in a local orthonormal basis $(\mathbf{e}_r, \mathbf{e}_\theta, \mathbf{e}_z)$. This is shown in Figure 2.11a. Within a Cartesian coordinate system the position vector $\mathbf{OM}$ reads

$$\mathbf{OM} = r\cos\theta\mathbf{e}_x + r\sin\theta\mathbf{e}_y + z\mathbf{e}_z$$

Derivation of $\mathbf{OM}$ with regard to the cylinder coordinates gives

$$\frac{\partial\mathbf{OM}}{\partial r} = \cos\theta\mathbf{e}_x + \sin\theta\mathbf{e}_y = \mathbf{e}_r$$

$$\frac{\partial\mathbf{OM}}{\partial\theta} = r(-\sin\theta\mathbf{e}_x + \cos\theta\mathbf{e}_y) = r\mathbf{e}_\theta$$

$$\frac{\partial\mathbf{OM}}{\partial z} = \mathbf{e}_z$$

Thus,

$$\frac{\partial\mathbf{e}_r}{\partial\theta} = \mathbf{e}_\theta; \quad \frac{\partial\mathbf{e}_\theta}{\partial r} = -\mathbf{e}_r$$

These derivatives of the base vectors need to be considered in the differential operators.

Scalar field $a(r, \theta, z, t)$

$$\operatorname{grad} a = \nabla a = \frac{\partial a}{\partial r}\mathbf{e}_r + \frac{1}{r}\frac{\partial a}{\partial\theta}\mathbf{e}_\theta + \frac{\partial a}{\partial z}\mathbf{e}_z$$

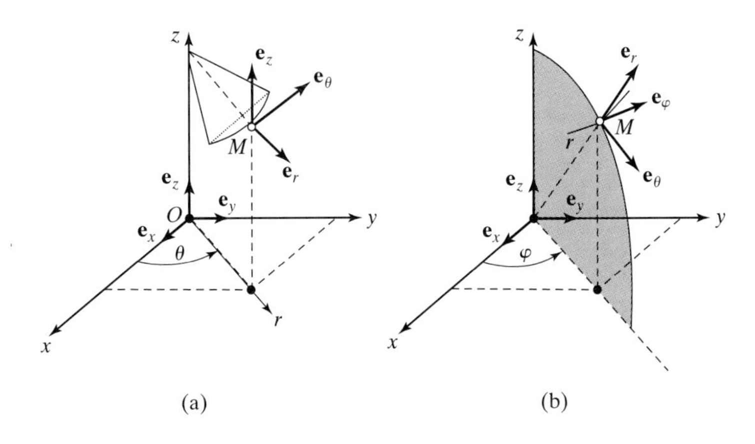

FIGURE 2.11: Coordinate systems: (a) cylinder coordinates; (b) spherical coordinates.

Vector field $\mathbf{b}(r, \theta, z, t)$

$$\mathbf{b}(r, \theta, z, t) = b_r(r, \theta, z, t)\mathbf{e}_r + b_\theta(r, \theta, z, t)\mathbf{e}_\theta + b_z(r, \theta, z, t)\mathbf{e}_z$$

$$\operatorname{div}\mathbf{b} = \nabla \cdot \mathbf{b} = \frac{1}{r}\frac{\partial}{\partial r}(r b_r) + \frac{1}{r}\frac{\partial b_\theta}{\partial \theta} + \frac{\partial b_z}{\partial z}$$

$$\begin{aligned} \operatorname{curl}\mathbf{b} &= \nabla \times \mathbf{b} \\ &= \left(\frac{1}{r}\frac{\partial b_z}{\partial \theta} - \frac{\partial b_\theta}{\partial z}\right)\mathbf{e}_r + \left(\frac{1}{r}\frac{\partial b_r}{\partial z} - \frac{\partial b_z}{\partial r}\right)\mathbf{e}_\theta + \left(\frac{1}{r}\frac{\partial (r b_\theta)}{\partial r} - \frac{1}{r}\frac{\partial b_r}{\partial \theta}\right)\mathbf{e}_z \end{aligned}$$

$$\operatorname{grad}\mathbf{b} = \nabla\mathbf{b} = \begin{pmatrix} \frac{\partial b_r}{\partial r} & \frac{1}{r}\left(\frac{\partial b_r}{\partial \theta} - b_\theta\right) & \frac{\partial b_r}{\partial z} \\ \frac{\partial b_\theta}{\partial r} & \frac{1}{r}\left(\frac{\partial b_\theta}{\partial \theta} + b_r\right) & \frac{\partial b_\theta}{\partial z} \\ \frac{\partial b_z}{\partial r} & \frac{1}{r}\frac{\partial b_z}{\partial \theta} & \frac{\partial b_z}{\partial z} \end{pmatrix}$$

We verify that $\operatorname{div}\mathbf{b} = \operatorname{tr}(\operatorname{grad}\mathbf{b})$.

Second-order symmetric tensor field $\mathbf{c}(r, \theta, z, t)$

$$\mathbf{c}(\mathbf{x}, t) = c_{ij}(r, \theta, z, t)\mathbf{e}_i \otimes \mathbf{e}_j$$

$$\begin{aligned} \operatorname{div}\mathbf{c} &= \nabla \cdot \mathbf{c} \\ &= \left(\frac{\partial c_{rr}}{\partial r} + \frac{1}{r}\frac{\partial c_{r\theta}}{\partial \theta} + \frac{\partial c_{rz}}{\partial z} + \frac{c_{rr} - c_{\theta\theta}}{r}\right)\mathbf{e}_r \\ &\quad + \left(\frac{\partial c_{\theta r}}{\partial r} + \frac{1}{r}\frac{\partial c_{\theta\theta}}{\partial \theta} + \frac{\partial c_{\theta z}}{\partial z} + 2\frac{c_{r\theta}}{r}\right)\mathbf{e}_\theta \\ &\quad + \left(\frac{\partial c_{zr}}{\partial r} + \frac{1}{r}\frac{\partial c_{z\theta}}{\partial \theta} + \frac{\partial c_{zz}}{\partial z} + \frac{c_{zr}}{r}\right)\mathbf{e}_z \end{aligned}$$

2.6.3 Spherical Coordinates

A point M is defined by the spherical coordinates r, θ, φ in a local orthonormal basis $(\mathbf{e}_r, \mathbf{e}_\theta, \mathbf{e}_\varphi)$. This is shown in Figure 2.11b. Within a Cartesian coordinate system the position vector $\mathbf{OM}$ reads

$$\mathbf{OM} = r\sin\theta\cos\varphi\mathbf{e}_x + r\sin\theta\sin\varphi\mathbf{e}_y + r\cos\theta\mathbf{e}_z$$

Derivation with respect to spherical coordinates gives

$$\frac{\partial \mathbf{OM}}{\partial r} = \sin\theta\cos\varphi\mathbf{e}_x + \sin\theta\sin\varphi\mathbf{e}_y + \cos\theta\mathbf{e}_z = \mathbf{e}_r$$

$$\frac{\partial \mathbf{OM}}{\partial \theta} = r(\cos\theta\cos\varphi\mathbf{e}_x + \cos\theta\sin\varphi\mathbf{e}_y - \sin\theta\mathbf{e}_z) = r\mathbf{e}_\theta$$

$$\frac{\partial \mathbf{OM}}{\partial \varphi} = r\sin\theta(-\sin\varphi\mathbf{e}_x + \cos\varphi\mathbf{e}_y) = r\sin\theta\mathbf{e}_\varphi$$

Thus,

$$\begin{array}{lll} \mathbf{e}_{r,r} = 0 & \mathbf{e}_{r,\theta} = \mathbf{e}_\theta & \mathbf{e}_{r,\varphi} = \sin\theta\mathbf{e}_\varphi \\ \mathbf{e}_{\theta,r} = 0 & \mathbf{e}_{\theta,\theta} = -\mathbf{e}_r & \mathbf{e}_{\theta,\varphi} = \cos\theta\mathbf{e}_\varphi \\ \mathbf{e}_{\varphi,r} = 0 & \mathbf{e}_{\varphi,r} = 0 & \mathbf{e}_{\varphi,\varphi} = -\sin\theta\mathbf{e}_r - \cos\theta\mathbf{e}_\theta \end{array}$$

where $\mathbf{e}_{\alpha,\beta} = \partial\mathbf{e}_\alpha/\partial\beta$.

Scalar field $a(r,\theta,\varphi,t)$

$$\operatorname{grad} a = \nabla a = \frac{\partial a}{\partial r}\mathbf{e}_r + \frac{1}{r}\frac{\partial a}{\partial \theta}\mathbf{e}_\theta + \frac{1}{r\sin\theta}\frac{\partial a}{\partial \varphi}\mathbf{e}_\varphi$$

Vector field $\mathbf{b}(r,\theta,\varphi,t)$

$$\mathbf{b}(r,\theta,\varphi,t) = b_r(r,\theta,\varphi,t)\mathbf{e}_r + b_\theta(r,\theta,\varphi,t)\mathbf{e}_\theta + b_\varphi(r,\theta,\varphi,t)\mathbf{e}_z$$

$$\operatorname{div}\mathbf{b} = \nabla\cdot\mathbf{b} = \frac{1}{r^2}\frac{\partial}{\partial r}(r^2 b_r) + \frac{1}{r\sin\theta}\frac{\partial}{\partial\theta}(\sin\theta\, b_\theta) + \frac{1}{r\sin\theta}\frac{\partial b_\varphi}{\partial\varphi}$$

$$\begin{aligned}
\operatorname{curl}\mathbf{b} &= \nabla\times\mathbf{b} \\
&= \frac{1}{r\sin\theta}\left(\frac{\partial}{\partial\theta}(\sin\theta\, b_\varphi) - \frac{\partial b_\theta}{\partial\varphi}\right)\mathbf{e}_r + \frac{1}{r}\left(\frac{1}{\sin\theta}\frac{\partial b_r}{\partial\varphi} - \frac{\partial}{\partial r}(r b_\varphi)\right)\mathbf{e}_\theta \\
&\quad + \frac{1}{r}\left(\frac{\partial(r b_\theta)}{\partial r} - \frac{\partial b_r}{\partial\theta}\right)\mathbf{e}_z
\end{aligned}$$

$$\operatorname{grad}\mathbf{b} = \nabla\mathbf{b} = \begin{pmatrix}
\frac{\partial b_r}{\partial r} & \frac{1}{r}\left(\frac{\partial b_r}{\partial\theta} - b_\theta\right) & \frac{1}{r}\left(\sin\theta\frac{\partial b_r}{\partial\varphi} - b_\varphi\right) \\
\frac{\partial b_\theta}{\partial r} & \frac{1}{r}\left(\frac{\partial b_\theta}{\partial\theta} + b_r\right) & \frac{1}{r}\left(\frac{1}{\sin\theta}\frac{\partial b_\theta}{\partial\varphi} - \cot\theta\, b_\varphi\right) \\
\frac{\partial b_\varphi}{\partial r} & \frac{1}{r}\frac{\partial b_\varphi}{\partial\theta} & \frac{1}{r}\left(\frac{1}{\sin\theta}\frac{\partial b_\varphi}{\partial\varphi} - \cot\theta\, b_\theta + b_r\right)
\end{pmatrix}$$

Second-order symmetric tensor field $\mathbf{c}(r,\theta,\varphi,t)$

$$\mathbf{c}(\mathbf{x},t) = c_{ij}(r,\theta,\varphi,t)\mathbf{e}_i\otimes\mathbf{e}_j$$

$$\begin{aligned}
\operatorname{div}\mathbf{c} &= \nabla\cdot\mathbf{c} \\
&= \left(\frac{\partial c_{rr}}{\partial r} + \frac{1}{r}\frac{\partial c_{r\theta}}{\partial\theta} + \frac{1}{r\sin\theta}\frac{\partial c_{r\varphi}}{\partial\varphi} + \frac{1}{r}(2c_{rr} - c_{\theta\theta} - c_{\varphi\varphi} + c_{r\theta}\cot\theta)\right)\mathbf{e}_r \\
&\quad + \left(\frac{\partial c_{\theta r}}{\partial r} + \frac{1}{r}\frac{\partial c_{\theta\theta}}{\partial\theta} + \frac{1}{r\sin\theta}\frac{\partial c_{\theta\varphi}}{\partial\varphi} + \frac{1}{r}[(c_{\theta\theta} - c_{\varphi\varphi})\cot\theta + 3c_{r\theta}]\right)\mathbf{e}_\theta \\
&\quad + \left(\frac{\partial c_{zr}}{\partial r} + \frac{1}{r}\frac{\partial c_{\varphi\theta}}{\partial\theta} + \frac{1}{r\sin\theta}\frac{\partial c_{\varphi\varphi}}{\partial\varphi} + \frac{1}{r}[3c_{r\varphi} + 2\cot\theta\, c_{\theta\varphi}]\right)\mathbf{e}_\varphi
\end{aligned}$$

PART TWO

MOMENTUM BALANCE, STRESS, AND STRESS STATES

CHAPTER 3

Momentum Balance and Stress

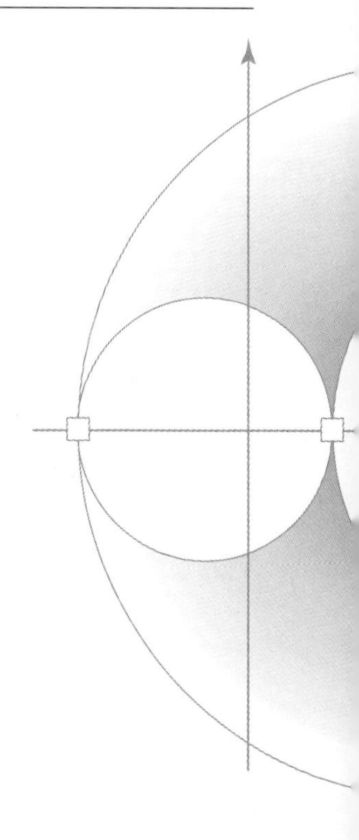

This and the next chapter deal with stresses. Beside the geometric description of deformation (whether finite or infinitesimal), the notion of stress is the second pillar of mechanics of material systems. Stresses are the consequence of external forces (volume and surface forces) applied to a material domain. However, in continuum mechanics, the existence of stresses as a tensor quantity derives from a physical law, the momentum balance. This physical law must be satisfied independent of the deformation that occurs in the solid. Therefore, this chapter is concerned with the hypothesis that allows the very definition of a stress tensor: the hypothesis of local contact forces. This hypothesis states that surface forces depend only on the normal at the boundary of the studied domain. It is at the basis of the very definition of the (Cauchy) stress tensor, and of the reduction formulas relating stresses to forces and moments in structural members. In combination with the global dynamic theorems of force and moment resultants, the local equilibrium equations (equation of motion) and the symmetry of the stress tensor are derived.

3.1 THE HYPOTHESIS OF LOCAL CONTACT FORCES

In continuum mechanics, it is usually assumed that a material domain (i.e., the structure or a structural element) is subjected to two kinds of external forces: volume forces and surface forces. This is sketched in Figure 3.1. We should note here that forces are usually described in the configuration in which they act. In general, this is the current (i.e., deformed) configuration.

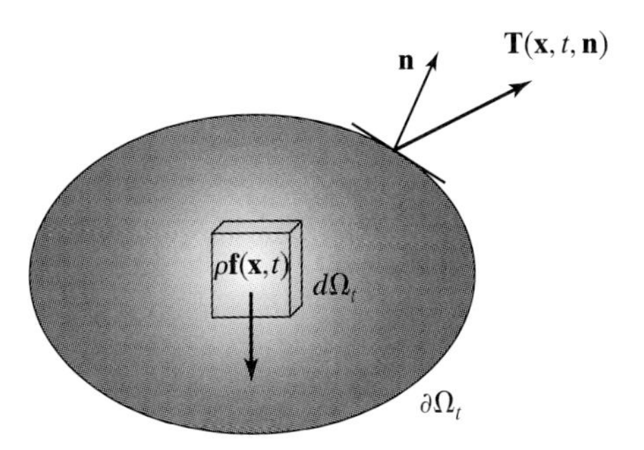

FIGURE 3.1: Definition of external forces: volume forces and surface forces.

3.1.1 Body Forces

The body (or volume) force acting on an elementary volume $d\Omega_t$ is

$$\text{in } \Omega_t : \mathbf{f}_v = \rho \mathbf{f}(\mathbf{x}, t)\, d\Omega_t \tag{3.1}$$

where ρ is the mass density (of dimension Mass/Length^3), and $\mathbf{f}(\mathbf{x}, t)$ is the volume force density per mass unit (of dimension Force/Mass = Length/Time^2). For instance, the most common volume force density is the one generated by gravity, for which the volume force density is the gravity vector $\mathbf{f} = \mathbf{g}$. The force density $\mathbf{f}(\mathbf{x}, t)$ will be assumed to depend only on position vector $\mathbf{x}$ of the material point in the current configuration, and on time t. Hence, nonlocal forces such as forces that depend on the distance between particles will not be considered here.

3.1.2 Surface Forces

Surface forces are of a different kind. They act on the surface $\partial\Omega_t$ that encloses the domain Ω_t:

$$\text{on } \partial\Omega_t : \mathbf{f}_s = \mathbf{T}(\mathbf{x}, t, \mathbf{n})\, da \tag{3.2}$$

where $\mathbf{T} = \mathbf{T}(\mathbf{x}, t, \mathbf{n})$ is the surface force density acting on the infinitesimal material surface da (of dimension Force/Length^2 = Mass/Length/Time^2). For instance, the surface force density exerted by a pressure p on the surface is $\mathbf{T} = -p\mathbf{n}$. In addition, it is common to assume that the surface force density $\mathbf{T}$ is not only defined by position vector $\mathbf{x}$ and time t; but also by the outward unit normal of the surface $\mathbf{n}$. This comes to assume that the surface forces acting on da are local contact forces exerted by adjacent material points. This hypothesis (and it is one), known as the hypothesis of local contact forces, excludes any action of forces that depend on the curvature (e.g., $\nabla\mathbf{n} = \text{grad}\ \mathbf{n}$) of surface $\partial\Omega$, as, for instance, forces due to capillary tension. We will see that this hypothesis is critical for the definition of a symmetric stress tensor (i.e., for its very existence [as a tensor quantity]).

3.2 MOMENTUM BALANCE

The momentum balance of some individual material points reads

$$\frac{d}{dt}\sum_i (m_i \mathbf{V}_i) = \sum_i \mathbf{f}_i^{\text{ext}} \tag{3.3}$$

$$\frac{d}{dt}\sum_i(\mathbf{x}_i \times m_i\mathbf{V}_i) = \sum_i(\mathbf{x}_i \times \mathbf{f}_i^{\text{ext}}) \tag{3.4}$$

where $m_i\mathbf{V}_i$ and $\mathbf{x}_i \times m_i\mathbf{V}_i$ denote the linear momentum and the angular momentum of a material point i of mass m_i and velocity vector $\mathbf{V}_i$ located by position vector $\mathbf{x}_i$ that is subjected to an external force $\mathbf{f}_i^{\text{ext}}$ and a moment $\mathbf{x}_i \times \mathbf{f}_i^{\text{ext}}$. The first equation states the conservation of the linear momentum, the second the conservation of the angular momentum for individual material points. Equations (3.3) and (3.4) apply to discrete material point systems. To derive the continuous counterparts, we replace the linear momentum and the angular momentum by the ones of the matter contained in the Representative Elementary Volume $d\Omega_t$ of continuum mechanics:

$$m_i\mathbf{V}_i \to \rho\mathbf{V}d\Omega_t; \quad \mathbf{x}_i \times m_i\mathbf{V}_i \to \mathbf{x} \times \rho\mathbf{V}d\Omega_t \tag{3.5}$$

In addition, the external forces acting on the continuous material domain Ω_t are the volume forces and surface forces given by (3.1) and (3.2). Hence, the instantaneous momentum balance of the whole matter contained in Ω_t can be written in the form

$$\frac{d}{dt}\int_{\Omega_t}\rho\mathbf{V}d\Omega_t = \int_{\Omega_t}\rho\mathbf{f}(\mathbf{x},t)\,d\Omega_t + \int_{\partial\Omega_t}\mathbf{T}(\mathbf{x},t,\mathbf{n})\,da \tag{3.6}$$

$$\frac{d}{dt}\int_{\Omega_t}(\mathbf{x}\times\rho\mathbf{V})\,d\Omega_t = \int_{\Omega_t}(\mathbf{x}\times\rho\mathbf{f}(\mathbf{x},t))\,d\Omega_t + \int_{\partial\Omega_t}(\mathbf{x}\times\mathbf{T}(\mathbf{x},t,\mathbf{n}))\,da \tag{3.7}$$

Equation (3.6) states that the instantaneous time derivative of the linear momentum of the whole matter contained in volume Ω_t is equal to the external forces (volume and surface forces) acting on this matter. Equation (3.7) states the same for the angular momentum.

Furthermore, since there is no mass supply to the elementary volume, $\rho d\Omega_t$ is constant throughout the deformation.[1] Application of the law of mass conservation allows rewriting the left-hand side of Eqs. (3.6) and (3.7) in the form

$$\frac{d}{dt}\int_{\Omega_t}\rho\mathbf{V}d\Omega_t = \int_{\Omega_t}\rho\frac{d\mathbf{V}}{dt}d\Omega_t = \int_{\Omega_t}\rho\boldsymbol{\gamma}d\Omega_t \tag{3.8}$$

$$\frac{d}{dt}\int_{\Omega_t}(\mathbf{x}\times\rho\mathbf{V})\,d\Omega_t = \int_{\Omega_t}(\mathbf{x}\times\rho\boldsymbol{\gamma})\,d\Omega_t \tag{3.9}$$

where $\boldsymbol{\gamma} = d\mathbf{V}/dt$. Equations (3.6), (3.8) and (3.7), (3.9) lead to a new formulation of the momentum balance, called the Dynamic Theorem.

[1] More precisely, the mass conservation in the elementary volume reads

$$dm = \rho(\mathbf{x},t)\,d\Omega_t = \rho_0(\mathbf{X})\,d\Omega$$

where ρ_0 is the mass density in the initial configuration (which is constant). Application of the volume transport formula (1.11) shows that

$$\rho_0/\rho = d\Omega_t/d\Omega = J = \det\mathbf{F}$$

THEOREM 1 of the dynamic resultant: The overall dynamic force is equal to the resultant of elementary body forces and surface forces:

$$\int_{\Omega_t} \rho\gamma \, d\Omega_t = \int_{\Omega_t} \rho\mathbf{f}(\mathbf{x}, t) \, d\Omega_t + \int_{\partial\Omega_t} \mathbf{T}(\mathbf{x}, t, \mathbf{n}) \, da \tag{3.10}$$

THEOREM 2 of the dynamic moment: The overall dynamic moment is equal to the resultant of the moments of elementary body forces and surface forces:

$$\int_{\Omega_t} (\mathbf{x} \times \rho\gamma) \, d\Omega_t = \int_{\Omega_t} (\mathbf{x} \times \rho\mathbf{f}(\mathbf{x}, t)) \, d\Omega_t + \int_{\partial\Omega_t} (\mathbf{x} \times \mathbf{T}(\mathbf{x}, t, \mathbf{n})) \, da \tag{3.11}$$

3.3 THE CAUCHY STRESS TENSOR

We now turn to a direct application of the theorem of dynamic resultant, deriving the action–reaction hypothesis for the vector of surface force density $\mathbf{T}(\mathbf{x}, t, \mathbf{n})$.

3.3.1 Action–Reaction Law

Consider the cylinder shown in Figure 3.2a. The two end sides are oriented by the unit outward normal vectors $\mathbf{n}$ and $-\mathbf{n}$. Applying the dynamic resultant theorem (3.10) leads to

$$\begin{aligned} \int_{\Omega_t} \rho(\gamma - \mathbf{f}) \, d\Omega_t &= \int_{\partial\Omega_t} \mathbf{T}(\mathbf{x}, t, \mathbf{n}) \, da \\ &= \int_a [\mathbf{T}(\mathbf{n}) - \mathbf{T}(-\mathbf{n})] \, da + \int_{\partial\Omega_t} \mathbf{T}(\mathbf{n} = \mathbf{e}_r) \, da \end{aligned} \tag{3.12}$$

where $\mathbf{T}(\mathbf{n})$ and $\mathbf{T}(-\mathbf{n})$ denote the surface force density vectors at the end sides of the cylinder, and $\mathbf{T}(\mathbf{n} = \mathbf{e}_r)$ the force density vector on the cylinder wall of surface $\partial\Omega_t$. Now, for $\partial\Omega_t \to 0$, we obtain the action–reaction law, sketched in Figure 3.2b:

$$\partial\Omega_t \to 0: \quad \mathbf{T}(\mathbf{n}) + \mathbf{T}(-\mathbf{n}) = 0 \tag{3.13}$$

or, equivalently

$$\mathbf{T}(-\mathbf{n}) = -\mathbf{T}(\mathbf{n}) \tag{3.14}$$

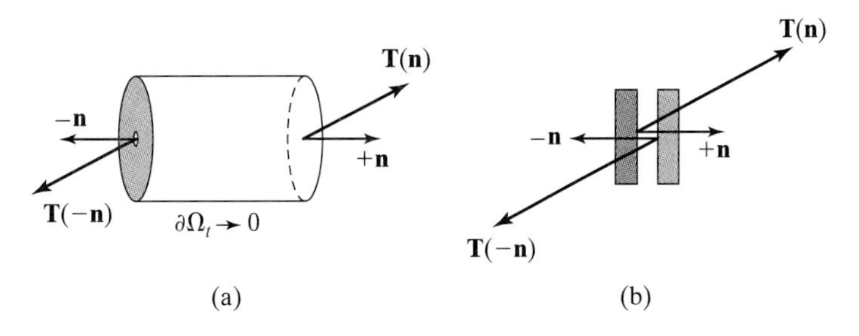

FIGURE 3.2: Action–reaction law (hypothesis): (a) on a cylinder with infinitesimal length; (b) continuity of the stress vector over a material surface.

Equation (3.13) describes the continuity of the surface traction vector $\mathbf{T}(\mathbf{n})$ at a material surface da. It is nothing but the hypothesis of local contact forces (3.2), which defines vector $\mathbf{T}(\mathbf{x}, t, \mathbf{n})$ at a material point with respect to the unit outward normal $\mathbf{n}$ [in other words, (3.2) or (3.13) do not apply to curvature dependent surface forces]: If the vector $\mathbf{n}$ is changed into its opposite $-\mathbf{n}$, $\mathbf{T}(\mathbf{n})$ is changed into $\mathbf{T}(-\mathbf{n}) = -\mathbf{T}(\mathbf{n})$. As we will see, this is a necessary condition for the existence of a symmetric stress tensor.

3.3.2 The Tetrahedron Lemma

In addition to the action–reaction law (which is actually a hypothesis since it is based on the assumption of local contact forces) let us consider the dynamic resultant theorem of the infinitesimal small tetrahedron in Figure 3.3, which has its three facets of surface S_j parallel to the coordinate planes. These oriented surfaces are related to the base surface S oriented by unit normal $\mathbf{n}$ by

$$S_j = S\,(\mathbf{n} \cdot \mathbf{e}_j) = S n_j \tag{3.15}$$

Applying the dynamic resultant theorem (3.10) to this tetrahedron reads

$$\begin{aligned} \int_{\Omega_t} \rho(\boldsymbol{\gamma} - \mathbf{f})\, d\Omega_t &= \int_{\partial\Omega_t} \mathbf{T}(\mathbf{x}, t, \mathbf{n})\, da \\ &= \mathbf{T}(\mathbf{n})S + \sum_{j=1,2,3} \mathbf{T}(-\mathbf{e}_j)S_j = \mathbf{T}(\mathbf{n})S + \sum_{j=1,2,3} \mathbf{T}(-\mathbf{e}_j)n_j S \end{aligned} \tag{3.16}$$

The left-hand side can be evaluated in the form

$$\int_{\Omega_t} \rho(\boldsymbol{\gamma} - \mathbf{f})\, d\Omega_t = \frac{hS}{3} O(\rho(\boldsymbol{\gamma} - \mathbf{f})) \tag{3.17}$$

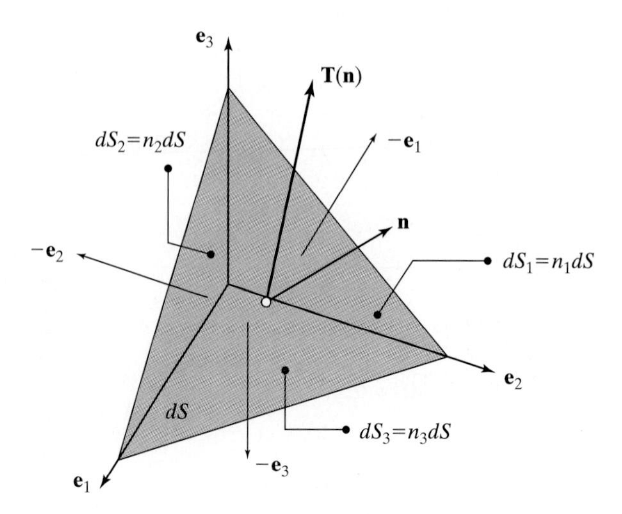

FIGURE 3.3: Tetrahedron lemma applied to stress vector $\mathbf{T}$.

where $hS/3$ is the volume of the tetrahedron[2] and $O(x)$ denotes the order of magnitude of the norm of quantity x. Hence, with $h \to 0$ (i.e., by degenerating the tetrahedron to a point), the left-hand side is zero. From (3.16), it follows that

$$\mathbf{T}(\mathbf{n}) + \sum_{j=1,2,3} \mathbf{T}(-\mathbf{e}_j)n_j = 0 \tag{3.18}$$

or, equivalently, using (3.14),

$$\mathbf{T}(\mathbf{n} = n_j\mathbf{e}_j) = \sum_{j=1,2,3} \mathbf{T}(\mathbf{e}_j)n_j \tag{3.19}$$

This is known as the tetrahedron lemma,[3] here applied to the stress vector $\mathbf{T}$. The tetrahedron lemma (3.19) defines a linear operator relating the stress vector $\mathbf{T}(\mathbf{x}, t, \mathbf{n})$ to $\mathbf{n}$. This operator is known as the Cauchy (or Eulerian) stress tensor $\boldsymbol{\sigma} = \boldsymbol{\sigma}(\mathbf{x}, t)$ of components σ_{ij}:

$$\mathbf{T}(\mathbf{x}, t, \mathbf{n}) = \boldsymbol{\sigma} \cdot \mathbf{n}; \quad T_j = \sigma_{ij}n_i; \quad \boldsymbol{\sigma} = \sigma_{ij}\mathbf{e}_i \otimes \mathbf{e}_j \tag{3.20}$$

We should note that the linear operator $\boldsymbol{\sigma} = \boldsymbol{\sigma}(\mathbf{x}, t)$ is independent of the normal $\mathbf{n}$, on which stress vector $\mathbf{T}(\mathbf{x}, t, \mathbf{n})$ depends. This underlines the importance of the hypothesis of local contact forces (3.2) for the very definition (3.20) of the stress $\boldsymbol{\sigma}$ as a tensor quantity.

Figure 3.4a (on the next page) gives an illustration of the physical significance of the Cauchy stress tensor $\boldsymbol{\sigma}$. Due to (3.20), the terms σ_{i1}, σ_{i2} and σ_{i3} $(i = 1, 2, 3)$ are the components of the surface force density $\mathbf{T}(\mathbf{x}, t, \mathbf{n} = \mathbf{e}_j) = T_i\mathbf{e}_i$ in the orthonormal basis $\mathbf{e}_i$ $(i = 1, 2, 3)$ acting at time t, per unit of surface area, on the material surface oriented by unit normal $\mathbf{n} = \mathbf{e}_j$ and located at point $\mathbf{x}$. In other words, the i-subscripts of σ_{ij} refer to the component i in the orthonormal basis of $\mathbf{T}$ built on the j-th material surface. This is shown in Figure 3.4b. For instance, the stress vector on the surface $\mathbf{e}_1$ reads

$$\mathbf{T}(\mathbf{e}_1) = \boldsymbol{\sigma} \cdot \mathbf{e}_1 = \sigma_{11}\mathbf{e}_1 + \sigma_{21}\mathbf{e}_2 + \sigma_{31}\mathbf{e}_3 \tag{3.21}$$

or, in matrix notation,

$$\begin{pmatrix} T_1 \\ T_2 \\ T_3 \end{pmatrix} = \begin{bmatrix} \sigma_{11} & \sigma_{12} & \sigma_{13} \\ \sigma_{21} & \sigma_{22} & \sigma_{23} \\ \sigma_{31} & \sigma_{32} & \sigma_{33} \end{bmatrix} \begin{pmatrix} 1 \\ 0 \\ 0 \end{pmatrix} = \begin{pmatrix} \sigma_{11} \\ \sigma_{21} \\ \sigma_{31} \end{pmatrix} \tag{3.22}$$

[2]The volume of the tetrahedron is 1/6 of that of the parallelepiped; that is (using the triple scalar product),

$$\Omega_t = \frac{1}{6}\mathbf{a} \cdot (\mathbf{b} \times \mathbf{c}) = \frac{1}{6}\,|\mathbf{a}|\,|\mathbf{b} \times \mathbf{c}| \cos\beta$$

where β is the angle between $\mathbf{a}$ and the product vector $\mathbf{b} \times \mathbf{c}$. In addition, $|\mathbf{b} \times \mathbf{c}|$ is twice the area S, and $|\mathbf{a}| \cos\beta$ is the altitude h of the tetrahedron. It follows $\Omega_t = hS/3$.

[3]For the purpose of completeness, the tetrahedron lemma states that if a relation of the form

$$\int_{\Omega_t} h(\mathbf{x}, t)\, d\Omega_t = -\int_{\partial\Omega_t} f(\mathbf{x}, t, \mathbf{n})\, da$$

holds for any volume Ω_t, then f can be written as a linear function of the n_i components of normal $\mathbf{n}$:

$$f(\mathbf{x}, t, \mathbf{n}) = f_i(\mathbf{x}, t)n_i$$

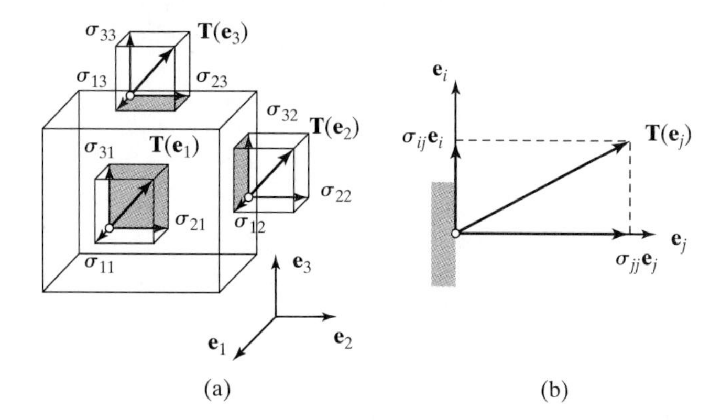

FIGURE 3.4: Cauchy stresses σ_{ij}: (a) components of the stress vector acting on the material surfaces of an elementary parallelepiped built on the coordinate axes; (b) the stress vector and its components acting on a material surface oriented by unit normal $\mathbf{e}_j$.

Exercise 20. As an accompanying example, we consider the two-layer soil system shown in Figure 3.5. The surface $y = 0$ is subjected to a constant pressure p. Specify the continuity conditions of the stress vector on the surface $\partial\Omega_t$ and at the interface $y = -h$.

The surface force density (= stress vector) reads

$$y = 0 : \mathbf{T}^d = -p\mathbf{n} = -p\mathbf{e}_y$$

where $\mathbf{n} = \mathbf{e}_y$ is the outward unit normal (see Figure 3.5), and superscript $(.)^d$ stands for prescribed data (here pressure p). From the hypothesis of normal contact forces, it follows that this stress vector is equal to the stress vector in the substratum, which reads

$$\mathbf{T}(\mathbf{e}_y) = \boldsymbol{\sigma} \cdot \mathbf{e}_y = \sigma_{xy}\mathbf{e}_x + \sigma_{yy}\mathbf{e}_y + \sigma_{zy}\mathbf{e}_z$$

At $y = 0$, it is $\mathbf{T}^d = \mathbf{T}(\mathbf{e}_y) = \boldsymbol{\sigma}(y = 0) \cdot \mathbf{e}_y$; thus from a comparison of the two previous equations,

$$y = 0 : \sigma_{yy} = -p; \quad \sigma_{xy} = \sigma_{zy} = 0$$

At the interface, the continuity of the stress vector reads

$$y = -h : \mathbf{T}(\mathbf{e}_y) + \mathbf{T}(-\mathbf{e}_y) = 0$$

where $\mathbf{T}(\mathbf{e}_y)$ is the stress vector at the interface in Layer 2:

$$\mathbf{T}(\mathbf{e}_y) = \sigma_{xy}^{(2)}\mathbf{e}_x + \sigma_{yy}^{(2)}\mathbf{e}_y + \sigma_{zy}^{(2)}\mathbf{e}_z$$

and $\mathbf{T}(-\mathbf{e}_y)$ the stress vector at the interface in Layer 1:

$$\mathbf{T}(-\mathbf{e}_y) = -\sigma_{xy}^{(1)}\mathbf{e}_x - \sigma_{yy}^{(1)}\mathbf{e}_y - \sigma_{zy}^{(1)}\mathbf{e}_z$$

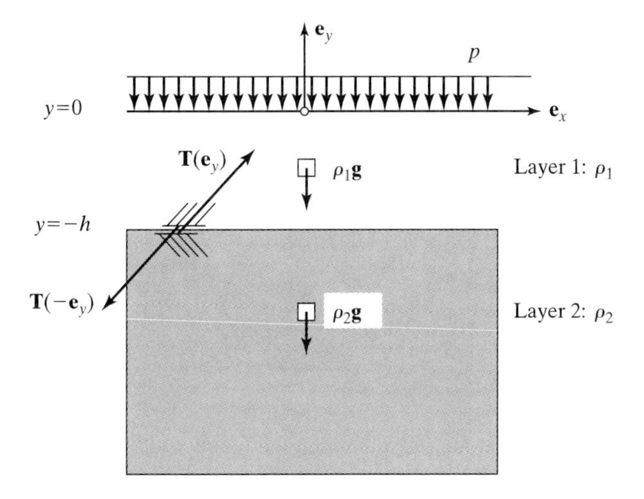

FIGURE 3.5: Two-layer soil substratum with different mass densities.

Hence, the continuity at the interface implies the continuity of the stress components:

$$\sigma_{xy}^{(1)} = \sigma_{xy}^{(2)}; \quad \sigma_{yy}^{(1)} = \sigma_{yy}^{(2)}; \quad \sigma_{zy}^{(1)} = \sigma_{zy}^{(2)}$$

Any other component of $\boldsymbol{\sigma}$, which is not involved in the continuity of the interface stress vector (for instance σ_{xx}), is not necessarily continuous along y. ■

3.3.3 Relation with Force and Moment Resultants: Reduction Formulas

The stress tensor $\boldsymbol{\sigma} = \boldsymbol{\sigma}(\mathbf{x}, t)$ is a local physical quantity defined by the hypothesis of local contact forces. It is useful to relate this local property with global force and moment quantities used (e.g., in structural mechanics of beams, plates, shells, etc.). These force and moment resultants are defined on a finite surface $\partial\Omega_t$. From (3.11), integration of the stress vector $\mathbf{T}(\mathbf{n})$ and of its moment $\mathbf{OM} \times \mathbf{T}(\mathbf{n})$, with $\mathbf{OM}$ the position vector locating any material point in the surface, yields

$$\mathcal{N} = \int_{\partial\Omega_t} \boldsymbol{\sigma} \cdot \mathbf{n}\, da; \quad \mathcal{M} = \int_{\partial\Omega_t} (\mathbf{OM} \times \boldsymbol{\sigma} \cdot \mathbf{n})\, da \tag{3.23}$$

where $\mathcal{N}$ is the vector of force resultants, and $\mathcal{M}$ the vector of moment resultants: Eq. (3.23) reduces the local stress distribution $\boldsymbol{\sigma} = \boldsymbol{\sigma}(\mathbf{x}, t)$ to some global force and moment resultants. Like the (Eulerian) stress tensor $\boldsymbol{\sigma}$, force vector $\mathcal{N}$ and moment vector $\mathcal{M}$ are defined in the current (i.e., deformed) configuration, where they are related by equilibrium to the external forces and moments applied to the structure.

Exercise 21. For a beam oriented in the $\mathbf{e}_x$-direction, determine the force and moment resultants.

In the section $a\mathbf{e}_x$ of the beam, any material point is located by the position vector $\mathbf{OM} = y\mathbf{e}_y + z\mathbf{e}_z$, with $y = 0, z = 0$ corresponding to the beam's axis of references (for instance, the "neutral axis"). The stress vector in this section reads

$$\mathbf{T}(\mathbf{e}_x) = \boldsymbol{\sigma} \cdot \mathbf{e}_x = \sigma_{xx}\mathbf{e}_x + \sigma_{yx}\mathbf{e}_y + \sigma_{zx}\mathbf{e}_z$$

Using $(3.23)_1$, the force resultants read

$$\mathcal{N} = N\mathbf{e}_x + V_y\mathbf{e}_y + V_z\mathbf{e}_z$$

where $N = \int_a \sigma_{xx} da$ is the axial force oriented in the $\mathbf{e}_x$-direction ($da = dy\,dz$); and $V_y = \int_a \sigma_{yx} da$ and $V_z = \int_a \sigma_{zx} da$ are the (in-plane) shear forces oriented respectively in the $\mathbf{e}_y$- and $\mathbf{e}_z$-directions.
In a similar way, we obtain the section moments. Use of $(3.23)_2$ in matrix notation reads

$$\begin{pmatrix} M_x \\ M_y \\ M_z \end{pmatrix} = \int_a \begin{pmatrix} 0 \\ y \\ z \end{pmatrix} \times \begin{pmatrix} \sigma_{xx} \\ \sigma_{yx} \\ \sigma_{zx} \end{pmatrix} da = \begin{pmatrix} \int_a (y\sigma_{zx} - z\sigma_{yx}) da \\ \int_a z\sigma_{xx} da \\ -\int_a y\sigma_{xx} da \end{pmatrix}$$

or, in tensor notation,

$$\mathcal{M} = M_x\mathbf{e}_x + M_y\mathbf{e}_y + M_z\mathbf{e}_z$$

where $M_x = \int_a (y\sigma_{zx} - z\sigma_{yx})\, da$ is the torsion moment around the $\mathbf{e}_x$-axis; and $M_y = \int_a z\sigma_{xx} da$ and $M_z = -\int_a y\sigma_{xx} da$ are the bending moments acting, respectively, around the $\mathbf{e}_y$- and $\mathbf{e}_z$-axes. ∎

3.4 EQUATION OF MOTION

With the stress tensor $\boldsymbol{\sigma}$ as linear operator in hand, we can return to the dynamic theorems.

3.4.1 The Dynamic Resultant Theorem and the Local Equilibrium Equation

Use of (3.20) in the global dynamic resultant theorem (3.10) gives

$$0 = \int_{\Omega_t} \rho(\mathbf{f} - \boldsymbol{\gamma})\, d\Omega_t + \int_{\partial\Omega_t} (\boldsymbol{\sigma} \cdot \mathbf{n})\, da \tag{3.24}$$

We now seek an expression that defines the momentum balance at the material point. To this end, we change the surface integral on the right hand side of (3.24) into a volume integral. From a mathematical point of view, this is achieved using the divergence theorem; that is,

$$\int_{\partial\Omega_t} (\boldsymbol{\sigma} \cdot \mathbf{n})\, da = \int_{\Omega_t} \operatorname{div} \boldsymbol{\sigma}\, d\Omega_t \tag{3.25}$$

From a more physical standpoint, the quantity described by (3.25) corresponds to an outflux of motion through the boundary $\partial\Omega_t$, which is balanced by the external supply of motion through the same boundary (i.e., by the external surface forces acting on $\partial\Omega_t$). In addition, substitution of (3.25) in (3.24) yields the dynamic resultant theorem in the form

$$0 = \int_{\Omega_t} [\operatorname{div} \boldsymbol{\sigma} + \rho(\mathbf{f} - \boldsymbol{\gamma})]\, d\Omega_t \tag{3.26}$$

Since the momentum balance must hold irrespective of any particular choice of domain Ω_t, it is not only the integral that is zero, but as well the integrand. This leads to the local equation of motion or equilibrium of the elementary volume $d\Omega_t$:

$$\operatorname{div} \boldsymbol{\sigma} + \rho(\mathbf{f} - \boldsymbol{\gamma}) = \mathbf{0} \tag{3.27}$$

or, in index notation in a Cartesian coordinate system,

$$\frac{\partial \sigma_{ij}}{\partial x_j} + \rho(f_i - \gamma_i) = 0 \tag{3.28}$$

where repeated subscript indicates summation. Equation (3.27) expresses the dynamic resultant theorem applied locally to the matter that constitutes, at time t, the elementary volume $d\Omega_t$, located at material point $\mathbf{x}$. It states that the sum of the elementary body forces $\rho \mathbf{f}\, d\Omega_t$ and of all the surface forces $\mathbf{T} da$ acting on the facets of the parallelepiped $d\Omega_t = dx_1 dx_2 dx_3$ is equal to the dynamic force $\rho \boldsymbol{\gamma}\, d\Omega_t$.

Exercise 22. Determine the equation of motion in the $\mathbf{e}_2$-direction, by considering the equilibrium of all forces acting on the elementary volume $d\Omega_t$ shown in Figure 3.4a.

The equilibrium equation in any direction is obtained by projecting the stress vectors of each surface and the external force vector together with the dynamic force on the considered direction. For the $\mathbf{e}_2$-direction, these forces are shown in Figure 3.6a. The equilibrium equation in the $\mathbf{e}_2$-direction reads

$$\begin{aligned} 0 = [\sigma_{22}^+ - \sigma_{22}^-]\, dx_1 dx_3 + [\sigma_{21}^+ - \sigma_{21}^-]\, dx_2 dx_3 + [\sigma_{23}^+ - \sigma_{23}^-]\, dx_1 dx_1 \\ + \rho(f_2 - \gamma_2)\, dx_1 dx_2 dx_3 \end{aligned}$$

Here, $\sigma_{2j}^- = \mathbf{T}(-\mathbf{e}_j) \cdot \mathbf{e}_2$ and $\sigma_{2j}^+ = \mathbf{T}(\mathbf{e}_j) \cdot \mathbf{e}_2$ denote the projection in the $\mathbf{e}_2$-direction of the stress vector $\mathbf{T}(\mathbf{n})$ of the surfaces at x_j and $x_j + dx_j$ oriented, respectively, by $\mathbf{n} = -\mathbf{e}_j$ and $\mathbf{n} = \mathbf{e}_j$. In addition, using a truncated series development (without summation on repeated subscripts),

$$\sigma_{2j}^+ = \sigma_{2j}^- + \frac{\partial \sigma_{2j}}{\partial x_j} dx_j$$

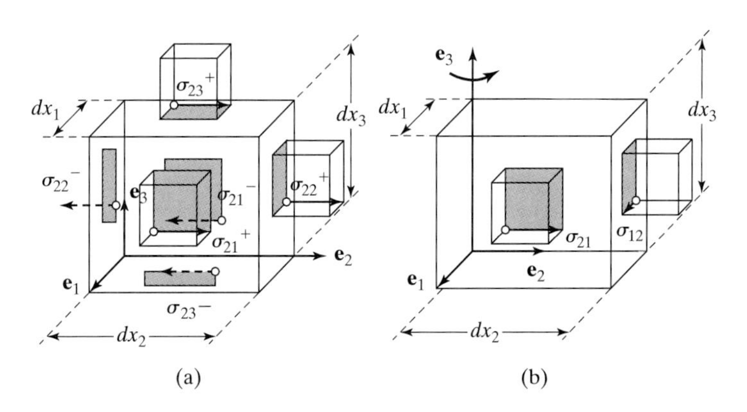

FIGURE 3.6: The dynamic theorems: (a) stress components in the $\mathbf{e}_2$-direction; (b) moment contributions around $\mathbf{e}_3$-axis.

the previous equilibrium equation can be rewritten in the form

$$0 = \left[\frac{\partial \sigma_{21}}{\partial x_1} + \frac{\partial \sigma_{22}}{\partial x_2} + \frac{\partial \sigma_{23}}{\partial x_3} + \rho(f_2 - \gamma_2)\right] dx_1 dx_2 dx_3$$

This is the sought equation of motion (3.28) in the $\mathbf{e}_2$-direction in a Cartesian coordinate frame. ■

Exercise 23. We consider the two-layer soil system shown in Figure 3.5. In each layer, the material is considered as homogeneous with different volume mass densities ρ_i $(i = 1, 2)$. Specify the equations of equilibrium for each layer, and determine the resulting stress components.

The body force in this "static" example (inertia effects neglected) is the volume force density

$$\rho_i \mathbf{f} = -\rho_i g \mathbf{e}_y$$

In addition, the assumption of homogeneity implies that the stresses depend only on the y-coordinate,

$$\boldsymbol{\sigma} = \boldsymbol{\sigma}(y)$$

For $\boldsymbol{\sigma} = \boldsymbol{\sigma}(y)$, the local equilibrium conditions (3.28) read

$$\mathbf{e}_x : \frac{\partial \sigma_{xy}(y)}{\partial y} = 0 \to \sigma_{xy} = C_{xy}$$

$$\mathbf{e}_y : \frac{\partial \sigma_{yy}(y)}{\partial y} - \rho_i g = 0 \to \sigma_{yy}(y) = \rho_i g y + C_{yy}$$

$$\mathbf{e}_z : \frac{\partial \sigma_{zy}(y)}{\partial y} = 0 \to \sigma_{zy} = C_{zy}$$

where C_{xy}, C_{yy}, C_{zy} are integration constants. They can be determined from the boundary and continuity conditions determined previously. In fact, using the surface condition,

$$y = 0 : \sigma_{yy} = -p; \quad \sigma_{xy} = \sigma_{zy} = 0$$

it follows that

$$C_{xy} = 0; \quad C_{yy} = -p; \quad C_{zy} = 0$$

and the known stress components in Layer 1 are

$$y \in (0, -h) : \sigma_{yy}^{(1)} = \rho_1 g y - p; \quad \sigma_{xy}^{(1)} = \sigma_{zy}^{(1)} = 0$$

Furthermore, the continuity at the interface $y = -h$,

$$\sigma_{xy}^{(1)} = \sigma_{xy}^{(2)}; \quad \sigma_{yy}^{(1)} = \sigma_{yy}^{(2)}; \quad \sigma_{zy}^{(1)} = \sigma_{zy}^{(2)}$$

implies the continuity of $\sigma_{yy}, \sigma_{xy}, \sigma_{zy}$:

$$\sigma_{yy}^{(2)} = \rho_2 g(y + h) - \rho_1 g h - p; \quad \sigma_{xy}^{(2)} = \sigma_{zy}^{(2)} = 0$$

Hence, without further information, only three of nine stress components can be determined. In matrix notation these are

$$(\sigma_{ij}) = \begin{bmatrix} ? & \sigma_{xy} & ? \\ ? & \sigma_{yy} & ? \\ ? & \sigma_{zy} & ? \end{bmatrix}$$

■

3.4.2 The Dynamic Moment Theorem and the Symmetry of the Stress Tensor

We have explored the global dynamic resultant theorem (3.10) to derive the local equation of motion (3.27) for the elementary volume $d\Omega_t$. We can proceed in the same manner with the global dynamic moment theorem (3.11). With (3.20), we rewrite (3.11) in the form

$$0 = \int_{\Omega_t} (\mathbf{x} \times \rho(\mathbf{f} - \gamma))\, d\Omega_t + \int_{\partial\Omega_t} (\mathbf{x} \times (\boldsymbol{\sigma} \cdot \mathbf{n}))\, da \tag{3.29}$$

Then, analogously to (3.25), we apply the divergence theorem to the last term, the moment flux integral, to transform the surface integral into a volume integral:

$$\int_{\partial\Omega_t} (\mathbf{x} \times (\boldsymbol{\sigma} \cdot \mathbf{n}))\, da = \int_{\Omega_t} (\mathbf{x} \times \operatorname{div} \boldsymbol{\sigma} \ + 2\boldsymbol{\Sigma}^{as})\, d\Omega_t \tag{3.30}$$

where $\boldsymbol{\Sigma}^{as}$ is the vector

$$2\boldsymbol{\Sigma}^{as} = (\sigma_{23} - \sigma_{32})\mathbf{e}_1 + (\sigma_{13} - \sigma_{31})\mathbf{e}_2 + (\sigma_{12} - \sigma_{21})\mathbf{e}_3 \tag{3.31}$$

Equation (3.29) becomes

$$0 = \int_{\Omega_t} [\mathbf{x} \times (\rho(\mathbf{f} - \gamma) + \operatorname{div} \boldsymbol{\sigma}) + 2\boldsymbol{\Sigma}^{as}]\, d\Omega_t \tag{3.32}$$

Use of the equation of motion (3.27) in (3.32) leads us to state the global dynamic moment theorem as follows:

$$0 = \int_{\Omega_t} 2\boldsymbol{\Sigma}^{as} d\Omega_t \tag{3.33}$$

Again, since this equation must hold for any volume Ω_t, it follows that $\boldsymbol{\Sigma}^{as} = 0$. Hence, from (3.31), this implies the symmetry of the strain tensor $\boldsymbol{\sigma}$:

$$\boldsymbol{\sigma} = {}^t\boldsymbol{\sigma}; \quad \sigma_{ij} = \sigma_{ji} \tag{3.34}$$

Exercise 24. Consider the balance of moments on the elementary parallelepiped $d\Omega_t$ shown in Figure 3.4. From the moment balance around the $\mathbf{e}_3$-axis, show that $\sigma_{12} = \sigma_{21}$.

The stress components contributing to the moment around the $\mathbf{e}_3$-axis are given in Figure 3.6b. The balance of moments reads

$$\sigma_{21} da_1 dx_1 - \sigma_{12} da_2 dx_2 = 0$$

where $da_1 = dx_2 dx_3$ [the surface on which $\mathbf{T}(\mathbf{e}_1)$ acts] and $da_2 = dx_1 dx_3$. Since $d\Omega_t = da_1 dx_1 = da_2 dx_2$, it follows that

$$(\sigma_{21} - \sigma_{12})\, d\Omega_t = 0$$

which proves the symmetry of the stress tensor in the Cartesian coordinate frame.

■

Exercise 25. Complete the components of the stress-tensor of the two-layer soil system of Figure 3.5. Is the problem well posed?

The symmetry of the stress tensor gives three additional stress components:

$$(\sigma_{ij}) = \begin{bmatrix} ? & \sigma_{xy} & ? \\ \sigma_{yx} = \sigma_{xy} & \sigma_{yy} & \sigma_{yz} = \sigma_{zy} \\ ? & \sigma_{zy} & ? \end{bmatrix}$$

The components that were not involved in the momentum balance and the boundary conditions remain undetermined: $\sigma_{xx}, \sigma_{xz} = \sigma_{zx}, \sigma_{zz}$. Their determination requires a constitutive material law. For instance, if we consider the material as isotropic (i.e., with the same material behavior in all directions), it follows for the given example that

$$\sigma_{xx}(y) = \sigma_{zz}(y)$$

This illustrates the necessity of a material law. ■

3.4.3 Principal Stresses and Principal Stress Directions

The symmetry of the stress tensor, expressed by (3.34), is the dynamic moment theorem applied to the elementary volume. This symmetry results from the absence of external moment couples, whether volume- or surface-related. Furthermore, it is this symmetry of the stress tensor that ensures the realness of its eigenvalues $\sigma_{J=1,2,3}$:

$$\boldsymbol{\sigma} \cdot \mathbf{u}_J = \sigma_J \mathbf{u}_J \tag{3.35}$$

These eigenvalues are called the principal stresses, and the associated eigenvectors $\mathbf{u}_J$ are the principal stress directions.

3.5 TRAINING SET: PRESSURE VESSEL FORMULA

As training set, consider the cylindrical vessel of inner radius R and thickness t, shown in Figure 3.7. The vessel is subjected to an internal pressure p (exerted, for example, by a liquid on the internal surface $r = R$). The outside surface is free of stresses. Body and inertia forces are neglected. The length L of the cylinder is much larger than the radius R (i.e., $R/L \ll 1$). We are interested in determining the stresses in the cylinder wall.

3.5.1 Cylinder Stress Components and Boundary Conditions

In a first step, it is instructive to determine the *a priori* shape of the stress tensor (i.e., the nonzero stress components of $\boldsymbol{\sigma}$). For the cylindrical problem at hand, the stress tensor is conveniently defined in cylinder coordinates of basis $(\mathbf{e}_r, \mathbf{e}_\theta, \mathbf{e}_z)$; and $\boldsymbol{\sigma}$ reads in matrix notation

$$(\sigma_{ij}) = \begin{bmatrix} \sigma_{rr} & \sigma_{r\theta} & \sigma_{rz} \\ & \sigma_{\theta\theta} & \sigma_{\theta z} \\ \text{sym.} & & \sigma_{zz} \end{bmatrix}$$

where σ_{rr} is the radial stress defined on a material facet oriented by outward unit $\mathbf{e}_r$. In turn, $\sigma_{\theta\theta}$ is the *hoop, tangential,* or *circumferential* stress. It is the normal

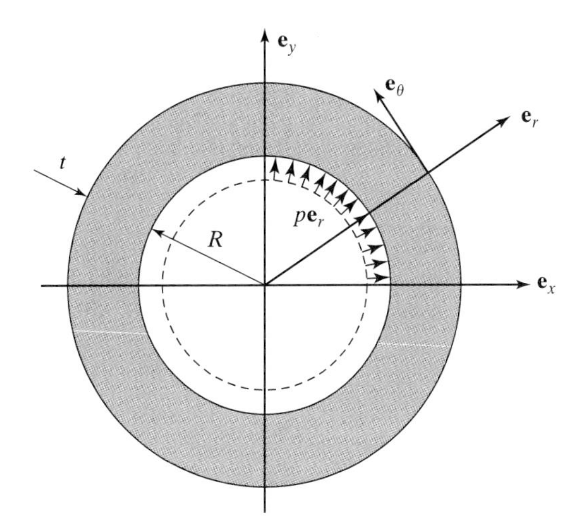

FIGURE 3.7: Training set: pressure vessel.

stress component on a longitudinal plane of the cylinder oriented by outward unit normal $\mathbf{e}_\theta$.

Exercise 26. In the cylinder coordinate frame, specify the force boundary conditions, and determine the nonzero stress components of $\boldsymbol{\sigma}$ in the vessel wall.

The boundary $\partial\Omega_t$ in the problem is composed of the inner surface $r = R$ and the outer surface $r = R + t$. At the inner surface, the imposed pressure reads

$$r = R : \mathbf{T}^d = p\mathbf{e}_r$$

In addition, the stress vector of the material on the inner surface reads

$$r = R : \mathbf{T}(\mathbf{n} = -\mathbf{e}_r) = \boldsymbol{\sigma} \cdot \mathbf{n} = -\boldsymbol{\sigma} \cdot \mathbf{e}_r$$

where $\mathbf{n} = -\mathbf{e}_r$ is the outward unit normal (note the sign). The stress vector at this surface can be developed in the form

$$r = R : \mathbf{T}(\mathbf{n} = -\mathbf{e}_r) = -\sigma_{rr}\mathbf{e}_r - \sigma_{\theta r}\mathbf{e}_\theta - \sigma_{zr}\mathbf{e}_z$$

A comparison of the stress vector components with the stress vector boundary condition $\mathbf{T}^d = p\mathbf{e}_r$ shows

$$r = R : \mathbf{T}^d = \mathbf{T} \leftrightarrow \sigma_{rr} = -p; \quad \sigma_{\theta r} = \sigma_{zr} = 0$$

In a similar fashion, we proceed at the boundary $r = R + t$, where

$$r = R + t : \mathbf{T}^d = 0$$

In turn, the outward normal to this surface is $\mathbf{n} = +\mathbf{e}_r$, and the stress vector at this surface reads here

$$r = R + t : \mathbf{T}(\mathbf{n} = \mathbf{e}_r) = \sigma_{rr}\mathbf{e}_r + \sigma_{\theta r}\mathbf{e}_\theta + \sigma_{zr}\mathbf{e}_z$$

Hence

$$r = R + t : \mathbf{T}^d = \mathbf{T} \leftrightarrow \sigma_{rr} = \sigma_{\theta r} = \sigma_{zr} = 0$$

From these boundary conditions, we see that $\sigma_{rr} = \sigma_{rr}(r)$, while $\sigma_{\theta r} = \sigma_{zr} = 0$ in the vessel wall. In return, with no boundary conditions in the $\mathbf{e}_\theta$- and $\mathbf{e}_z$-directions in hand, nothing can be concluded on the values of the hoop stress, $\sigma_{\theta\theta}$, and the longitudinal stress, σ_{zz}. However, with respect to the structural dimension $R/L \ll 1$, the shear stress $\sigma_{z\theta}$ is zero. Finally, with respect to the rotational symmetry of the cylinder problem (structure + boundary conditions), $\sigma_{\theta\theta} = \sigma_{\theta\theta}(r)$. In summary, the stress tensor in the vessel wall is of the diagonal form:

$$r \in (R, R+t) : \boldsymbol{\sigma} = \sigma_{rr}(r)\mathbf{e}_r \otimes \mathbf{e}_r + \sigma_{\theta\theta}(r)\mathbf{e}_\theta \otimes \mathbf{e}_\theta + \sigma_{zz}\mathbf{e}_z \otimes \mathbf{e}_z$$

and satisfies the boundary conditions

$$\sigma_{rr}(r = R) = -p; \quad \sigma_{rr}(r = R + t) = 0$$

This is one necessary condition that stress field $\boldsymbol{\sigma}$ needs to satisfy. ■

3.5.2 Momentum Balance in Cylinder Coordinates

The second condition that stress $\boldsymbol{\sigma}$ needs to satisfy is the equation of motion. In contrast to (3.27), the equation of motion (3.28) is restricted to Cartesian coordinates and does not apply to cylinder coordinates. Avoiding the introduction of curvilinear spatial operators (see Section 2.6.2), we can determine the momentum balance equation from equilibrium considerations on an infinitesimal cylinder wall element.

Exercise 27. Determine the equation of motion in the $\mathbf{e}_r$- and $\mathbf{e}_\theta$-directions, by considering the equilibrium of all forces acting on an elementary volume $d\Omega_t$ in a cylinder coordinate frame.

The stress components acting on the representative elementary cylinder volume $d\Omega_t = rdr\, d\theta\, dz$ in the $\mathbf{e}_r$- and $\mathbf{e}_\theta$-directions are given in Figure 3.8. Not considered in this figure is the shear stress $\sigma_{\theta r} = \sigma_{r\theta}$, which was found to be zero. Proceeding as in the Cartesian case, the balance of forces in the $\mathbf{e}_r$-direction reads

$$\mathbf{e}_r : 0 = \left[\sigma_{rr} + \frac{\partial \sigma_{rr}}{\partial r} dr\right](r + dr)d\theta - \sigma_{rr} r d\theta - \sigma_{\theta\theta} dr \sin d\theta$$

Linearizing this equation (i.e., $\sin d\theta = d\theta$ and $dr^2 \to 0$) with respect to the infinitesimal order of the involved quantities leads to the momentum balance equation:

$$\mathbf{e}_r : 0 = \frac{\partial \sigma_{rr}}{\partial r} + \frac{1}{r}(\sigma_{rr} - \sigma_{\theta\theta})$$

Analogously, we obtain the equilibrium equation in the $\mathbf{e}_\theta$-direction from Figure 3.8 (for $\sigma_{r\theta} = 0$):

$$\mathbf{e}_\theta : 0 = \left[\sigma_{\theta\theta} + \frac{\partial \sigma_{\theta\theta}}{\partial \theta} d\theta\right] dr \cos d\theta - \sigma_{\theta\theta} dr$$

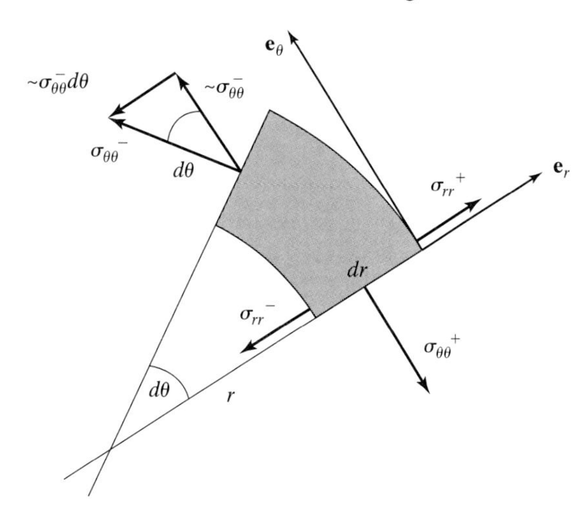

FIGURE 3.8: Stress components acting on an infinitesimal cylinder wall element in a rotation symmetrical problem.

Hence, after linearization ($\cos d\theta = 1$),

$$\mathbf{e}_\theta : \frac{\partial \sigma_{\theta\theta}}{\partial \theta} = 0$$

For the purpose of completeness, let us add the momentum balance in the $\mathbf{e}_z$-direction, which, due to $R/L \ll 1$ (absence of shear stresses) and in the static case, reads

$$\mathbf{e}_z : \frac{\partial \sigma_{zz}}{\partial z} = 0$$

■

3.5.3 One Statically Admissible Stress Field

In addition to the boundary conditions $[\sigma_{rr}(r = R) = -p; \quad \sigma_{rr}(r = R+t) = 0]$, the sought stress field needs to satisfy the previously determined momentum balance equation. We have already seen that $\sigma_{rr} = \sigma_{rr}(r)$ and $\sigma_{\theta\theta} = \sigma_{\theta\theta}(r)$. Use of these stress components in the momentum balance equation in the $\mathbf{e}_r$-direction leads to the following ordinary differential equation (ODE):

$$\mathbf{e}_r : 0 = \frac{\partial \sigma_{rr}(r)}{\partial r} + \frac{1}{r}[\sigma_{rr}(r) - \sigma_{\theta\theta}(r)]$$

Hence, any combination of $\sigma_{rr}(r)$ and $\sigma_{\theta\theta}(r)$ that satisfies this differential equation and the boundary conditions is a statically admissible solution.

Exercise 28. In a first approximation, assume that the hoop stress is constant in the vessel wall (i.e., $\sigma_{\theta\theta} = \text{const}$). Determine the radial stress $\sigma_{rr}(r)$, and show that the resulting stress field is statically admissible.

The solution of the ODE for $\sigma_{\theta\theta} = \text{const}$ reads

$$\sigma_{rr}(r) = \frac{A}{r} + B$$

where A, B are integration constants, to be determined from the boundary conditions:

$$\sigma_{rr}(r = R) = -p = \frac{A}{R} + B$$

$$\sigma_{rr}(r = R + t) = 0 = \frac{A}{R + t} + B$$

This leads to

$$\left[\begin{array}{cc} \frac{1}{R} & 1 \\ \frac{1}{R+t} & 1 \end{array}\right] \left(\begin{array}{c} A \\ B \end{array}\right) = \left(\begin{array}{c} -p \\ 0 \end{array}\right) \rightarrow \left(\begin{array}{c} A \\ B \end{array}\right) = -p\frac{R}{t}\left(\begin{array}{c} R+t \\ 1 \end{array}\right)$$

and to the stresses

$$\sigma_{rr} = p\frac{R}{t}\left(1 - \frac{R+t}{r}\right); \quad \sigma_{\theta\theta} = p\frac{R}{t}$$

■

This was a long way to determine the well-known vessel formula $\sigma_{\theta\theta} = pR/t$, but it illustrates well the required steps to derive a stress field solution for a given force problem. We should note here that this vessel formula is based on the critical assumption of a constant hoop stress $\sigma_{\theta\theta}$ over the cylinder thickness t. It therefore appears that the found solution is restricted to thin wall cylinders, with $t/R \ll 1$.

Furthermore, we could have obtained the vessel formula by applying the reduction formula $(3.23)_1$ to the cylinder vessel.

Exercise 29. Determine the vessel formula by applying the reduction formula to the cylinder vessel.

We first note that the total momentum supplied by pressure p to the circle is zero, since

$$\int_A p\mathbf{n}da = 0$$

This integral holds for any closed surface A subjected to a constant internal pressure. We therefore apply the reduction element (3.23) to the half-circle, shown in Figure 3.9:

$$\begin{aligned} 2V\mathbf{e}_\theta &= \int_{\theta=0}^{\theta=\pi} p\mathbf{e}_r R d\theta = pR\int_{\theta=0}^{\theta=\pi} (\cos\theta\mathbf{e}_x + \sin\theta\mathbf{e}_y)\, d\theta \\ &= pR[\sin\theta\mathbf{e}_x - \cos\theta\mathbf{e}_y]_{\theta=0}^{\theta=\pi} = 2pR\mathbf{e}_y = 2pR\mathbf{e}_\theta \end{aligned}$$

In addition, if we note $\langle\sigma_{\theta\theta}\rangle = \frac{1}{t}\int_t \sigma_{\theta\theta}dr$, the mean traction in the cylinder wall, such that $V = \langle\sigma_{\theta\theta}\rangle\, t$, we obtain the pressure vessel formula:

$$\langle\sigma_{\theta\theta}\rangle = \frac{pR}{t}$$

We should note the difference of the solution obtained here and the one obtained by direct integration based on the assumption of a constant hoop stress, $\langle\sigma_{\theta\theta}\rangle$ being the mean value over the cylinder wall thickness t.

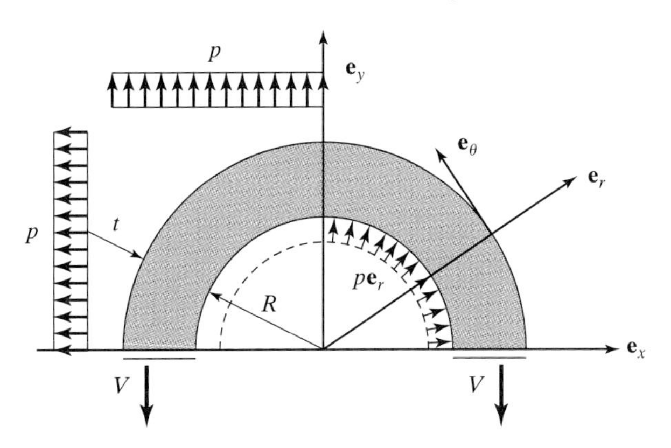

FIGURE 3.9: Vessel formula determined by application of the reduction formula to the half-circle. ■

From this last remark, it appears that the found solution based on the assumption $\sigma_{\theta\theta}$ = const is not unique, as any stress field satisfying the equilibrium condition and the boundary condition is an admissible solution of the problem. We will see later that it requires a constitutive law relating stresses and strains to obtain a unique solution.

CHAPTER 4

Stress and Stress States

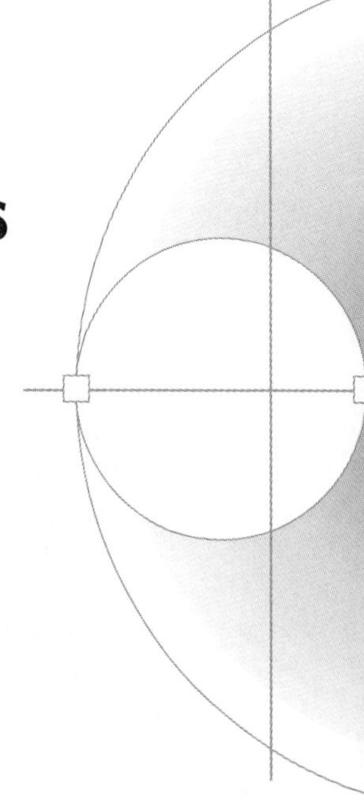

This chapter is devoted to stress states. The starting point is the definition of a statically admissible stress field: a symmetric stress tensor that satisfies the continuity condition of the stress vector at any material surface, the momentum balance equation, and the force boundary conditions. By "static" we mean that inertia effects will not be considered here. Through the stress vector, we introduce the stresses into the Mohr plane. In this Mohr stress plane, the notion of a strength criterion is introduced that encloses admissible stress states. Several strength criteria are discussed: tension cutoff, Tresca criterion, and Mohr–Coulomb criterion.

4.1 STATICALLY ADMISSIBLE STRESS FIELD

In Chapter 3, we introduced the stress tensor $\boldsymbol{\sigma} = \boldsymbol{\sigma}(\mathbf{x}, t)$ through the hypothesis of normal contact forces (3.20) with (3.14). This hypothesis ensures the continuity of the stress vector at any material surface within the material domain Ω_t. In domain Ω_t, this local stress quantity satisfies the equation of motion (3.27) (i.e., the theorem of dynamic resultant) and the symmetry condition (3.34) (i.e., the theorem of dynamic resultant). Disregarding inertia forces (i.e., $\rho\gamma d\Omega_t = \mathbf{0}$), these conditions read

$$\text{in } \Omega_t : \left\{ \begin{array}{l} \mathbf{T} = \boldsymbol{\sigma} \cdot \mathbf{n} \\ \operatorname{div} \boldsymbol{\sigma} + \rho \mathbf{f} = \mathbf{0} \\ \boldsymbol{\sigma} = {}^t\boldsymbol{\sigma} \end{array} \right\} \tag{4.1}$$

In addition, on the boundary $\partial\Omega_{\mathbf{T}^d} \in \partial\Omega_t$, where a stress vector $\mathbf{T}^d$ is prescribed (the superscript d stands for given *data*), the stress tensor $\boldsymbol{\sigma}$ satisfies the (surface) force boundary condition:

$$\text{on } \partial\Omega_{\mathbf{T}^d} : \mathbf{T}^d = \mathbf{T}(\mathbf{n}) = \boldsymbol{\sigma} \cdot \mathbf{n} \tag{4.2}$$

A stress field $\boldsymbol{\sigma} = \boldsymbol{\sigma}(\mathbf{x}, t)$ that satisfies relations (4.1) and (4.2) is called a *statically admissible stress field.*[1] This notion is independent of the material law obeyed by the matter contained in Ω_t.

4.2 PRACTICAL STRESS PLANE AND STRESS SPACE QUANTITIES

4.2.1 Surface Tension and Shear Stress

The definition of the stress tensor $\mathbf{T} = \boldsymbol{\sigma} \cdot \mathbf{n}$ on a material surface da oriented by (outward) unit normal $\mathbf{n}$ allows for a straightforward interpretation of the stress components acting on the oriented surface $\mathbf{n}da$:

$$\mathbf{T} = \boldsymbol{\sigma} \cdot \mathbf{n} = \sigma\mathbf{n} + \tau\mathbf{t} \tag{4.3}$$

The normal and tangential components, σ and τ, are sketched in Figure 4.1a. The normal stress component $\sigma = \mathbf{n}\cdot\boldsymbol{\sigma}\cdot\mathbf{n}$ is the surface traction. Following its orientation (i.e., using the mechanics sign convention), $\sigma > 0$ stands for tension, and $\sigma < 0$ for compression exerted by the normal stress on the surface. Things are somehow different for the shear stress $\tau = \mathbf{t} \cdot \boldsymbol{\sigma} \cdot \mathbf{n}$, oriented in the tangential direction $\mathbf{t}$. Its intensity is defined by

$$|\tau| = \sqrt{(\boldsymbol{\sigma} \cdot \mathbf{n})^2 - (\mathbf{n} \cdot \boldsymbol{\sigma} \cdot \mathbf{n})^2} = \sqrt{\mathbf{T}^2 - \sigma^2} \tag{4.4}$$

In turn, the direction of the in-plane shear stress is given as $\tau\mathbf{t} = |\tau|\text{sign}(\tau)\mathbf{t}$.

Exercise 30. Consider the stress vector on the surface oriented by $\mathbf{n} = \mathbf{e}_1$. Determine the normal and the tangential stress component, (σ, τ) and the corresponding directions $(\mathbf{n}, \mathbf{t})$.

We have seen that the stress vector on $\mathbf{n} = \mathbf{e}_1$ reads [i.e., (3.21)]

$$\mathbf{T}(\mathbf{e}_1) = \boldsymbol{\sigma} \cdot \mathbf{e}_1 = \sigma_{11}\mathbf{e}_1 + \sigma_{21}\mathbf{e}_2 + \sigma_{31}\mathbf{e}_3$$

Use in (4.4) yields

$$\sigma = \mathbf{n} \cdot \boldsymbol{\sigma} \cdot \mathbf{n} = \sigma_{11}; \quad |\tau| = \sqrt{\sigma_{21}^2 + \sigma_{31}^2}$$

The direction of the shear stress $\mathbf{t} = t_2\mathbf{e}_2 + t_2\mathbf{e}_2$ is given by vector algebra:

$$\text{sign}\,(\tau)\mathbf{t} = \frac{\sigma_{21}\mathbf{e}_2 + \sigma_{31}\mathbf{e}_3}{\sqrt{\sigma_{21}^2 + \sigma_{31}^2}}$$

We should note here that σ and $|\tau|$ can be considered as invariants of the surface stress vector built on the material plane $\mathbf{n} = \mathbf{e}_1$. ■

[1] We have already seen in Section 2.5.1 the definition of a kinematically admissible displacement field, satisfying the displacement boundary conditions on $\partial\Omega_{\boldsymbol{\xi}^d}$, and the total boundary is

$$\partial\Omega_t = \partial\Omega_{\boldsymbol{\xi}^d} \cup \partial\Omega_{\mathbf{T}^d}$$

The static admissible stress field satisfying (4.1) and (4.2) is the counterpart on the stress side of the notion of a kinematically admissible displacement field. As we will see later, these elementary notions form the core of analytical and approximated solution strategies for mechanics problems.

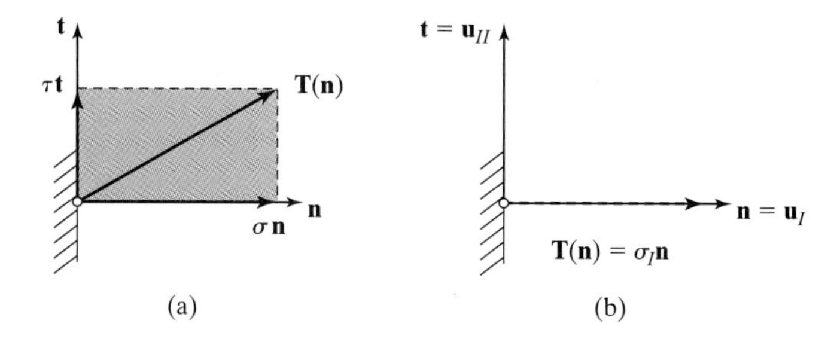

FIGURE 4.1: Stress vector $\mathbf{T}(\mathbf{n})$: (a) definition of stress vector components; (b) stress vector on a plane oriented by a principal stress axis.

4.2.2 Principal Stresses

We have seen in Section 3.4.3 that the symmetry of the stress tensor $\boldsymbol{\sigma} = {}^t\boldsymbol{\sigma}$ implies the existence of real eigenvalues of the stress tensor. In the space defined by the principal stress directions $\mathbf{u}_I, \mathbf{u}_{II}, \mathbf{u}_{III}$, the stress tensor reads

$$\boldsymbol{\sigma} = \sigma_I \mathbf{u}_I \otimes \mathbf{u}_I + \sigma_{II} \mathbf{u}_{II} \otimes \mathbf{u}_{II} + \sigma_{III} \mathbf{u}_{III} \otimes \mathbf{u}_{III} \tag{4.5}$$

By convention, we consider $\sigma_I \geq \sigma_{II} \geq \sigma_{III}$. The stress vector built on a material surface oriented in the principal direction $\mathbf{n} = \mathbf{u}_J$ is

$$\mathbf{T}(\mathbf{n} = \mathbf{u}_J) = \boldsymbol{\sigma} \cdot \mathbf{u}_J = \sigma_J \mathbf{u}_J \tag{4.6}$$

Equation (4.6) is the same equation as (3.35) (i.e., the very definition of the eigenvectors and eigenvalues of the stress tensor). However, (4.6) with (4.3) reveal the physical nature of the eigenvalue problem (3.35): It is the stress vector with only normal stress component σ_J on the surface oriented in one of the principal stress directions $\mathbf{u}_J$. This is illustrated in Figure 4.1b.

4.2.3 The Hydrostatic Pressure and Other Stress Invariants

One particular eigenvalue stress state is the hydrostatic stress state. The hydrostatic stress state is defined by a pressure p exerted in all directions ($\sigma_I = \sigma_{II} = \sigma_{III} = -p$), thus:

$$\boldsymbol{\sigma} = -p\mathbf{1}; \quad \sigma_{ij} = -p\delta_{ij}; \quad (\sigma_{ij}) = \begin{bmatrix} -p & & \\ & -p & \\ & & -p \end{bmatrix} \tag{4.7}$$

The particularity of this hydrostatic pressure is that it applies in all directions (i.e., $\mathbf{T} = -p\mathbf{n}$), and consequently any normal is a principal stress direction. In other words, this hydrostatic stress state is independent of the direction: It is a stress invariant.

More generally, a stress invariant is a scalar quantity expressed as a function of the stress components σ_{ij}, which does not change when the basis of the regarded 3D-system change (i.e., circular permutation of subscripts). These stress invariants can be constructed as polynomial functions of $\boldsymbol{\sigma}$:

$$I_1 = \operatorname{tr} \boldsymbol{\sigma} = \sigma_{ii} \tag{4.8}$$

$$I_2 = \frac{1}{2}\text{tr}\,(\boldsymbol{\sigma} \cdot \boldsymbol{\sigma}) = \frac{1}{2}\sigma_{ij}\sigma_{ji} \tag{4.9}$$

$$I_3 = \frac{1}{3}\text{tr}\,(\boldsymbol{\sigma} \cdot \boldsymbol{\sigma} \cdot \boldsymbol{\sigma}) = \frac{1}{3}\sigma_{ij}\sigma_{jk}\sigma_{ki} \tag{4.10}$$

where repeated subscripts indicates summation.

Exercise 31. Express the stress invariants I_1, I_2, I_3 as a function of the principal stresses $\sigma_I \geq \sigma_{II} \geq \sigma_{III}$. Determine the stress invariants of the hydrostatic stress state and the corresponding hydrostatic axis in the principal stress space defined by $(\mathbf{u}_I, \mathbf{u}_{II}, \mathbf{u}_{III})$.

Use of (4.5) in (4.8) to (4.10) gives

$$I_1 = \sigma_I + \sigma_{II} + \sigma_{III}; \quad I_2 = \frac{1}{2}(\sigma_I^2 + \sigma_{II}^2 + \sigma_{III}^2); \quad I_3 = \frac{1}{3}(\sigma_I^3 + \sigma_{II}^3 + \sigma_{III}^3)$$

For the hydrostatic stress state $\sigma_I = \sigma_{II} = \sigma_{III} = -p$, the invariants are

$$I_1 = -3p; \quad I_2 = \frac{3p^2}{2}; \quad I_3 = -p^3$$

Finally, in the principal stress space, the hydrostatic stress state is defined along an axis going through the origin:

$$\mathbf{n} = \frac{\sqrt{3}}{3}(\mathbf{u}_I + \mathbf{u}_{II} + \mathbf{u}_{III})$$

The axis defined by $\mathbf{n}$ is called the hydrostatic axis. The stress vector of a hydrostatic stress state built on any plane oriented by this hydrostatic axis is

$$\mathbf{T}(\mathbf{n}) = \sigma_m \mathbf{n} = \frac{\sqrt{3}}{3}[\sigma_I \mathbf{u}_I + \sigma_{II}\mathbf{u}_{II} + \sigma_{III}\mathbf{u}_{III}]$$

where $\sigma_m = I_1/3$ is the mean volume stress (i.e., in a pure hydrostatic stress state $\sigma_m = -p$). ■

The invariants of the stress tensor play an important role in the description of the specific behavior of a certain class of materials, which is referred to as *isotropic* material behavior. A material behavior is said to be *isotropic* if the behavior is the same in all directions. An immediate consequence of this isotropy is that the behavior can be described in terms of the stress invariants I_1, I_2, I_3. Let $f = f(\boldsymbol{\sigma})$ be a function describing a material behavior that is determined by stress $\boldsymbol{\sigma}$. In the case of an isotropic material behavior, this function is a scalar function depending on the invariants

$$f = f(\boldsymbol{\sigma}) = f(I_1, I_2, I_3) \tag{4.11}$$

4.2.4 Stress Deviator

The first stress invariant I_1 represents the volume, spherical or hydrostatic part of the stress tensor. The other component of stress tensor $\boldsymbol{\sigma}$ is the deviator part, denoted $\mathbf{s}$. By definition, this stress deviator has a zero volume component (i.e., $\text{tr}\,\mathbf{s} = 0$); hence,

$$\mathbf{s} = \boldsymbol{\sigma} - \frac{I_1}{3}\mathbf{1} = \boldsymbol{\sigma} - \sigma_m \mathbf{1} \tag{4.12}$$

where $\sigma_m = I_1/3$ is the mean (volume) stress. The stress deviator has the same principal directions (eigenvectors) as stress tensor $\boldsymbol{\sigma}$, and the principal values (eigenvalues) therefore read (show this)

$$s_J = \sigma_J - \sigma_m = \frac{1}{3}(2\sigma_J - \sigma_K - \sigma_L) \tag{4.13}$$

In addition, the invariants[2] of $\mathbf{s}$ are

$$J_1 = \text{tr}\,\mathbf{s} = 0; \quad J_2 = \frac{1}{2}\text{tr}\,(\mathbf{s}\cdot\mathbf{s}) = \frac{1}{2}s_{ij}s_{ji}; \quad J_3 = \frac{1}{3}\text{tr}\,(\mathbf{s}\cdot\mathbf{s}\cdot\mathbf{s}) = \frac{1}{3}s_{ij}s_{jk}s_{ki} \tag{4.14}$$

Exercise 32. Consider the surface oriented by the hydrostatic axis in the principal stress/principal stress deviator space defined by $(\mathbf{u}_I, \mathbf{u}_{II}, \mathbf{u}_{III})$. Determine the normal and tangential components of a stress vector $\mathbf{T}(\mathbf{n})$, built on this oriented surface, shown in Figure 4.2.

The stress vector on the surface oriented by $\mathbf{n}$ reads

$$\mathbf{T}(\mathbf{n}) = \sigma_{\text{oct}}\mathbf{n} + \tau_{\text{oct}}\mathbf{t}$$

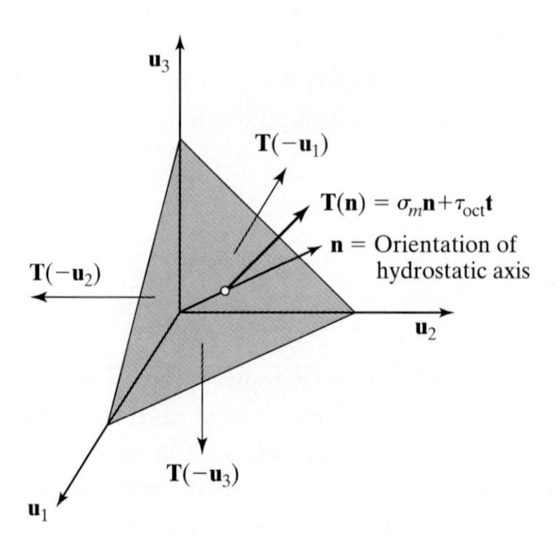

FIGURE 4.2: Octrahedral stresses: geometric–physical interpretation of in the principal stress space.

[2]The invariants of the stress tensor and of the stress deviator are related by

$$I_2 = J_2 + I_1^2/6; \quad I_3 = J_3 + 2I_1J_2/3 + I_1^3/27$$

This shows that only three invariants exist for the second-order stress tensor.

In addition, recalling the tetrahedron lemma (3.19), this stress vector is equal to

$$\mathbf{T}(\mathbf{n}) = \sum_{J=I,II,III} \mathbf{T}(\mathbf{u}_J) n_J$$

where $\mathbf{T}(\mathbf{u}_J)$ is the stress vector built on the principal stress planes oriented by $\mathbf{u}_J$, and the components of the normal orienting the hydrostatic axis are $n_J = \sqrt{3}/3$; thus, according to (4.6),

$$\mathbf{T}(\mathbf{n}) = \frac{\sqrt{3}}{3} \sum_{J=I,II,III} \sigma_J \mathbf{u}_J$$

A combination of the two previous equations leads to

$$\mathbf{T}(\mathbf{n}) = \sigma_{\text{oct}} \mathbf{n} + \tau_{\text{oct}} \mathbf{t} = \frac{\sqrt{3}}{3} (\sigma_I \mathbf{u}_I + \sigma_{II} \mathbf{u}_{II} + \sigma_{III} \mathbf{u}_{III})$$

The normal stress on this plane is the mean stress:

$$\sigma_{\text{oct}} = \sigma_m = I_1/3 = (\sigma_I + \sigma_{II} + \sigma_{III})/3$$

while (4.4), (4.13), and (4.14) show that the tangential stress is determined by the principal deviator stress components s_J:

$$\tau_{\text{oct}} \mathbf{t} = \frac{\sqrt{3}}{3} \sum_{J=I,II,III} [\sigma_J - \sigma_m] \mathbf{u}_J = \frac{\sqrt{3}}{3} \sum_{J=I,II,III} s_J \mathbf{u}_J$$

respectively, by the second stress deviator invariant:

$$|\tau_{\text{oct}}| = \frac{\sqrt{3}}{3} \sqrt{s_I^2 + s_{II}^2 + s_{III}^2} = \sqrt{\frac{2J_2}{3}}$$

This gives a geometric–physical interpretation of the invariants I_1 and J_2: They define the normal and the shear components, σ_{oct} and $|\tau_{\text{oct}}|$, of the stress vector $\mathbf{T}(\mathbf{n})$ built in the principal stress space on the plane oriented by the hydrostatic axis. These stress components are referred to as octrahedral stresses. ■

Last, we should note that the set of three invariants of the stress tensor is not restricted to the one introduced here by (4.8)–(4.10) through a polynomial argument. Other sets of invariants are commonly used for the description of isotropic material behavior. They can be obtained by algebraic operations using the set of invariants introduced here.

4.3 THE MOHR STRESS PLANE

We have already seen the imminent role of the stress vector $\mathbf{T}(\mathbf{n}) = \boldsymbol{\sigma} \cdot \mathbf{n}$ for describing stress states. We will now seek for the corresponding graphical representation in the Mohr stress plane. We consider a stress vector on a material plane oriented

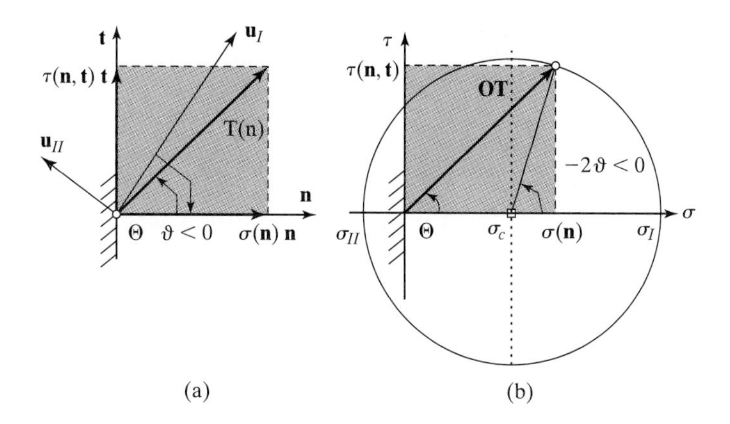

FIGURE 4.3: Stress vector representation in (a) the material plane; (b) in the Mohr stress plane.

by unit vector $\mathbf{n}$. Let $\mathbf{u}_I$ and $\mathbf{u}_{II}$ be the principal directions of the stresses in this plane (and $\mathbf{u}_{III}$ the normal to this plane). Proceeding as in Section 2.3, we note $\vartheta = \vartheta(\mathbf{u}_I, \mathbf{n})$, the angle between the principal stress axis $\mathbf{u}_I$ and the normal $\mathbf{n}$. We recall (2.27):

$$\mathbf{n} = \cos\vartheta\,\mathbf{u}_I + \sin\vartheta\,\mathbf{u}_{II}; \quad \mathbf{t} = -\sin\vartheta\,\mathbf{u}_I + \cos\vartheta\,\mathbf{u}_{II} \tag{4.15}$$

The normal and the tangential component of the stress vector $\mathbf{T}(\mathbf{n}) = \sigma\mathbf{n} + \tau\mathbf{t}$ are related to the principal stresses:

$$\sigma = \mathbf{n}\cdot\boldsymbol{\sigma}\cdot\mathbf{n} = \sigma_I\cos^2\vartheta + \sigma_{II}\sin^2\vartheta = \frac{\sigma_I + \sigma_{II}}{2} + \frac{\sigma_I - \sigma_{II}}{2}\cos 2\vartheta \tag{4.16}$$

$$\tau = \mathbf{t}\cdot\boldsymbol{\sigma}\cdot\mathbf{n} = -(\sigma_I - \sigma_{II})\sin\vartheta\cos\vartheta = \frac{\sigma_I - \sigma_{II}}{2}\sin(-2\vartheta) \tag{4.17}$$

As in Section 2.3, we obtain for the second-order stress tensor the two Mohr coordinates (σ, τ) of the stress vector representing, respectively, the normal stress σ and the shear stress τ. In the Mohr stress plane, this stress point is situated on a circle of center

$$\sigma_c = \frac{1}{2}(\sigma_I + \sigma_{II}); \quad \tau = 0 \tag{4.18}$$

and of radius

$$R = \frac{1}{2}(\sigma_I - \sigma_{II}) \tag{4.19}$$

The graphical representation in the material plane and the Mohr plane is given in Figure 4.3. A summary of the links between the Material plane and the Mohr plane is given in Table 4.1.

Here we assumed that the normal $\mathbf{u}_{III}$ to the material plane $(\mathbf{n}, \mathbf{t})$ is a principal direction of the stress vector associated with the principal stress σ_{III}. There are three Mohr circles corresponding to this principal stress material space associated with the different (principal stress) material basis:

Quantity	Material Plane	Mohr Plane
Basis	$(\mathbf{n}, \mathbf{t})$	(σ, τ)
Stress Vector	$\mathbf{T}(\mathbf{n}) = \sigma\mathbf{n} + \tau\mathbf{t}$	$\mathbf{OT} = (\ \sigma \quad \tau\ )$
Angle Θ	$\Theta(\mathbf{n}, \mathbf{T}(\mathbf{n}))$	$\Theta(\mathbf{n}, \mathbf{OT})$
Angle ϑ	$\vartheta(\mathbf{u}_I, \mathbf{n})$	$-2\vartheta(\mathbf{u}_I, \mathbf{n})$

TABLE 4.1: Relations between quantities in the material plane and the Mohr plane.

$$(\mathbf{u}_I \times \mathbf{u}_{II}) : \sigma_c = \frac{1}{2}(\sigma_I + \sigma_{II}); \quad R = \frac{1}{2}(\sigma_I - \sigma_{II}) \tag{4.20}$$

$$(\mathbf{u}_{II} \times \mathbf{u}_{III}) : \sigma_c = \frac{1}{2}(\sigma_{II} + \sigma_{III}); \quad R = \frac{1}{2}(\sigma_{II} - \sigma_{III}) \tag{4.21}$$

$$(\mathbf{u}_I \times \mathbf{u}_{III}) : \sigma_c = \frac{1}{2}(\sigma_I + \sigma_{III}); \quad R = \frac{1}{2}(\sigma_I - \sigma_{III}) \tag{4.22}$$

The three circles are presented in Figure 4.4. There is one maximum circle enclosing the other smaller ones. With the convention $\sigma_I \geq \sigma_{II} \geq \sigma_{III}$, this is the circle defined by (4.22) of radius $R = \frac{1}{2}(\sigma_I - \sigma_{III})$. In addition, due to the very definition of the principal stresses expressed by (4.6), it is readily understood that any stress vector on a plane oriented by a normal $\mathbf{n}$ is enclosed in this maximum Mohr stress circle, defined by the principal stresses. This maximum circle is called *THE* Mohr circle.

By way of application, we illustrate the Mohr representation of some common stress states:

Hydrostatic Stress State.

With $\boldsymbol{\sigma} = -p\mathbf{1}$ and $\sigma_I = \sigma_{II} = \sigma_{III} = -p$ the three Mohr circles degenerate to a point in the Mohr plane (Figures 4.5a and b).

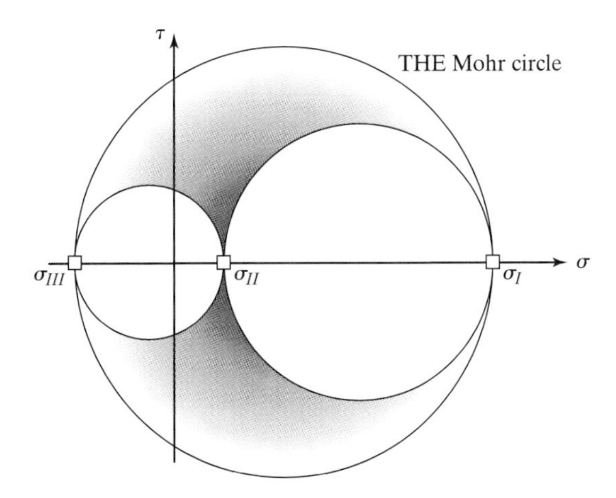

FIGURE 4.4: The Mohr circle and the principal mohr circles in the Mohr stress plane $(\sigma \times \tau)$.

Uniaxial Tension.

The uniaxial tension is defined by

$$\boldsymbol{\sigma} = \sigma \mathbf{e}_z \otimes \mathbf{e}_z \tag{4.23}$$

where $\mathbf{e}_z$ is the direction of the stress application of intensity $\sigma > 0$. Obviously, the principal stresses are $\sigma_I = \sigma; \sigma_{II} = \sigma_{III} = 0$. This stress state corresponds to two circles with center and radius $\sigma_c = R = \sigma/2$, and a zero-radius circle at the origin (Figures 4.5c and d).

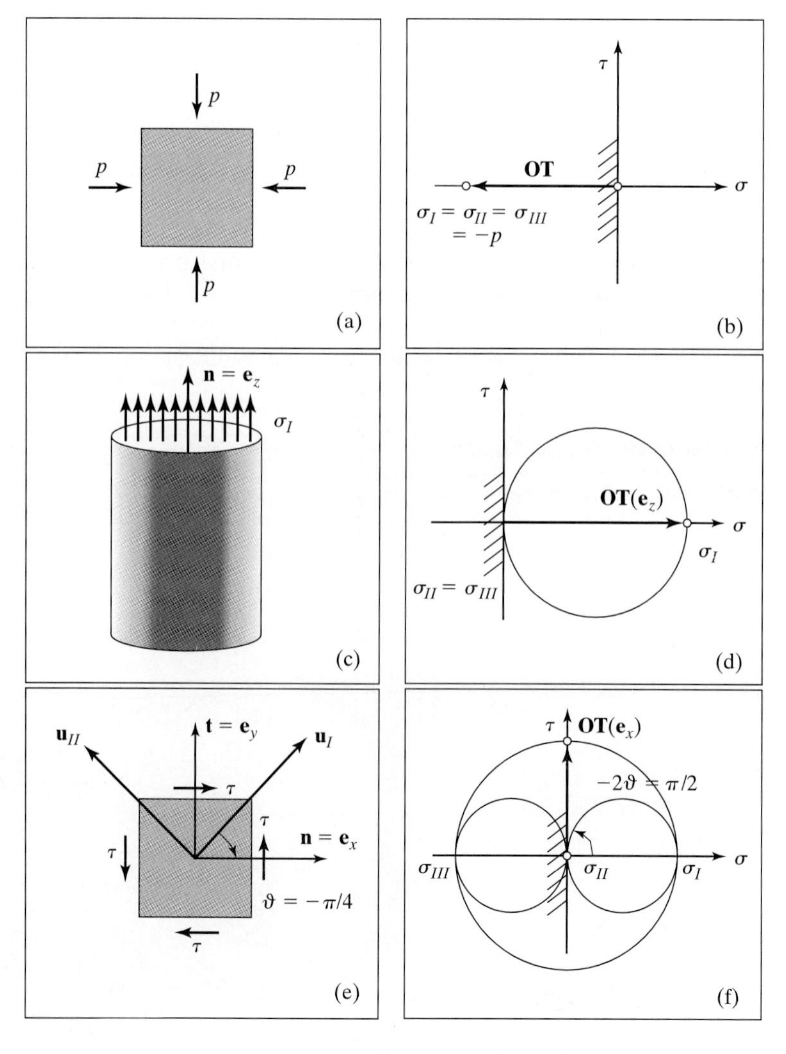

FIGURE 4.5: Remarkable stress states in material and Mohr plane: (a)–(b) hydrostatic stress state; (c)–(d) uniaxial tension; (e)–(f) simple shear stress state.

Simple Shear.

The simple shear is defined by (*cf.* Figure 4.5e)

$$\boldsymbol{\sigma} = \tau[\mathbf{e}_y \otimes \mathbf{e}_x + \mathbf{e}_x \otimes \mathbf{e}_y] \tag{4.24}$$

For a stress vector built on the surface oriented by the normal $\mathbf{n} = \mathbf{e}_x$, the stress vector is given by

$$\mathbf{T}(\mathbf{n} = \mathbf{e}_x) = \boldsymbol{\sigma} \cdot \mathbf{n} = \tau \mathbf{t} = \tau \mathbf{e}_y \tag{4.25}$$

This corresponds to the stress vector $\mathbf{OT}(\mathbf{e}_x) = (\ 0 \quad \tau\ )$ in the Mohr stress plane.[3] The principal stresses are $\sigma_I = \tau; \sigma_{II} = 0; \sigma_{III} = -\tau$ and the corresponding eigenvalues $\mathbf{u}_{I,III} = \frac{\sqrt{2}}{2}[\mathbf{e}_x + \mathbf{e}_y]; \mathbf{u}_{III} = \mathbf{e}_z$. The three Mohr circles are given in Figure 4.5f.

Plane Stress State.

The plane stress state is defined by (*cf.* Figure 4.6a)

$$\mathbf{T}(\mathbf{n} = \mathbf{e}_z) = \boldsymbol{\sigma} \cdot \mathbf{n} = 0 \tag{4.26}$$

From (3.21), this implies

$$\sigma_{xz} = \sigma_{yz} = \sigma_{zz} = 0; \quad (\sigma_{ij}) = \begin{bmatrix} \sigma_{xx} & \sigma_{xy} & 0 \\ & \sigma_{yy} & 0 \\ \text{sym} & & 0 \end{bmatrix} \tag{4.27}$$

The principal stresses are the principal stresses in the in-plane $(\mathbf{e}_x, \mathbf{e}_y)$, material plane, and $\sigma_{zz} = 0$. The three Mohr circles for this plane stress condition are given in Figure 4.6b. The plane stress assumption is a common assumption in classical plate and shell theories.

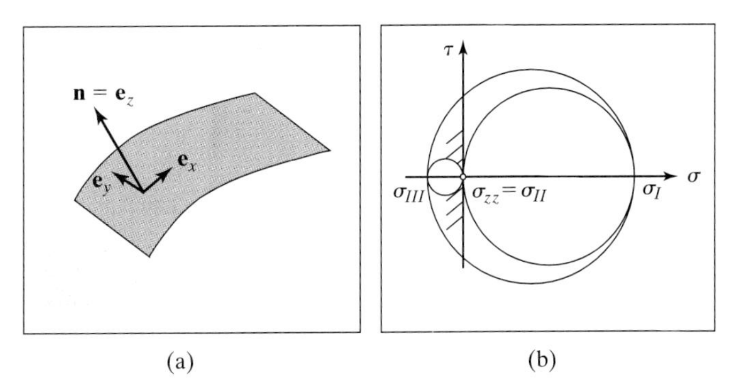

FIGURE 4.6: Plane stress state: (a) in the material plane; (b) in the Mohr stress plane.

[3]The determination of principal stresses and principal directions is similar to the one applied in Section 2.5.3 dealing with torsion deformation.

4.4 ADMISSIBLE STRESS STATES AND STRENGTH CRITERION

So far we have only been concerned with the definition of a statically admissible stress field, satisfying the momentum balance (4.1) and the boundary conditions (4.2). These conditions are independent of the material behavior of the matter. However, with the physical significance of the stresses in hand, we can now move on to link the stresses to material strength properties. Later, we will see how to link these limit stress states to strains.

Exercise 33. Consider a material with a finite tensile strength, denoted f_t'. For the previously considered stress states of hydrostatic pressure, uniaxial tension, and simple shear, determine the maximum admissible stress that can be applied. Give an illustration of this "tension cutoff" criterion in the Mohr plane, and determine the stress direction in which the tension strength is reached.
The strength criterion reads

$$f(\sigma_I) = \sigma_I - f_t' \leq 0 \tag{4.28}$$

with σ_I the maximum principal stress. The stress direction in which this strength is reached is the principal stress axis $\mathbf{u}_I$.
For the hydrostatic pressure it is $\sigma_I = -p < f_t'$. The tension strength criterion does not restrict the pressure application. In turn, in the case of the direct tension test, with $\sigma_I = \sigma$ the applied stress, it is obviously $\sigma_{\max} = f_t'$, and $\mathbf{u}_I = \mathbf{e}_z$. Last, in the case of the simple shear, with $\sigma_I = \tau$, the maximum shear stress is $\tau_{\max} = f_t'$, and $\mathbf{u}_I = \frac{\sqrt{2}}{2}[\mathbf{e}_x + \mathbf{e}_y]$. Figure 4.7 illustrates the tension cutoff criterion in the Mohr plane.

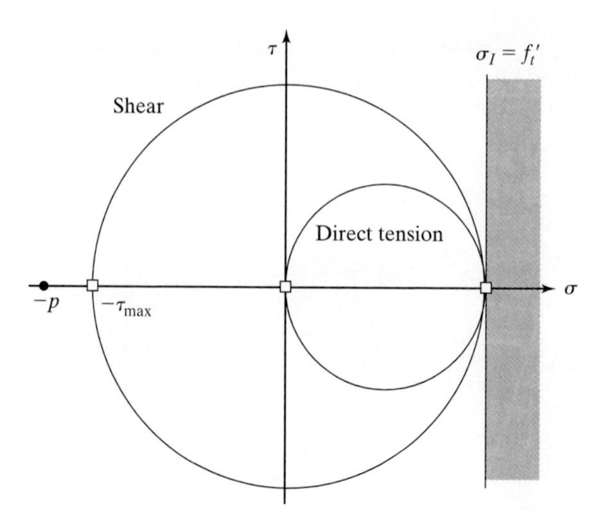

FIGURE 4.7: Tension cutoff criterion and limit stress states of simple shear and direct tension.

■

The example of the tension strength criterion illustrates the main features of strength-of-material criteria:

1. A material strength domain, which defines admissible stress states in the form of a strength criterion:
$$\forall \boldsymbol{\sigma}: \quad f(\boldsymbol{\sigma}) \leq 0 \tag{4.29}$$
This strength domain is determined by material tests. From physical evidence, all stress states must lie within the strength domain of the material. It goes without saying that the stress fields must be statically admissible.
2. A material plane, on which the limit stress state is reached.

4.4.1 The Tresca Criterion

The Tresca criterion reads as a function of the principal stresses:

$$f(\boldsymbol{\sigma}) = \sigma_I - \sigma_{III} - \sigma_0 \leq 0 \tag{4.30}$$

with σ_0 a stress threshold. In the Mohr plane (see Figure 4.4), $\sigma_I - \sigma_{III}$ represents twice the maximum radius R of the Mohr circle. It does not define an individual lower or upper bound of σ_I or σ_{III}; and the Tresca criterion can be written in the alternative form:
$$f(\boldsymbol{\sigma}) = |\tau| - \sigma_0/2 \leq 0 \tag{4.31}$$
with $|\tau| = \sqrt{\mathbf{T}^2 - \sigma^2}$ according (4.4). This one-parameter strength criterion requires one material test to determine the *shear strength* parameter $\sigma_0/2$.

Exercise 34. Study the following material tests carried out on a Tresca material (i.e., a material of which the strength domain is defined by the Tresca strength criterion): uniaxial tension, uniaxial compression, and simple shear. Determine the material plane on which the strength criterion (4.31) is reached.
In uniaxial tension,

$$\boldsymbol{\sigma} = f_t' \mathbf{e}_z \otimes \mathbf{e}_z; \quad \sigma_I = f_t' = \sigma_0; \quad \sigma_{III} = 0$$

With respect to the direction of stress application $\mathbf{e}_z$, the maximum shear in uniaxial tension is reached on a plane oriented by $\vartheta(\mathbf{u}_I, \mathbf{n}) = \mp\pi/4$.
In uniaxial compression,

$$\boldsymbol{\sigma} = -f_c' \mathbf{e}_z \otimes \mathbf{e}_z; \quad \sigma_I = 0; \quad \sigma_{III} = -f_c' = -\sigma_0$$

The plane on which the maximum shear is reached is oriented by $\vartheta(\mathbf{u}_{III}, \mathbf{n}) = \pm\pi/4$. Hence, one of the underlying concepts of the Tresca criterion is the symmetric strength behavior in tension and compression ($f_t'/f_c' = 1$). In other words, the Tresca criterion is independent of the mean stress σ_m but depends only on the deviator stresses. Such materials are referred to as (hydrostatic) pressure insensitive. The Tresca criterion is one strength criterion that models pressure insensitive behavior.
Finally, in simple shear,

$$\boldsymbol{\sigma} = \tau_{\max}[\mathbf{e}_y \otimes \mathbf{e}_x + \mathbf{e}_x \otimes \mathbf{e}_y]; \quad |\tau_{\max}| = \sigma_0/2$$

The direction of shear stress application corresponds to the in-plane unit tangential vector of the material plane of maximum shear.

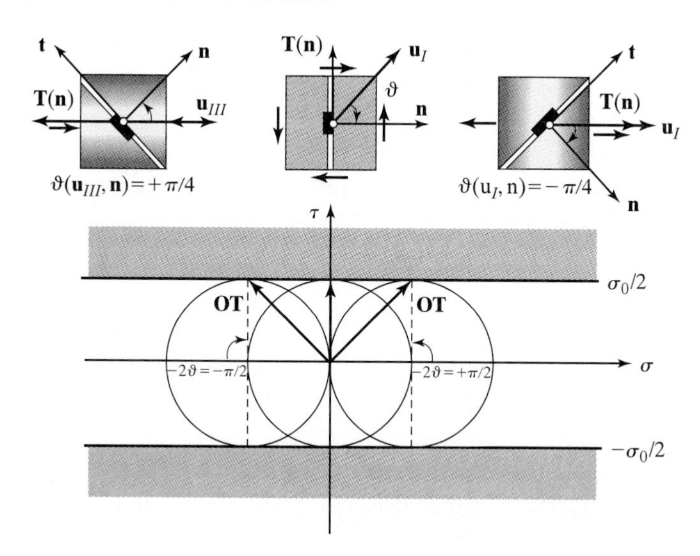

FIGURE 4.8: The Tresca criterion applied to uniaxial compression, simple shear, and uniaxial tension.

The limit stress states and the corresponding material planes of maximum shear are given in Figure 4.8. ■

We have used the strength criterion together with well-defined material test stress states. However, strength criteria also apply on the structural level, provided that the considered stress field is statically admissible.

Exercise 35. Consider the semi-infinite half-space subjected to a pressure p over a length B (corresponding, e.g., to a foundation).

In the subdomains defined in Figure 4.9a, we consider as stress field

$$\text{Zone } 1 : \boldsymbol{\sigma}^{(1)} = -p[\mathbf{e}_x \otimes \mathbf{e}_x + \mathbf{e}_y \otimes \mathbf{e}_y + \mathbf{e}_z \otimes \mathbf{e}_z]$$

$$\text{Zone } 2 : \boldsymbol{\sigma}^{(2)} = -p[\mathbf{e}_y \otimes \mathbf{e}_y + \mathbf{e}_z \otimes \mathbf{e}_z]$$

Determine the maximum admissible pressure $p_{\max}$ for a Tresca material. We verify that this stress field is statically admissible, as it satisfies the boundary conditions at $x = 0$,

$$x = 0;\quad y \in B : \mathbf{T}(\mathbf{n} = -\mathbf{e}_x) = -\boldsymbol{\sigma}^{(1)} \cdot \mathbf{e}_x = \mathbf{T}^d = p\mathbf{e}_x$$

$$x = 0;\quad y \notin B : \mathbf{T}(\mathbf{n} = -\mathbf{e}_x) = -\boldsymbol{\sigma}^{(2)} \cdot \mathbf{e}_x = \mathbf{T}^d = 0$$

and the continuity condition at the interface between the two zones (e.g., on the right interface):

$$\mathbf{T}(\mathbf{n}^{(1)} = \mathbf{e}_y) + \mathbf{T}(\mathbf{n}^{(2)} = -\mathbf{e}_y) = 0$$

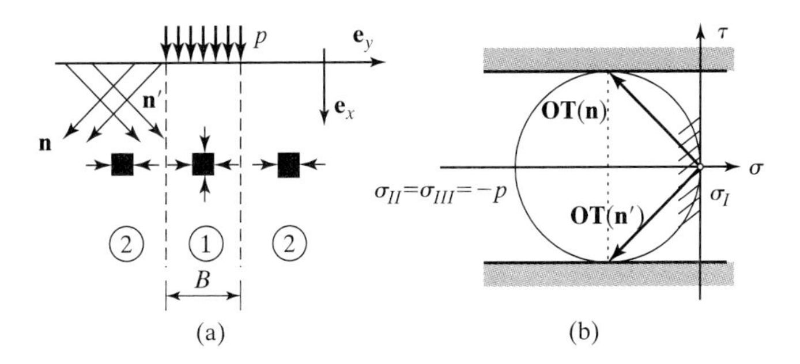

FIGURE 4.9: The Tresca criterion applied to a soil substratum subjected to foundation loading: (a) material–structural plane; (b) Mohr stress plane.

The stress state in zone 1 corresponds to a hydrostatic stress state, and the Mohr circles therefore degenerate to a point at $\sigma_I = \sigma_{II} = \sigma_{III} = -p$. In turn, in zone 2 it is

$$\sigma_I = 0; \quad \sigma_{II} = \sigma_{III} = -p$$

Hence, the maximum Mohr circle (see Figure 4.9b), in combination with the Tresca criterion (4.30), yields a maximum value of the applied pressure:

$$f(\boldsymbol{\sigma}) = 0 \leftrightarrow p_{\max} = \sigma_0$$

The material planes corresponding to the maximum shear defined by the Tresca criterion are inclined to the vertical (down) axis by $\vartheta = \mp\pi/4$. This is shown in Figure 4.9a. ■

4.4.2 The Mohr–Coulomb Criterion

The Mohr–Coulomb criterion reads

$$f(\boldsymbol{\sigma}) = |\tau| + \sigma \tan\varphi - c \leq 0 \tag{4.32}$$

with $|\tau| = \sqrt{\mathbf{T}^2 - \sigma^2}, \sigma = \mathbf{n} \cdot \boldsymbol{\sigma} \cdot \mathbf{n}$. This strength criterion has two material parameters: cohesion c, and friction angle φ. Hence, two independent material tests [i.e., two points in the $(\sigma \times \tau)$-plane] are required to determine the strength parameters of this two-parameter model. The Mohr–Coulomb criterion includes the Tresca criterion for $\varphi = 0 \to c = \sigma_0/2$, and the tension cutoff criterion for $\varphi = \pi/2 \to c \cot\varphi = f_t'$.

Exercise 36. Study the limit stress states of uniaxial tension, uniaxial compression and the simple shear for a Mohr–Coulomb material. Determine the material plane on which the strength criterion is reached.

In the Mohr stress plane, the Mohr–Coulomb strength criterion is represented by two inclined straight lines defined by $f(\boldsymbol{\sigma}) = 0 : |\tau| = -\sigma\tan\varphi + c$. In this plane, the maximum admissible stresses are situated on the Mohr circle that has as

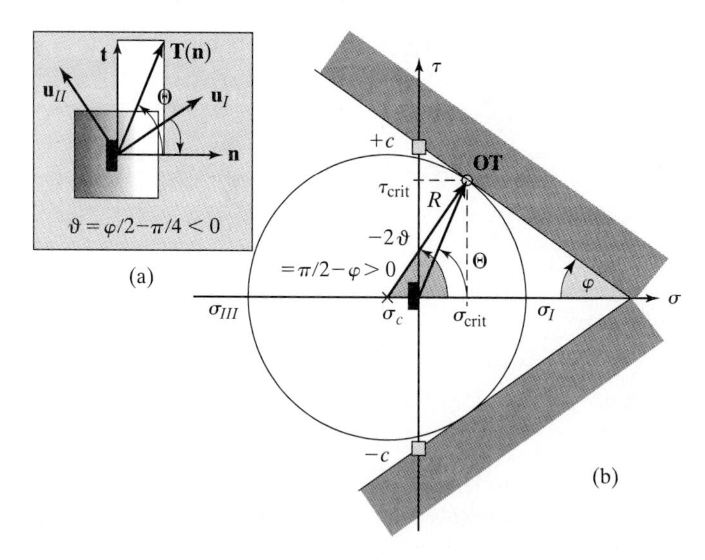

FIGURE 4.10: The Mohr–Coulomb criterion: (a) material plane on which the strength criterion is reached; (b) the strength criterion in the Mohr plane.

tangent the inclined straight lines. This is sketched in Figure 4.10b. From geometric considerations, the Mohr coordinates of the stress vector $\mathbf{OT}$ on the Mohr circle touching the Mohr–Coulomb criterion are[4]

$$\sigma_{\text{crit}} = \sigma_c + R\cos(\pi/2 - \varphi) = \sigma_c + R\sin\varphi$$

$$\tau_{\text{crit}} = R\sin(\pi/2 - \varphi) = R\cos\varphi$$

with σ_c and R the center and radius of the maximum Mohr circle:

$$\sigma_c = \frac{\sigma_I + \sigma_{III}}{2}; \quad R = \frac{\sigma_I - \sigma_{III}}{2}$$

Use in the Mohr–Coulomb criterion (4.32) gives

$$f(\boldsymbol{\sigma}) = 0 : \sigma_I(1 + \sin\varphi) - \sigma_{III}(1 - \sin\varphi) - 2c\cos\varphi = 0 \tag{4.33}$$

In addition, in the Mohr plane the angle between the principal stress axis $\mathbf{u}_I$ and the normal $\mathbf{n}$ of the plane, on which the stress vector $\mathbf{T} = \sigma_{\text{crit}}\mathbf{n} + \tau_{\text{crit}}\mathbf{t}$ is built, is $\pi/2 - \varphi$; thus in the material plane (Figure 4.10a)

$$-2\vartheta(\mathbf{u}_I, \mathbf{n}) = \pi/2 - \varphi \rightarrow \vartheta(\mathbf{u}_I, \mathbf{n}) = \varphi/2 - \pi/4$$

Finally, the angle $\Theta(\mathbf{n}, \mathbf{T}(\mathbf{n})) = \Theta(\mathbf{n}, \mathbf{OT}(\mathbf{n}))$ can directly be extracted from the Mohr plane.

---

[4]Recall

$$\sin(a - b) = \sin a \cos b - \cos a \sin b$$
$$\cos(a - b) = \cos a \cos b + \sin a \sin b$$

Application to uniaxial tension gives

$$f(\boldsymbol{\sigma} = f_t'\mathbf{e}_z \otimes \mathbf{e}_z) = 0; \quad \sigma_I = f_t' > 0; \quad \sigma_{III} = 0$$

$$f_t' = \frac{2c\cos\varphi}{1+\sin\varphi}$$

In uniaxial compression,

$$f(\boldsymbol{\sigma} = -f_c'\mathbf{e}_z \otimes \mathbf{e}_z) = 0; \quad \sigma_I = 0; \quad \sigma_{III} = -f_c' < 0$$

$$f_c' = \frac{2c\cos\varphi}{1-\sin\varphi}$$

In simple shear,

$$f(\boldsymbol{\sigma} = \tau_{\max}[\mathbf{e}_y \otimes \mathbf{e}_x + \mathbf{e}_x \otimes \mathbf{e}_y]) = 0; \quad \sigma_I = \tau_{\max}; \quad \sigma_{III} = -\tau_{\max}$$

$$|\tau_{\max}| = c\cos\varphi$$

The Mohr–Coulomb criterion is a typical example of a pressure sensitive strength criterion that captures the asymmetric strength in compression and tension through the friction angle φ:

$$\frac{f_t'}{f_c'} = \frac{1-\sin\varphi}{1+\sin\varphi} \leftrightarrow \varphi = \arcsin\left(\frac{f_c' - f_t'}{f_c' + f_t'}\right)$$

The limit stress states and the corresponding material planes, on which the strength criterion (4.32) is reached, are given in Figure 4.11.
Compare these results with those obtained for the Tresca criterion given in Figure 4.8. ■

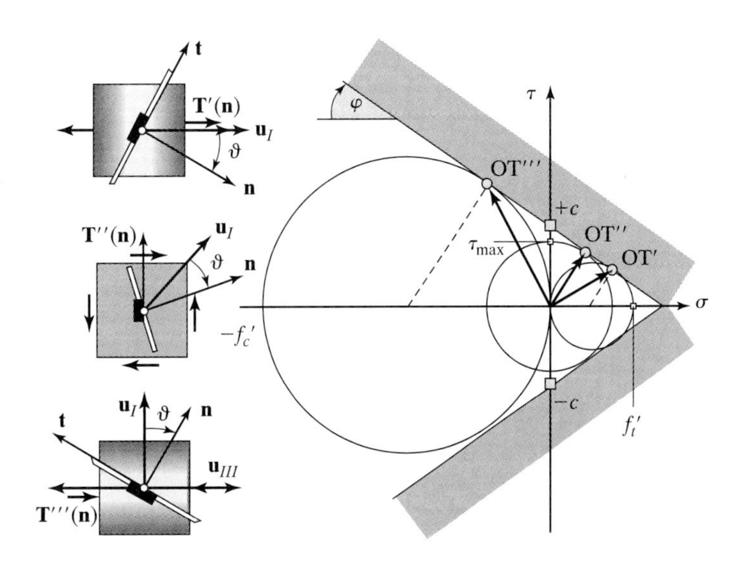

FIGURE 4.11: The Mohr–Coulomb criterion applied to uniaxial tension, simple shear, and uniaxial compression.

4.5 TRAINING SET: EXCAVATION PIT

Consider the excavation pit of height H sketched in Figure 4.12. We consider a (x, y, z)-coordinate system, in which x gives the distance from the surface ($x > 0$ down) and in which y separates the wall of the excavation from the pit ($y \leq 0$). The studied structure is subjected to gravity forces. In a step-by-step fashion, we want to study the conditions for statically admissible stress fields and determine, by means of a material strength criterion, the critical depth H_{crit} that the structure can sustain without violating static equilibrium and the local material strength.

4.5.1 Statically Admissible Stress Fields

According to (4.1) and (4.2), the stresses $\boldsymbol{\sigma}$ in the studied domain must satisfy the equilibrium equation, the continuity of the stress vector at any material plane, and the stress vector boundary conditions. For purpose of analysis, we divide the studied domain into three subdomains (Figure 4.12).

Exercise 37. We restrict ourselves to stress solutions that are functions of x alone:

$$\boldsymbol{\sigma} = \boldsymbol{\sigma}(x)$$

Develop in detail the conditions that the stress field needs to satisfy to be statically admissible.

The only force in the problem is the body force $\rho \mathbf{f} = \rho g \mathbf{e}_x$, with $\mathbf{e}_x$ the vertical axis. The momentum balance equation reads

$$\operatorname{div} \boldsymbol{\sigma} + \rho g \mathbf{e}_x = \mathbf{0}$$

Since $\boldsymbol{\sigma} = \boldsymbol{\sigma}(x)$, the momentum balance equation is satisfied for

$$\frac{\partial \sigma_{xx}}{\partial x} + \rho g = 0 \leftrightarrow \sigma_{xx} = -\rho g x + C_x$$

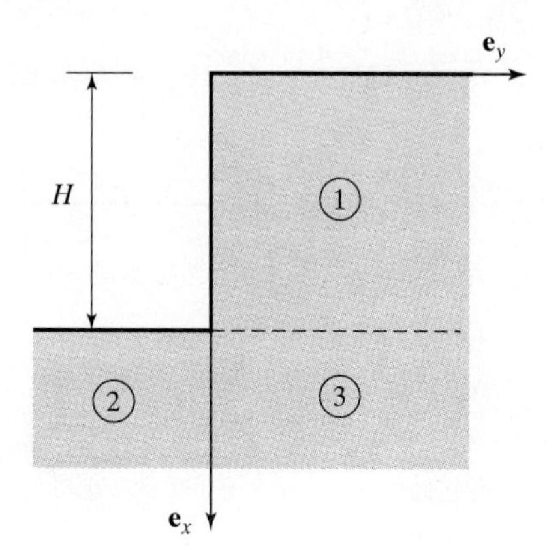

FIGURE 4.12: Training set: excavation pit.

$$\frac{\partial \sigma_{yx}}{\partial x} = 0 \leftrightarrow \sigma_{yx} = C_y$$

$$\frac{\partial \sigma_{zx}}{\partial x} = 0 \leftrightarrow \sigma_{zx} = C_z$$

where C_x, C_y, C_z are integration constants to be determined in each zone by the appropriate boundary and continuity conditions.
The boundary and continuity conditions in the three zones are

1. **Boundary condition of zone 1** ($0 \leq x \leq H;\ y \geq 0$): The boundary conditions are the ones on the free surfaces at $x = 0$, and $y = 0$:

$$x = 0 : \mathbf{T}(\mathbf{n} = -\mathbf{e}_x) = -\boldsymbol{\sigma}^{(1)} \cdot \mathbf{e}_x = 0 \leftrightarrow \sigma_{xx}^{(1)} = \sigma_{yx}^{(1)} = \sigma_{zx}^{(1)} = 0$$

$$y = 0 : \mathbf{T}(\mathbf{n} = -\mathbf{e}_y) = -\boldsymbol{\sigma}^{(1)} \cdot \mathbf{e}_y = 0 \leftrightarrow \sigma_{xy}^{(1)} = \sigma_{yy}^{(1)} = \sigma_{zy}^{(1)} = 0$$

From the first boundary condition it follows that

$$C_x^{(1)} = C_y^{(1)} = C_z^{(1)} = 0$$

Thus

$$0 \leq x \leq H;\ y \geq 0 : \sigma_{xx}^{(1)} = -\rho g x; \quad \sigma_{yx}^{(1)} = \sigma_{zx}^{(1)} = 0$$

In addition, the second boundary condition leads to

$$\sigma_{yy}^{(1)}(x) = 0; \quad \sigma_{zy}^{(1)}(x) = 0$$

Hence, from the six stress components, we determined five; the normal stress in the z-direction, remains undetermined:

$$\sigma_{zz}^{(1)} = \sigma_{zz}^{(1)}(x)$$

2. **Boundary condition of zone 2** ($x \geq H;\ y \leq 0$): The stress-free surface condition at $x = H$ and $y \leq 0$ reads

$$x = H;\ \ y \leq 0 : \mathbf{T}(\mathbf{n} = -\mathbf{e}_x) = -\boldsymbol{\sigma}^{(2)} \cdot \mathbf{e}_x = 0 \leftrightarrow \sigma_{xx}^{(2)} = \sigma_{yx}^{(2)} = \sigma_{zx}^{(2)} = 0$$

This gives

$$C_x^{(2)} = \rho g H; \quad C_y^{(2)} = C_z^{(2)} = 0$$

$$x \geq H;\ y \leq 0 : \sigma_{xx}^{(2)} = -\rho g (x - H); \quad \sigma_{yx}^{(2)} = \sigma_{zx}^{(2)} = 0$$

3. **Continuity condition between zones 1 and 3:**

$$x = H;\ y \geq 0 : \mathbf{T}(\mathbf{n}^{(3)} = -\mathbf{e}_x) + \mathbf{T}(\mathbf{n}^{(1)} = +\mathbf{e}_x) = 0$$

$$\leftrightarrow \sigma_{xx}^{(1)} = \sigma_{xx}^{(3)}; \quad \sigma_{yx}^{(1)} = \sigma_{yx}^{(3)}; \quad \sigma_{zx}^{(1)} = \sigma_{zx}^{(3)}$$

Thus, with the solution of zone 1 in hand, we obtain the following stress components in zone 3:

$$x \geq H;\ y \geq 0 : \sigma_{xx}^{(3)} = -\rho g x; \quad \sigma_{yx}^{(3)} = \sigma_{zx}^{(3)} = 0$$

4. Continuity condition between zones 2 and 3:

$$y = 0; \quad x \geq H : \mathbf{T}(\mathbf{n}^{(2)} = \mathbf{e}_y) + \mathbf{T}(\mathbf{n}^{(3)} = -\mathbf{e}_y) = 0$$

$$\leftrightarrow \sigma_{xy}^{(2)} = \sigma_{xy}^{(3)}; \quad \sigma_{yy}^{(2)} = \sigma_{yy}^{(3)}; \quad \sigma_{zy}^{(2)} = \sigma_{zy}^{(3)}$$

In addition to σ_{zz}, $\sigma_{yy} = \sigma_{yy}(x)$ and $\sigma_{zy} = \sigma_{zy}(x)$ remain undetermined in zones 2 and 3. However, with respect to the assumed infinite extension of the system in the z-direction, we can set the shear stress $\sigma_{zy}^{(2)} = \sigma_{zy}^{(3)} = 0$.

We summarize the results:

$$\text{Zone 1} : (\sigma_{ij}) = \begin{bmatrix} \sigma_{xx}^{(1)} = -\rho g x & 0 & 0 \\ & 0 & 0 \\ \text{sym} & & \sigma_{zz}^{(1)} = \sigma_{zz}^{(1)}(x) \end{bmatrix}$$

$$\text{Zone 2} : (\sigma_{ij}) = \begin{bmatrix} \sigma_{xx}^{(2)} = -\rho g (x - H) & 0 & 0 \\ & \sigma_{yy}^{(2)} = \sigma_{yy}^{(2)}(x) & 0 \\ \text{sym} & & \sigma_{zz}^{(2)} = \sigma_{zz}^{(2)}(x) \end{bmatrix}$$

$$\text{Zone 3} : (\sigma_{ij}) = \begin{bmatrix} \sigma_{xx}^{(3)} = -\rho g x & 0 & 0 \\ & \sigma_{yy}^{(3)} = \sigma_{yy}^{(3)}(x) & 0 \\ \text{sym} & & \sigma_{zz}^{(3)} = \sigma_{zz}^{(3)}(x) \end{bmatrix}$$

At this stage, we are left with unknown functions $\sigma_{zz} = \sigma_{zz}(x)$ in all three zones, and with unknown functions $\sigma_{yy} = \sigma_{yy}(x)$ in zones 2 and 3. This stress indeterminacy needs to be closed by a closure scheme, based on the introduction of constitutive relations linking (directly or indirectly) the involved statically admissible stresses. Clearly, the conditions of a statically admissible stress field is a prerequisite for the application of any constitutive law. ■

4.5.2 A Critical Excavation Height

Among the statically admissible stress fields and without referring to displacements and strains, we can choose any stress tensor of the form derived here above. For instance, let us consider the following stress field:

$$\begin{aligned} \boldsymbol{\sigma}^{(1)} &= \boldsymbol{\sigma}^{(3)} = -\rho g x \mathbf{e}_x \otimes \mathbf{e}_x \\ \boldsymbol{\sigma}^{(2)} &= -\rho g (x - H) \mathbf{e}_x \otimes \mathbf{e}_x \end{aligned}$$

This stress field corresponds to a uniaxial compression stress, with $\sigma_{xx}^{(1)} = \sigma_{xx}^{(3)} = -\rho g x$ the applied stress. The Mohr plane representation of this stress state was given in Figure 4.8 for a Tresca material, and in Figure 4.11 for a Mohr–Coulomb criterion. However, the previously defined statically admissible stress field is unbounded, as it increases (in zones 2 and 3) linearly with x. Such an unbounded stress state—albeit statically admissible—is not compatible with the finite strength domain, $f(\boldsymbol{\sigma}) \leq 0$, of strength criteria with finite shear strength (as the Tresca criterion or the Mohr–Coulomb strength criterion). Indeed, when the strength criterion is reached, the value of σ_{xx} would be fixed to a constant value, equal to σ_0, and beyond a height of $H \geq \sigma_0/\rho g$ the stress would be constant (i.e., $\partial\sigma_{xx}/\partial x = 0$). This would violate the equilibrium condition $\partial\sigma_{xx}/\partial x = -\rho g$; and the stress field below $H \geq \sigma_0/\rho g$ would cease to be statically admissible. Therefore, this solution is not admissible.

Exercise 38. Consider the following stress field[5]:

$$\text{Zone 1} : \boldsymbol{\sigma}^{(1)} = -\rho g x \mathbf{e}_x \otimes \mathbf{e}_x$$

$$\text{Zone 2} : \boldsymbol{\sigma}^{(2)} = -\rho g(x - H)\mathbf{e}_x \otimes \mathbf{e}_x - \rho g x(\mathbf{e}_y \otimes \mathbf{e}_y + \mathbf{e}_z \otimes \mathbf{e}_z)$$

$$\text{Zone 3} : \boldsymbol{\sigma}^{(3)} = -\rho g x \mathbf{e}_x \otimes \mathbf{e}_x - \rho g x(\mathbf{e}_y \otimes \mathbf{e}_y + \mathbf{e}_z \otimes \mathbf{e}_z)$$

Give a representation of this stress field in the Mohr stress plane. Check that this stress field is bounded. Explore the limit stress states for the Tresca criterion, and determine the orientation of the critical material surfaces of the excavation pit, on which the strength criterion is reached.

The principal stresses are

$$\text{Zone 1} : \sigma_I = \sigma_{II} = 0; \quad \sigma_{III} = -\rho g x$$

$$\text{Zone 2} : \sigma_I = -\rho g(x - H); \sigma_{II} = \sigma_{III} = -\rho g x$$

$$\text{Zone 3} : \sigma_I = \sigma_{II} = \sigma_{III} = -\rho g x$$

The Mohr plane representation of the stress states is given in Figure 4.13. In all three zones, the Mohr circles have a finite extension varying between $R = 0$ and $R = \rho g H/2$:

$$\text{Zone 1} : \tau(0 \leq x \leq H) = \rho g x/2$$

$$\text{Zone 2} : \tau(x \geq H) = \rho g H/2$$

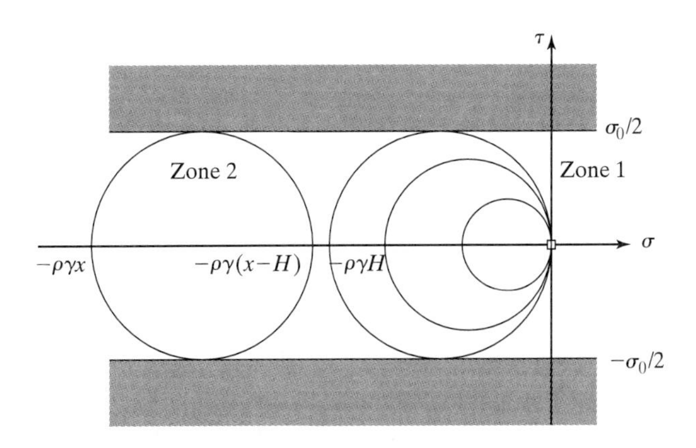

FIGURE 4.13: The Mohr plane with Tresca criterion of the excavation pit.

[5]Check that this stress field is statically admissible.

$$\text{Zone } 3 : \tau(x \geq H) = 0$$

Hence, the stress state remains statically admissible even when it reaches the strength criterion.

Consider the Tresca criterion (4.31), which limits the shear stress $|\tau_{\max}| \leq \sigma_0/2$. For the maximum shear stress in zone 2, it turns out that the Tresca criterion defines the critical excavation height:

$$f(\boldsymbol{\sigma}) = 0 \leftrightarrow H_{\text{crit}} = \frac{\sigma_0}{\rho g}$$

This maximum shear stress is reached on planes oriented by $\vartheta(\mathbf{e}_1, \mathbf{n}) = \mp\pi/4$. ■

Exercise 39. Repeat this exercise by considering a Mohr–Coulomb strength criterion. ■

4.6 PROBLEM SET: TRIAXIAL TEST AND LIMIT LOAD OF A CIRCULAR FOUNDATION

Part I: Triaxial Test: Consider a cylinder sample (radius R, height H) made of a homogeneous material between the two rigid plates of a material testing machine (see Figure 4.14). The lower surface $z = 0$ is fixed. Subjected to a force F, the upper plateau of the testing machine at $z = H$ can be moved down in the vertical direction (i.e., $\mathbf{F} = -F\mathbf{e}_z, F > 0$). The contact between the sample and the plateaus of the machine is assumed frictionless. In addition to the vertical force, a pressure p is exerted on the cylinder wall. Inertia and body forces are neglected. Throughout the exercise, we assume $F > \pi R^2 p$.

1. For the problem in hand, specify precisely the conditions that statically admissible stress fields need to satisfy. Determine one (among all possible statically admissible stress fields) that is constant in the sample. In the Mohr plane $(\sigma \times \tau)$, give a graphical representation of the chosen (constant) stress field. In both the Mohr plane and the material plane, determine the surface and the corresponding stress vector, where the shear stress is maximum.
2. The material is a Mohr–Coulomb material, such that the shear stress τ and the normal stress σ are related by the inequality

$$|\tau| \leq c - \sigma \tan\varphi$$

   (c = cohesion, φ = friction angle). For the chosen constant stress field, determine a relation linking the external actions F and p when the Mohr–Coulomb strength criterion is reached. Determine the corresponding critical surfaces and the stress vector on these surfaces, in both the Mohr plane and the material plane.

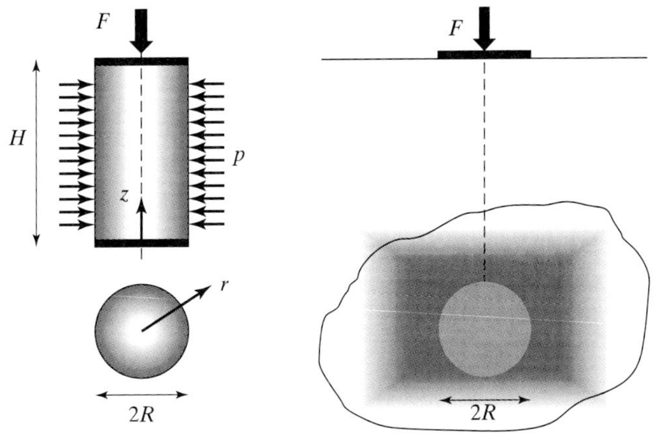

FIGURE 4.14: Problem set: from the Triaxial test to the limit load of a circular foundation on a Mohr–Coulomb soil substrate.

Part II: Circular Foundation on a Mohr–Coulomb Soil: Consider now a circular foundation of radius R on a semi-infinite half-space representing a homogeneous soil substratum (see Figure 4.14). The foundation is subjected in its center to a vertical force of intensity F.

1. With the results from Part I, determine one statically admissible stress field.
2. The soil substratum is assumed to obey the Mohr–Coulomb criterion. Determine the load bearing capacity F of this structural system. In the material plane, represent the orientation of the critical material surfaces, on which the Mohr–Coulomb criterion is reached.

4.6.1 Part I: Triaxial Test

a-1. Statically Admissible Stress Field A statically admissible stress field is a stress field that satisfies (i) the force boundary conditions, (ii) the momentum balance, and (iii) the symmetry of the stress tensor. For the triaxial test at hand, the boundary conditions are as follows:

1. On the cylinder surface:

$$r = R : \mathbf{T}^d = -p\mathbf{e}_r = \mathbf{T}(\mathbf{n} = \mathbf{e}_r) = \boldsymbol{\sigma} \cdot \mathbf{e}_r \leftrightarrow \sigma_{rr}(r = R) = -p$$

2. Frictionless contact at plateaus:

$$z = 0 \text{ and } z = H : \mathbf{T}(\mathbf{n} = \mathbf{e}_z) \cdot \mathbf{t} = 0 \leftrightarrow \left\{ \begin{array}{c} \mathbf{t} = \mathbf{e}_r : \sigma_{rz} = 0 \\ \mathbf{t} = \mathbf{e}_\theta : \sigma_{\theta z} = 0 \end{array} \right\}$$

3. On the upper plateau (application of reduction formula):

$$z = H : \mathcal{N}^d = -F\mathbf{e}_z = \int_{A=\pi R^2} \boldsymbol{\sigma} \cdot \mathbf{e}_z da = \mathbf{e}_z \int_{A=\pi R^2} \sigma_{zz} da$$

Note that it cannot *a priori* be concluded that $\sigma_{zz} = -F/A$, since σ_{zz} may not be constant over the plateau section. In turn, if we note $\langle \sigma_{zz} \rangle = \frac{1}{\pi R^2} \int_{A=\pi R^2} \sigma_{zz} da$ the average vertical stress, the force boundary condition can be rewritten in the form:

$$z = H : \langle \sigma_{zz} \rangle = -\frac{F}{\pi R^2}$$

Given the rotational symmetry of the triaxial test, it follows that it is $\sigma_{\theta r} = 0$, and the stress tensor is of the diagonal form

$$\boldsymbol{\sigma} = \sigma_{rr} \mathbf{e}_r \otimes \mathbf{e}_r + \sigma_{\theta\theta} \mathbf{e}_\theta \otimes \mathbf{e}_\theta + \sigma_{zz} \mathbf{e}_z \otimes \mathbf{e}_z$$

This stress tensor satisfies the boundary conditions and is symmetrical:

$$\boldsymbol{\sigma} = {}^t\boldsymbol{\sigma}$$

Finally, neglecting body forces, the stress tensor $\boldsymbol{\sigma}$ in the material sample must satisfy the following momentum balance equations (cylinder coordinates):

$$\text{in } \Omega : \operatorname{div} \boldsymbol{\sigma} = 0 : \left\{ \begin{array}{r} \mathbf{e}_r : \frac{\partial \sigma_{rr}}{\partial r} + \frac{1}{r}[\sigma_{rr} - \sigma_{\theta\theta}] = 0 \\ \mathbf{e}_\theta : \frac{\partial \sigma_{\theta\theta}}{\partial \theta} = 0 \\ \mathbf{e}_z : \frac{\partial \sigma_{zz}}{\partial z} = 0 \end{array} \right\}$$

a-2. Choice of Constant Stress Field and Mohr Representation

$$\frac{\partial \sigma_{rr}}{\partial r} = 0 : \sigma_{rr} = \sigma_{\theta\theta} = -p; \quad \sigma_{zz} \equiv \langle \sigma_{zz} \rangle = -\frac{F}{\pi R^2}$$

$$F > \pi R^2 p \rightarrow \sigma_{zz} < \sigma_{rr} = \sigma_{\theta\theta} < 0$$

The principal stresses:

$$\sigma_I = \sigma_{II} = -p > \sigma_{III} = -\frac{F}{\pi R^2}$$

In the Mohr stress plane, the Mohr circle corresponding to stress vectors built on surfaces oriented by $\mathbf{n} = \mathbf{e}_r$ and $\mathbf{n} = \mathbf{e}_\theta$ degenerates to a point situated at $\sigma_I = \sigma_{II} = -p$. The Mohr circle has as center

$$\sigma_c = \frac{1}{2}(\sigma_I + \sigma_{III}) = -\frac{1}{2}\left(p + \frac{F}{\pi R^2}\right)$$

and radius

$$\mathcal{R} = \frac{1}{2}(\sigma_I - \sigma_{III}) = \frac{1}{2}\left(-p + \frac{F}{\pi R^2}\right)$$

This is shown in Figure 4.15b.

In the Mohr plane, we determine the maximum shear $\max \tau = \mathcal{R}$. The plane, on which the maximum shear occurs, is a cone oriented by $\mathbf{n} = \frac{\sqrt{2}}{2}[\mathbf{e}_r - \mathbf{e}_z]$ (see graphical determination displayed in Figure 4.15).

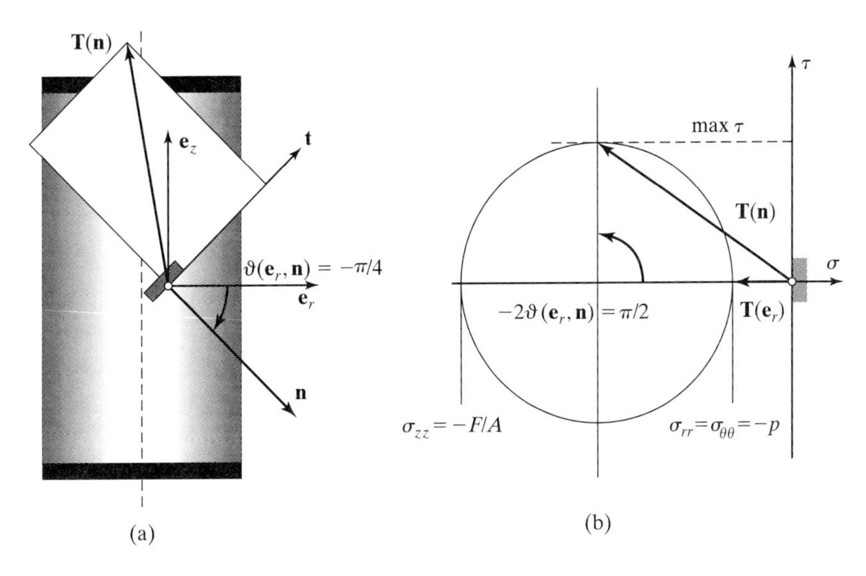

FIGURE 4.15: Triaxial test and maximum shear in (a) material plane; (b) Mohr stress plane.

b. Mohr–Coulomb Criterion The Mohr–Coulomb limit stress state is shown in Figure 4.16b in the Mohr stress plane. The minimum and maximum shear stress is given by

$$\tau = \pm \mathcal{R} \sin(\pi/2 - \varphi) = \pm \frac{\sigma_I - \sigma_{III}}{2} \cos\varphi$$

and the corresponding normal stress is

$$\sigma = \sigma_c + \mathcal{R}\cos(\pi/2 - \varphi) = \frac{\sigma_I + \sigma_{III}}{2} + \frac{\sigma_I - \sigma_{III}}{2} \sin\varphi$$

where $\mathcal{R}$ denotes the radius of the Mohr circle. Use in the Mohr–Coulomb criterion gives

$$\frac{\sigma_I - \sigma_{III}}{2} \cos\varphi \leq c - \frac{1}{2}[\sigma_I + \sigma_{III} + (\sigma_I - \sigma_{III})\sin\varphi]\tan\varphi$$

Letting $\sigma_I = -p$ and $\sigma_{III} = -F/\pi R^2$, we obtain

$$F \leq \left(\frac{2c\cos\varphi}{1-\sin\varphi} + p\frac{1+\sin\varphi}{1-\sin\varphi}\right)\pi R^2$$

$$p \geq -\frac{2c\cos\varphi}{1+\sin\varphi} + \frac{F}{\pi R^2}\frac{1-\sin\varphi}{1+\sin\varphi}$$

The orientation of the surfaces, on which the Mohr–Coulomb material reaches the strength, is defined by angle $\vartheta = \mp(\pi/4 - \varphi/2)$ ("$-$" for maximum shear, "$+$" for minimum shear); see Figure 4.16. For instance, the unit normal of the surface on which the shear stress is maximum reads

$$\mathbf{n} = \cos\vartheta \mathbf{e}_r + \sin\vartheta \mathbf{e}_z$$

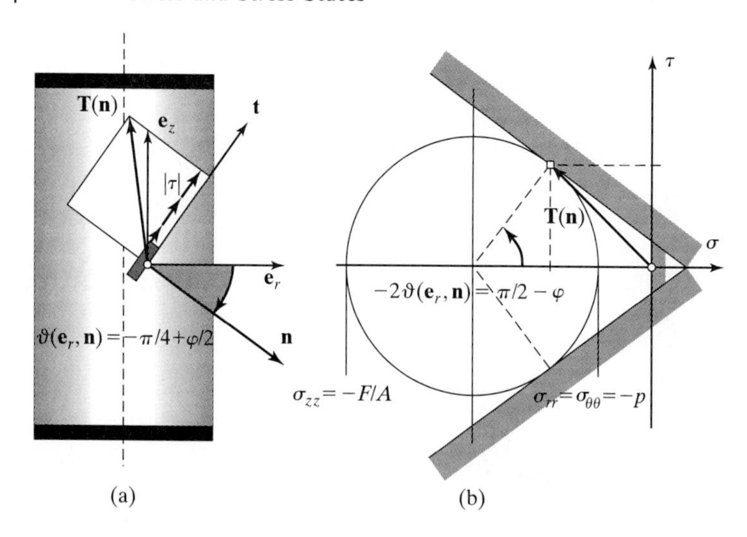

FIGURE 4.16: Triaxial test on a Mohr–Coulomb material in (a) the material plane; (b) Mohr stress plane.

with

$$\cos\vartheta = \cos(\varphi/2 - \pi/4) = \frac{\sqrt{2}}{2}\Big[\cos(\varphi/2) + \sin(\varphi/2)\Big]$$

$$\sin\vartheta = \sin(\varphi/2 - \pi/4) = \frac{\sqrt{2}}{2}\Big[\sin(\varphi/2) - \cos(\varphi/2)\Big]$$

4.6.2 Part II: Limit Load of a Circular Foundation

a. Statically Admissible Stress Field We consider two zones:

1. Zone 1 below the circular foundation $r \in [0, R]$, in which we assume the statically admissible stress field from the triaxial test:

$$\text{Zone 1: } \boldsymbol{\sigma}^{(1)} = \frac{-F}{\pi R^2}\mathbf{e}_z \otimes \mathbf{e}_z - p[\mathbf{e}_r \otimes \mathbf{e}_r + \mathbf{e}_\theta \otimes \mathbf{e}_\theta]$$

2. Zone 2 for $r > R$. The stress field must satisfy the boundary conditions at the surface $z = 0$:

$$\mathbf{T}^d(\mathbf{n} = \mathbf{e}_z) = \boldsymbol{\sigma}^{(2)} \cdot \mathbf{e}_z = 0$$

In addition, $\boldsymbol{\sigma}^{(1)}$ and $\boldsymbol{\sigma}^{(2)}$ must satisfy the stress vector continuity over the interface between Zones 1 and 2:

$$r = R : \mathbf{T}^{(1)}(\mathbf{n} = \mathbf{e}_r) = \mathbf{T}^{(2)}(\mathbf{n} = -\mathbf{e}_r)$$

Thus, with $\mathbf{T}^{(1)}(\mathbf{n} = \mathbf{e}_r) = -p\mathbf{e}_r$, we choose the following constant stress field in Zone 2:

$$\text{Zone 2: } \boldsymbol{\sigma}^{(2)} = -p[\mathbf{e}_r \otimes \mathbf{e}_r + \mathbf{e}_\theta \otimes \mathbf{e}_\theta]$$

Note that the hoop stress $\sigma_{\theta\theta}^{(2)} = -p$ comes from the momentum balance equation for $\sigma_{rr} = \text{const}$, and the stress field is statically admissible.

b. Limit Load The Mohr representations of the stress fields, $\boldsymbol{\sigma}^{(1)}$ and $\boldsymbol{\sigma}^{(2)}$, are given in Figure 4.17. The Mohr representation of $\boldsymbol{\sigma}^{(1)}$ is identical with the one displayed in Figure 4.16b. In addition, the Mohr representation of $\boldsymbol{\sigma}^{(2)}$ corresponds to a circle, with center and radius

$$\sigma_c^{(2)} = -p/2; \quad \mathcal{R} = p/2$$

and

$$\sigma_I^{(2)} = 0 > \sigma_{II}^{(2)} = \sigma_{III}^{(2)} = -p$$

Proceeding as in the first part of this exercise, the maximum pressure in Zone 2 for a Mohr–Coulomb criterion is restricted by

$$\text{Zone 2: } p \leq \max p = 2c\frac{\cos\varphi}{1-\sin\varphi}$$

In turn, the force–pressure relation developed for the triaxial test still holds for Zone 1, that is,

$$\text{Zone 1: } F \leq \left(\frac{2c\cos\varphi}{1-\sin\varphi} + p\frac{1+\sin\varphi}{1-\sin\varphi}\right)\pi R^2$$

Finally, use of the limit pressure $\max p$ in the previous equation leads to the limit load:

$$F \leq \max F = \frac{4c\cos\varphi}{(1-\sin\varphi)^2}\pi R^2$$

For $F = \max F$, the stresses in both zones meet the Mohr–Coulomb strength criterion. The orientation of the surfaces, on which the strength criterion is reached, is shown in Figure 4.18 (obtained from exploitation of the Mohr circles in Figure 4.17).

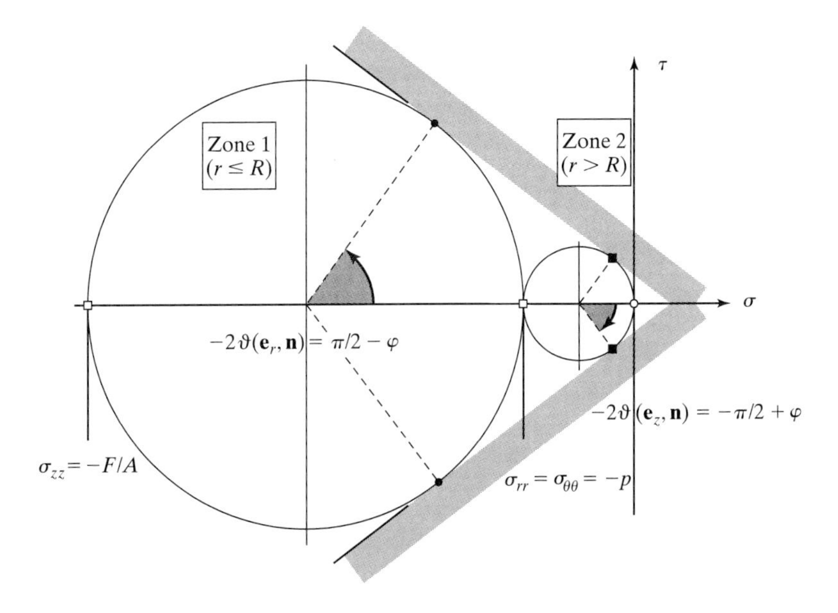

FIGURE 4.17: Mohr representation of the two-zone stress field. Note that in Zone 1, $\vartheta(\mathbf{e}_r, \mathbf{n}) = -\pi/4 + \varphi$, while in Zone 2, $\vartheta = \vartheta(\mathbf{e}_z, \mathbf{n}) = \pi/4 - \varphi$.

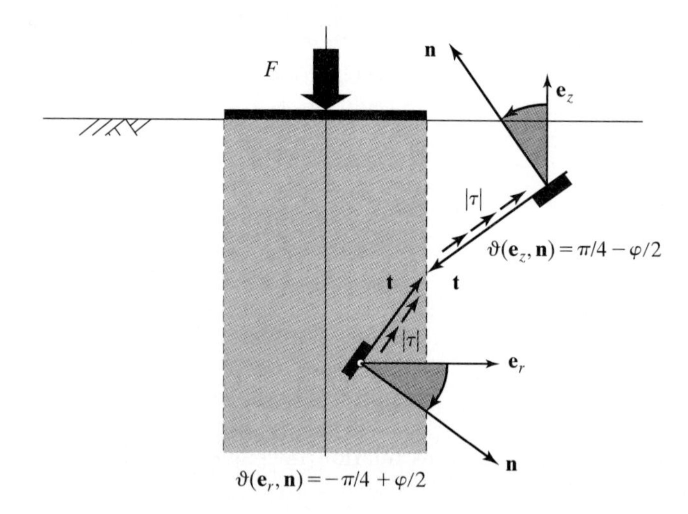

FIGURE 4.18: Material planes on which the Mohr–Coulomb criterion is reached ($\tau = +|\tau|$ in Zone 1, $\tau = -|\tau|$ in Zone 2).

4.7 PROBLEM SET: STATICALLY ADMISSIBLE STRESS FIELDS OF A BEAM

We want to study the stress fields of the beam of length L oriented along the x-axis and of square section $2a \times 2a$ shown in Figure 4.19. At $x = 0$, the section (centered around O) is in contact without friction with a fixed rigid plate. At the other end, at $x = L$, the section (centered around O') is subjected to a moment of intensity M around the z-axis and an axial force N in the x-direction. The moment and force are generated by means of a rigid plate to which the beam end surface remains (frictionless) attached. The other surfaces at $y = \pm a$ and at $z = \pm a$ are stress free.

1. **Pure Bending**: We first consider the pure bending case (i.e., $N = 0$). We restrict ourselves to uniaxial stress solutions of the form

$$\boldsymbol{\sigma} = \sigma(y)\mathbf{e}_x \otimes \mathbf{e}_x$$

   (a) Specify the conditions that function $\sigma(y)$ in the beam needs to satisfy in order to be statically admissible. Specify all boundary conditions.

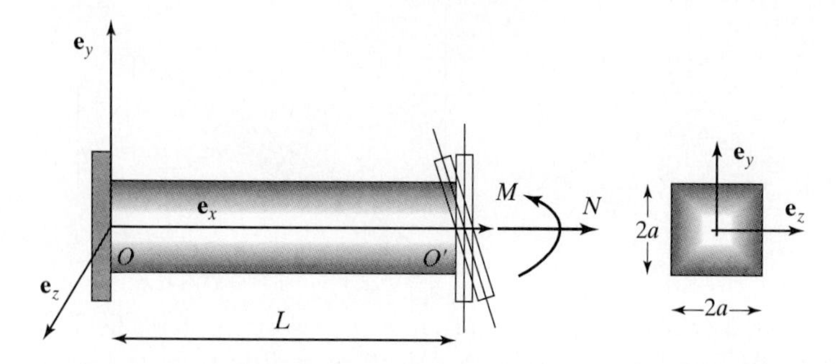

FIGURE 4.19: Problem set: beam subjected to combined bending and axial load.

(b) Among all statically admissible stress fields—as previously defined—consider the following two stress fields, $\boldsymbol{\sigma}^{(1)}$ and $\boldsymbol{\sigma}^{(2)}$, defined by functions $\sigma^{(1)}(y)$ and $\sigma^{(2)}(y)$:

$$\text{Field 1} : \sigma^{(1)}(y) = \sigma_1 \frac{y}{a}$$

$$\text{Field 2} : \sigma^{(2)}(y) = \sigma_2 \text{sign}\,(y/a) = \left\{ \begin{array}{c} \sigma_2 \text{ if } 0 < Y \leq a \\ -\sigma_2 \text{ if } -a \leq Y < 0 \end{array} \right\}$$

Determine the constants σ_1 and σ_2 as a function of M. Represent the stress fields $\boldsymbol{\sigma}^{(1)}$ and $\boldsymbol{\sigma}^{(2)}$ in the Mohr stress plane. Determine the material surfaces subjected to the maximum shear, and express this maximum shear as a function of the moment M.

(c) The material of the beam obeys the Tresca strength criterion, with σ_0 as uniaxial tension strength. For the stress fields, $\boldsymbol{\sigma}^{(1)}$ and $\boldsymbol{\sigma}^{(2)}$, determine the maximum values of the moment M [i.e., $M_1 = \max M(\boldsymbol{\sigma}^{(1)})$ and $M_1 = \max M(\boldsymbol{\sigma}^{(1)})$] for which the Tresca criterion is satisfied. Compare the values.

2. **Combined Bending and Axial Load**: In addition to the bending moment M, consider an axial force N. Give a modified expression of $\sigma^{(2)}(y)$ so that the stress field $\boldsymbol{\sigma}^{(2)}$ remains statically admissible under combined axial load and bending. Represent this new stress field in the Mohr stress plane. In the generalized load plane $[N \times M]$, determine the strength domain in which any load point N and M is situated, when the Tresca criterion is respected within the structure.

4.7.1 Pure Bending

a. Statically Admissible Stress Field Momentum balance ($\rho \mathbf{f} = 0$) is satisfied, since

$$\text{div}\,\boldsymbol{\sigma} = \text{tr}\left[\text{grad}\left(\sigma(y)\mathbf{e}_x \otimes \mathbf{e}_x\right)\right] = 0$$

$$\text{grad}\left(\sigma(y)\mathbf{e}_x \otimes \mathbf{e}_x\right) = \frac{\partial \sigma(y)}{\partial x} \mathbf{e}_x \otimes \mathbf{e}_x \otimes \mathbf{e}_x = 0$$

Symmetry of stress tensor:

$$\boldsymbol{\sigma} = {}^t\boldsymbol{\sigma} = \sigma(y)\mathbf{e}_x \otimes \mathbf{e}_x$$

Boundary conditions:

1. At beam surface ($z = \pm a, y = \pm a$):

$$\mathbf{T}^d(\mathbf{n} = \pm \mathbf{e}_z) = \pm \boldsymbol{\sigma} \cdot \mathbf{e}_z = 0$$

2. At $x = L$ (application of reduction elements):

$$\mathcal{N} = \int_{\partial \Omega_t} \boldsymbol{\sigma} \cdot \mathbf{e}_x da = N\mathbf{e}_x + V_y \mathbf{e}_y + V_z \mathbf{e}_z = 0 \leftrightarrow N = V_y = V_z = 0$$

$$\mathcal{M} = \int_{\partial \Omega_t} (\mathbf{OM} \times \boldsymbol{\sigma} \cdot \mathbf{e}_x) da = \mathcal{M}^d = M\mathbf{e}_z$$

With $\mathbf{OM} = y\mathbf{e}_y + z\mathbf{e}_z$ and $\boldsymbol{\sigma} \cdot \mathbf{e}_x = \sigma(y)\mathbf{e}_x$:

$$\begin{pmatrix} M_x \\ M_y \\ M_z \end{pmatrix} = \int_{A=4a^2} \begin{pmatrix} 0 \\ y \\ z \end{pmatrix} \times \begin{pmatrix} \sigma(y) \\ 0 \\ 0 \end{pmatrix} da = \begin{pmatrix} 0 \\ 0 \\ -\int y\sigma(y)da \end{pmatrix}$$

Thus

$$\mathcal{M}^d = M\mathbf{e}_z = \left[-2a \int_{y=-a}^{y=+a} y\sigma(y)\, dy\right] \mathbf{e}_z$$

b. Stress Fields and Mohr Representation

Linear stress field:

$$M = -2\sigma_1 \int_{y=-a}^{y=+a} y^2 dy = -\frac{4a^3\sigma_1}{3} \rightarrow \sigma_1 = -\frac{3M}{4a^3}$$

Block stress field:

$$M = -2a\sigma_2 \left[\int_{y=0}^{y=+a} ydy - \int_{y=-a}^{y=0} ydy\right] = -2\sigma_2 a^3 \rightarrow \sigma_2 = -\frac{M}{2a^3}$$

Mohr representation:

$$\mathbf{T}^{(1)}(\mathbf{n} = \mathbf{e}_x) = -\frac{3M}{4a^3}\left(\frac{y}{a}\right)\mathbf{e}_x; \quad \mathbf{T}^{(2)}(\mathbf{n} = \mathbf{e}_x) = -\frac{M}{2a^3}\text{sign}\left(\frac{y}{a}\right)\mathbf{e}_x$$

The Mohr circles for the two stress fields are displayed in Figures 4.20 and 4.21, together with the surface on which the maximum shear occurs. It follows that

$$\max\tau = \begin{Bmatrix} \max\sigma^{(1)}/2 = \frac{3}{8}M/a^3 \\ \max\sigma^{(2)}/2 = \frac{1}{4}M/a^3 \end{Bmatrix}$$

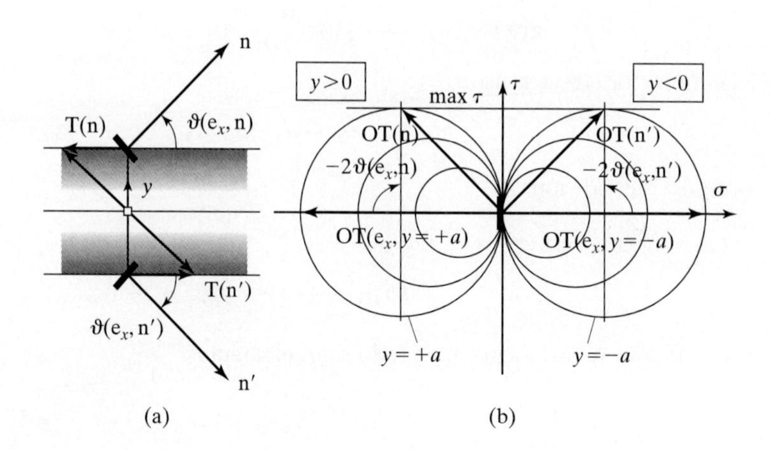

FIGURE 4.20: Stress field 1: representation in (a) the physical plane; (b) the Mohr plane.

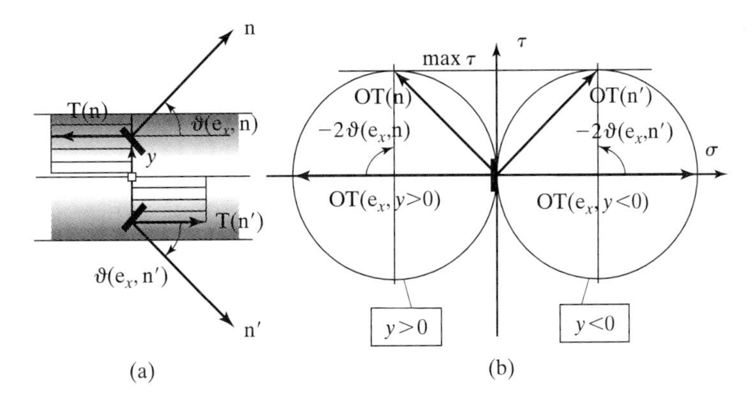

FIGURE 4.21: Stress field 2: representation in (a) the physical plane; (b) the Mohr plane.

In both cases, the surfaces where the maximum shear is reached are oriented by

$$\mathbf{n} = \frac{\sqrt{2}}{2}[\mathbf{e}_x + \text{sign}\ (y/a)\,\mathbf{e}_y]$$

c. Tresca Criterion

$$|\tau| \leq \sigma_0/2$$

Thus,

$$\max M^{(1)} = \frac{4}{3}\sigma_0 a^3 \leq \max M^{(2)} = 2\sigma_0 a^3$$

The difference is readily understood from the stress-vector profiles over y displayed in Figures 4.20a and 4.21a.

4.7.2 Combined Bending and Axial Load

The following modified stress field is statically admissible:

$$\boldsymbol{\sigma}^{(2)} = \left[\sigma_2 \text{sign}\left(\frac{y}{a}\right) + C\right]\mathbf{e}_x \otimes \mathbf{e}_x$$

It shifts the normal stress resultant by a constant factor:

$$\begin{aligned}\mathcal{N}^d &= N\mathbf{e}_x = \int_{\partial\Omega_t} \boldsymbol{\sigma}^{(2)} \cdot \mathbf{e}_x da = \left[\int_{\partial\Omega_t} \sigma_2 \text{sign}\left(\frac{y}{a}\right) da + \int_{\partial\Omega_t} C da\right]\mathbf{e}_x \\ &= 4Ca^2\mathbf{e}_x \to C = \frac{N}{4a^2}\end{aligned}$$

The moment remains constant:

$$\mathcal{M}^d = M\mathbf{e}_z = \int_{\partial\Omega_t} (\mathbf{OM} \times \boldsymbol{\sigma} \cdot \mathbf{e}_x)\, da = -2\sigma_2 a^3 \mathbf{e}_z \to \sigma_2 = -\frac{M}{2a^3}$$

The stress vector in $-a \leq y \leq +a$ reads

$$\mathbf{T}(\mathbf{n} = \mathbf{e}_x) = \left[-\frac{M}{2a^3} \operatorname{sign}\left(\frac{y}{a}\right) + \frac{N}{4a^2} \right] \mathbf{e}_x$$

The Mohr representation is given in Figure 4.22b. For $N > 0$, the maximum shear stress occurs in the tension zone, $y/a < 0$ (see Figure 4.22a). For a Tresca material, it is restricted to

$$|\tau| = \max \frac{\sigma_2}{2} = \frac{1}{2}\left(\frac{M}{2a^3} + \frac{N}{4a^2}\right) \leq \frac{\sigma_0}{2}$$

In the $(M \times N)$-plane, the Tresca criterion restricts the admissible normal and bending moments to

$$\frac{M}{2\sigma_0 a^3} + \frac{N}{4\sigma_0 a^2} \leq 1$$

This is shown in Figure 4.23.

Note that only a part of the beam section reaches the material strength limit (see Figure 4.22). It is therefore expected that the actual strength domain of the section is still larger than the one displayed in Figure 4.23.

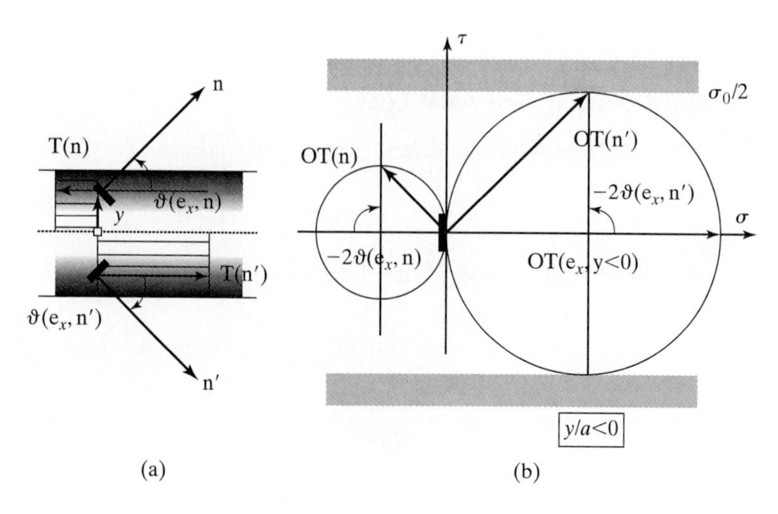

FIGURE 4.22: Combined bending and axial load for one statically admissible stress field: (a) physical plane; (b) Mohr plane.

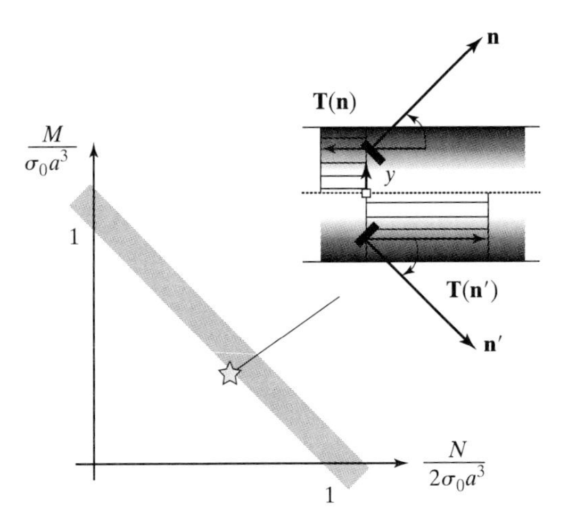

FIGURE 4.23: M-N interaction curve for a rectangular section, determined for a statically admissible stress field and the Tresca strength criterion.

4.8 PROBLEM SET: WHY SANDCASTLES FALL

We want to study the stress fields in a dry and a humid sandpile, idealized as an inclined semi-infinite half-space oriented at an angle α to the horizontal (see Figure 4.24). We choose an x-z coordinate system, in which z gives the distance from the surface of the pile ($z > 0$ down) and x gives the distance parallel to the surface (infinite extension in the y-direction). The sandpile is subjected to its deadweight (volume mass density ρ, and $\mathbf{g}$ the earth acceleration vector), and static evolutions are assumed.

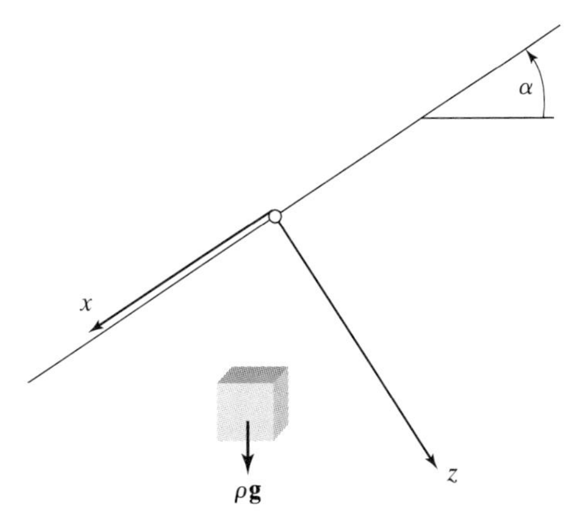

FIGURE 4.24: Problem set: Mohr–Coulomb's problem—idealized problem of a sandpile.

1. **Dry Sandpile—The Mohr–Coulomb Result**: We restrict ourselves to solutions that are functions of z alone; that is,

$$\boldsymbol{\sigma} = \boldsymbol{\sigma}(z)$$

Furthermore, the sand behavior is assumed isotropic.

   (a) Determine precisely the conditions that stress field $\boldsymbol{\sigma}$ needs to satisfy in order to be statically admissible (S.A.). Determine the nonzero stress components of $\boldsymbol{\sigma}$, and give a precise of the stress components of which the value is not given by static equilibrium (Statically Admissible-stress conditions).

   (b) For a given distance $z > 0$ from the surface, represent the previously determined stress state in the Mohr stress plane. In this plane, indicate the angle α.

   (c) We want to provide the critical angle $\alpha \leq \max\alpha$, by considering that the material in the sandpile obeys the (dry sand) Mohr–Coulomb criterion:

$$|\tau| + \sigma \tan\varphi \leq 0$$

   where τ is the tangential stress across some plane interior to the sandpile, σ is the normal stress across the same plane, and $\tan\varphi$ is the internal friction angle. Show the criterion in the Mohr space, and determine the critical value of α at which the material reaches the Mohr–Coulomb criterion.

2. **Humid Sandpile**[6]: Consider now a sandpile in which a normal adhesive stress s_A is exerted across every plane, in addition to whatever other stresses may exist due to body forces. This adhesive stress introduces a normal force between pairs of contiguous particles that allows the sandpile to support a finite shear stress (i.e., τ), even in the limit of zero applied compressive stresses (i.e., $\sigma = 0$). The maximum shear stress, in this case, is $\max|\tau| = s_A \tan\varphi$.

   (a) Propose a modified Mohr–Coulomb criterion, which for $s_A = 0$ gives the dry sand Mohr–Coulomb criterion.

   (b) In comparison with the dry sand criterion, how does the Mohr plane representation change in the case of a humid sandpile? Determine the critical angle at which the material reaches the humid sand failure criterion. In comparison with the dry sandpile, does $\max\alpha$ increase or decrease? Conclude by suggesting how sandcastles fall.

4.8.1 Dry Sandpile

a. Statically Admissible Stress Field

$$\operatorname{div}\boldsymbol{\sigma} + \rho\mathbf{f} = 0$$

$$\boldsymbol{\sigma} = {}^t\boldsymbol{\sigma}$$

[6] Inspired by Halsey, T.C. and Levine, A.J., (1998). "How Sandcastles Fall," *Physical Review Letters*, Vol. 80, No. 14, 3141–3144.

Deadweight:

$$\rho \mathbf{f} = \rho \mathbf{g} = \rho g[\cos\alpha \mathbf{e}_z + \sin\alpha \mathbf{e}_x]$$

Momentum Balance:

$$\mathbf{e}_x : \frac{\partial \sigma_{xz}(z)}{\partial z} + \rho g \sin\alpha = 0; \quad \left(\frac{\partial \sigma_{xx}(z)}{\partial x} = \frac{\partial \sigma_{xy}(z)}{\partial y} = 0\right)$$

$$\mathbf{e}_y : \frac{\partial \sigma_{yz}(z)}{\partial z} = 0; \quad \left(\frac{\partial \sigma_{yx}(z)}{\partial x} = \frac{\partial \sigma_{yy}(z)}{\partial y} = 0\right)$$

$$\mathbf{e}_z : \frac{\partial \sigma_{zz}(z)}{\partial z} + \rho g \cos\alpha = 0; \quad \left(\frac{\partial \sigma_{zx}(z)}{\partial x} = \frac{\partial \sigma_{zy}(z)}{\partial y} = 0\right)$$

Boundary Conditions:

$$z = 0 : \mathbf{T}^d(\mathbf{n} = -\mathbf{e}_z) = -\boldsymbol{\sigma} \cdot \mathbf{e}_z = \sigma_{xz}\mathbf{e}_x + \sigma_{yz}\mathbf{e}_y + \sigma_{zz}\mathbf{e}_z = \mathbf{0}$$

$$\sigma_{xz}(z = 0) = \sigma_{yz}(z = 0) = \sigma_{zz}(z = 0) = 0$$

It follows that

$$\sigma_{xz}(z) = \sigma_{zx}(z) = -\rho g z \sin\alpha$$

$$\sigma_{yz}(z) = \sigma_{zy}(z) = 0$$

$$\sigma_{zz}(z) = -\rho g z \cos\alpha$$

Infinite extension in the y-direction implies no shear stress in $(y \times x)$-plane:

$$\sigma_{yx}(z) = \sigma_{xy} = 0$$

Isotropic behavior:

$$\sigma_{xx}(z) = \sigma_{yy}(z) = C(z)$$

Thus

$$(\boldsymbol{\sigma}) = \begin{bmatrix} C(z) & 0 & -\rho g z \sin\alpha \\ & C(z) & 0 \\ \text{sym} & & -\rho g z \cos\alpha \end{bmatrix}$$

with $\sigma_{xx}(z) = \sigma_{yy}(z) = C(z)$ undetermined.

b. Mohr Representation We build the stress vectors on a plane oriented by

- $\mathbf{n} = \mathbf{e}_x$:

$$\mathbf{T}(\mathbf{n} = \mathbf{e}_x) = C(z)\mathbf{e}_x - \rho g z \sin\alpha \mathbf{e}_z$$

- $\mathbf{n} = \mathbf{e}_y$:

$$\mathbf{T}(\mathbf{n} = \mathbf{e}_y) = C(z)\mathbf{e}_y$$

- $\mathbf{n} = \mathbf{e}_z$:

$$\mathbf{T}(\mathbf{n} = \mathbf{e}_z) = -\rho g z \sin\alpha \mathbf{e}_x - \rho g z \cos\alpha \mathbf{e}_z$$

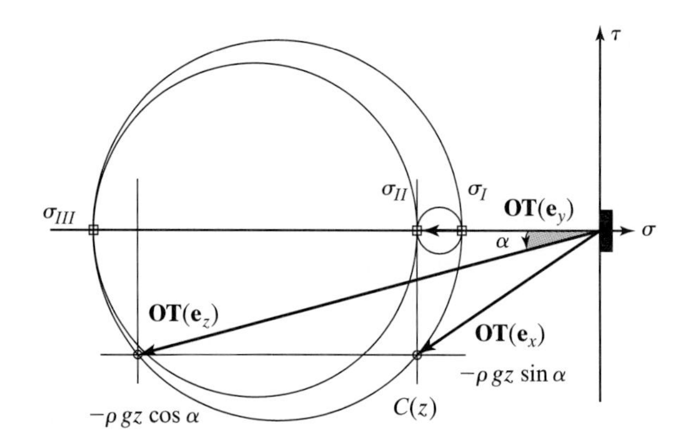

FIGURE 4.25: Mohr representation for $C(z) > -\rho g z \cos \alpha$.

The stress vector $\mathbf{T}(\mathbf{n} = \mathbf{e}_y)$ is a principal stress vector, while the other two have a normal stress and a shear component. We distinguish two cases: The case $C(z) > -\rho g z \cos \alpha$ is shown in Figure 4.25, and the case $C(z) < -\rho g z \cos \alpha$ in Figure 4.26, together with the angle α.

c. Mohr–Coulomb Criterion From Figures 4.25 and 4.26 it appears that the case $C(z) < -\rho g z \cos \alpha$ gives a smaller loading angle α. This loading angle must be smaller than the critical angle $\max \alpha$, which corresponds to the case when the stress vector $\mathbf{OT}(\mathbf{e}_z)$ becomes tangent to the Mohr circle in Figure 4.26. In this case, the angle α coincides with the friction angle φ. This is shown in Figure 4.27. Thus,

$$\alpha \leq \max \alpha = \varphi$$

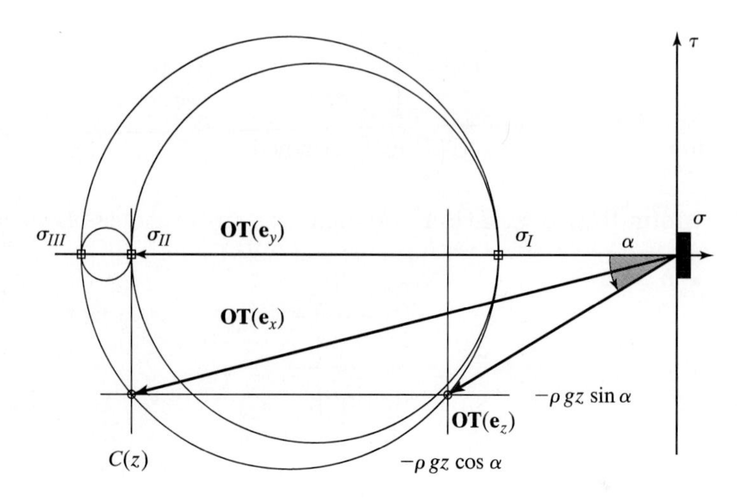

FIGURE 4.26: Mohr representation for $C(z) < -\rho g z \cos \alpha$.

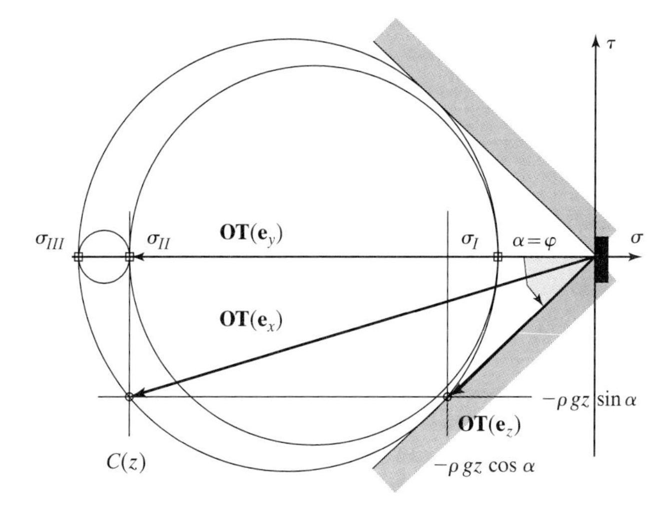

FIGURE 4.27: Mohr–Coulomb's limit case: $\alpha = \max \alpha = \varphi$.

4.8.2 Humid Sandpile

a. Modified Mohr–Coulomb Criterion The adhesive force plays the role of a cohesion, such that

$$\sigma = 0 : |\tau| - s_A \tan \varphi \leq 0$$

Therefore,

$$|\tau| + (\sigma - s_A) \tan \varphi \leq 0$$

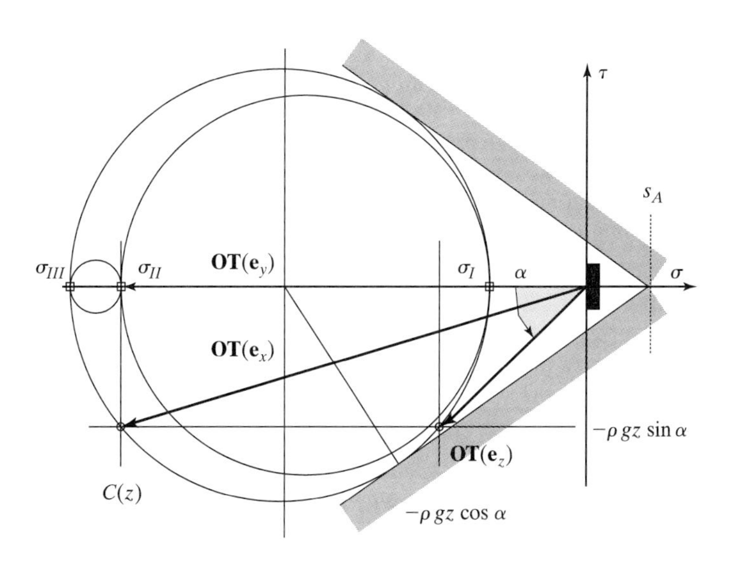

FIGURE 4.28: Mohr–Coulomb criterion with adhesive forces related to humidity.

b. Mohr Representation The modified Mohr–Coulomb criterion is shown in Figure 4.28: The criterion is shifted along the σ-axis by s_A. Instead of shifting the criterion, we work with an effective stress tensor:

$$\boldsymbol{\sigma}' = \boldsymbol{\sigma} - s_A \mathbf{1}$$

This operation does not change the principal stress axis. Therefore, a Mohr representation in the $((\sigma - s_A) \times \tau)$-plane can be given, together with an effective Mohr–Coulomb criterion:

$$|\tau| + \sigma' \tan\varphi \leq 0$$

This gives the same Mohr circles as shown in Figure 4.27, when replacing σ by $\sigma' = \sigma - s_A$. This is shown in Figure 4.29. Note, however, that the effective stress concerns only the normal stress, and not the shear stress. Consequently, the angle α' in this figure is not equal to the inclination angle α. This is readily understood from the effective stress vector:

$$\mathbf{T}'(\mathbf{n} = \mathbf{e}_z) = -\rho g z \sin\alpha \mathbf{e}_x - (\rho g z \cos\alpha + s_A)\mathbf{e}_z$$

In turn, this effective friction angle α' is equal to (see Figure 4.29)

$$\tan\alpha' = \tan\varphi = \frac{\rho g z \sin\alpha}{\rho g z \cos\alpha + s_A} = \tan\alpha \left(1 + \frac{s_A \cos\alpha}{\rho g z}\right)^{-1}$$

Solving this equation for α gives the maximum inclination of the sandpile, which increases with s_A increasing. Also note that the greater z the smaller α becomes, which approaches at a certain depth the critical angle of the dry sand Mohr–Coulomb problem. This suggests that humid sandpiles fail in depth.

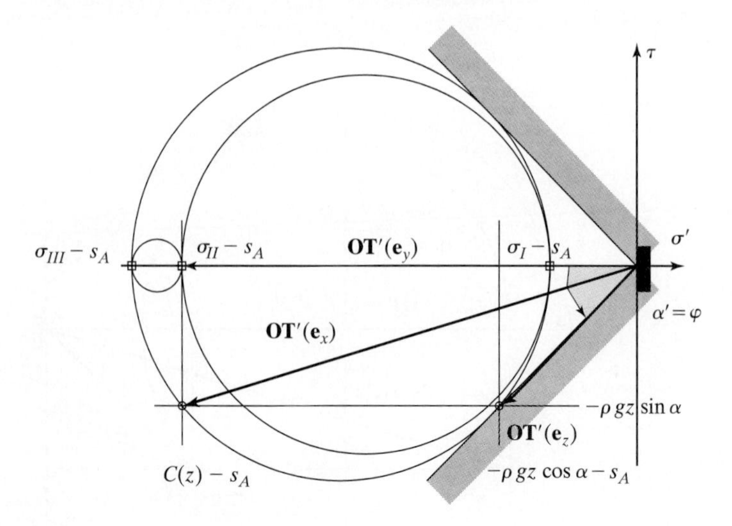

FIGURE 4.29: The Mohr–Coulomb criterion in the effective Mohr stress plane.

PART THREE

ELASTICITY AND ELASTICITY BOUNDS

CHAPTER 5

Thermoelasticity

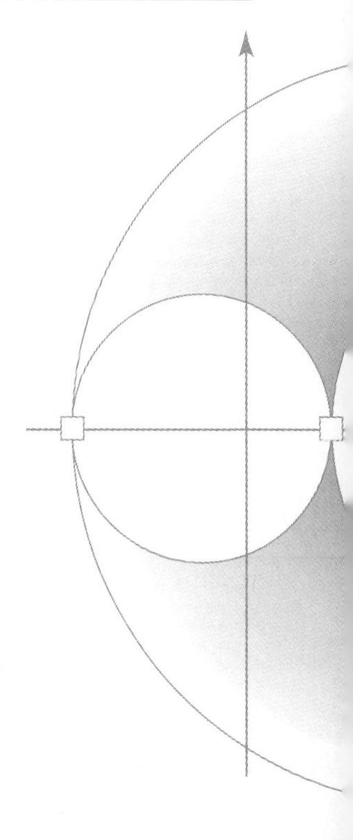

This and the next chapter deal with elastic behavior of continuous material systems. Elasticity is the first material law we study. This chapter is concerned with the description of thermoelastic behavior on the material level. The starting point is the study of a 1D think model of thermoelasticity. With a focus on the energy side of the thermoelastic response, the first law and the second law of thermodynamics are derived. Within this energy approach, we extend the 1D model to the general 3D case, and derive the stress equation of state for general and isotropic thermoelasticity. The material tests that allow the determination of the material properties of the linear isotropic thermoelastic material law are discussed: bulk modulus, shear modulus, and thermal dilatation coefficient. With the constitutive equations in hand, we move on to continuous elastic material systems. Direct solving methods for problems involving linear isotropic material systems are presented; the displacement method and the stress method.

5.1 THE NECESSITY OF MATERIAL LAWS

In the application sections (the training sets) of the previous chapters, we have encountered the stress indeterminacy. This stress indeterminacy results from the fact that there are more unknown in the problem than equations to solve. Indeed, the unknowns of the continuous problem at hand are the six stress components $\sigma_{ij} = \sigma_{ji}$, the six strain components $\varepsilon_{ij} = \varepsilon_{ji}$, and the three displacements ξ_i.

This gives a total of 15 unknowns. On the other hand, there are three equilibrium equations (3.27):

$$\operatorname{div} \boldsymbol{\sigma} + \rho \mathbf{f} = \mathbf{0} \tag{5.1}$$

and six strain-displacement relations [i.e., (1.42) or (2.8)[1]]:

$$\varepsilon = \frac{1}{2} \left(\operatorname{grad} \boldsymbol{\xi} + {}^t\operatorname{grad} \boldsymbol{\xi} \right) \tag{5.2}$$

Thus, there is a total of nine equations for 15 unknowns; six equations are missing for the problem to be well posed. The missing equations need to be provided by relations linking (statically admissible) stresses and (kinematically admissible) strains: This is the first motivation for material laws.

5.2 THE NOTION OF ELASTICITY AND ELASTICITY POTENTIAL

Elasticity is the most classical material law. It links the six stress components to the six strain components through a one-to-one relation,

$$\boldsymbol{\sigma} \rightleftharpoons \boldsymbol{\varepsilon} : \boldsymbol{\sigma} = \boldsymbol{\sigma}(\boldsymbol{\varepsilon}); \quad \boldsymbol{\varepsilon} = \boldsymbol{\varepsilon}(\boldsymbol{\sigma}) \tag{5.3}$$

5.2.1 1D Think Model

To motivate the forthcoming developments, we study the response of the spring of initial unit length $l = 1$ $[L]$ shown in Figure 5.1a. The application of an external stress σ to the system, represented in the 1D think model (e.g., a force per unit surface) leads to a relative length increase $\varepsilon = \Delta l/[L]$. Since there is no other physical phenomenon that induces displacement (as, for instance, a temperature variation), there exists a unique relation linking σ to ε:

$$\sigma \rightleftharpoons \varepsilon : \sigma = \sigma(\varepsilon); \quad \varepsilon = \varepsilon(\sigma) \tag{5.4}$$

We have a look on the energy side of the elastic response: At a given externally supplied stress σ, which is in equilibrium with the force in the spring, the work that is incrementally supplied from the outside to the system equals $\sigma d\varepsilon$. It actually represents the strain energy due to the work supplied from the outside to the spring.[2] This incremental external work supply is entirely stored in the spring in form of the free energy $\psi = \psi(\varepsilon)$ as mechanical work that can be recovered later on; thus,

$$\sigma d\varepsilon = d\psi \tag{5.5}$$

Equation (5.5) must hold irrespective of the sign of $\sigma d\varepsilon$. This implies the strict reversibility of the internal energy changes $d\psi$. In addition, use of $\psi = \psi(\varepsilon)$ in (5.5) yields

$$\left(\sigma - \frac{\partial \psi(\varepsilon)}{\partial \varepsilon} \right) d\varepsilon = 0 \tag{5.6}$$

---

[1] In all what follows, we will assume the linear deformation theory.

[2] In fact, the total work supplied to the system is the force $F = \sigma A$ multiplied with the displacement Δl (i.e., $F \times \Delta l$ of dimension $[E] = L^2 M T^{-2}$) or, equivalently, $\sigma \varepsilon V$, where V is the volume. It follows that the incrementally supplied work per unit volume is $\sigma d\varepsilon$. It is the work done by the externally supplied force, which is in equilibrium with the internal force in the spring, along an infinitesimal change of length of the spring. The theorem behind this rather intuitive derivation will be developed in Chapter 6, dealing with the theorem of virtual work rate.

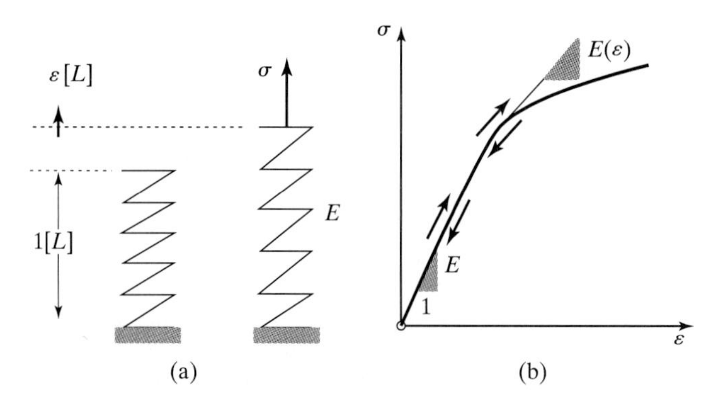

FIGURE 5.1: (a) 1D think model of elasticity. (b) Stress-strain relation.

The study of the 1D think model reveals the existence of an elastic potential $\psi = \psi(\varepsilon)$ that links the stress σ to strain ε. This link is provided by the *equation of state*:

$$\sigma = \sigma(\varepsilon) = \frac{\partial\psi(\varepsilon)}{\partial\varepsilon} \tag{5.7}$$

From Figure 5.1a, this elastic potential can be expressed as

$$\psi(\varepsilon) = \frac{1}{2}E\varepsilon^2 \rightarrow \sigma = \frac{\partial\psi(\varepsilon)}{\partial\varepsilon} = E\varepsilon \tag{5.8}$$

where E denotes the stiffness of the spring:

$$E = \frac{\partial\sigma}{\partial\varepsilon} = \frac{\partial^2\psi(\varepsilon)}{\partial\varepsilon^2} \tag{5.9}$$

If the relation linking stress and strains is linear, we speak of a linear elastic behavior. For the 1D think model, this implies for the spring stiffness E to be constant. In turn, if $E = E(\varepsilon)$, the elastic behavior is nonlinear. In this case, (5.9) defines the tangent elastic stiffness. This is sketched in Figure 5.1b.

Exercise 40. Consider an initial prestress σ_0 in the spring shown in Figure 5.1a. Express the free energy of the system, and derive the equation of state for stress σ.

Prestressing means that there exists an initial energy stored in the system:

$$\psi(\varepsilon) = \sigma_0\varepsilon + \frac{1}{2}E\varepsilon^2$$

Substitution in (5.6) yields

$$[\sigma - (\sigma_0 + E\varepsilon)]\,d\varepsilon = 0$$

Thus

$$\sigma = \frac{\partial\psi(\varepsilon)}{\partial\varepsilon} = \sigma_0 + E\varepsilon$$

■

5.2.2 1D Thermodynamics of Thermoelasticity

A material law describes a change of internal energy. Elasticity is associated with a *reversible* change of this energy. In addition to the mechanical loading, we want to study the change in internal energy caused by a temperature variation $\theta = T - T_0$. This change of temperature is often caused by an external heat supply, which we noted Q. In this non-isothermal case, the change in internal energy, dU, is due to the work or strain energy $dW = \sigma d\varepsilon$ and the heat supply dQ. This is the first law of thermodynamics:

$$dU = dW + dQ = \sigma d\varepsilon + dQ \tag{5.10}$$

The first law describes the conservation of energy in all of its forms. For instance, in an undeformable system ($d\varepsilon = 0$), the change in internal energy is only caused by heat supply from the exterior, $dU = dQ$. On the contrary, if the change in internal energy is only caused by the energy supply in form of mechanical work, $dU = \sigma d\varepsilon$, the evolutions are called adiabatic: No heat is supplied or can escape from the system. Note clearly the difference with respect to isothermal evolutions, in which no temperature variation occurs (i.e., $dT = 0 \to \theta = 0$). From these considerations, it appears that the physical quantities in (5.10) (i.e., the change in internal energy dU, work dW, and heat dQ) are not sufficient to describe all thermal conditions of the thermo-mechanical problem. It requires another physical quantity, the entropy S, and another conservation law, the second law of thermodynamics.

The entropy S expresses the quality of the energy stored in the system. The second law defines the sense of development of the entropy S. Indeed, the second law states that the change of internal entropy dS is always greater than or equal to the external entropy supply. This external supply is due to heat supply. For a uniformly heated system (homogeneous system), the law reads

$$dS \geq \frac{dQ}{T} \tag{5.11}$$

with T the absolute temperature. Substitution of the first law in (5.11) yields

$$\varphi dt = \sigma d\varepsilon - (dU - TdS) \geq 0 \tag{5.12}$$

Remarkably, Eq. (5.12) states that part of the energy provided during time dt in the form of work to the system (i.e., the strain energy $dW = \sigma d\varepsilon$) is not entirely stored in the system in form of mechanical work recoverable later on. This part of the energy is irreversibly dissipated. By contrast, reversible processes are nondissipative. It follows for thermoelastic reversible processes:

$$\varphi dt = 0 \Leftrightarrow dQ = TdS \tag{5.13}$$

In addition, Eqs. (5.10), (5.12), and (5.13) suggest the introduction of an energy that expresses the maximum capacity of the system to do work:

$$\psi = U - TS \tag{5.14}$$

Energy ψ is called the free energy, or the Helmholtz energy. We have already used this free energy to study the isothermal elastic deformation in Section 5.2.1. In turn, in the considered non-isothermal evolution ($dT \neq 0$), use of (5.14) in (5.12) yields

$$\varphi dt = \sigma d\varepsilon - SdT - d\psi = 0 \tag{5.15}$$

The previous equation specifies the energy conservation and entropy conservation of thermoelastic evolutions.

For the 1D (uniformly) heated spring system, the energy state is described by strain ε and temperature variation $\theta = T - T_0$; thus, formally in terms of the free energy ψ

$$\psi = \psi(\varepsilon, \theta = T - T_0) \tag{5.16}$$

The arguments of the free energy are called *state variables*: They describe the energy states of the elementary system (here the spring system). In the case of reversible processes, the values of these state variables can be controlled from the outside (i.e., the relative length variation ε and temperature T can be prescribed from the outside to the system). They are therefore also called *external state variables*.

Use of (5.16) in (5.15) yields

$$\varphi dt = \left(\sigma - \frac{\partial\psi}{\partial\varepsilon}\right) d\varepsilon - \left(S + \frac{\partial\psi}{\partial T}\right) dT = 0 \tag{5.17}$$

The partial derivatives of free energy $\psi = \psi(\varepsilon, T)$ in (5.17) are carried out at constant values of the variable not involved in the derivation. From a physical standpoint, this means that the physical evolution of strain increment $d\varepsilon$ and temperature increment dT can occur independent of each other. We also say that ε and T constitute a normal set of state variables; thus, from (5.17),

$$\sigma = \left(\frac{\partial\psi}{\partial\varepsilon}\right)_T; \quad S = -\left(\frac{\partial\psi}{\partial T}\right)_\varepsilon \tag{5.18}$$

Equations (5.18) are the 1D thermoelastic state equations for stress σ and entropy S in the elementary system.[3] In the case of a linear thermoelastic behavior, σ

[3] We have a quick look at the dimension of the involved quantities: Stress σ has the dimension

$$[\sigma] = L^{-2}F = LMT^{-2}$$

where F stands for the dimension of force (e.g., of unit, Newton), L for a length dimension (e.g., of unit, meter), M for the dimension of mass (e.g., of unit, kilogram), T for the dimension of time (e.g., of unit, second); the volume heat Q has as dimension:

$$[Q] = L^{-3}E = L^{-2}F$$

where $E = LF$ stands for the heat (or energy) dimension (e.g., of unit, Joule). E has the same dimension as a mechanical moment. The entropy S is considered here also as a volume quantity of dimension

$$[S] = E\Theta^{-1}L^{-3} = FL^{-2}\Theta^{-1}$$

(with Θ the temperature dimension, of unit Kelvin). Therefore, in the continuum approach, the volume density of the entropy S represents from a dimensional standpoint a stress per unit temperature. We always have to take care to use a consistent unit system. For instance, if we consider for stress σ as unit MPa = MN/m^2, then the appropriate choice of units for heat is MJ, and for the volume heat capacity C it is MJ/m^3/K.

and S depend linearly on the state variables ε and T. This linearity, known as *physical linearization*, can be checked from the outside. Within the hypothesis of physical linearization, the free energy $\psi = \psi(\varepsilon, T)$ around a stress free ($\sigma_0 = 0$) and isothermal reference state can be written in the form (quadratic expansion)

$$\psi = \psi(\varepsilon, \theta = T - T_0) = \frac{1}{2}\left(E\varepsilon^2 - 2\alpha E\theta\varepsilon - \frac{\mathcal{C}\theta^2}{T_0}\right) \tag{5.19}$$

Substituting this quadratic expression into (5.18), the thermoelastic equations of state are linear with regard to the external variables:

$$\sigma = \sigma(\varepsilon, \theta) = \left(\frac{\partial \psi}{\partial \varepsilon}\right)_T = E\varepsilon - \alpha E\theta \tag{5.20}$$

$$S = S(\varepsilon, \theta) = -\left(\frac{\partial \psi}{\partial T}\right)_\varepsilon = \frac{\mathcal{C}\theta}{T_0} + \alpha E\varepsilon \tag{5.21}$$

The dependence of stress σ on temperature variation θ is the well-known phenomenon of thermal expansion, with α the thermal dilatation coefficient (of dimension $[\alpha] = \Theta^{-1}$): In a stress-free experiment ($\sigma = 0$), a temperature rise $\theta > 0$ leads to the thermal dilatation $\alpha\theta$. In return, if this expansion is restrained ($\varepsilon = 0$), a compressive stress $\sigma = -\alpha E\theta$ is generated in the spring. Remarkably, the consequence of this thermomechanical cross-effect is that the entropy S also depends on strain ε. This thermomechanical coupling is due to the (assumed) existence of a thermoelastic potential $\psi = \psi(\varepsilon, T)$. Its very existence implies the Maxwell symmetry:

$$\left(\frac{\partial \sigma}{\partial T}\right)_\varepsilon = -\left(\frac{\partial S}{\partial \varepsilon}\right)_T = \frac{\partial^2 \psi}{\partial \varepsilon \partial T} = -\alpha E \tag{5.22}$$

Finally, $\mathcal{C}$ is the volume heat capacity (of dimension $[\mathcal{C}] = L^{-2}F\Theta^{-1} = L^{-1}MT^{-2}\Theta^{-1}$)[4]: It is the heat required to increase the temperature in the elementary system by one temperature unit.

Exercise 41. Consider the 1D spring system within an adiabatic chamber that avoids any loss of heat to the outside (i.e., $dQ = 0$). A tension stress $\sigma > 0$ is applied. Does the temperature within this closed system increase or decrease? For an isothermal experimental set-up determine the required heat supply from the outside to maintain isothermal conditions (i.e., $dT = 0$). Determine the elastic stiffness that is measured respectively in an adiabatic and in an isothermal experimental set-up.

1. **Adiabatic Temperature Change**: From Eq. (5.13), it follows that for adiabatic conditions

$$\varphi dt = 0 \Leftrightarrow dQ = TdS = 0$$

Use of the entropy equation of state (5.21) leads to:

$$TdS = 0 : dT = -\frac{\alpha E d\varepsilon}{\mathcal{C}/T_0}$$

[4]Typical values for $\mathcal{C}$ are as follows: Concrete: $\sim$ 2.4 MJ/m^3/K; Steel: $\sim$ 3.5 MJ/m^3/K; Wood: $\sim$ 1.3 MJ/m^3/K; Soil: $\sim$ 2.0 MJ/m^3/K.

Finally, replacing strain $d\varepsilon$ by the expression obtained from the stress equation of state (5.20) yields

$$dT = -\frac{\alpha d\sigma}{\mathcal{C}/T_0 + \alpha^2 E}$$

Hence, with $\alpha \geq 0$, the temperature in adiabatic tension ($d\sigma > 0$) decreases, while it increases in a compression experiment. Note that this change of temperature is reversible.

2. **Isothermal Heat Supply**: The required heat to maintain the temperature constant, $T = T_0$, is obtained from (5.13) and (5.21):

$$\varphi dt = 0 \Leftrightarrow dQ = T_0 dS = \alpha E T_0 d\varepsilon = \alpha T_0 d\sigma$$

Hence, in tension ($d\sigma > 0$), heat must be supplied from the outside ($dQ > 0$) to balance the intrinsic cooling tendency due to the tension stress application. This heat is called the latent heat of deformation. In the thermoelastic model, this latent heat results from the Maxwell symmetry (5.22).

3. **Isothermal and Adiabatic Elastic Stiffness**: The stiffness that is measured in an isothermal experiment is the stiffness E. In turn, in an adiabatic experiment, the use of the temperature variation in (5.20) leads to

$$\sigma = \bar{E}\varepsilon = E\left(1 + \frac{\alpha^2 E}{\mathcal{C}/T_0}\right)\varepsilon$$

Hence, the elastic stiffness that is measured in an adiabatic experiment is the adiabatic stiffness $\bar{E} \geq E$. This underlines the importance of specifying the thermal conditions when dealing with thermoelastic material properties (whether experimentally or on paper). ■

5.2.3 Linear 3D Thermoelasticity

To extend the 1D equations to the 3D case, it suffices to replace the scalar quantities by their tensorial counterparts (i.e., $\varepsilon \rightarrow \varepsilon_{ij}$ and $\sigma \rightarrow \sigma_{ij}$). For instance, the zero dissipation of thermoelasticity expressed by (5.15) reads in the 3D case:

$$\varphi dt = \boldsymbol{\sigma} : d\boldsymbol{\varepsilon} - SdT - d\psi = 0 \tag{5.23}$$

where $\boldsymbol{\sigma} : d\boldsymbol{\varepsilon} = \text{tr}\,(\boldsymbol{\sigma} \cdot d\boldsymbol{\varepsilon}) = \sigma_{ij} d\varepsilon_{ji}$ is the strain energy supplied in form of work to the elementary volume $d\Omega$. In the continuum model, the Helmholtz energy ψ is considered a volume quantity. Within the framework of physical linearization, the free energy of thermoelasticity reads in the 3D case (*quadratic* expansion):

$$\psi = \psi(\boldsymbol{\varepsilon}, \theta = T - T_0) = \frac{1}{2}\boldsymbol{\varepsilon} : \mathbf{C} : \boldsymbol{\varepsilon} - \mathbf{C} : \boldsymbol{\alpha} : \boldsymbol{\varepsilon}\theta - \frac{1}{2}\frac{\mathcal{C}\theta^2}{T_0} \tag{5.24}$$

or, equivalently, in tensor components

$$\psi = \psi(\varepsilon_{ij}, \theta = T - T_0) = \frac{1}{2}\varepsilon_{ij} C_{ijkl} \varepsilon_{kl} - C_{ijkl}\alpha_{kl}\varepsilon_{ij}\theta - \frac{1}{2}\frac{\mathcal{C}\theta^2}{T_0} \tag{5.25}$$

Use of (5.24) in (5.23) leads to the 3D state equations of *linear* thermoelasticity. These are as follows:

- The stress equation of state:

$$\boldsymbol{\sigma} = \boldsymbol{\sigma}(\boldsymbol{\varepsilon}, \theta) = \left(\frac{\partial \psi}{\partial \boldsymbol{\varepsilon}}\right)_T = \mathbf{C} : \boldsymbol{\varepsilon} - \mathbf{C} : \boldsymbol{\alpha}\theta \tag{5.26}$$

$$\sigma_{ij} = \sigma_{ij}(\varepsilon_{ij}, \theta) = \left(\frac{\partial \psi}{\partial \varepsilon_{ij}}\right)_T = C_{ijkl}\varepsilon_{kl} - C_{ijkl}\alpha_{kl}\theta \tag{5.27}$$

- The entropy equation of state:

$$S = S(\boldsymbol{\varepsilon}, \theta) = -\left(\frac{\partial \psi}{\partial T}\right)_{\boldsymbol{\varepsilon}} = \frac{\mathcal{C}\theta}{T_0} + \mathbf{C} : \boldsymbol{\alpha} : \boldsymbol{\varepsilon} \tag{5.28}$$

$$S = S(\varepsilon_{ij}, \theta) = -\left(\frac{\partial \psi}{\partial T}\right)_{\varepsilon_{ij}} = \frac{\mathcal{C}\theta}{T_0} + C_{ijkl}\alpha_{kl}\varepsilon_{ij} \tag{5.29}$$

The physical significance of the 3D thermoelastic properties is the same as in the 1D case; only the tensorial order is changed. For instance, $\boldsymbol{\alpha}$ is the second-order tensor of thermal dilatation coefficient, relating in a stress-free experiment the temperature variation θ to the strains:

$$\boldsymbol{\sigma} = 0 \rightarrow \boldsymbol{\varepsilon} = \boldsymbol{\alpha}\theta; \quad \varepsilon_{ij} = \alpha_{ij}\theta \tag{5.30}$$

We note here that due to the symmetry of the strain tensor $\varepsilon_{ij} = \varepsilon_{ji}$, the tensor of thermal dilation coefficient is also symmetric (i.e., $\alpha_{ij} = \alpha_{ji}$). Hence, there are six thermal dilatation coefficients to be determined from experiments.

In an isothermal experiment, the fourth-order stiffness tensor $\mathbf{C} = C_{ijkl}\mathbf{e}_i \otimes \mathbf{e}_j \otimes \mathbf{e}_k \otimes \mathbf{e}_l$ relates the strains to stresses:

$$\theta = 0 \rightarrow \boldsymbol{\sigma} = \mathbf{C} : \boldsymbol{\varepsilon}; \quad \sigma_{ij} = C_{ijkl}\varepsilon_{kl} \tag{5.31}$$

Without any further information, the possible permutations of the subscripts of C_{ijkl} imply *a priori* $3^4 = 81$ unknown (isothermal) elastic stiffness coefficients. However, the symmetry of the stress tensor ($\boldsymbol{\sigma} = {}^t\boldsymbol{\sigma}$) and the strain tensor ($\boldsymbol{\varepsilon} = {}^t\boldsymbol{\varepsilon}$), that is,

$$\sigma_{ij} = C_{ijkl}\varepsilon_{kl}; \quad \sigma_{ji} = C_{jikl}\varepsilon_{kl} \tag{5.32}$$

$$\sigma_{ij} = C_{ijkl}\varepsilon_{kl}; \quad \sigma_{ij} = C_{ijlk}\varepsilon_{lk} \tag{5.33}$$

implies the following symmetry relations:

$$C_{ijkl} = C_{jikl} = C_{ijlk} = C_{jilk} \tag{5.34}$$

This symmetry reduces the number of elastic unknowns from 81 to $6 \times 6 = 36$ components that can be presented in the following matrix form:

$$\begin{pmatrix} \sigma_{11} \\ \sigma_{22} \\ \sigma_{33} \\ \sigma_{12} = \\ \sigma_{21} \\ \sigma_{13} = \\ \sigma_{31} \\ \sigma_{23} = \\ \sigma_{32} \end{pmatrix} = \begin{bmatrix} C_{1111} & C_{1122} & C_{1133} & \begin{array}{c} C_{1112}= \\ C_{1121} \end{array} & \begin{array}{c} C_{1113}= \\ C_{1131} \end{array} & \begin{array}{c} C_{1123}= \\ C_{1132} \end{array} \\ C_{2211} & C_{2222} & C_{2233} & \begin{array}{c} C_{2212}= \\ C_{2221} \end{array} & \begin{array}{c} C_{2213}= \\ C_{2231} \end{array} & \begin{array}{c} C_{2223}= \\ C_{2232} \end{array} \\ C_{3311} & C_{3322} & C_{3333} & \begin{array}{c} C_{3312}= \\ C_{3321} \end{array} & \begin{array}{c} C_{3313}= \\ C_{3331} \end{array} & \begin{array}{c} C_{3323}= \\ C_{3332} \end{array} \\ \begin{array}{c} C_{1211}= \\ C_{2111} \end{array} & \begin{array}{c} C_{1222}= \\ C_{2122} \end{array} & \begin{array}{c} C_{1233}= \\ C_{2133} \end{array} & \begin{array}{c} C_{1212}= \\ C_{2112}= \\ C_{1221}= \\ C_{2121} \end{array} & \begin{array}{c} C_{1213}= \\ C_{2113}= \\ C_{1231}= \\ C_{2131} \end{array} & \begin{array}{c} C_{1223}= \\ C_{2123}= \\ C_{1232}= \\ C_{2132} \end{array} \\ \begin{array}{c} C_{1311}= \\ C_{3111} \end{array} & \begin{array}{c} C_{1322}= \\ C_{3122} \end{array} & \begin{array}{c} C_{1333}= \\ C_{3133} \end{array} & \begin{array}{c} C_{1312}= \\ C_{3112}= \\ C_{1321}= \\ C_{3121} \end{array} & \begin{array}{c} C_{1313}= \\ C_{3113}= \\ C_{1331}= \\ C_{3131} \end{array} & \begin{array}{c} C_{1323}= \\ C_{3123}= \\ C_{1332}= \\ C_{3132} \end{array} \\ \begin{array}{c} C_{2311}= \\ C_{3211} \end{array} & \begin{array}{c} C_{1322}= \\ C_{3122} \end{array} & \begin{array}{c} C_{2333}= \\ C_{3233} \end{array} & \begin{array}{c} C_{2312}= \\ C_{3212}= \\ C_{2321}= \\ C_{3221} \end{array} & \begin{array}{c} C_{2313}= \\ C_{3213}= \\ C_{2331}= \\ C_{3231} \end{array} & \begin{array}{c} C_{2323}= \\ C_{3223}= \\ C_{2332}= \\ C_{3232} \end{array} \end{bmatrix} \begin{pmatrix} \varepsilon_{11} \\ \varepsilon_{22} \\ \varepsilon_{33} \\ \varepsilon_{12} = \\ \varepsilon_{21} \\ \varepsilon_{13} = \\ \varepsilon_{31} \\ \varepsilon_{23} = \\ \varepsilon_{32} \end{pmatrix} \tag{5.35}$$

In addition, the quadratic form of $\frac{1}{2}\boldsymbol{\varepsilon} : \mathbf{C} : \boldsymbol{\varepsilon} = \frac{1}{2}\varepsilon_{ij}C_{ijkl}\varepsilon_{kl}$ in the expression of free energy (5.24) or (5.25) entails the following remarkable (Maxwell) symmetry:

$$C_{ijkl} = \frac{\partial \sigma_{ij}}{\partial \varepsilon_{kl}} = \frac{\partial}{\partial \varepsilon_{kl}}\left(\frac{\partial \psi}{\partial \varepsilon_{ij}}\right); \quad C_{klij} = \frac{\partial \sigma_{kl}}{\partial \varepsilon_{ij}} = \frac{\partial}{\partial \varepsilon_{ij}}\left(\frac{\partial \psi}{\partial \varepsilon_{kl}}\right) \tag{5.36}$$

Thus,

$$C_{ijkl} = C_{klij} = \frac{\partial^2 \psi}{\partial \varepsilon_{ij} \partial \varepsilon_{kl}} \tag{5.37}$$

The symmetry condition (5.37) that results from an energy argument [in the same way as (5.9) in the 1D case] renders the 6×6 matrix, defined by (5.35), symmetric. This symmetry condition reduces the unknown isothermal elastic constants to $(6 + 1) \times 6/2 = 21$. Hence, it suffices to specify the energy expression for particular material configurations to obtain the expression of the elastic stiffness tensor.

5.3 ISOTROPIC LINEAR THERMOELASTIC MATERIAL PROPERTIES

An isotropic behavior is defined as a behavior that does not depend on specific directions. From an energy point of view, this means that the elastic potential reads as a function of the strain invariants (have a look back at the polynomial construction of invariants in Section 4.2.3):

$$\operatorname{tr} \boldsymbol{\varepsilon} = \varepsilon_{ii} \tag{5.38}$$

$$\frac{1}{2}\operatorname{tr}(\boldsymbol{\varepsilon} \cdot \boldsymbol{\varepsilon}) = \frac{1}{2}\varepsilon_{ij}\varepsilon_{ji} \tag{5.39}$$

$$\frac{1}{3}\operatorname{tr}(\boldsymbol{\varepsilon} \cdot \boldsymbol{\varepsilon} \cdot \boldsymbol{\varepsilon}) = \frac{1}{3}\varepsilon_{ij}\varepsilon_{jk}\varepsilon_{ki} \tag{5.40}$$

In the framework of physical linearization and the associated quadratic expansion (5.24) of free energy ψ, only the linear and the quadratic invariants, (5.38) and

(5.39), are required to describe the energy states of the linear isotropic thermoelastic material:

$$\psi(\varepsilon, \theta = T - T_0) = \frac{\lambda}{2}(\operatorname{tr}\varepsilon)^2 + 2G\left(\frac{1}{2}\operatorname{tr}(\varepsilon \cdot \varepsilon)\right) - (3\lambda + 2G)\alpha(\operatorname{tr}\varepsilon)\,\theta - \frac{1}{2}\frac{C\theta^2}{T_0} \tag{5.41}$$

where λ and G are two elastic constants associated with the elastic energy contribution of the first and second strain invariant. They are called Lamé constants. Substitution of (5.41) in (5.26) and (5.28) yields the following:

- The linear isotropic thermoelastic stress equation of state:

$$\boldsymbol{\sigma} = \left(\frac{\partial \psi}{\partial \varepsilon}\right)_T = \lambda(\operatorname{tr}\varepsilon)\,\mathbf{1} + 2G\varepsilon - (3\lambda + 2G)\alpha\theta\mathbf{1} \tag{5.42}$$

$$\sigma_{ij} = \left(\frac{\partial \psi}{\partial \varepsilon_{ij}}\right)_T = \lambda\varepsilon_{kk}\delta_{ij} + 2G\varepsilon_{ij} - (3\lambda + 2G)\alpha\theta\delta_{ij} \tag{5.43}$$

- The linear isotropic thermoelastic entropy equation of state:

$$S = -\left(\frac{\partial \psi}{\partial T}\right)_{\varepsilon} = \frac{C\theta}{T_0} + (3\lambda + 2G)\alpha \operatorname{tr}\varepsilon \tag{5.44}$$

$$S = -\left(\frac{\partial \psi}{\partial T}\right)_{\varepsilon_{ij}} = \frac{C\theta}{T_0} + (3\lambda + 2G)\alpha\varepsilon_{kk} \tag{5.45}$$

In the isotropic case, the $21 + 6 = 27$ thermoelastic constants (21 isothermal elastic constants + 6 thermal dilatation coefficients) reduce to three: the isothermal Lamé constants, λ and G, and the isotropic thermal dilatation coefficient, α. Hence, three material tests are required to determine the three thermoelastic constants.

5.3.1 Bulk Modulus

Exercise 42. It is known from fluid mechanics that the pressure p is related to the relative volume variation $\Delta V/V$. In the linear range, this can be expressed by

$$p = -K\frac{\Delta V}{V}$$

where K is the bulk modulus. Determine the relation that links the bulk modulus to the isothermal Lamé constants, λ and μ.

The hydrostatic stress state is defined by $\boldsymbol{\sigma} = -p\mathbf{1}$ of mean stress $\sigma_m = I_1/3 = -p$ (*cf.* Figure 5.2a). In addition, from (5.42 on next page), the mean stress is related to the isothermal relative volume variation $\operatorname{tr}\varepsilon = \Delta V/V$ by

$$\sigma_m = \frac{1}{3}\operatorname{tr}\boldsymbol{\sigma} = \left(\lambda + \frac{2}{3}G\right)\operatorname{tr}\varepsilon = K\operatorname{tr}\varepsilon$$

■

Hence, the bulk modulus K that relates the first invariant of the stress tensor to the first invariant of the strain tensor reads

$$K = \lambda + \frac{2}{3}G \tag{5.46}$$

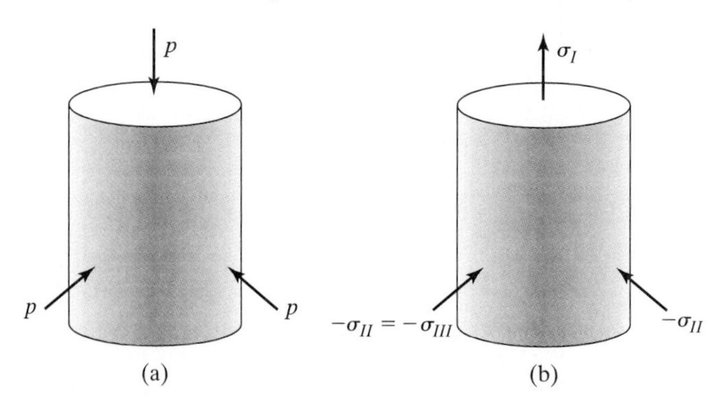

FIGURE 5.2: Elastic material tests: (a) hydrostatic test; (b) triaxial test set-up.

5.3.2 Shear Modulus

The shear modulus G relates the shear stress σ_{ij} $(i \neq j)$ to the shear strain ε_{ij} $(i \neq j)$. These shear strains represent the angular variation $\gamma(\mathbf{e}_i, \mathbf{e}_j)$, or half-distortion, between initially orthonormal axis $\mathbf{e}_i \perp \mathbf{e}_j$ (have a look back at Section 2.2.1):

$$i \neq j : \sigma_{ij} = 2G\varepsilon_{ij} = 2G\gamma(\mathbf{e}_i, \mathbf{e}_j) \tag{5.47}$$

However, experiments with prescribed distortion are difficult to perform.

Exercise 43. Consider the triaxial material test set-up sketched in Figure 5.2b. Show that the shear modulus can be determined in this shear stress-free material test by

$$\sigma_I - \sigma_{III} = 2G(\varepsilon_I - \varepsilon_{III}) \tag{5.48}$$

where σ_I, σ_{III}, and $\varepsilon_I, \varepsilon_{III}$ are principal stresses and principal strains, respectively. Give an interpretation in the Mohr planes of stresses and strains.

The stress state in this triaxial test is given by (see 5.2b)

$$\boldsymbol{\sigma} = \sigma_I \mathbf{u}_I \otimes \mathbf{u}_I + \sigma_{III}(\mathbf{u}_{II} \otimes \mathbf{u}_{II} + \mathbf{u}_{III} \otimes \mathbf{u}_{III})$$

with $\sigma_I > \sigma_{II} = \sigma_{III}$ the principal stresses associated with the principal directions $\mathbf{u}_J$ $(J = 1, 2, 3)$. Due to the isotropic nature of the stress-strain relation, the principal directions of the stress tensor $\boldsymbol{\sigma} \cdot \mathbf{u}_J = \sigma_J \mathbf{u}_J$ [defined by Eq. (4.6)] and the principal directions of the strain tensor $\boldsymbol{\varepsilon} \cdot \mathbf{u}_J = \varepsilon_J \mathbf{u}_J$ [defined by Eq. (2.9)] coincide. The same applies to the principal directions of the stress deviator $\mathbf{s}$ and the strain deviator $\mathbf{e} = \boldsymbol{\varepsilon} - \frac{1}{3}\mathrm{tr}\,(\boldsymbol{\varepsilon})\,1$. The constitutive relation linking stress deviator and strain deviator follows from (5.42) and (5.46):

$$\begin{aligned} \mathbf{s} &= \boldsymbol{\sigma} - \sigma_m 1 = \lambda\,(\mathrm{tr}\,\boldsymbol{\varepsilon})\,1 + 2G\boldsymbol{\varepsilon} - (\lambda + 2G/3)\,(\mathrm{tr}\,\boldsymbol{\varepsilon})\,1 \\ &= 2G\,(\boldsymbol{\varepsilon} - (\mathrm{tr}\,\boldsymbol{\varepsilon})\,1/3) = 2G\mathbf{e} \end{aligned}$$

Hence, the shear modulus G relates the stress deviator $\mathbf{s}$ to the strain deviator. For the triaxial stress state in hand, the stress deviator reads (have a look back at

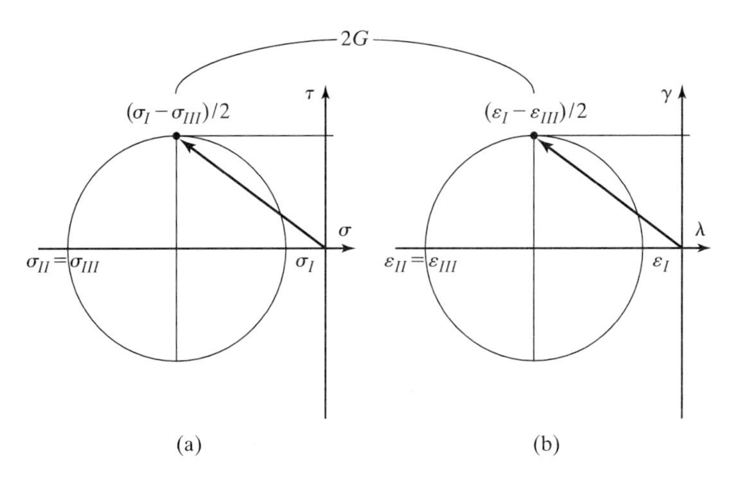

FIGURE 5.3: Shear modulus determination from a triaxial test: (a) in the Mohr stress plane; (b) in the Mohr strain plane.

Section 4.2.4)

$$(s_{ij}) = (\sigma_I - \sigma_{III}) \begin{pmatrix} 2/3 & & \\ & -1/3 & \\ & & -1/3 \end{pmatrix}$$

and the strain deviator

$$(e_{ij}) = (\varepsilon_I - \varepsilon_{III}) \begin{pmatrix} 2/3 & & \\ & -1/3 & \\ & & -1/3 \end{pmatrix}$$

Therefore, since the stress quantities, $\boldsymbol{\sigma}$ and $\mathbf{s}$, and the strain quantities, $\boldsymbol{\varepsilon}$ and $\mathbf{e}$, have the same principal directions, it follows that

$$s_{ij} = 2Ge_{ij} \rightarrow \sigma_I - \sigma_{III} = 2G(\varepsilon_I - \varepsilon_{III})$$

which is (5.48): The triaxial test gives access to the experimental determination of the shear modulus G—from a shear stress-free experiment. Figure 5.3 displays the triaxial stress and strain state in the Mohr stress plane and the Mohr strain plane, respectively. The shear modulus G relates the maximum shear stress component $\max \tau = \frac{1}{2}(\sigma_I - \sigma_{III})$ to the maximum distortion $\max 2\gamma = \varepsilon_I - \varepsilon_{III}$. ■

5.3.3 Thermal Dilatation Coefficient

The thermal dilatation coefficient is determined from a stress-free experiment, in which the temperature is varied. From (5.42), it follows that

$$\boldsymbol{\sigma} = 0 \rightarrow \lambda\,(\operatorname{tr}\boldsymbol{\varepsilon})\,1 + 2G\boldsymbol{\varepsilon} = (3\lambda + 2G)\alpha\theta 1 = 0 \rightarrow \operatorname{tr}\boldsymbol{\varepsilon} = 3\alpha\theta \tag{5.49}$$

Due to the isotropic nature of the thermal dilatation (i.e., $\boldsymbol{\alpha} = \alpha\mathbf{1}$), it is sufficient to measure the thermal expansion in one direction. The thermal dilatation coefficient is the linear dilatation caused by a unit temperature variation $\theta = 1$ K.

5.3.4 Young's Modulus and Poisson Ratio

It is useful to specify the relations that link the isothermal elastic properties, λ and G, to the Young's modulus E and the Poisson ratio ν:

$$E = G\frac{3\lambda + 2G}{\lambda + G}; \quad \nu = \frac{\lambda}{2(\lambda + G)} \tag{5.50}$$

$$\lambda = \frac{\nu E}{(1+\nu)(1-2\nu)}; \quad G = \frac{E}{2(1+\nu)} \tag{5.51}$$

$$K = \lambda + \frac{2}{3}G = \frac{E}{3(1-2\nu)} \tag{5.52}$$

Among these properties, two are sufficient to describe the linear isotropic (isothermal) elastic behavior.

Exercise 44. The fourth-order isotropic linear elastic stiffness tensor can be written in the form

$$\mathbf{C} = K\mathbf{1} \otimes \mathbf{1} + 2G\left(\mathbf{I} - \frac{1}{3}\mathbf{1} \otimes \mathbf{1}\right) \tag{5.53}$$

and its inverse, the fourth-order isotropic linear elastic compliance tensor:

$$\mathbf{\Lambda} = \mathbf{C}^{-1} = \frac{1+\nu}{E}\mathbf{I} - \frac{\nu}{E}\mathbf{1} \otimes \mathbf{1} \tag{5.54}$$

where $\mathbf{I}$ is the fourth-order unit tensor, and $\mathbf{1}$ the second-order unit tensor. Give a matrix representation of (5.53) and (5.54). ■

Equation (5.54) allows us to invert the stress equation of state (5.42):

$$\boldsymbol{\varepsilon} = \mathbf{\Lambda} : \boldsymbol{\sigma} = \frac{1+\nu}{E}\boldsymbol{\sigma} - \frac{\nu}{E}(\operatorname{tr}\boldsymbol{\sigma})\mathbf{1} + \alpha\theta\mathbf{1} \tag{5.55}$$

$$\varepsilon_{ij} = \Lambda_{ijkl}\sigma_{kl} = \frac{1+\nu}{E}\sigma_{ij} - \frac{\nu}{E}\sigma_{kk}\delta_{ij} + \alpha\theta\delta_{ij} \tag{5.56}$$

Last, a stability requirement is that the quadratic form of the elastic volume density $\frac{1}{2}\boldsymbol{\varepsilon} : \mathbf{C} : \boldsymbol{\varepsilon} = \frac{1}{2}\varepsilon_{ij}C_{ijkl}\varepsilon_{kl}$ in (5.24) is positive definite. This reads

$$\boldsymbol{\varepsilon} : \mathbf{C} : \boldsymbol{\varepsilon} \geq 0 \tag{5.57}$$

In the isotropic case, it is readily understood from (5.41) that this implies the following stability restrictions for the elastic properties:

$$G \geq 0; \quad K \geq 0 \tag{5.58}$$

or, equivalently,

$$E \geq 0; \quad -1 \leq \nu \leq 0.5 \tag{5.59}$$

This stability criterion can be understood from the 1D think model, with which we have started (*cf.* Figure 5.1a): It excludes a negative spring stiffness (which is physically senseless).

5.3.5 An Instructive Exercise: Dimensional Analysis of the Boussinesq Problem

We consider the Boussinesq problem displayed in Figure 5.4: A half-space is subjected to a point load P directed in the z-direction. We are interested in determining the displacement at the surface of the half-space (i.e., at $z = 0$). The problem is complex and requires some refined mathematical tools for comprehensively solving this challenging elasticity problem, which we will develop later (see Section 5.6). However, by means of dimensional analysis, a solution can be found on the basis of the elements of linear isotropic elasticity developed to this point. By dimensional analysis, we mean to make use of the dimensional homogeneity of the physical problem at hand. While the physical basis of dimensional analysis can be found in classical textbooks on the topic or chapters in fluid mechanics textbooks,[5] we develop here this powerful technique through an instructive exercise, which gives an inside in the physical background of the elasticity constants.

The problem[6] we want to solve is the surface displacement w due to an applied concentrated force P. A simple check of dimensions of the two quantities, w and P, underlines the necessity of a material law: w has dimension of length, say L, and force P has a force dimension, say F. Therefore, the sought solution, which relates w and P, is a dimension function relating the two base dimensions L and F. In the case of an isotropic elastic material behavior, the material constants are the bulk modulus K and the shear modulus G, which, in addition to the distance r from the point of load application, will determine the surface displacement:

$$w = f(P, r, K, G) \tag{5.60}$$

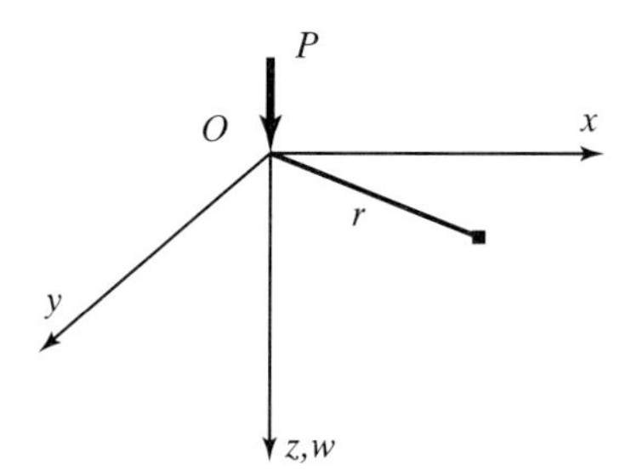

FIGURE 5.4: Boussinesq problem: half-space subjected to a concentrated force.

If (5.60) is a physically meaningful description of the problem, then w and f must have the same dimension (i.e., length L). Since the dimension of function f results

[5]See, for instance, Barenblatt, G. I. (1996). *Scaling, Self-Similarity, and Intermediate Asymptotics.* Cambridge Texts in Applied Mathematics, Cambridge University Press, Cambridge, UK, 386 pages; (in particular Chapter 0–1, pp. 1–63).

Huerre, P. (1997). *Mécanique des Fluides. Tome I. Cours,* Départment de Mécanique, Ecole Polytechnique, France. [in French]; (in particular, Chapter V, pp. 107–140).

Sonin, A. A. (1997). *The Physical Basis of Dimensional Analysis.* Department of Mechanical Engineering, MIT, Cambridge, MA, 52 pages.

[6]Inspired by the presentation of Borodich, F. M. (1998). "Similarity Methods in Hertz Contact Problems and Their Relations with the Meyer Hardness Test." *Technical Report TR/MAT/FMB/98–98,* Glasgow, Caledonian University, UK.

from the dimensions of its arguments (i.e., P, r, K, G), certain algebraic relations between these physical quantities must exist. This is the underlying idea of dimensional analysis, which explores the dimensional homogeneity of physical problems. For the problem at hand, the quantities involved in (5.60) have the following dimensions in a LF-base dimension system:

$$[w] = [r] = L; \quad [P] = F; \quad [K] = [G] = L^{-2}F \tag{5.61}$$

We note that two out of the five involved quantities are dimensionally independent, while the other three can be expressed as power functions of the former. This suggests a combination of the different dimension functions to derive three dimensionless quantities. For instance, if we choose r and G as dimensionally independent quantities, we express the remaining three in the following dimensionless form:

$$\Pi = \frac{w}{r}; \quad \Pi_1 = \frac{P}{Gr^2}; \quad \Pi_2 = \frac{K}{G} \tag{5.62}$$

Furthermore, if physically meaningful, (5.60) must hold irrespective of any particular choice of system of units (i.e., meter m, foot ft, ... for L; Newton N, Dyne dyn, Pound-force lbf, ... for F; and so on), the previous dimensional analysis, or more precisely the pi-theorem,[7] implies that relation (5.60) must reduce to a relation $\Pi = \mathcal{F}(\Pi_1, \Pi_2)$ between the three dimensionless quantities:

$$\Pi = \frac{w}{r} = \mathcal{F}\left(\Pi_1 = \frac{P}{Gr^2}, \Pi_2 = \frac{K}{G}\right) \tag{5.63}$$

Hence, by means of dimensional analysis, we arrived at reducing the relation between five dimensional quantities to one between three dimensionless quantities. The second invariant, which expresses the relation between the bulk modulus and the shear modulus, is readily understood from (5.50) to (5.52): It is only a function of the Poisson ratio:

$$\Pi_2 = \frac{K}{G} = \frac{2}{3}\frac{1+\nu}{1-2\nu} \tag{5.64}$$

This gives an interesting interpretation of the Poisson ratio: It expresses a bulk-to-shear modulus ratio. Since we can always construct new invariants as power functions of previously identified dimensionless numbers, we can replace the bulk-to-shear modulus ratio in (5.63) by ν. Finally, since the material is *linear* elastic, the displacement w must be a linear function of the applied load P, respectively of the first invariant Π_1; that is,

$$\mathcal{F}\left(\Pi_1 = \frac{P}{Gr^2}, \nu\right) = \Pi_1 \mathcal{F}_2(\nu) \tag{5.65}$$

Last, with the two mechanical arguments (5.64) and (5.65), the surface displacement solution is of the form

$$w = \frac{P}{Gr}\mathcal{F}_2(\nu) \tag{5.66}$$

[7] The pi-theorem derives its name from Buckingham's naming of the dimensionless variables in his 1914 paper: Buckingham, E. (1914). "On Physically Similar Systems. Illustrations of the Use of Dimensional Analysis." *Physical Review*, Vol. 4, 345–376.

Function $\mathcal{F}_2(\nu)$ remains undetermined and requires a refined solving method to be determined, which is developed in a forthcoming section. It will turn out to be $\mathcal{F}_2(\nu) = (1-\nu)/2\pi$.

5.4 DIRECT SOLVING METHODS OF ELASTIC PROBLEMS

We have formulated the thermoelastic material law on the material level of the elementary volume $d\Omega$ within the hypothesis of infinitesimal deformation. To close this chapter, we develop the steps for solving problems involving thermoelastic material systems. This will be done within the framework of small perturbation.

5.4.1 Small Perturbation

The small perturbation assumption includes two hypotheses:

1. The hypothesis of infinitesimal deformation, introduced in Chapter 2:

$$\forall \mathbf{X}: \quad \|\text{Grad}\,\boldsymbol{\xi}\| \ll 1 \rightarrow \boldsymbol{\varepsilon} = \frac{1}{2}\left(\text{grad}\,\boldsymbol{\xi} + {}^t\,\text{grad}\,\boldsymbol{\xi}\right) \tag{5.67}$$

2. The hypothesis of small displacement:

$$\forall \mathbf{X}: \quad \frac{\|\boldsymbol{\xi}\|}{\mathcal{L}_c} \ll 1 \tag{5.68}$$

where $\mathcal{L}_c$ is a characteristic length of the considered problem.

The infinitesimal deformation assumption (5.67) allows merging the initial configuration with the current one in all spatial derivations. The addition of the small displacement hypothesis (5.68) defines a first-order strain-displacement theory:

$$\boldsymbol{\xi}(\mathbf{X}) = \boldsymbol{\xi}(\mathbf{x} - \boldsymbol{\xi}) \simeq \boldsymbol{\xi}(\mathbf{x}) - \boldsymbol{\xi}(\mathbf{x}) \cdot \text{grad}\,\boldsymbol{\xi}(\mathbf{x}) \simeq \boldsymbol{\xi}(\mathbf{x}) \tag{5.69}$$

Hence, in this first-order strain-displacement theory, all physical quantities are expressed with respect to a fixed initial configuration. For instance, the small perturbation hypothesis implies that

$$\rho(\mathbf{x}, t) \simeq \rho_0(\mathbf{X}) \tag{5.70}$$

where ρ_0 is the mass density in the initial (undeformed) configuration, assumed to be constant.

5.4.2 Governing Equations

The governing equations are the momentum balance equation, the boundary conditions, and the material law.

Equilibrium Equation.

Within the hypothesis of small perturbations, the static momentum balance equation reads

$$\text{in } \Omega : \operatorname{div} \boldsymbol{\sigma} + \rho_0 \mathbf{f} = 0 \tag{5.71}$$

or, in index notation in a Cartesian coordinate system,

$$\frac{\partial \sigma_{ij}}{\partial X_j} + \rho_0(\mathbf{X}) f_i = 0 \tag{5.72}$$

Equation (5.72) needs to be compared with the exact expression (3.28): Within the hypothesis of small perturbations, the spatial derivatives are carried out with respect to the initial configuration, and the body force is defined on the initial fixed configuration.

Boundary Conditions.

The boundary conditions are composed of the stress and displacement boundary conditions:

$$\text{on } \partial\Omega = \partial\Omega_{\boldsymbol{\xi}^d} \cup \partial\Omega_{\mathbf{T}^d} : \left\{ \begin{array}{l} \text{on } \partial\Omega_{\mathbf{T}^d} : \boldsymbol{\sigma} \cdot \mathbf{n} = \mathbf{T}^d \\ \text{on } \partial\Omega_{\boldsymbol{\xi}^d} : \boldsymbol{\xi} = \boldsymbol{\xi}^d \end{array} \right\} \tag{5.73}$$

where d stands for given *d*ata. We should note here that displacements and stresses can only be simultaneously prescribed at the same point on boundary $\partial\Omega$, if they apply in different directions; for instance,

$$\mathbf{T} \cdot \mathbf{n} = \mathbf{T}^d \cdot \mathbf{n} = \sigma^d; \quad \boldsymbol{\xi} \cdot \mathbf{t} = \xi_t^d; \quad \mathbf{n} \cdot \mathbf{t} = 0 \tag{5.74}$$

with σ^d the prescribed normal stress and ξ_t^d the prescribed tangential displacement.

In addition, in order for the problem to be linear, unilateral contact conditions and friction at the interface are disregarded. The contact is perfect and without friction:

$$\boldsymbol{\xi} \cdot \mathbf{n} = 0; \quad \mathbf{T} \cdot \mathbf{t} = 0 \tag{5.75}$$

This is a necessary condition for the response of the system to be path independent.

Linear Thermoelastic Material Law.

The last equation is the constitutive law relating stress and strain. Hence, for isotropic linear thermoelastic material systems,

$$\boldsymbol{\sigma} = \boldsymbol{\sigma}_0 + \lambda (\operatorname{tr} \boldsymbol{\varepsilon}) \mathbf{1} + 2G\boldsymbol{\varepsilon} - (3\lambda + 2G)\alpha\theta \mathbf{1} \tag{5.76}$$

Here, $\boldsymbol{\sigma}_0$ is a possible initial stress state. In addition, in what follows we consider the temperature field $\theta = \theta^d(\mathbf{X})$ as given in the studied domain Ω. Its determination from the heat equation and conduction laws goes beyond the scope of this chapter. Hence, the unknown of the problems are the stress field $\boldsymbol{\sigma}$, the strain field $\boldsymbol{\varepsilon}$, and the displacement field $\boldsymbol{\xi}$.

5.4.3 Theorem of Superposition

An immediate consequence of the problem stated previously is the theorem of superposition. It states that the solution fields $(\boldsymbol{\sigma}, \boldsymbol{\varepsilon}, \boldsymbol{\xi})_I$ and $(\boldsymbol{\sigma}, \boldsymbol{\varepsilon}, \boldsymbol{\xi})_{II}$ associated with load cases $(\mathbf{T}^d, \boldsymbol{\xi}^d, \rho_0 \mathbf{f})_I$ and $(\mathbf{T}^d, \boldsymbol{\xi}^d, \rho_0 \mathbf{f})_{II}$ can be superposed:

$$\lambda_I \begin{pmatrix} \mathbf{T}^d \\ \boldsymbol{\xi}^d \\ \rho_0 \mathbf{f} \end{pmatrix}_I + \lambda_{II} \begin{pmatrix} \mathbf{T}^d \\ \boldsymbol{\xi}^d \\ \rho_0 \mathbf{f} \end{pmatrix}_{II} = \lambda_I \begin{pmatrix} \boldsymbol{\sigma} \\ \boldsymbol{\varepsilon} \\ \boldsymbol{\xi} \end{pmatrix}_I + \lambda_{II} \begin{pmatrix} \boldsymbol{\sigma} \\ \boldsymbol{\varepsilon} \\ \boldsymbol{\xi} \end{pmatrix}_{II} \tag{5.77}$$

where λ_I and λ_{II} are load factors. We should note again the importance of the linearity of the boundary conditions. In fact, in the case of a unilateral contact (e.g., Hertz contact problem), the theorem of superposition does not apply.

5.4.4 Displacement Method

The displacement method consists of considering the displacement field $\boldsymbol{\xi} = \boldsymbol{\xi}(\mathbf{X})$ as principal unknown. Starting from the unknown displacement field, the strain tensor $\boldsymbol{\varepsilon} = \boldsymbol{\varepsilon}[\boldsymbol{\xi}(\mathbf{X})]$ according (5.67) is determined. Substitution in the material law (5.76) yields the stress field $\boldsymbol{\sigma} = \boldsymbol{\sigma}\{\boldsymbol{\varepsilon}[\boldsymbol{\xi}(\mathbf{X})]\}$, which needs to satisfy the equilibrium equation (5.71).[8] Solving this differential equation, and using the boundary conditions to determine the integration constants, leads to the sought solution. This direct solution method is summarized by (5.78).

$$\begin{array}{cccc} \boldsymbol{\xi} \rightarrow & \boldsymbol{\varepsilon} = \frac{1}{2}\left(\operatorname{grad} \boldsymbol{\xi} + {}^t \operatorname{grad} \boldsymbol{\xi}\right) & & \boldsymbol{\sigma} = \boldsymbol{\sigma}_0 + \lambda\,(\operatorname{tr} \boldsymbol{\varepsilon})\,\mathbf{1} + 2G\boldsymbol{\varepsilon} \\ & & & \downarrow \\ & & & \operatorname{div} \boldsymbol{\sigma} + \rho_0 \mathbf{f} = \mathbf{0} \\ & & & \downarrow \\ & & & \text{Integration} \\ & & & \downarrow \\ & \begin{array}{c}\text{Boundary Conditions} \\ \text{on } \partial\Omega = \partial\Omega_{\boldsymbol{\xi}^d} \cup \partial\Omega_{\mathbf{T}^d}\end{array} & \rightarrow & \text{Solution} \end{array} \tag{5.78}$$

Exercise 45. Consider the soil layer of height H, shown in Figure 5.5. At the surface $x = 0$, a uniform pressure p is prescribed. At $x = H$ the soil layer is attached to an undeformable rigid surface. The material system is assumed to be homogeneous, without weight, initially stress free, and at constant temperature $T = T_0$. Determine the displacement on the surface, and the stresses.

Given the semi-infinite nature of the problem, the displacement field is of the form

$$\boldsymbol{\xi} = u(X)\mathbf{e}_x$$

---

[8]Substitution of the isothermal linear elastic material law in the momentum balance equation gives

$$(\lambda + \mu)\operatorname{grad}(\operatorname{div} \boldsymbol{\xi}) + \mu \operatorname{div}(\operatorname{grad} \boldsymbol{\xi}) + \rho_0 \mathbf{f} = 0$$

with $\operatorname{div} \boldsymbol{\xi} = \varepsilon_{ii} = \operatorname{tr} \boldsymbol{\varepsilon}$ (see Chapter 2). The previous equation is known as Navier's equation. In Cartesian coordinates it reads

$$(\lambda + \mu)\frac{\partial}{\partial X_i}\left(\frac{\partial \xi_j}{\partial X_J}\right) + \nabla^2 \xi_i + \rho_0 f_i = 0$$

where ∇^2 is the Laplacian operator.

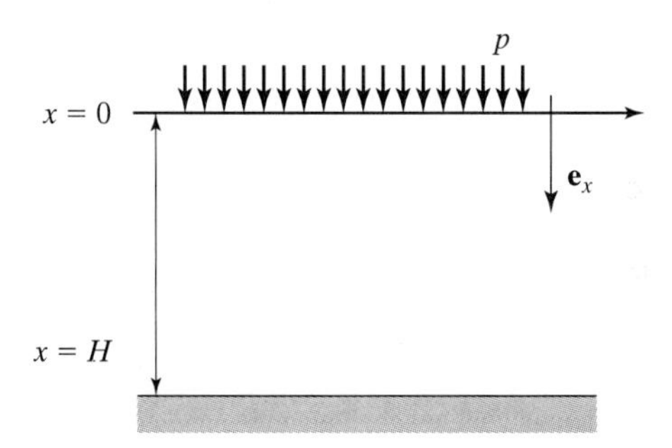

FIGURE 5.5: Soil layer subjected to a uniform load.

From (5.67), the strain tensor reads

$$\boldsymbol{\varepsilon} = \frac{\partial u(X)}{\partial X}\mathbf{e}_x \otimes \mathbf{e}_x$$

The hypothesis of small perturbation implies here that

$$\left|\frac{\partial u(X)}{\partial X}\right| \ll 1; \qquad \frac{|u(X)|}{H} \ll 1$$

Use of $\boldsymbol{\varepsilon}$ in constitutive law (5.76) yields ($\theta = 0$)

$$(\boldsymbol{\sigma}) = \begin{pmatrix} \lambda + 2G & 0 & 0 \\ & \lambda & 0 \\ \text{sym} & & \lambda \end{pmatrix} \frac{\partial u}{\partial X}$$

The equilibrium condition reads here

$$\frac{\partial \sigma_{xx}}{\partial x} = (\lambda + 2G)\frac{\partial^2 u}{\partial X^2} = 0$$

Thus,

$$\sigma_{xx} = \text{const}$$

The value of σ_{xx} follows from the stress boundary condition at $x = 0$:

$$\text{on } \partial\Omega_{\mathbf{T}^d} \leftrightarrow x = 0 : \boldsymbol{\sigma} \cdot \mathbf{n} = \mathbf{T}^d = p\mathbf{e}_x$$

Hence, with $\mathbf{n} = -\mathbf{e}_x$,

$$\sigma_{xx} = -p$$

It follows that

$$\frac{\partial u}{\partial X} = \frac{-p}{(\lambda + 2G)}$$

and after integration

$$u(X) = \frac{-p}{(\lambda + 2G)}X + \text{Cst}$$

The integration constant Cst is determined from the displacement boundary condition at $x = H$:

$$\text{on } \partial\Omega_{\boldsymbol{\xi}^d} : \boldsymbol{\xi} = \boldsymbol{\xi}^d = 0 \leftrightarrow u(X = H) = 0$$

Thus,

$$u(X) = \frac{-p}{(\lambda + 2\mu)}(X - H)$$

We should note that this solution is only valid within the solution domain $X \in [0, H]$. Finally, the stresses are

$$\sigma_{yy} = \sigma_{zz} = -\frac{\lambda}{\lambda + 2G}p = -\frac{\nu}{1-\nu}p \geq \sigma_{xx} = -p$$

$\nu/(1-\nu)$ can be determined in an odometer test. ■

5.4.5 Stress Method

The stress method consists of considering the stress field $\boldsymbol{\sigma} = \boldsymbol{\sigma}(\mathbf{X})$ as principal unknown. The starting point is the construction of a statically admissible stress field as defined in Chapter 4 (see Section 4.1). This statically admissible stress field is applied in the constitutive equation (5.55) to determine the strain field. By integration, the displacement field is obtained. The stress method can be summarized as follows:

$$\begin{array}{cccc}
\boldsymbol{\sigma} \rightarrow & \begin{array}{c}\text{Statically Admissible}\\ \text{in } \Omega : \operatorname{div}\boldsymbol{\sigma} + \rho_0\mathbf{f} = \mathbf{0}\\ \text{on } \partial\Omega_{\mathbf{T}^d} : \boldsymbol{\sigma}\cdot\mathbf{n} = \mathbf{T}^d\end{array} & \rightarrow & \boldsymbol{\varepsilon} = \frac{1+\nu}{E}[\boldsymbol{\sigma} - \boldsymbol{\sigma}_0] - \frac{\nu}{E}[\operatorname{tr}(\boldsymbol{\sigma} - \boldsymbol{\sigma})]1 \\
 & & & \downarrow \\
 & & & \boldsymbol{\varepsilon} = \frac{1}{2}\left(\operatorname{grad}\boldsymbol{\xi} + {}^t\operatorname{grad}\boldsymbol{\xi}\right) \\
 & & & \downarrow \\
 & & & \text{Integration} \\
 & & & \downarrow \\
 & \begin{array}{c}\text{Kinematically Admissible}\\ \text{on } \partial\Omega_{\boldsymbol{\xi}^d} : \boldsymbol{\xi} = \boldsymbol{\xi}^d\end{array} & \rightarrow & \text{Solution}
\end{array} \tag{5.79}$$

5.5 TRAINING SET: FROM THE CYLINDER TUBE TO DEEP TUNNELING

This training set is concerned with the analysis of the elastic equilibrium of some cylinder tubes under pressure. The starting point is the elastic analysis of a cylinder subjected to an external and an internal pressure: It is the necessary elastic extension of the training set of Chapter 3, dealing with admissible stress states in a pressure vessel (pressure vessel formula). Then we show how this elastic solution can be used for analyzing elastic stress and displacement that occur in deep tunneling, a tube being driven into a natural prestressed continuum.

5.5.1 Elastic Equilibrium of a Cylinder Tube

We want to study the elastic equilibrium of the cylinder tube of internal radius R_0 and external radius $R_1 = R_0 + t$ (with t the cylinder wall thickness). This cylinder of height ℓ and oriented along $0z$ is subjected at its internal and external wall to the pressures p_0 and p_1, respectively (see Figure 5.6a). On both end sides of the cylinder, at $z = 0$ and $z = \ell$, the expansion of the tube is restrained in the z-direction. The tube is made of a homogeneous isotropic linear elastic material. Body forces are neglected.

Exercise 46. Specify the boundary conditions. What can be said about the components of the displacement field $\boldsymbol{\xi}$ in the tube?

The boundary $\partial\Omega$ is composed of the force boundary $\partial\Omega_{\mathbf{T}^d}$ and the displacement boundary $\partial\Omega_{\boldsymbol{\xi}^d}$. Part of the stress boundary conditions have already been specified in Section 3.5.1, which we recall:

$$\text{on } \partial\Omega_{\mathbf{T}^d} : \left\{ \begin{array}{c} r = R_0 : \mathbf{T}^d = p_0\mathbf{e}_r \\ r = R1 : \mathbf{T}^d = -p_1\mathbf{e}_r \end{array} \right\}$$

In addition, at $z = 0$ and $z = \ell$, a zero displacement is prescribed in the z-direction:

$$\text{on } \partial\Omega_{\boldsymbol{\xi}^d} : \left\{ \begin{array}{c} z = 0 : \boldsymbol{\xi} \cdot (-\mathbf{e}_z) = \xi_z^d = 0 \\ z = \ell : \boldsymbol{\xi} \cdot (+\mathbf{e}_z) = \xi_z^d = 0 \end{array} \right\}$$

Furthermore, the displacements in the radial and circumferential direction, ξ_r and ξ_θ, are not restrained. Since the contact within the hypothesis of small perturbation

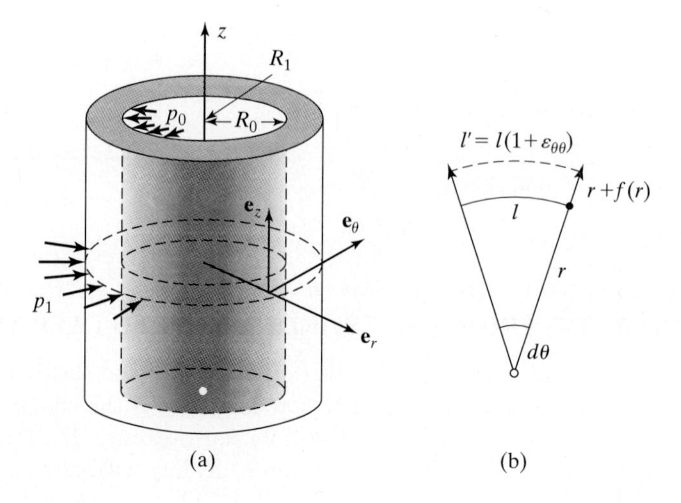

FIGURE 5.6: Training set: (a) cylinder tube subjected to internal and external pressure; (b) geometric significance of hoop strain $\varepsilon_{\theta\theta}$.

is frictionless, it follows from (5.75) that the tangential components of the stress vector built on $z = 0$ and $z = \ell$ are zero. This contact condition reads here

$$\text{Contact}: \left\{ \begin{array}{ll} z = 0: & \begin{array}{c} \mathbf{T}(\mathbf{n} = -\mathbf{e}_z) \cdot \mathbf{e}_r = -\sigma_{zr} = 0 \\ \mathbf{T}(\mathbf{n} = -\mathbf{e}_z) \cdot \mathbf{e}_\theta = -\sigma_{z\theta} = 0 \end{array} \\ z = \ell: & \begin{array}{c} \mathbf{T}(\mathbf{n} = \mathbf{e}_z) \cdot \mathbf{e}_r = \sigma_{zr} = 0 \\ \mathbf{T}(\mathbf{n} = \mathbf{e}_z) \cdot \mathbf{e}_\theta = \sigma_{z\theta} = 0 \end{array} \end{array} \right\}$$

The displacement field in cylinder coordinates reads

$$\boldsymbol{\xi} = \xi_r \mathbf{e}_r + \xi_\theta \mathbf{e}_\theta + \xi_z \mathbf{e}_z$$

From the displacement boundary conditions on $\partial\Omega_{\boldsymbol{\xi}^d}$, it follows that $\xi_z = 0$ in the cylinder. In addition, the rotational symmetry of the problem implies $\xi_\theta = 0$. Thus, the displacement field reduces to

$$\boldsymbol{\xi} = \xi_r \mathbf{e}_r$$

Last, with a similar reasoning, the radial displacement is a function of only r,

$$\xi_r = f(r)$$

■

The only unknown displacement of the problem is the 1D radial displacement function $f(r)$. This suggests the use of the displacement method for the elastic equilibrium analysis.

Exercise 47. Using the previous results, determine the stress field $\boldsymbol{\sigma}$ by application of the linear elastic material law in the cylinder tube.

The displacement method is summarized in (5.78). For the cylinder tube at hand, the method consists of the following steps:

1. The starting point of the displacement method is the radial displacement $\xi_r = f(r)$.
2. We need to determine the strain components. From the displacement components and the associated symmetry, it follows that

$$\begin{aligned} \varepsilon_{ij} &= 0 \text{ for } i \neq j \text{ (no distortion)} \\ \varepsilon_{zz} &= 0 \text{ since } \xi_z = 0 \end{aligned}$$

   Hence, we are left with the radial and hoop strain components, ε_{rr} and $\varepsilon_{\theta\theta}$. The radial strain is the linear dilatation in the r−direction, that is,

$$\varepsilon_{rr} = \frac{f(r + dr) - f(r)}{dr} = \frac{df}{dr} = f'$$

   In addition, the hoop strain $\varepsilon_{\theta\theta}$ represents the relative length increase of the circumferential arch-length (see Figure 5.6b), thus

$$\varepsilon_{\theta\theta} = \frac{\Delta l}{l} = \frac{[r + f(r)]\, d\theta - r d\theta}{r d\theta} = \frac{f(r)}{r}$$

This leads to the strain tensor

$$(\varepsilon_{ij}) = \begin{pmatrix} f' & 0 & 0 \\ & f/r & 0 \\ \text{sym} & & 0 \end{pmatrix}$$

The relative volume variation, $\operatorname{tr}\boldsymbol{\varepsilon}$, therefore reads

$$\operatorname{tr}\boldsymbol{\varepsilon} = f' + f/r$$

3. Use of the strain tensor $\boldsymbol{\varepsilon}$ in the isotropic elastic state equation (5.76) leads to ($\boldsymbol{\sigma}_0 = 0, \theta = 0$)

$$\boldsymbol{\sigma} = \lambda\,(\operatorname{tr}\boldsymbol{\varepsilon})\,\mathbf{1} + 2G\boldsymbol{\varepsilon} : \left\{ \begin{array}{l} \sigma_{rr} = (\lambda + 2G)f' + \lambda f/r \\ \sigma_{\theta\theta} = \lambda f' + (\lambda + 2G)f/r \\ \sigma_{zz} = \lambda(f' + f/r) \\ \text{other } \sigma_{ij} = 0 \end{array} \right\}$$

4. The equilibrium equations for the axisymmetrical problem were derived in Section 3.5.3. Since $f = f(r)$, the equilibrium equations in the $\mathbf{e}_\theta$-direction, and the $\mathbf{e}_z$-direction are satisfied:

$$\frac{\partial \sigma_{\theta\theta}}{\partial \theta} = 0; \quad \frac{\partial \sigma_{zz}}{\partial z} = 0$$

We are left with the radial balance equation:

$$\mathbf{e}_r : \frac{\partial \sigma_{rr}}{\partial r} + \frac{1}{r}\left(\sigma_{rr} - \sigma_{\theta\theta}\right) = 0$$

With

$$\frac{\partial \sigma_{rr}}{\partial r} = \frac{\partial}{\partial r}\left[(\lambda + 2G)f' + \lambda f/r\right] = (\lambda + 2G)f'' + \lambda f'/r - \lambda f/r^2$$

it follows that

$$\mathbf{e}_r : (\lambda + 2G)[f'' + f'/r - f/r^2] = 0$$

5. The solution of the problem consists in solving the differential equation[9]:

$$f'' + f'/r - f/r^2 = 0 \rightarrow f = Ar + \frac{B}{r}$$

[9] Use MATLAB, Maple, Mathematica, and so on. The manual solution can be achieved when rewriting the ODE in the form

$$f'' + \frac{f'}{r} - \frac{f}{r^2} = \frac{\partial}{\partial r}\left(f' + \frac{f}{r}\right) = \frac{\partial}{\partial r}\left[\frac{1}{r}\frac{\partial}{\partial r}(rf)\right] = 0$$

leading to the sought solution:

$$f = Ar + \frac{B}{r}$$

We should also note that the differential equation can be written in the form

$$\frac{\partial}{\partial r}\left(f' + \frac{f}{r}\right) = \frac{\partial}{\partial r}\left(\operatorname{tr}\boldsymbol{\varepsilon}\right)$$

More generally, in the absence of body forces, Navier's equation reduces to

$$\nabla^2\left(\operatorname{tr}\boldsymbol{\varepsilon}\right) = 0$$

A, B are integration constants. Substitution of the solution in the strain-displacement relations leads to:

$$\varepsilon_{rr} = f' = A - \frac{B}{r^2}; \quad \varepsilon_{\theta\theta} = \frac{f}{r} = A + \frac{B}{r^2}$$

and

$$\operatorname{tr} \boldsymbol{\varepsilon} = f' + f/r = 2A$$

The stress components read

$$\sigma_{rr} = \lambda \operatorname{tr} \boldsymbol{\varepsilon} + 2G\varepsilon_{rr} = 2A(\lambda + G) - 2G\frac{B}{r^2}$$

$$\sigma_{\theta\theta} = \lambda \operatorname{tr} \boldsymbol{\varepsilon} + 2G\varepsilon_{\theta\theta} = 2A(\lambda + G) + 2G\frac{B}{r^2}$$

$$\sigma_{zz} = \lambda \operatorname{tr} \boldsymbol{\varepsilon} = 2\lambda A$$

6. The integration constants A, B are determined from the boundary conditions:

$$r = R_0 : \quad \mathbf{T}^d = p_0 \mathbf{e}_r \quad = \boldsymbol{\sigma} \cdot (\mathbf{n} = -\mathbf{e}_r) = -\sigma_{rr}(r = R_0)\mathbf{e}_r = -\left[2A(\lambda + G) - 2G\frac{B}{R_0^2}\right] \mathbf{e}_r$$

$$r = R_1 : \quad \mathbf{T}^d = -p_1 \mathbf{e}_r \quad = \boldsymbol{\sigma} \cdot (\mathbf{n} = \mathbf{e}_r) = \sigma_{rr}(r = R_1)\mathbf{e}_r = \left[2A(\lambda + G) - 2G\frac{B}{R_1^2}\right] \mathbf{e}_r$$

Thus

$$\begin{bmatrix} -2(\lambda + G) & \frac{2\mu}{R_0^2} \\ 2(\lambda + G) & \frac{2\mu}{R_1^2} \end{bmatrix} \begin{pmatrix} A \\ B \end{pmatrix} = \begin{pmatrix} p_0 \\ -p_1 \end{pmatrix}$$

$$A = \frac{p_0 R_0^2 - p_1 R_1^2}{2(\lambda + G)(R_1^2 - R_0^2)}; \quad B = \frac{(p_0 - p_1) R_0^2 R_1^2}{2G(R_1^2 - R_0^2)}$$

7. The solution reads

$$\xi_r = p_0 \frac{R_0^2}{R_1^2 - R_0^2}\left(\frac{r}{2(\lambda + G)} + \frac{R_1^2}{2Gr}\right) - p_1 \frac{R_1^2}{R_1^2 - R_0^2}\left(\frac{r}{2(\lambda + G)} + \frac{R_0^2}{2Gr}\right)$$

$$\sigma_{rr} = p_0 \frac{R_0^2}{R_1^2 - R_0^2}\left(1 - \frac{R_1^2}{r^2}\right) - p_1 \frac{R_1^2}{R_1^2 - R_0^2}\left(1 - \frac{R_0^2}{r^2}\right)$$

$$\sigma_{\theta\theta} = p_0 \frac{R_0^2}{R_1^2 - R_0^2}\left(1 + \frac{R_1^2}{r^2}\right) - p_1 \frac{R_1^2}{R_1^2 - R_0^2}\left(1 + \frac{R_0^2}{r^2}\right)$$

$$\sigma_{zz} = \frac{\lambda}{\lambda + G}\left[\frac{p_0 R_0^2 - p_1 R_1^2}{R_1^2 - R_0^2}\right] = 2\nu \left[\frac{p_0 R_0^2 - p_1 R_1^2}{R_1^2 - R_0^2}\right]$$

where (5.50) has been used. ■

5.5.2 Case Study: The Pressure Vessel Revisited. Elastic Yield Limit

For $p_1 = 0$, the example corresponds to the pressure vessel studied in Chapter 3.

Exercise 48. For $p_1 = 0$ and a Tresca material (shear strength $\sigma_0/2$), determine the maximum pressure $\max p_0$ corresponding to the elastic limit stress state. Show where the strength criterion is reached in the cylinder, and determine the orientation of the critical surface on which the maximum shear for this elastic yield limit occurs.

We let $p_1 = 0$ in the general solution:

$$p_1 = 0 : (\sigma_{ij}) = \varpi \begin{pmatrix} 1 - \frac{R_1^2}{r^2} & & \\ & 1 + \frac{R_1^2}{r^2} & \\ & & 2\nu \end{pmatrix}; \quad \varpi = p_0 \frac{R_0^2}{R_1^2 - R_0^2}$$

The previous expression allows for an interesting representation of the elastic stress states in the Mohr plane.[10] The Mohr circles corresponding to the stresses in the $(\mathbf{e}_r \times \mathbf{e}_\theta)$-plane (i.e., cylinder wall, see Fig. 5.7a) vary only in radius $1 \leq \mathcal{R}/\varpi \leq (R_1/R_0)^2$ around a fixed center:

$$\sigma_c = \frac{\sigma_{\theta\theta} + \sigma_{rr}}{2} = \varpi; \quad \mathcal{R} = \frac{\sigma_{\theta\theta} - \sigma_{rr}}{2} = \varpi \frac{R_1^2}{r^2}$$

The Mohr circles shown in Figure 5.7b are normalized with regard to ϖ. The figure also shows the Tresca strength criterion:

$$f = \max(\sigma_{\theta\theta} - \sigma_{rr}) - \sigma_0 = 2\varpi \frac{R_1^2}{R_0^2} - \sigma_0 \leq 0$$

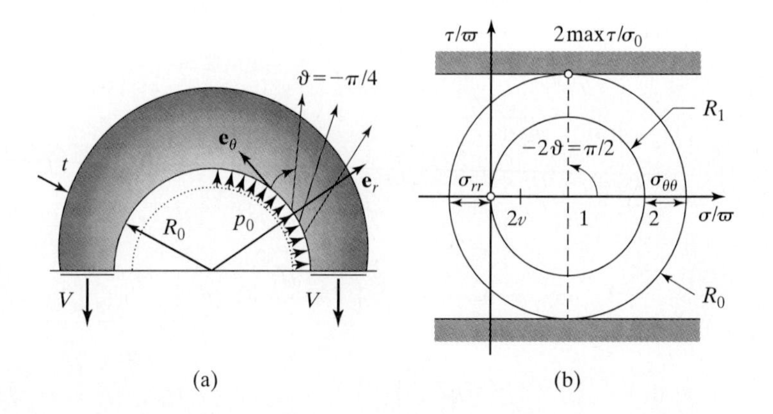

FIGURE 5.7: Pressure vessel revisited: (a) material system with elastic yield surface directions; (b) Mohr plane with Tresca strength criterion.

[10] We assume that $\sigma_I \geq \sigma_{II} = \sigma_{zz} = 2\nu\varpi \geq \sigma_{III}$.

Thus,

$$f = 0 \leftrightarrow \frac{\max p_0}{\sigma_0/2} = 1 - \frac{R_0^2}{R_1^2}$$

The elastic strength domain (yield locus) is reached on the inner cylinder wall. The orientation of the maximum shear in the cylinder $(\mathbf{e}_r \times \mathbf{e}_\theta)$-wall is shown in Figure 5.7a. ■

Last, with the solution in hand, we can revisit the pressure vessel formula derived in Chapter 3 from averaging the elastic hoop stress:

$$\begin{aligned} \langle \sigma_{\theta\theta} \rangle &= \frac{1}{R_1 - R_0} \int_{R_0}^{R_1} \sigma_{\theta\theta} dr = \varpi \int_{R_0}^{R_1} \left(1 + \frac{R_1^2}{r^2}\right) dr \\ &= \frac{p_0 R_0^2}{(R_1 - R_0)(R_1^2 - R_0^2)} \left[r - \frac{R_1^2}{r} \right]_{r=R_0}^{r=R_1} = \frac{p_0}{R_0(R_1 - R_0)} \end{aligned}$$

If we let $t = R_1 - R_0$, the vessel formula is obtained.[11]

5.5.3 Case Study: Deep Tunneling

Consider an initial natural stress state $\boldsymbol{\sigma} = \boldsymbol{\sigma}_0$ in a continuum. In this continuum, a tunnel of radius R_0 is driven, excavating in $r \leq R$ the (geological) prestressed material. This is the general stress picture of tunneling. In deep tunneling, the variation of the initial prestress due to body forces can be neglected. Assuming a hydrostatic initial prestress reads

$$\boldsymbol{\sigma}_0 = -p_0 \mathbf{1}$$

Exercise 49. Determine the stress state after tunneling in the continuum surrounding the tunnel. The material is assumed isotropic linear elastic. For purpose of analysis, divide the studied domain into two subdomains (Figure 5.8): the excavation zone 1, and the outer zone 2.

The problem consists of two load cases:

1. **Initial Prestress**: At the interface of zones 1 and 2, the stress continuity implies a radial outward-oriented pressure p_0 exerted from zone 1 on zone 2. The load case and the corresponding system response can be written in the form

$$\begin{pmatrix} \mathbf{T}^d = p_0 \mathbf{e}_r \\ \boldsymbol{\xi}^d = 0 \\ \rho_0 \mathbf{f} = 0 \end{pmatrix}_I \rightarrow \begin{pmatrix} \boldsymbol{\sigma}_0 = -p_0 \mathbf{1} \\ \boldsymbol{\varepsilon} = 0 \\ \boldsymbol{\xi} = 0 \end{pmatrix}_I$$

[11] For thin cylinders, the following approximation can be made:

$$R_1 = R_0(1 + \epsilon); \quad \epsilon = t/R_0 \ll 1$$

Substitution in the general expression of the hoop stress (derived for thick cylinders) gives

$$\epsilon = t/R_0 \ll 1 \rightarrow \sigma_{\theta\theta} \sim \frac{p_0}{\delta R_0^2}$$

Show this.

By definition, the natural prestressing related to deformation at some geological time scales is associated with a zero displacement. This is, however, a matter of choice of the initial reference configuration.

2. **Tunneling**: After excavation, the stress vector on the tunnel wall is zero, and the pressure p_0 is released by excavation. In addition, with excavation the displacement of the surrounding continuum is activated. The resulting (incremental) solicitation of excavation corresponds to imposing $\mathbf{T}^d = -p_0\mathbf{e}_r$ on the tunnel wall. This load case and the corresponding system response can be written in the form

$$\begin{pmatrix} \mathbf{T}^d = -p_0\mathbf{e}_r \\ \boldsymbol{\xi}^d = 0 \\ \rho_0\mathbf{f} = 0 \end{pmatrix}_{II} \rightarrow \begin{pmatrix} \boldsymbol{\sigma} \\ \boldsymbol{\varepsilon} \\ \boldsymbol{\xi} \end{pmatrix}_{II}$$

The solution for this load case can be obtained from the general linear and elastic solution of the cylinder. The tunnel in the continuum can be viewed as an infinite thick cylinder, $R_1 \rightarrow \infty$, with a pressure $p_1 = 0$. The stress solution for this load case reads[12]

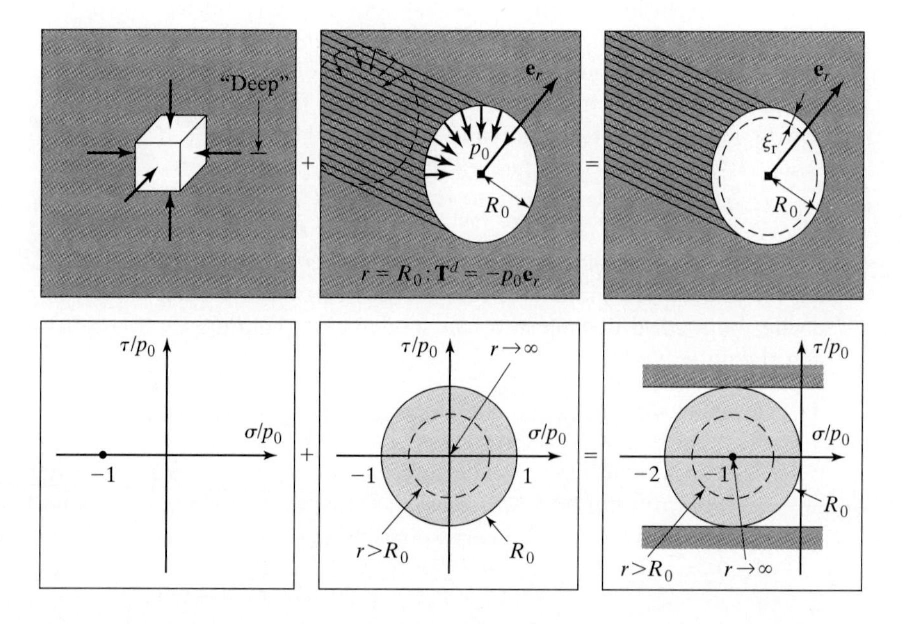

FIGURE 5.8: The theorem of superposition applied to deep tunneling in an elastic continuum: physical systems and corresponding Mohr planes.

[12]For instance, the asymptotic case $R \rightarrow \infty$ for σ_{rr} is developed as follows:

$$\lim_{R_1\rightarrow\infty} \sigma_{rr} = p_0\frac{R_0^2}{r^2}\lim_{R_1\rightarrow\infty}\left(\frac{r^2 - R_1^2}{R_1^2 - R_0^2}\right) = -p_0\frac{R_0^2}{r^2}$$

$$R_1 \to \infty : (\sigma_{ij})_{II} = p_0 \begin{pmatrix} \frac{R_0^2}{r^2} & & \\ & -\frac{R_0^2}{r^2} & \\ & & 0 \end{pmatrix}$$

and the displacement

$$\boldsymbol{\xi}_{II} = \xi_r \mathbf{e}_r = p_0 \frac{R_0^2}{2Gr} \mathbf{e}_r$$

Given the overall linearity of the problem (i.e., of deformation, boundary and continuity condition, material law, etc.), the theorem of superposition (5.77) can be applied to the two load cases. The stress solution for the deep tunneling problem then reads

$$\boldsymbol{\sigma} = \boldsymbol{\sigma}_I + \boldsymbol{\sigma}_{II} = -p_0 \mathbf{1} + p_0 \frac{R_0^2}{r^2} [\mathbf{e}_r \otimes \mathbf{e}_r - \mathbf{e}_\theta \otimes \mathbf{e}_\theta]$$

or, equivalently,

$$(\sigma_{ij}) = -p_0 \begin{pmatrix} 1 - \frac{R_0^2}{r^2} & & \\ & 1 + \frac{R_0^2}{r^2} & \\ & & 1 \end{pmatrix}$$

This theorem of superposition applied to the deep tunneling situation is sketched in Figure 5.8. In the Mohr plane the deep tunneling corresponds to an expansion of the hydrostatic stress point (the center of all Mohr circles) with Mohr circle radius $R = p_0 (R_0/r)^2$. The maximum shear stress occurs at the tunnel wall at $r = R_0$. For a Tresca material, the maximum admissible initial stress is equal to the shear stress:

$$f = 0 : p_0 = \frac{\sigma_0}{2}$$

The Mohr circles of this tunneling problem are also given in Figure 5.8. ■

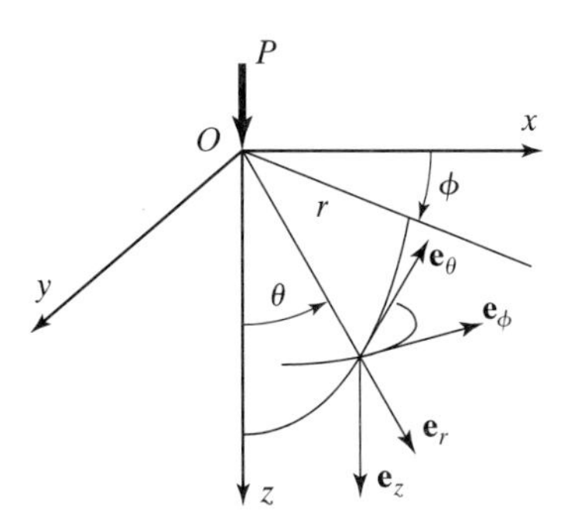

FIGURE 5.9: The Boussinesq problem: a half-space in spherical coordinates subjected to a concentrated force.

5.6 PROBLEM SET: THE BOUSSINESQ PROBLEM—REFINED ANALYSIS

We consider a semi-infinite half-space of a linear elastic material, characterized by the bulk modulus and shear modulus, K and G. At the origin of coordinates O, the half-space is subjected to a vertical force of intensity P directed along the z-axis. The gravity will not be considered. The aim of this exercise is to derive the displacement and the stress solution of this classical elasticity problem known as the Boussinesq problem. In spherical coordinates (see Figure 5.9) the displacement solution is sought for in the form

$$\boldsymbol{\xi} = u_r \mathbf{e}_r + v_\theta \mathbf{e}_\theta + w_\phi \mathbf{e}_\phi$$

1. **Dimensional Analysis**: By means of dimensional analysis, show that

$$u_r = \frac{F}{Gr}\mathcal{U}(\nu, \theta); \quad v_\theta = \frac{F}{Gr}\mathcal{V}(\nu, \theta); \quad w_\phi = 0$$

   where $\mathcal{U}$ and $\mathcal{V}$ are dimensionless functions that depend only on the Poisson ratio ν and angle θ.

2. **Displacement Method in Spherical Coordinates**: We want to determine the solution of the problem using the displacement method:

   (a) Derive the expressions of the strain and the stress components as a function of $\mathcal{U}(\nu, \theta)$ and $\mathcal{V}(\nu, \theta)$ and their derivatives $\mathcal{U}'$and $\mathcal{V}'$ with respect to θ.

   (b) Express the boundary conditions that the functions $\mathcal{U}$ and $\mathcal{V}$ need to satisfy. In particular, derive the condition that ensures the static equilibrium of any half-sphere of radius r centered at the origin.

   (c) For a stress field for which $\sigma_{i\phi} = \sigma_{\phi i} = 0$ for $i \neq \phi$, the equilibrium equations reduce to (see Section 2.6.3)

$$\frac{\partial \sigma_{rr}}{\partial r} + \frac{1}{r}\frac{\partial \sigma_{r\theta}}{\partial \theta} + \frac{1}{r}\left(2\sigma_{rr} - \sigma_{\theta\theta} - \sigma_{\phi\phi} + \sigma_{r\theta}\cot\theta\right) = 0$$

$$\frac{\partial \sigma_{\theta r}}{\partial r} + \frac{1}{r}\frac{\partial \sigma_{\theta\theta}}{\partial \theta} + \frac{1}{r}\left(\sigma_{\theta\theta} - \sigma_{\phi\phi}\right)\cot\theta + 3\sigma_{r\theta} = 0$$

   Derive, in spherical coordinates, the displacement field solution and the stress field solution.

3. **The Uniform Pressure Disk Problem**: We consider now, instead of a concentrated force, a vertical uniform pressure of intensity p applied over a disk of radius R centered at the origin. In cylindrical coordinates (ρ, α, z) derive an integral form for the vertical stress σ_{zz} as a function of ρ and z. Give a closed form expression of σ_{zz} for the points belonging to the Oz-axis $(\rho = 0)$. Finally, determine the vertical settlement w_0 at the origin of coordinates.

5.6.1 Dimensional Analysis

For an isotropic material, the problem is axisymmetric and therefore does not depend on ϕ. It follows $w_\phi = 0$. The displacement solution depends then on the applied load P, the material properties K and G, and the spherical coordinates r and θ:

$$u_r = f_r(P, K, G, r, \theta); \quad v_\theta = f_\theta(P, K, G, r, \theta)$$

The previous equation can be physically relevant only if *dimensionally* homogenous (that is, if u_r and f_r, and if v_θ and f_θ have the same physical dimension, here length L). The physical dimension of functions f_r and f_θ result from the physical dimensions of their arguments. Dimensional analysis explores how the combination of the arguments can result in an expression that has dimension of a length. The physical dimension of any physical quantity Q can be expressed through its dimension function $[Q]$, which is a power function of the fundamental dimensions. Using a LMT-base dimension system, where L stands for length, M for mass, and T for time, the dimension function reads

$$[Q] = L^\alpha M^\beta T^\gamma$$

where α, β and γ are exponents. The dimension functions of the different quantities involved in the problem read

$$[u_r] = [v_\theta] = [r] = L; \quad [P] = LMT^{-2}; \quad [K] = [G] = L^{-1}MT^{-2}; \quad [\theta] = 1$$

These dimension functions are conveniently summarized in the form of the exponent matrix of dimensions, in which the exponents α, β and γ form the columns. If we note that u_r and v_θ are independent variables, we can analyze the dimensional problem independently. The exponent matrix related to u_r reads

	$[u_r]$	$[P]$	$[G]$	$[K]$	$[r]$	$[\theta]$
L	1	1	−1	−1	1	0
M	0	1	1	1	0	0
T	0	−2	−2	−2	0	0

The number of dimensionally independent quantities is obtained by determining the rank of the exponent matrix of dimensions,[13] here two. If we choose G and r as the two dimensionally independent quantities, we construct from the four remaining dimensionally dependant variables the following four invariants:

$$\Pi = \frac{u_r}{r}; \quad \Pi_1 = \frac{P}{Gr^2}; \quad \Pi_2 = \frac{K}{G}; \quad \Pi_3 = \theta$$

In order for the physical problem to be dimensionally consistent, the solution is a relation that links these four dimensionless quantities:

$$\Pi = \frac{u_r}{r} = \mathrm{U}\left(\Pi_1 = \frac{P}{Gr^2}, \, \Pi_2 = \frac{K}{G}, \, \Pi_3 = \theta\right)$$

Finally, we add the following two mechanical considerations to the pure dimensional considerations: (i) The bulk-to-shear modulus ratio is a unique function of the Poisson ratio (i.e., $\frac{K}{G} = \frac{2}{3}\frac{1+\nu}{1-2\nu}$), and we can therefore replace Π_2 by the elasticity

[13]The rank of a matrix is the dimension of the biggest nonsingular matrix that can be extracted from it, and that corresponds to the maximum number of linearly independent columns.

invariant ν; (ii) the material is *linear* elastic, which means that the displacement u_r and the applied load P are linearly related; that is, in terms of the invariant relation,

$$\mathrm{U}(\Pi_1, \nu, \theta) = \Pi_1 \times \mathcal{U}(\nu, \theta)$$

With these considerations, which equally apply to v_θ, we arrive at the sought relations:

$$u_r = \frac{F}{Gr}\mathcal{U}(\nu, \theta) \quad v_\theta = \frac{F}{Gr}\mathcal{V}(\nu, \theta)$$

5.6.2 Displacement Method in Spherical Coordinates

We consider the displacement field,

$$\boldsymbol{\xi} = u_r(r, \theta)\,\mathbf{e}_r + v_\theta(r, \theta)\,\mathbf{e}_\theta$$

in which u_r and v_θ are given by the previously derived dimensionless forms.

a. Strain and Stress Components The strain components in spherical coordinates read (the grad operator in spherical coordinates can be found in Section 2.6.3)

$$\begin{aligned}
\varepsilon_{rr} &= \frac{\partial u_r}{\partial r} = -\frac{P}{Gr^2}\mathcal{U}; \\
\varepsilon_{\theta\theta} &= \frac{1}{r}\left(\frac{\partial v_\theta}{\partial \theta} + u_r\right) = \frac{P}{Gr^2}\left(\mathcal{U} + \mathcal{V}'\right); \\
\varepsilon_{\phi\phi} &= \frac{v_\theta}{r}\cot\theta + \frac{u_r}{r} = \frac{P}{Gr^2}\left(\mathcal{U} + \mathcal{V}\cot\theta\right); \\
\varepsilon_{r\theta} &= \varepsilon_{\theta r} = \frac{1}{2}\left[r\frac{\partial}{\partial r}\left(\frac{v_\theta}{r}\right) + \frac{1}{r}\frac{\partial u_r}{\partial \theta}\right] = \frac{P}{Gr^2}\left[\frac{1}{2}\mathcal{U}' - \mathcal{V}\right] \\
\text{other } \varepsilon_{ij} &= 0
\end{aligned}$$

where $\mathcal{U}' = d\mathcal{U}/d\theta$ and $\mathcal{V}' = d\mathcal{V}/d\theta$. It is instructive to note that the strain components are all linearly related to $\Pi_1 = P/Gr^2$. This is, of course, due to the linearity of the problem. We also note the relation

$$\varepsilon_{kk} = \varepsilon_{rr} + \varepsilon_{\theta\theta} + \varepsilon_{\phi\phi} = \frac{P}{Gr^2}\left(\mathcal{U} + \frac{(\mathcal{V}\sin\theta)'}{\sin\theta}\right)$$

The stress components are obtained by application of the linear constitutive equations of elasticity:

$$\sigma_{ij} = \left(K - \frac{2}{3}G\right)\varepsilon_{kk}\delta_{ij} + 2G\varepsilon_{ij}$$

If we let $K/G - 2/3 = 2\nu/(1-2\nu)$, the stress components read

$$
\begin{aligned}
\sigma_{rr} &= \frac{2P}{r^2}\left[\frac{\nu}{1-2\nu}\left(\mathcal{U}+\frac{(\mathcal{V}\sin\theta)'}{\sin\theta}\right)-\mathcal{U}\right]\\
\sigma_{\theta\theta} &= \frac{2P}{r^2}\left[\frac{\nu}{1-2\nu}\left(\mathcal{U}+\frac{(\mathcal{V}\sin\theta)'}{\sin\theta}\right)+\mathcal{U}+\mathcal{V}'\right]\\
\sigma_{\phi\phi} &= \frac{2P}{r^2}\left[\frac{\nu}{1-2\nu}\left(\mathcal{U}+\frac{(\mathcal{V}\sin\theta)'}{\sin\theta}\right)+\mathcal{U}+\mathcal{V}\cot\theta\right]\\
\sigma_{r\theta} &= \sigma_{\theta r} = \frac{2P}{r^2}\left[\frac{1}{2}\mathcal{U}'-\mathcal{V}\right] \quad \text{other } \sigma_{ij}=0
\end{aligned}
$$

b. Conditions of Kinematically Compatibility and of Statically Admissibility Except for the singularity at $r=0$ (which corresponds to the point of load application), the plane $\theta=\pi/2$ is stress free. Therefore, $\sigma_{\theta\theta}\left(\theta=\frac{\pi}{2}\right)=\sigma_{r\theta}\left(\theta=\frac{\pi}{2}\right)=0$, that is,

$$
\mathcal{U}+\mathcal{V}'\,|_{\theta=\frac{\pi}{2}}=0;\quad \frac{1}{2}\mathcal{U}'-\mathcal{V}\,|_{\theta=\frac{\pi}{2}}=0
$$

In addition, static equilibrium of any half sphere of radius r centered at the origin requires

$$
2\pi r^2\int_0^{\pi/2}(\sigma_{rr}\cos\theta-\sigma_{r\theta}\sin\theta)\sin\theta\,d\theta=-P
$$

resulting in

$$
\int_0^{\pi/2}\left[\frac{\nu}{1-2\nu}\left(\mathcal{U}+\frac{(\mathcal{V}\sin\theta)'}{\sin\theta}\right)-\mathcal{U}\right]\cos\theta\sin\theta\,d\theta-\int_0^{\pi/2}\left[\frac{1}{2}\mathcal{U}'-\mathcal{V}\right]\sin^2\theta\,d\theta=-\frac{1}{4\pi}
$$

c. Displacement and Stress Solution A substitution of the previously obtained stress expressions into the equilibrium equations yields, after some transformation,

$$
4(1-\nu)\mathcal{U}\sin\theta-(1-2\nu)(\mathcal{U}'\sin\theta)'+4(1-\nu)(\mathcal{V}\sin\theta)'=0
$$

$$
(3-4\nu)\mathcal{U}'+2(1-\nu)\left(\frac{(\mathcal{V}\sin\theta)'}{\sin\theta}\right)'=0
$$

Using the first of the preceding differential equations, the boundary condition $\mathcal{U}+\mathcal{V}'\,|_{\theta=\frac{\pi}{2}}=0$ reduces to

$$
\mathcal{U}''\,|_{\theta=\frac{\pi}{2}}=0
$$

An integration of the second differential equation leads to

$$
(\mathcal{V}\sin\theta)'=\frac{\sin\theta}{2(1-\nu)}\left[C_1-(3-4\mathcal{V})\mathcal{U}\right]
$$

where C_1 is an integration constant. Substitution of this result in the former equation provides a differential equation for $\mathcal{U}$:

$$
2\mathcal{U}\sin\theta+(\mathcal{U}'\sin\theta)'=\frac{2C_1}{1-2\nu}\sin\theta
$$

The solution of this differential equation, which meets the boundary condition $\mathcal{U}'' \,|_{\theta=\frac{\pi}{2}}= 0$, is[14]

$$\mathcal{U} = \frac{C_1}{1-2\nu} + C_2 \cos\theta$$

where C_2 is a second integration constant. If we integrate the second of the differential equations, resulting from the equilibrium equations, and use the solution for $\mathcal{U}$, we obtain:

$$\mathcal{V}\sin\theta = \frac{C_1}{1-2\nu}(\cos\theta - 1) + \frac{C_2\,(3-4\nu)}{8\,(1-\nu)}(\cos 2\theta - 1)$$

The integration constant has been chosen here so that the solution remains valid for $\theta \to 0$. The integration constants C_1 and C_2 can now be determined using the remaining boundary conditions. In particular, the boundary condition

$$\frac{1}{2}\mathcal{U}' - \mathcal{V} \,|_{\theta=\frac{\pi}{2}}= 0$$

yields

$$C_2 = -\frac{4\,(1-\nu)}{(1-2\nu)^2}\,C_1$$

Using this result and expressing the remaining integral condition of the static equilibrium of any half-sphere centered at the origin, some further calculation yields

$$C_1 = -\frac{(1-2\nu)^2}{4\pi}$$

Finally, collecting the preceding results, the displacement solution reads

$$\begin{aligned} u_r &= \frac{P}{4\pi G r}\left[4\,(1-\nu)\cos\theta - (1-2\nu)\right] \\ v_\theta &= \frac{P}{4\pi G r}\left[\frac{1-2\nu}{1+\cos\theta} - (3-4\nu)\right]\sin\theta \end{aligned}$$

Last, if we note

$$\mathcal{U} + \frac{(\mathcal{V}\sin\theta)'}{\sin\theta} = -\frac{1}{2\pi}(1-2\nu)\cos\theta$$

the stress solution is obtained in the form

$$\begin{aligned} \sigma_{rr} &= \frac{P}{2\pi r^2}\left[(1-2\nu) - 2\,(2-\nu)\cos\theta\right] \\ \sigma_{\theta\theta} &= \frac{P}{2\pi r^2}(1-2\nu)\frac{\cos^2\theta}{1+\cos\theta} \\ \sigma_{\phi\phi} &= \frac{P}{2\pi r^2}(1-2\nu)\left[\cos\theta - \frac{1}{1+\cos\theta}\right] \\ \sigma_{r\theta} &= \sigma_{\theta r} = \frac{P}{2\pi r^2}(1-2\nu)\frac{\sin\theta\cos\theta}{1+\cos\theta} \end{aligned}$$

[14]The general solution of the differential equation includes a third term of the form $C_3\,(1-\cos\theta)\,\mathrm{arg\,tanh}\,(\cos\theta)$, where C_3 is another integration constant. This term cancels out when considering the specific boundary condition of the problem.

5.6.3 The Uniform Pressure Disk Problem

The solution consists, in a first step, of a change from spherical coordinates to cylindrical coordinates. If we note

$$\mathbf{e}_z = \cos\theta\,\mathbf{e}_r - \sin\theta\,\mathbf{e}_\theta$$

the vertical stress is obtained in the form

$$\begin{aligned}\sigma_{zz} &= \begin{pmatrix} \cos\theta & -\sin\theta & 0 \end{pmatrix} \begin{pmatrix} \sigma_{rr} & \sigma_{\theta r} & \sigma_{\phi r} \\ \sigma_{r\theta} & \sigma_{\theta\theta} & \sigma_{\phi\theta} \\ \sigma_{r\phi} & \sigma_{\theta\phi} & \sigma_{\phi\phi} \end{pmatrix} \begin{pmatrix} \cos\theta \\ -\sin\theta \\ 0 \end{pmatrix} \\ &= \cos^2\theta\,\sigma_{rr} - 2\sin\theta\cos\theta\,\sigma_{r\theta} + \sin^2\theta\,\sigma_{\theta\theta}\end{aligned}$$

Furthermore, if we note that $z = r\cos\theta$, and making use of the previous sections, the expression for the vertical stress σ_{zz} induced by a vertical infinitesimal force dP located at the origin is

$$\sigma_{zz} = -\frac{3dP}{2\pi}\frac{z^3}{r^5}$$

Due to the linearity of the employed elasticity model, we can use the superposition theorem in order to find the vertical stress due to a pressure p that is uniformly applied over a disk lying on the surface $z = 0$. Indeed, we can sum up the individual effects of infinitesimal forces $dP = pdA$ that apply over the infinitesimal area $dS = \mathrm{r}\,d\alpha\,d\mathrm{r}$, $0 < \mathrm{r} < R$, $0 < \alpha < 2\pi$ forming the disk.

Let r be the distance between a point of polar angle α, located at the surface $z = 0$ at a distance r from the origin, and a point of polar radius ρ and zero polar angle, located at depth z. Simple geometric considerations show that

$$r = \left(z^2 + \rho^2 + \mathrm{r}^2 - 2\rho\mathrm{r}\cos\alpha\right)^{1/2}$$

From the superposition theorem and the preceding expression of σ_{zz} for the point load problem, we derive

$$\sigma_{zz} = -\frac{3p}{\pi}z^3 \int_0^{\pi}\int_0^{R} \frac{\mathrm{r}\,d\alpha\,d\mathrm{r}}{\left(z^2 + \rho^2 + \mathrm{r}^2 - 2\rho\mathrm{r}\cos\alpha\right)^{5/2}}$$

Due to the symmetry of the problem, this result must hold irrespective of the polar angle of the considered point. For $\rho = 0$, the preceding expression therefore reduces to

$$\sigma_{zz}(\rho = 0, z) = -3p\int_0^{R/z} \frac{\bar{\mathrm{r}}\,d\bar{\mathrm{r}}}{\left(1 + \bar{\mathrm{r}}^2\right)^{5/2}}$$

where $\bar{\mathrm{r}} = \mathrm{r}/z$, and, after integration,

$$\sigma_{zz}(\rho = 0, z) = -p\left[1 - \frac{1}{\left[1 + (R/z)^2\right]^{3/2}}\right]$$

Alternatively, a surface point corresponds to $z = 0$ in cylinder coordinates and to $\theta = \pi/2$ in spherical coordinates. Hence, the surface vertical displacement

$w\left(\rho, z=0\right)$ induced by an infinitesimal vertical force dP located at the origin is equal to $-v_\theta\left(r=\rho, \theta=\frac{\pi}{2}\right)$:

$$w\left(\rho=r, z=0\right)=\left(\frac{1-\nu}{2\pi}\right)\frac{dP}{Gr}$$

Employing again the superposition theorem, in a similar way as for the derivation of σ_{zz}, the vertical displacement due to a uniform vertical pressure p applying over a disk of radius R is obtained in the form:

$$w\left(\rho, z=0\right)=\left(\frac{1-\nu}{2\pi}\right)\frac{2p}{G}\int_0^\pi\int_0^R\frac{\mathrm{r}\,d\alpha\,d\mathrm{r}}{\left(\rho^2+\mathrm{r}^2-2\rho\mathrm{r}\cos\alpha\right)^{1/2}}$$

The settlement w_0 at the origin of coordinates corresponds to the value of w at $\rho=0$. Introducing the Young modulus E with $G=\frac{E}{2(1+\nu)}$, it reads

$$w_0=2pR\,\frac{1-\nu^2}{E}$$

5.7 PROBLEM SET: ELASTIC BENDING OF A BEAM

We consider a beam of length L along the x-axis and section A (see Figure 5.10). At $x=0$, the section (centered around O) is in contact without friction with a fixed rigid plate. At the other end, at $x=L$, the section (centered around O') fixed on a frictionless rigid plate is subjected to a rotation of intensity $\vartheta \ll 1$ around the z-axis. The surface of the beam is stress free. The beam is composed of a homogeneous isotropic linear elastic material defined by Young's modulus E and Poisson ratio ν.

1. **Direct Solving Method**: We consider a longitudinal stress field of the form

$$\boldsymbol{\sigma}=\sigma(y)\mathbf{e}_x\otimes\mathbf{e}_x$$

   (a) Develop in detail all conditions that $\boldsymbol{\sigma}$ needs to satisfy to be statically admissible.

   (b) Determine the analytical solution of the elastic beam bending problem (displacement solution $\boldsymbol{\xi}$, stress solution $\boldsymbol{\sigma}$), and show that the curvature is

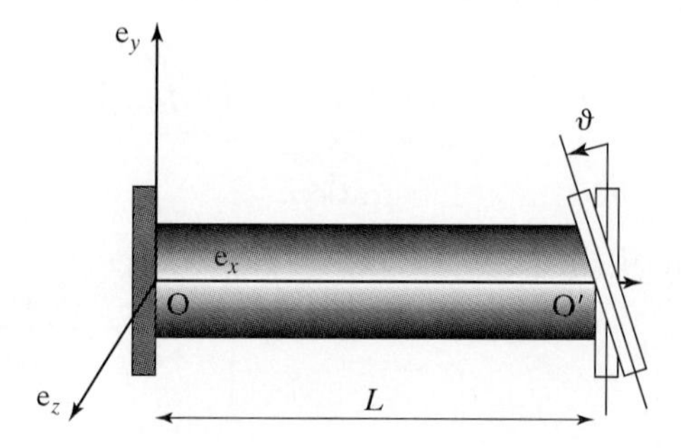

FIGURE 5.10: Problem set: elastic bending of a beam.

related to the bending moment M_z by a constant

$$\vartheta/L = M_z/C$$

Determine C.

2. **Variational Method**[15]: We consider a statically admissible stress field of the form

$$\boldsymbol{\sigma}'(y) = [\sigma(y) + m\delta\sigma]\mathbf{e}_x \otimes \mathbf{e}_x$$

where $\sigma(y)$ is the unknown solution of the elastic bending problem. By using the stress based variational method, determine $\sigma(y)$. (Note that this part can be solved independent of the first part of this exercise.)

5.7.1 Direct Solving Method: Stress Approach

Statically Admissible Stress Field

Equilibrium Condition and Symmetry of Stress Tensor:

$$\operatorname{div}\boldsymbol{\sigma} = 0 \rightarrow \frac{\partial\sigma(y)}{\partial x} = 0; \quad \boldsymbol{\sigma} = {}^t\boldsymbol{\sigma}$$

Force Boundary Conditions on $\partial\Omega_{\mathbf{T}^d}$:

1. Stress-free surfaces (oriented by $\mathbf{n} = \mathbf{e}_y$ and $\mathbf{n} = \mathbf{e}_z$)

$$\mathbf{T}^d = \mathbf{0} \equiv \mathbf{T}(\mathbf{n}) = \boldsymbol{\sigma} \cdot \mathbf{n} \rightarrow \sigma_{yy} = \sigma_{zz} = 0$$

2. Frictionless contact at $x = 0$ and $x = L$:

$$T_y^d = 0 \equiv \mathbf{T}(\mathbf{n} = \mathbf{e}_x) \cdot \mathbf{e}_y = \sigma_{yx} = 0$$

$$T_z^d = 0 \equiv \mathbf{T}(\mathbf{n} = \mathbf{e}_x) \cdot \mathbf{e}_z = \sigma_{zx} = 0$$

Note that the missing boundary condition in the x-direction at $x = 0$ and $y = 0$ is a displacement boundary condition (i.e., $\partial\Omega_{\boldsymbol{\xi}^d}$).

Elastic State Equation

$$\boldsymbol{\varepsilon} = \frac{1+\nu}{E}\boldsymbol{\sigma} - \frac{\nu}{E}(\operatorname{tr}\boldsymbol{\sigma})\,\mathbf{1} = \frac{\sigma(y)}{E}\begin{bmatrix} 1 & 0 & 0 \\ & -\nu & 0 \\ \text{sym} & & -\nu \end{bmatrix}$$

Displacement Field (by Integration)

$$\varepsilon_{xx} = \frac{\partial\xi_x}{\partial x} = \frac{\sigma(y)}{E} \rightarrow \xi_x = \frac{\sigma(y)}{E}x + C_x(y, z)$$

[15]The exercise provides a link between the exact elastic solution methods and the variational methods developed in Chapter 6.

$$\varepsilon_{yy} = \frac{\partial \xi_y}{\partial y} = -\nu \frac{\sigma(y)}{E} \rightarrow \xi_y = -\frac{\nu}{E} \int \sigma(y)\, dy + C_y(x,z)$$

$$\varepsilon_{yy} = \frac{\partial \xi_z}{\partial z} = -\nu \frac{\sigma(y)}{E} \rightarrow \xi_z = -\frac{\nu \sigma(y)}{E} z + C_z(x,y)$$

where $C_x = C_x(y,z)$, $C_y = C_y(x,z)$ and $C_z = C_z(x,y)$ are integration functions, which depend on the other two coordinates that were not involved in the integration.

Displacement Boundary Conditions $\partial\Omega_{\boldsymbol{\xi}^d}$

$$\text{on } x = 0 : \xi_x^d = 0 \equiv \xi_x(x=0) = C_x(y,z)$$

$$\text{on } x = L : \xi_x^d(y) = -\vartheta y \equiv \xi_x(x=L) = \frac{\sigma(y)}{E} L$$

It follows that

$$\sigma(y) = -E\frac{\vartheta y}{L}$$

Further Deformation Considerations

Displacement symmetry condition:

$$z = 0 : \xi_z = 0 = C_z(x,y)$$

Zero shear strain condition $\varepsilon_{xy} = 0$:

$$2\varepsilon_{xy} = \frac{\partial \xi_x}{\partial y} + \frac{\partial \xi_y}{\partial x} = 0$$

Thus,

$$\frac{\partial \xi_x}{\partial y} = \frac{\partial}{\partial y}\left(\frac{\sigma(y)}{E} x\right) = -\frac{\vartheta x}{L} = -\frac{\partial \xi_y}{\partial x} = -\frac{\partial C_y(x,z)}{\partial x}$$

$$\rightarrow C_y(x.z) = \frac{\vartheta x^2}{2L} + c_y(z)$$

Zero shear strain condition $\varepsilon_{yz} = 0$:

$$2\varepsilon_{yz} = \frac{\partial \xi_y}{\partial z} + \frac{\partial \xi_z}{\partial y} = 0$$

Thus, recalling $C_z(x,y) = 0$,

$$\frac{\partial \xi_y}{\partial z} = \frac{\partial C_y(x,z)}{\partial z} = -\frac{\partial \xi_z}{\partial y} = -\frac{\nu \vartheta z}{L}$$

$$\rightarrow C_y(x,z) = -\frac{\nu \vartheta z^2}{2L} + c_y(x)$$

From a comparison of the two expressions obtained for $C_y(x,z)$, it follows that

$$C_y(x,z) = \frac{\vartheta}{2L}\left(x^2 - \nu z^2\right)$$

Solution $(\boldsymbol{\sigma}, \boldsymbol{\xi})$

$$\boldsymbol{\sigma} = -E\frac{\vartheta y}{L}\mathbf{e}_x \otimes \mathbf{e}_x$$

$$\xi_x = -\vartheta y \frac{x}{L}; \quad \xi_y = \frac{\vartheta}{2L}\left(x^2 + \nu(y^2 - z^2)\right); \quad \xi_z = \nu\frac{\vartheta}{L}yz$$

Reduction Formulas

$$\begin{aligned}\mathcal{M} &= \int_A \mathbf{OM} \times \mathbf{T}(\mathbf{n} = \mathbf{e}_x)\, da = \int_A \begin{pmatrix} 0 \\ y \\ z \end{pmatrix} \times \begin{pmatrix} \sigma(y) \\ 0 \\ 0 \end{pmatrix} da \\ &= \int_A \begin{pmatrix} 0 \\ z\sigma(y) \\ -y\sigma(y) \end{pmatrix} da = \underbrace{\int_A z\sigma(y)\, da}_{M_y} \mathbf{e}_y + \underbrace{-\int_A y\sigma(y)\, da}_{M_z} \mathbf{e}_z\end{aligned}$$

Recalling that $\int_A z\sigma(y)dz\, dy = 0$, it follows that

$$\mathcal{M} = M_z\mathbf{e}_z = \frac{\vartheta}{L}E\int_A y^2 dy\, dz\, \mathbf{e}_z = \frac{\vartheta}{L}EI_z\mathbf{e}_z$$

where I is the inertia moment around the z-axis. It follows the section constitutive law

$$\frac{\vartheta}{L} = \frac{M_z}{EI_z}$$

5.7.2 Stress-Based Variational Method

With the same arguments employed in the first part of the exercise, the stress field $\boldsymbol{\sigma}'$ is statically admissible. We therefore can employ it in the stress-based variational method:

$$\mathcal{E}_{\text{com}}(\boldsymbol{\sigma} = \sigma(y)\mathbf{e}_x \otimes \mathbf{e}_x) = \min_{\boldsymbol{\sigma}' \text{ SA}} \mathcal{E}_{\text{com}}(\boldsymbol{\sigma}' = (\sigma(y) + m\delta\sigma)\mathbf{e}_x \otimes \mathbf{e}_x)$$

The theorem of minimum complementary energy states here that $\sigma(y)$ is a minimum among all statically admissible stress solutions $\sigma' = \sigma(y) + m\delta\sigma$, which is obtained for $m \to 0$, irrespective of the value of $\delta\sigma$. Therefore,

$$\forall\delta\sigma : \frac{\partial}{\partial m}\left[\mathcal{E}_{\text{com}}(\boldsymbol{\sigma}' = (\sigma(y) + m\delta\sigma)\mathbf{e}_x \otimes \mathbf{e}_x)\right]_{m=0} = 0$$

We determine $\mathcal{E}_{\text{com}}(\boldsymbol{\sigma}') = W^*(\boldsymbol{\sigma}') - \Phi^*(\boldsymbol{\sigma}')$:

1. Internal complementary energy:

$$\begin{aligned}W^*(\boldsymbol{\sigma}') &= \int_\Omega \frac{1}{2E}\left[-\nu(\operatorname{tr}\boldsymbol{\sigma}')^2 + (1+\nu)\operatorname{tr}(\boldsymbol{\sigma}' \cdot \boldsymbol{\sigma}')\right] d\Omega \\ &= \int_\Omega \frac{1}{2E}(\sigma(y) + m\delta\sigma)^2 d\Omega = \int_A \frac{L}{2E}(\sigma(y)^2 + 2\sigma(y)m\delta\sigma + m^2\delta\sigma)\, da\end{aligned}$$

where we made use that $\boldsymbol{\sigma}' = \boldsymbol{\sigma}'(y)$.

2. External work by prescribed displacements on $\partial\Omega_{\boldsymbol{\xi}^d}$:

$$\Phi^*(\boldsymbol{\sigma}') = \int_{\partial\Omega_{\boldsymbol{\xi}^d}} \mathbf{T}' \cdot \boldsymbol{\xi}^d da = -\int_{A,x=L} (\sigma(y) + m\delta\sigma)\vartheta y \, da$$

Thus,

$$\mathcal{E}_{\text{com}}(\boldsymbol{\sigma}') = \int_A \left[\frac{L}{2E}(\sigma(y)^2 + 2\sigma(y)m\delta\sigma + m^2\delta\sigma) + (\sigma(y) + m\delta\sigma)\vartheta y \right] da$$

Stationary of complementary energy delivers

$$\frac{\partial \mathcal{E}_{\text{com}}(m)}{\partial m} = 0 : \quad = \int_A \left[\frac{L}{2E}(2\sigma(y)\delta\sigma + 2m\delta\sigma) + \delta\sigma\vartheta y \right] da$$
$$= \int_A \delta\sigma \left[\frac{L}{E}(\sigma(y) + m) + \vartheta y \right] da = 0$$

Application of the theorem of minimum complementary energy finally gives

$$\forall \delta\sigma : \frac{\partial \mathcal{E}_{\text{com}}(m)}{\partial m}|_{m=0} = 0 : \sigma(y) = -E\frac{\vartheta y}{L}$$

CHAPTER 6

The Theorem of Virtual Work and Variational Methods in Elasticity

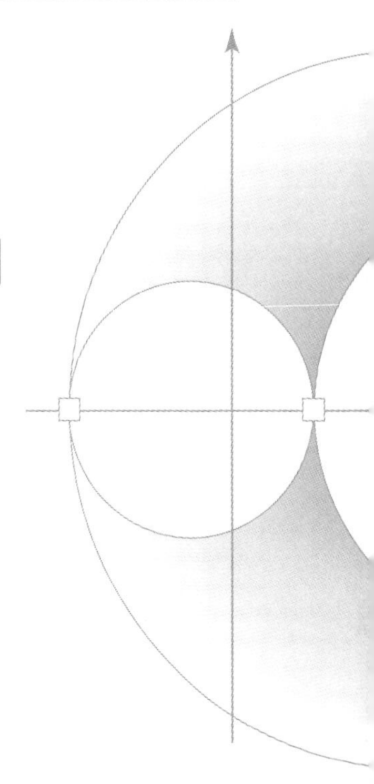

This second chapter on elasticity deals with variational methods. The starting point is the general theorem of virtual work, the weak (or integral) form of the local equation of motion. For a given elastic material system, it is shown that the application of this theorem with (1) a kinematically admissible displacement field, and (2) a statically admissible stress field leads, respectively, to an upper and a lower bound of the exact elastic solution. This exact solution is the one for which simultaneously the stress field is statically admissible, and the displacement field kinematically admissible. In turn, upper and lower approximations can be derived based on energy minimum theorems: the theorem of minimum potential energy, and the theorem of minimum complementary energy. By way of application (training set), the energy theorems will be applied to the determination of the effective material properties of heterogeneous material systems.

6.1 THEOREM OF VIRTUAL WORK

For clarity, we will adopt the hypothesis of small perturbation, as outlined in Section 5.4.1.

6.1.1 The Theorem

In the absence of inertia forces, the work done by the external forces is equal to the internal work:

$$\mathcal{W}_{\text{ext}} = \mathcal{W}_{\text{int}} \tag{6.1}$$

The external work is caused by volume and surface forces, which do work along a displacement $\boldsymbol{\xi}$. With the notations of Chapter 3, the external work reads

$$\mathcal{W}_{\text{ext}} = \int_{\Omega} \boldsymbol{\xi} \cdot (\rho \mathbf{f})\, d\Omega + \int_{\partial\Omega} \boldsymbol{\xi} \cdot \mathbf{T} da \tag{6.2}$$

The first term expresses the external work by volume forces $\rho \mathbf{f} d\Omega$, the second the external work by surface tractions $\mathbf{T}$ at the boundary $\partial\Omega$. In the case of a rigid body movement or an undeformable medium, the internal work is zero, and consequently $\mathcal{W}_{\text{ext}} = 0$. Moreover, with definition (3.20) of the stress tensor $\mathbf{T}(\mathbf{n}) = \boldsymbol{\sigma} \cdot \mathbf{n}$, the second term on the right-hand side can be developed in the form

$$\begin{aligned} \int_{\partial\Omega} \boldsymbol{\xi} \cdot \mathbf{T} da &= \int_{\partial\Omega} (\boldsymbol{\xi} \cdot \boldsymbol{\sigma}) \cdot \mathbf{n} da = \int_{\Omega} \operatorname{div} (\boldsymbol{\xi} \cdot \boldsymbol{\sigma})\, d\Omega \\ &= \int_{\Omega} (\operatorname{grad} \boldsymbol{\xi} : \boldsymbol{\sigma} + \boldsymbol{\xi} \cdot \operatorname{div} \boldsymbol{\sigma})\, d\Omega \\ &= \int_{\Omega} (\boldsymbol{\varepsilon} : \boldsymbol{\sigma} + \boldsymbol{\xi} \cdot (-\rho \mathbf{f}))\, d\Omega \end{aligned} \tag{6.3}$$

where the divergence theorem, the symmetry of both the strain and the stress tensor,[1] and the quasi-static momentum balance equation ($\operatorname{div} \boldsymbol{\sigma} = -\rho \mathbf{f}$) have been used. Rewriting (6.3) in the form of (6.1) leads to the theorem of virtual work:

THEOREM 3 of virtual work: The work supplied from the outside in the form of body forces and surface forces is equal to the internal strain energy:

$$\underbrace{\int_{\Omega} \boldsymbol{\xi} \cdot (\rho \mathbf{f}) d\Omega + \int_{\partial\Omega} \boldsymbol{\xi} \cdot \mathbf{T} da}_{\mathcal{W}_{\text{ext}}} = \underbrace{\int_{\Omega} \boldsymbol{\sigma} : \boldsymbol{\varepsilon} d\Omega}_{\mathcal{W}_{\text{int}}} \tag{6.4}$$

This theorem is independent of the material law linking the stress $\boldsymbol{\sigma}$ to the strain $\boldsymbol{\varepsilon}$. It is just the dual form (or weak form) of the local momentum balance equation.

Exercise 50. Show that the theorem of virtual work can be derived starting from the local momentum balance.

We introduce separately (i) the stress tensor $\boldsymbol{\sigma}$ satisfying the local momentum balance and the boundary condition

$$\forall \boldsymbol{\sigma} : \left\{ \begin{array}{l} \text{in } \Omega : \operatorname{div} \boldsymbol{\sigma} + \rho \mathbf{f} = 0 \\ \text{on } \partial\Omega : \mathbf{T}(\mathbf{n}) = \boldsymbol{\sigma} \cdot \mathbf{n} \end{array} \right\}$$

[1] $\operatorname{grad} \boldsymbol{\xi} : \boldsymbol{\sigma} = (\boldsymbol{\varepsilon} + \boldsymbol{\omega}) : \boldsymbol{\sigma} = \boldsymbol{\varepsilon} : \boldsymbol{\sigma}$.

and (*ii*) the displacement field $\boldsymbol{\xi}'$ defining the linearized strain tensor

$$\varepsilon' = \frac{1}{2}(\operatorname{grad}\boldsymbol{\xi}' + {}^t\operatorname{grad}\boldsymbol{\xi}')$$

Multiplying the local momentum balance with $\boldsymbol{\xi}'$ and integrating over Ω gives

$$\begin{aligned} 0 &= \int_\Omega \left[\boldsymbol{\xi}' \cdot (\operatorname{div}\boldsymbol{\sigma} + \rho\mathbf{f})\right] d\Omega \\ &= \int_\Omega \left[-\operatorname{grad}\boldsymbol{\xi}' \cdot \boldsymbol{\sigma}\right] d\Omega + \int_{\partial\Omega} \left[\boldsymbol{\xi}' \cdot (\boldsymbol{\sigma} \cdot \mathbf{n})\right] da + \int_\Omega \left[\boldsymbol{\xi}' \cdot \rho\mathbf{f}\right] d\Omega \\ &= -\int_\Omega \boldsymbol{\sigma} : \varepsilon' \, d\Omega + \int_{\partial\Omega} \left[\boldsymbol{\xi}' \cdot \mathbf{T}\right] da + \int_\Omega \left[\boldsymbol{\xi}' \cdot \rho\mathbf{f}\right] d\Omega \end{aligned}$$

We note again that $\boldsymbol{\sigma}$ and ε' are not related by any material law. We keep this in mind for the forthcoming developments. ■

6.1.2 Application to Heterogeneous Material Systems: The Hill Lemma

One advantage of using an integral form of the momentum balance in the form of the theorem of virtual work is that it relates the external work supplied from the outside to a material system or structure to the sum of the internal strain energy of the different components in the structure. The externally supplied work, therefore, gives access to the overall internal energy changes that are intimately related to the material behavior. Conversely, the theorem provides a means to upscale the micromechanical behavior of heterogeneous material systems to the macroscopic

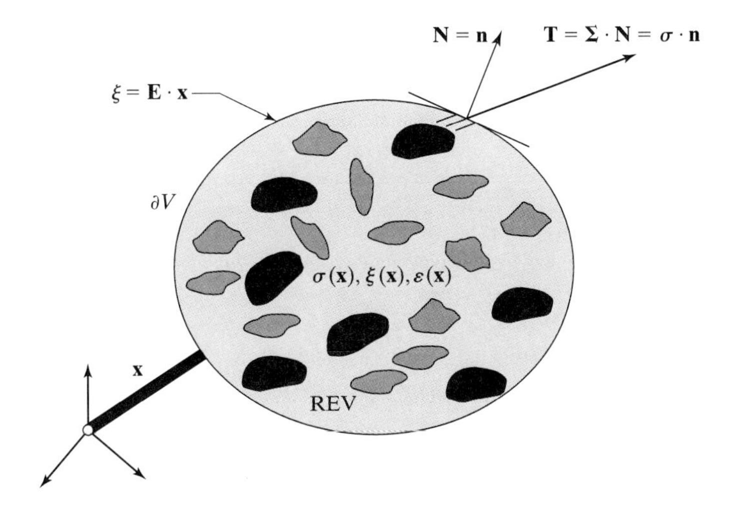

FIGURE 6.1: Heterogeneous material system: representative equivalent homogeneous elementary volume with regular stress or displacement boundary conditions.

representative elementary volume (REV)[2] (see Figure 6.1), where the material appears as homogeneous in a statistical sense. This is the focus of homogenization theories. The premise of homogenization theories is that the macroscopic stress and strain quantities are volume averages of local stress and strain quantities:

$$\mathbf{\Sigma} = \langle \boldsymbol{\sigma} \rangle = \frac{1}{V}\int_V \boldsymbol{\sigma}(\mathbf{x})\, dV; \quad \mathbf{E} = \langle \boldsymbol{\varepsilon} \rangle = \frac{1}{V}\int_V \boldsymbol{\varepsilon}(\mathbf{x})\, dV \tag{6.5}$$

where $\langle q \rangle$ denotes the volume average of the spatially varying quantity $q(\mathbf{x})$: $\boldsymbol{\sigma}(\mathbf{x})$ and $\boldsymbol{\varepsilon}(\mathbf{x})$ are spatially varying microscopic stress and strain quantities, and $\mathbf{\Sigma} = \langle \boldsymbol{\sigma} \rangle$ and $\mathbf{E} = \langle \boldsymbol{\varepsilon} \rangle$ are the equivalent homogeneous stress and strain tensors defined on the macroscopic level of the REV $d\Omega = V$.

What we are interested in is the macroscopic internal work in the REV as a function of $\mathbf{\Sigma}$ and $\mathbf{E}$ that results from the microscopic internal work in the heterogeneous material system; that is,

$$\mathcal{W}_{\text{int}}(\mathbf{\Sigma}, \mathbf{E}) = \int_V \boldsymbol{\sigma}(\mathbf{x}) : \boldsymbol{\varepsilon}(\mathbf{x})\, dV \tag{6.6}$$

To find this expression, we use the theorem of virtual work rate (6.4).

We first have a look at the stress side of the problem. If we note that stress $\mathbf{\Sigma}$ is, by its very definition, constant in V, any stress vector built on the boundary ∂V of the REV is $\mathbf{T}(\mathbf{N}) = \mathbf{\Sigma} \cdot \mathbf{N}$, where $\mathbf{N}$ is the unit outward normal to ∂V. This force boundary condition is a first link between the macroscopic stress and the microscopic stress, if we prescribe on the heterogeneous material system enclosed in V the macroscopic stress vector as stress boundary condition $\mathbf{t}^d$ (see Figure 6.1):

$$\text{on } \partial V : \mathbf{t}^d = \mathbf{\Sigma} \cdot \mathbf{N} \equiv \boldsymbol{\sigma} \cdot \mathbf{n} \tag{6.7}$$

We assume here that the macroscopic stress vector $\mathbf{T}(\mathbf{N}) = \mathbf{\Sigma} \cdot \mathbf{N}$ and the microscopic stress vector $\mathbf{t}(\mathbf{n}) = \boldsymbol{\sigma}(\mathbf{x}) \cdot \mathbf{n}$ on ∂V act on the same nominal surface $dA = da$ oriented by the normal $\mathbf{N} = \mathbf{n}$. Since $\mathbf{\Sigma}$ is constant on the boundary ∂V, we therefore assume that $\boldsymbol{\sigma} \cdot \mathbf{n}$ is also constant on ∂V. Furthermore, the microscopic stress $\boldsymbol{\sigma}(\mathbf{x})$ satisfies all features to be statically admissible; that is, the boundary condition (6.7), the continuity of the stress vector at any material plane in the heterogeneous material volume, $\mathbf{t}(\mathbf{n}) + \mathbf{t}(-\mathbf{n}) = 0$, the symmetry of the stress tensor, $\boldsymbol{\sigma} = {}^t\boldsymbol{\sigma}$, and the static equilibrium equations in V, which are $\operatorname{div} \boldsymbol{\sigma} = 0$ when disregarding volume forces. In this case, taking into account the symmetry of $\mathbf{\Sigma} = {}^t\mathbf{\Sigma}$, the external work can be developed in the form:

$$\begin{aligned} \mathcal{W}_{\text{ext}} &= \int_{\partial V} (\boldsymbol{\sigma} \cdot \mathbf{n}) \cdot \boldsymbol{\xi} da = \int_{\partial V} (\mathbf{\Sigma} \cdot \mathbf{N}) \cdot \boldsymbol{\xi} da = \mathbf{\Sigma} : \int_{\partial V} (\boldsymbol{\xi} \otimes \mathbf{N})\, da \\ &= \mathbf{\Sigma} : \int_V \boldsymbol{\varepsilon} dV = \mathbf{\Sigma} : \langle \boldsymbol{\varepsilon} \rangle V \end{aligned} \tag{6.8}$$

where we made use of the divergence theorem.[3] Then use of (6.8) in the theorem of virtual work (6.4) delivers the sought relation (6.6):

[2]Have a quick look back at Section 1.1.

[3]In this development, we use

$$(\mathbf{N} \cdot \mathbf{\Sigma}) \cdot \boldsymbol{\xi} = (\boldsymbol{\xi} \otimes \mathbf{N}) : \mathbf{\Sigma}$$

$$\mathcal{W}_{\text{ext}} = \boldsymbol{\Sigma} : \langle \boldsymbol{\varepsilon} \rangle V \equiv \int_V \boldsymbol{\sigma}(\mathbf{x}) : \boldsymbol{\varepsilon}(\mathbf{x})\, dV = \mathcal{W}_{\text{int}} \tag{6.9}$$

Note again that the stress $\boldsymbol{\sigma}$ and the strain $\boldsymbol{\varepsilon} = \frac{1}{2}(\text{grad}\,\boldsymbol{\xi} + {}^t\text{grad}\,\boldsymbol{\xi})$ are not related by any material law. The only requirement is that the stress is statically admissible, satisfying the regular stress boundary condition (6.7).

Expression (6.9) was derived under the assumption of a prescribed regular stress field everywhere on ∂V, which provides a link between the macroscopic and microscopic stress quantities. Analogously, we seek a second link involving a kinematically admissible displacement field that links the two scales. The macroscopic strain $\mathbf{E}$ is by its very definition constant in V. This means that the displacement on the boundary ∂V is linearly related to $\mathbf{E}$. This provides a second link, now between the macroscopic strain and the microscopic displacement, which we consider as prescribed on the boundary ∂V (see Figure 6.1):

$$\text{on } \partial V : \boldsymbol{\xi}^d = \mathbf{E} \cdot \mathbf{x} \tag{6.10}$$

In this case, the external work reads[4]

$$\begin{aligned} \mathcal{W}_{\text{ext}} &= \int_{\partial V} \boldsymbol{\xi}^d \cdot (\boldsymbol{\sigma} \cdot \mathbf{n}) \cdot da = \mathbf{E} : \int_{\partial V} (\mathbf{x} \otimes \boldsymbol{\sigma}) \cdot \mathbf{n} da \\ &= \mathbf{E} : \int_V \text{div}\,(\mathbf{x} \otimes \boldsymbol{\sigma})\, dV = \mathbf{E} : \int_V \boldsymbol{\sigma} dV = \mathbf{E} : \langle \boldsymbol{\sigma} \rangle V \end{aligned} \tag{6.11}$$

Finally, use of (6.11) in the theorem of virtual work (6.4) gives a second relation for (6.6):

$$\mathcal{W}_{\text{ext}} = \mathbf{E} : \langle \boldsymbol{\sigma} \rangle V \equiv \int_V \boldsymbol{\sigma}(\mathbf{x}) : \boldsymbol{\varepsilon}(\mathbf{x})\, dV = \mathcal{W}_{\text{int}} \tag{6.12}$$

In summary, provided regular stress boundary conditions (6.7) or regular displacement boundary conditions (6.10), the volume average of the microscopic internal work is strictly equivalent to the product of the volume averages of stress and strain:

$$\langle \boldsymbol{\sigma} : \boldsymbol{\varepsilon} \rangle = \langle \boldsymbol{\sigma} \rangle : \langle \boldsymbol{\varepsilon} \rangle \tag{6.13}$$

---

which is similar to relation (1.50), $(\mathbf{a} \otimes \mathbf{b}) \cdot \mathbf{c} = (\mathbf{b} \cdot \mathbf{c})\mathbf{a}$. Furthermore, we consider the following development:

$$\begin{aligned} \int_{\partial V} (\boldsymbol{\xi} \otimes \mathbf{N})\, da &= \int_{\partial V} (\boldsymbol{\xi} \otimes \mathbf{1}) \cdot \mathbf{N} da \\ &= \int_V \text{div}\,(\boldsymbol{\xi} \otimes \mathbf{1})\, dV = \int_V \text{tr}\,(\text{grad}\,(\boldsymbol{\xi} \otimes \mathbf{1}))\, dV \\ &= \int_V \left[(\text{grad}\,(\boldsymbol{\xi}) \otimes \mathbf{1}) \cdot \mathbf{1} \right] dV = \int_V \text{grad}\,(\boldsymbol{\xi})\, dV \end{aligned}$$

[4] Note that,

$$\text{div}\,(\mathbf{x} \otimes \boldsymbol{\sigma}) = \text{tr}\,(\text{grad}\,(\mathbf{x} \otimes \boldsymbol{\sigma})) = \text{tr}\,(\boldsymbol{\sigma} \otimes \text{grad}\,(\mathbf{x}) + \text{grad}\,(\boldsymbol{\sigma}) \otimes \mathbf{x}) = \text{tr}\,(\boldsymbol{\sigma} \otimes \mathbf{1}) + \text{div}\,(\boldsymbol{\sigma}) \otimes \mathbf{x}$$

Then, if we note that $\text{div}\,(\boldsymbol{\sigma}) = 0$ (equilibrium in the absence of body forces), and that $\text{tr}\,(\boldsymbol{\sigma} \otimes \mathbf{1}) = (\boldsymbol{\sigma} \otimes \mathbf{1}) \cdot \mathbf{1} = \boldsymbol{\sigma}$, it follows that

$$\text{div}\,(\boldsymbol{\sigma} \otimes \mathbf{x}) = \boldsymbol{\sigma}$$

which we used in (6.11).

This remarkable result (relating the volume integral of a product to the product of two volume integrals) is known as the Hill lemma. We keep this lemma in mind in forthcoming exercises dealing with heterogeneous material systems.

6.1.3 Potential Energy and Complementary Energy

Employing the split of the boundary $\partial\Omega = \partial\Omega_{\boldsymbol{\xi}^d} \cup \partial\Omega_{\mathbf{T}^d}$, we can rewrite the left-hand side of (6.4) in the form

$$\mathcal{W}_{\text{ext}} = \int_\Omega \boldsymbol{\xi} \cdot (\rho \mathbf{f})\, d\Omega + \int_{\partial\Omega_{\mathbf{T}^d}} \boldsymbol{\xi} \cdot \mathbf{T}^d da + \int_{\partial\Omega_{\boldsymbol{\xi}^d}} \boldsymbol{\xi}^d \cdot \mathbf{T} da \tag{6.14}$$

where $\mathbf{T}^d$ are prescribed surface forces on $\partial\Omega_{\mathbf{T}^d}$, and $\boldsymbol{\xi}^d$ are prescribed displacements on $\partial\Omega_{\boldsymbol{\xi}^d}$. The previous expression suggests the following separation of the external work:

1. The external work due to prescribed body and surface forces:

$$\Phi(\boldsymbol{\xi}) = \int_\Omega \boldsymbol{\xi} \cdot (\rho \mathbf{f})\, d\Omega + \int_{\partial\Omega_{\mathbf{T}^d}} \boldsymbol{\xi} \cdot \mathbf{T}^d da \tag{6.15}$$

2. The external work due to prescribed displacements:

$$\Phi^*(\boldsymbol{\sigma}) = \int_{\partial\Omega_{\boldsymbol{\xi}^d}} \boldsymbol{\xi}^d \cdot \mathbf{T} da = \Phi^*(\boldsymbol{\sigma}) = \int_{\partial\Omega_{\boldsymbol{\xi}^d}} \boldsymbol{\xi}^d \cdot (\boldsymbol{\sigma} \cdot \mathbf{n})\, da \tag{6.16}$$

With this separation, which will turn out quite useful in further developments, the theorem of virtual work (6.4) is rewritten in the form

$$\underbrace{\Phi(\boldsymbol{\xi}) + \Phi^*(\boldsymbol{\sigma})}_{\mathcal{W}_{\text{ext}}} = \underbrace{\int_\Omega \boldsymbol{\sigma} : \boldsymbol{\varepsilon} d\Omega}_{\mathcal{W}_{\text{int}}} \tag{6.17}$$

In a similar fashion, but restricted to elastic behavior, we can separate the internal work $\mathcal{W}_{\text{int}}$ on the right-hand side of (6.17) in an energy potential associated with strain $\boldsymbol{\varepsilon}$, and a second part associated with stress $\boldsymbol{\sigma}$. To this end, let

$$\boldsymbol{\sigma} : \boldsymbol{\varepsilon} = \psi(\boldsymbol{\varepsilon}) + \psi^*(\boldsymbol{\sigma}) \tag{6.18}$$

where $\psi(\boldsymbol{\varepsilon})$ is Helmholtz free energy volume density, introduced in Section 5.2.3, and $\psi^*(\boldsymbol{\sigma})$ is referred to as complementary elastic energy. Integrating (6.18) yields

$$\mathcal{W}_{\text{int}} = \int_\Omega \boldsymbol{\sigma} : \boldsymbol{\varepsilon} d\Omega = W(\boldsymbol{\varepsilon}) + W^*(\boldsymbol{\sigma}) \tag{6.19}$$

Here, $W(\boldsymbol{\varepsilon}) = \int_\Omega \psi(\boldsymbol{\varepsilon}) d\Omega$ is the overall free energy of the system. In turn, $\psi^*(\boldsymbol{\sigma})$ and $W^*(\boldsymbol{\sigma}) = \int_\Omega \psi^*(\boldsymbol{\sigma}) d\Omega$ are the complementary energy potentials complementing the internal work, respectively, on a material level [energy volume density $\psi^*(\boldsymbol{\sigma})$] and on a structural level. With (6.19), separating works associated with displacements and stresses, we can rewrite the theorem of virtual work in the form

$$\underbrace{W(\boldsymbol{\varepsilon}) - \Phi(\boldsymbol{\xi})}_{\mathcal{E}_{\text{pot}}(\boldsymbol{\xi})} + \underbrace{W^*(\boldsymbol{\sigma}) - \Phi^*(\boldsymbol{\sigma})}_{\mathcal{E}_{\text{com}}(\boldsymbol{\sigma})} = 0 \tag{6.20}$$

$\mathcal{E}_{\text{pot}}(\boldsymbol{\xi})$ is the elastic potential energy of the material system, and $\mathcal{E}_{\text{com}}$ the complementary elastic energy. Their physical significance will be revealed in forthcoming developments of the application of the theorem of virtual work in variational methods in elasticity.

6.1.4 Convexity

Convexity is an important mathematical property, in particular for variational problems based on energy minimization. We briefly recall the general definition. For a scalar function $f(x)$, convexity means that the tangent to the curve $f(x)$ is always smaller than the secant connecting two points x and x', as illustrated in Figure 6.2. This convexity of a differentiable scalar function in $[x, x']$ reads

$$\frac{\partial f}{\partial x}|_x (x' - x) \leq f(x') - f(x) \tag{6.21}$$

In addition, for a twice differentiable function, the convexity condition reads

$$\frac{\partial^2 f}{\partial x^2} \geq 0 \tag{6.22}$$

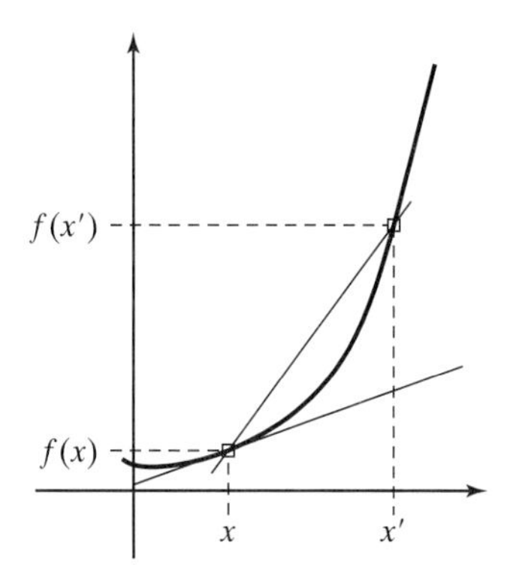

FIGURE 6.2: Convexity of a scalar function.

Exercise 51. Explore the convexity conditions for the 1D elasticity potential $\psi(\varepsilon)$ introduced in Chapter 5 (have a look back at Section 5.2.1).

For the elastic energy $\psi(\varepsilon) = \frac{1}{2}E\varepsilon^2$ of the 1D elastic think model defined by Eq. (5.8), the convexity condition (6.21) reads

$$\begin{aligned} \frac{\partial \psi}{\partial \varepsilon}(\varepsilon' - \varepsilon) &\leq \psi(\varepsilon') - \psi(\varepsilon) \\ \sigma(\varepsilon' - \varepsilon) &\leq \psi(\varepsilon') - \psi(\varepsilon) \end{aligned}$$

where the stress equation of state (5.7) has been used. Equation (6.22), reads here

$$E = \frac{\partial^2 \psi}{\partial \varepsilon^2} \geq 0$$

This relation is the stability requirement of the 1D isothermal elastic material behavior. ■

6.2 VARIATIONAL METHOD I: THEOREM OF MINIMUM POTENTIAL ENERGY

6.2.1 1D Think Model

Consider the one-parameter structural system shown in Figure 6.3a, on which a force F^d is prescribed. The material behavior in this structure is linear elastic. The applied load leads to a displacement x at the point of load application, and a strain ε in the material system. The potential energy of this elementary elastic system reads

$$\mathcal{E}_{\text{pot}} = W(\varepsilon) - \Phi(x) \tag{6.23}$$

where $W(\varepsilon = x/L)$ is the internal elastic energy stored in the 1D system of section A and length L. The counterpart of $W(\varepsilon = x/L)$ on the material level is the free energy $\psi(\varepsilon)$ (see Section 5.2.1), which corresponds to the area under the stress-strain curve shown in Figure 6.3b. In turn, $\Phi(x)$ is the potential energy of the external force F^d applied to the system. For the one-parameter system in hand, $W(\varepsilon)$ and $\Phi(x)$ read

$$W(\varepsilon) = \psi(\varepsilon)AL = \frac{1}{2}E\varepsilon^2 AL; \quad \Phi(x) = F^d x \tag{6.24}$$

We should note that $\varepsilon = x/L$ can be seen as the condition for x to be kinematically admissible.

Let us now assume that the prescribed force F^d leads to an additional infinitesimal displacement dx, provoking an additional external work increment,

$$d\mathcal{W}_{\text{ext}} = F^d dx \tag{6.25}$$

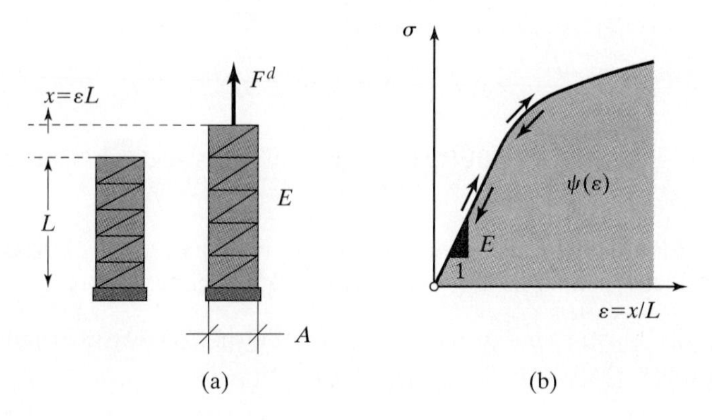

FIGURE 6.3: 1D elastic material system: (a) one-parameter think model; (b) elasticity potential at the material level.

Due to the assumed elasticity of the material, this external work increment is entirely stored in the material system as elastic energy dW, which can be recovered later. We have seen in Chapter 5 that this nondissipative behavior is expressed by the combination of the first and second laws, which reads at a structural level and under isothermal conditions:

$$d\mathcal{D} = \int_{\Omega=AL} \varphi dt d\Omega = d\mathcal{W}_{\text{ext}} - dW \equiv 0 \tag{6.26}$$

where $d\mathcal{D}$ denotes the overall dissipation in Ω, which for the given reversible behavior is zero. Let us now substitute (6.23) with (6.24) in (6.26), that is, $d\mathcal{E}_{\text{pot}} = dW - \left(F^d dx + dF^d x\right)$:

$$d\mathcal{D} = -d\mathcal{E}_{\text{pot}} + dF^d x \equiv 0 \tag{6.27}$$

The previous equation shows that, at constant prescribed force ($dF^d = 0$), the change in potential energy is zero. Furthermore, since the principal unknown of the potential energy is the displacement x, it follows that

$$d\mathcal{D} = -\frac{\partial \mathcal{E}_{\text{pot}}(x)}{\partial x} dx \equiv 0 \to \forall dx; \quad \frac{\partial \mathcal{E}_{\text{pot}}(x)}{\partial x} = 0 \tag{6.28}$$

Thus, with (6.24),

$$d\mathcal{D} = -\frac{\partial}{\partial x}\left(\frac{1}{2}E\left(\frac{x}{L}\right)^2 AL - F^d x\right) dx \equiv 0 \to \forall dx; \quad \frac{x}{L} = \frac{F^d}{EA} \tag{6.29}$$

The elastic solution $x/L = F^d/EA$ corresponds to the stationary of the potential energy. Clearly, the very existence of this stationary condition is due to the absence of dissipative mechanisms, $d\mathcal{D} \equiv 0$ (i.e., in the absence of damping, yielding, or other irreversible phenomena). But we also note that $\mathcal{E}_{\text{pot}}$ in (6.29) is a convex function of its argument x. This convexity means that the displacement solution x realizes actually a minimum of the potential energy $\mathcal{E}_{\text{pot}}$. It is then intuitively understood that the potential energy $\mathcal{E}_{\text{pot}}(x')$ associated with any other displacement x' is greater than or equal to the minimum potential energy $\mathcal{E}_{\text{pot}}(x)$:

$$\mathcal{E}_{\text{pot}}(x) = W(\varepsilon) - \Phi(x) \leq W(\varepsilon') - \Phi(x') = \mathcal{E}_{\text{pot}}(x') \tag{6.30}$$

Equation (6.30) forms the basis of displacement based variational formulation using the theorem of minimum potential energy.

Exercise 52. We consider two displacements x and x'; the first corresponds to the solution of the structural problem, the second not. By applying the theorem of virtual work, derive the energy minimum (6.30).

For the two displacements, Eq. (6.1) reads here

$$\begin{aligned} F^d x &= \sigma\varepsilon AL \\ F^d x' &= \sigma\varepsilon' AL \end{aligned}$$

where the left-hand side represents external work, and the right-hand side the internal work. A combination of the previous equations yields

$$F^d(x'-x)=\sigma(\varepsilon'-\varepsilon)AL$$

or, equivalently, using the stress equation of state,

$$F^d(x'-x)=\frac{\partial\psi}{\partial\varepsilon}(\varepsilon'-\varepsilon)AL$$

Finally, a substitution of the convexity condition of ψ (see Section 6.1.4) yields

$$F^d(x'-x)=\frac{\partial\psi}{\partial\varepsilon}(\varepsilon'-\varepsilon)AL\leq[\psi(\varepsilon')-\psi(\varepsilon)]AL$$

which is (6.30). This formulation is not restricted to *linear* elastic material behavior. In fact, only the theorem of virtual work, the stress state equation, and the convexity condition of the elastic potential were applied. ■

6.2.2 3D Displacement-Based Variational Formulation

The previous study of the one-parameter structural system highlights the key ingredients of displacement-based variational formulations:

1. Among all kinematically admissible displacement solutions, the displacement solution minimizes the potential energy of the system. The potential energy is a convex function of its argument, the kinematically admissible displacement; and the potential energy is the difference between the overall internal free energy of the system, and the external energy supplied in the form of work by prescribed forces to the system.
2. Any other displacement solution leads to a greater potential energy. From an energy point of view, such a solution is an *upper (energy) bound.*

The focus of this section is the extension of the 1D formulation to 3D elastic problems. To this end, we apply the theorem of virtual work (6.17) to the solution field $\boldsymbol{\xi}$, and another *kinematically admissible* displacement field, say $\boldsymbol{\xi}'$:

$$\underbrace{\int_\Omega \boldsymbol{\xi}\cdot(\rho\mathbf{f})\,d\Omega+\int_{\partial\Omega_{\mathbf{T}^d}}\boldsymbol{\xi}\cdot\mathbf{T}^d da}_{\Phi(\boldsymbol{\xi})}+\underbrace{\int_{\partial\Omega_{\boldsymbol{\xi}^d}}\boldsymbol{\xi}^d\cdot\mathbf{T}da}_{\Phi^*(\boldsymbol{\sigma})}=\int_\Omega\boldsymbol{\sigma}:\boldsymbol{\varepsilon}d\Omega \tag{6.31}$$

$$\underbrace{\int_\Omega \boldsymbol{\xi}'\cdot(\rho\mathbf{f})\,d\Omega+\int_{\partial\Omega_{\mathbf{T}^d}}\boldsymbol{\xi}'\cdot\mathbf{T}^d da}_{\Phi(\boldsymbol{\xi}')}+\underbrace{\int_{\partial\Omega_{\boldsymbol{\xi}^d}}\boldsymbol{\xi}^d\cdot\mathbf{T}da}_{\Phi^*(\boldsymbol{\sigma})}=\int_\Omega\boldsymbol{\sigma}:\boldsymbol{\varepsilon}'d\Omega \tag{6.32}$$

Proceeding as in the 1D case (see Exercise 52), subtracting (6.31) from (6.32) yields

$$\underbrace{\int_\Omega (\boldsymbol{\xi}'-\boldsymbol{\xi})\cdot(\rho\mathbf{f})\,d\Omega+\int_{\partial\Omega_{\mathbf{T}^d}}(\boldsymbol{\xi}'-\boldsymbol{\xi})\cdot\mathbf{T}^d da}_{\Phi(\boldsymbol{\xi}')-\Phi(\boldsymbol{\xi})}=\int_\Omega\boldsymbol{\sigma}:(\boldsymbol{\varepsilon}'-\boldsymbol{\varepsilon})\,d\Omega \tag{6.33}$$

or, equivalently, using the stress equation of state, $\boldsymbol{\sigma} = \partial\psi/\partial\boldsymbol{\varepsilon}$,

$$\Phi(\boldsymbol{\xi}') - \Phi(\boldsymbol{\xi}) = \int_\Omega \frac{\partial\psi}{\partial\boldsymbol{\varepsilon}} : (\boldsymbol{\varepsilon}' - \boldsymbol{\varepsilon})\, d\Omega \tag{6.34}$$

We note that only the external work associated with prescribed volume forces and surface traction on $\partial\Omega_{\mathbf{T}^d}$ enters the displacement-based variational method. Indeed, since we restrict ourselves to kinematically admissible displacement fields, the external work term $\Phi^*(\boldsymbol{\sigma})$ associated with prescribed displacements on $\partial\Omega_{\boldsymbol{\xi}^d}$ is the same in (6.31) and (6.32) and skips out when subtracting the latter from the former.[5] In addition, the convexity condition (6.21) for the 3D elasticity potential reads

$$\frac{\partial\psi}{\partial\boldsymbol{\varepsilon}} : (\boldsymbol{\varepsilon}' - \boldsymbol{\varepsilon}) \le \psi(\boldsymbol{\varepsilon}') - \psi(\boldsymbol{\varepsilon}) \tag{6.35}$$

Finally, use of (6.35) in (6.34) yields the displacement based variational inequality:

$$\underbrace{\int_\Omega (\boldsymbol{\xi}' - \boldsymbol{\xi}) \cdot (\rho\mathbf{f}) d\Omega + \int_{\partial\Omega_{\mathbf{T}^d}} (\boldsymbol{\xi}' - \boldsymbol{\xi}) \cdot \mathbf{T}^d da}_{\Phi(\boldsymbol{\xi}') - \Phi(\boldsymbol{\xi})} \le \underbrace{\int_\Omega [\psi(\boldsymbol{\varepsilon}') - \psi(\boldsymbol{\varepsilon})]\, d\Omega}_{W(\boldsymbol{\varepsilon}') - W(\boldsymbol{\varepsilon})} \tag{6.36}$$

or, equivalently, in form of (6.30),

$$\underbrace{W(\boldsymbol{\varepsilon}) - \Phi(\boldsymbol{\xi})}_{\mathcal{E}_{\text{pot}}(\boldsymbol{\xi})} \le \underbrace{W(\boldsymbol{\varepsilon}') - \Phi(\boldsymbol{\xi}')}_{\mathcal{E}_{\text{pot}}(\boldsymbol{\xi}')} \tag{6.37}$$

Equation (6.37) is the theorem of minimum potential energy:

> THEOREM 4 of minimum potential energy: The potential energy $\mathcal{E}_{\text{pot}}(\boldsymbol{\xi})$ of the elastic solution $\boldsymbol{\xi}$ realizes an energy minimum with regard to the potential energy $\mathcal{E}_{\text{pot}}(\boldsymbol{\xi}')$ of any approximation of the solution by a kinematically admissible displacement field $\boldsymbol{\xi}' = \boldsymbol{\xi}^d$ on $\partial\Omega_{\boldsymbol{\xi}^d}$:
>
> $$\mathcal{E}_{\text{pot}}(\boldsymbol{\xi}) = \min_{\boldsymbol{\xi}' = \boldsymbol{\xi}^d \text{ on } \partial\Omega_{\boldsymbol{\xi}^d}} [W(\boldsymbol{\varepsilon}') - \Phi(\boldsymbol{\xi}')] \tag{6.38}$$
>
> with
>
> $$\mathcal{E}_{\text{pot}}(\boldsymbol{\xi}') = \int_\Omega \psi(\boldsymbol{\varepsilon}')\, d\Omega - \left[\int_\Omega \boldsymbol{\xi}' \cdot (\rho\mathbf{f})\, d\Omega + \int_{\partial\Omega_{\mathbf{T}^d}} \boldsymbol{\xi}' \cdot \mathbf{T}^d da\right] \tag{6.39}$$

[5] Indeed, for the same stress $\boldsymbol{\sigma}$, and $\boldsymbol{\xi}'^d = \boldsymbol{\xi}^d$ (as prescribed from the outside), it is

$$\int_{\partial\Omega_{\boldsymbol{\xi}^d}} (\boldsymbol{\xi}'^d - \boldsymbol{\xi}^d) \cdot \mathbf{T} da = 0$$

This theorem holds for any elastic behavior, whether linear or nonlinear. The stress $\boldsymbol{\sigma}' = f(\boldsymbol{\varepsilon}')$ that can be derived by means of an elastic constitutive relation from the strain $\boldsymbol{\varepsilon}' = \boldsymbol{\varepsilon}'(\boldsymbol{\xi}')$ is not required to be statically admissible.[6]

The theorem of minimum potential energy sets out a formidable optimization problem: To seek kinematically admissible displacement fields which come closest to the minimum potential energy (i.e., the solution of the elasticity problem).

In the derivation of the displacement-based variational formulation (6.37) we assumed the same stress $\boldsymbol{\sigma}$ in the work expressions (6.31) and (6.32), associated with the $(\boldsymbol{\xi}, \boldsymbol{\varepsilon})$- and $(\boldsymbol{\xi}, \boldsymbol{\varepsilon})'$-solution. Let us now consider that stress $\boldsymbol{\sigma}$ in (6.33) is related by a linear elastic constitutive law to $\boldsymbol{\varepsilon}$ and $\boldsymbol{\varepsilon}'$, respectively,

$$\int_\Omega (\boldsymbol{\xi}' - \boldsymbol{\xi}) \cdot (\rho \mathbf{f})\, d\Omega + \int_{\partial\Omega_{\mathbf{T}^d}} (\boldsymbol{\xi}' - \boldsymbol{\xi}) \cdot \mathbf{T}^d da = \int_\Omega \boldsymbol{\varepsilon} : \mathbf{C} : (\boldsymbol{\varepsilon}' - \boldsymbol{\varepsilon})\, d\Omega \tag{6.40}$$

$$\int_\Omega (\boldsymbol{\xi}' - \boldsymbol{\xi}) \cdot (\rho \mathbf{f})\, d\Omega + \int_{\partial\Omega_{\mathbf{T}^d}} (\boldsymbol{\xi}' - \boldsymbol{\xi}) \cdot \mathbf{T}^d da = \int_\Omega \boldsymbol{\varepsilon}' : \mathbf{C} : (\boldsymbol{\varepsilon}' - \boldsymbol{\varepsilon})\, d\Omega \tag{6.41}$$

A subtraction of (6.40) from (6.41) yields

$$0 = \int_\Omega (\boldsymbol{\varepsilon}' - \boldsymbol{\varepsilon}) : \mathbf{C} : (\boldsymbol{\varepsilon}' - \boldsymbol{\varepsilon})\, d\Omega \tag{6.42}$$

This equation shows that $\boldsymbol{\varepsilon}' = \boldsymbol{\varepsilon}$ is the solution of the problem. It is on this line that the uniqueness of the elasticity solution is proven.

6.2.3 Displacement Method: Application to Linear Isotropic Elastic Material Systems

We consider an isotropic elastic material system. The local material behavior is defined by the elastic potential in functions of the two strain invariants of $\boldsymbol{\varepsilon}'$ (have a look back at Section 5.3):

$$\psi(\boldsymbol{\varepsilon}') = \frac{\lambda}{2}(\operatorname{tr} \boldsymbol{\varepsilon}')^2 + G \operatorname{tr}(\boldsymbol{\varepsilon}' \cdot \boldsymbol{\varepsilon}') \tag{6.43}$$

where λ and G are the Lamé constants. The internal energy in an isotropic elastic material system therefore reads

$$W(\boldsymbol{\varepsilon}') = \int_\Omega \left(\frac{\lambda}{2}(\operatorname{tr} \boldsymbol{\varepsilon}')^2 + G \operatorname{tr}(\boldsymbol{\varepsilon}' \cdot \boldsymbol{\varepsilon}') \right) d\Omega \tag{6.44}$$

In turn, the external energy supplied to the system in form of volume forces and prescribed surface forces $\mathbf{T}^d$ is

$$\Phi(\boldsymbol{\xi}') = \int_\Omega \boldsymbol{\xi}' \cdot (\rho \mathbf{f})\, d\Omega + \int_{\partial\Omega_{\mathbf{T}^d}} \boldsymbol{\xi}' \cdot \mathbf{T}^d da \tag{6.45}$$

[6] We replaced indeed the stress $\boldsymbol{\sigma} = \partial\psi/\partial\boldsymbol{\varepsilon}$ in (6.33) by inequality (6.35) based on the convexity of the free energy ψ. This dispenses us to consider any association of $\boldsymbol{\sigma}' = f(\boldsymbol{\xi}')$ with the force boundary conditions and the local stress equilibrium condition.

where $\boldsymbol{\xi}'$ is a kinematically admissible displacement field. Equations (6.44) and (6.45) enter the theorem of the minimum of the potential energy (6.38). We should note that the isotropic elastic material parameters λ and G need not be constant in the structural domain Ω (i.e., homogeneous); for instance, they may depend on the position vector $\mathbf{x}$. This is the case of heterogeneous material systems composed of isotropic phases.

Exercise 53. We consider the heterogeneous material system shown in Figure 6.4. On the boundary of the material system, a pressure p is prescribed. The behavior of the materials composing the solid is linear elastic. In a first approximation, we consider a displacement field of the form

$$\boldsymbol{\xi}' = a\mathbf{x}$$

where a is a constant, and $\mathbf{x}$ the position vector. Give an upper energy bound for the exact potential energy $\mathcal{E}_{\text{pot}}(\boldsymbol{\xi})$ of the system.

We need the strain $\boldsymbol{\varepsilon}'$ associated with $\boldsymbol{\xi}'$:

$$\boldsymbol{\varepsilon}' = \frac{\partial \boldsymbol{\xi}'}{\partial \mathbf{x}} = a\mathbf{1}; \quad \operatorname{tr}\boldsymbol{\varepsilon}' = 3a; \quad \operatorname{tr}(\boldsymbol{\varepsilon}' \cdot \boldsymbol{\varepsilon}') = 3a^2$$

Therefore, the volume variation over the entire volume can be expressed by

$$\Delta\Omega' = \int_{\Omega} \operatorname{tr}\boldsymbol{\varepsilon}' d\Omega = 3a\Omega$$

Use in (6.44) yields

$$\begin{aligned} W(\boldsymbol{\varepsilon}') &= \int_{\Omega} \left[\frac{\lambda(\mathbf{x})}{2} (\operatorname{tr}\boldsymbol{\varepsilon}')^2 + G(\mathbf{x}) \operatorname{tr}(\boldsymbol{\varepsilon}' \cdot \boldsymbol{\varepsilon}') \right] d\Omega \\ &= \int_{\Omega} \left[\frac{\lambda(\mathbf{x})}{2} 9a^2 + G(\mathbf{x}) 3a^2 \right] d\Omega = \frac{9a^2}{2} \int_{\Omega} \left[\lambda(\mathbf{x}) + \frac{2}{3} G(\mathbf{x}) \right] d\Omega \end{aligned}$$

Recalling (5.46),

$$K(\mathbf{x}) = \lambda(\mathbf{x}) + \frac{2}{3} G(\mathbf{x})$$

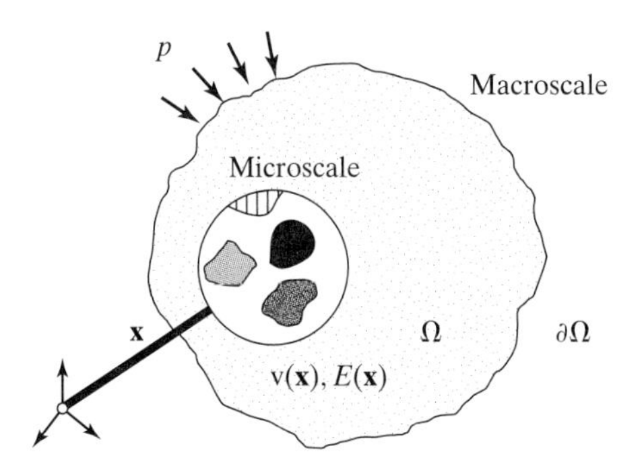

FIGURE 6.4: Bulk modulus of a heterogeneous material system.

the elastic energy stored in the system reads

$$W(\varepsilon') = \frac{9a^2}{2} \int_\Omega K(\mathbf{x})\, d\Omega$$

In a second step, we determine the external work (6.45) provided by forces. Neglecting volume forces, while noting $\mathbf{T}^d = -p\mathbf{n}$, we have

$$\Phi(\boldsymbol{\xi}') = \int_{\partial\Omega_{\mathbf{T}^d}} \boldsymbol{\xi}' \cdot \mathbf{T}^d da = -p \int_{\partial\Omega_{\mathbf{T}^d}} \boldsymbol{\xi}' \cdot \mathbf{n} da$$

We note that $\boldsymbol{\xi}' \cdot \mathbf{n}$ is the normal displacement to the boundary, caused by the pressure application. Noting that $\partial\Omega_{\mathbf{T}^d} = \partial\Omega$, the closed surface integral of $\boldsymbol{\xi}' \cdot \mathbf{n}$ represents the volume increase $\Delta\Omega$ due to pressure application:

$$\int_{\partial\Omega_{\mathbf{T}^d} = \partial\Omega} \boldsymbol{\xi}' \cdot \mathbf{n} da = \oint \boldsymbol{\xi}' \cdot \mathbf{n} da = \Delta\Omega'$$

Furthermore, since $\Delta\Omega' = 3a\Omega$, we obtain

$$\Phi(\boldsymbol{\xi}') = -p\Delta\Omega' = -3ap\Omega$$

The potential energy associated with the approximation $\boldsymbol{\xi}'$ therefore reads

$$\mathcal{E}_{\text{pot}}(\boldsymbol{\xi}') = W(\varepsilon') - \Phi(\boldsymbol{\xi}') = \frac{9a^2}{2} \int_\Omega K(\mathbf{x})\, d\Omega + 3pa\Omega$$

For a given prescribed pressure, this potential has a minimum for

$$\frac{\partial \mathcal{E}_{\text{pot}}(\boldsymbol{\xi}')}{\partial a}\Big|_{p=\text{const}} = 9a \int_\Omega K(\mathbf{x})\, d\Omega + 3p\Omega = 0$$

Thus

$$a = -\frac{p\Omega}{3 \int_\Omega K(\mathbf{x})\, d\Omega} = -\frac{p}{3\langle K(\mathbf{x})\rangle}$$

where $\langle K(\mathbf{x})\rangle$ is the volume average

$$\langle K(\mathbf{x})\rangle = \frac{1}{\Omega} \int_\Omega K(\mathbf{x})\, d\Omega$$

According to the minimum potential energy theorem, the energy $\mathcal{E}_{\text{pot}}(\boldsymbol{\xi}')$ is an upper bound of the (yet) unknown exact potential energy $\mathcal{E}_{\text{pot}}(\boldsymbol{\xi})$:

$$\mathcal{E}_{\text{pot}}(\boldsymbol{\xi}) \leq \mathcal{E}_{\text{pot}}(\boldsymbol{\xi}') = -\frac{p^2\Omega}{2\langle K(\mathbf{x})\rangle}$$

■

6.2.4 Relation with Displacement-Based Finite Element Method

The displacement-based variational method is at the origin of the finite element method, in which the continuous displacement is replaced by a discretization of the form

$$\xi' = \sum_{i=0}^{N+1} q_i N_i(x_j) \tag{6.46}$$

where $N_i(x_j)$ are the basis functions (also called shape or trial functions), and q_i are discrete displacements or nodal displacements of a finite element. The elastic energy of a linear elastic finite element can be written in the form

$$W(q_i, q_j) = \frac{1}{2} k_{ij} q_i q_j \tag{6.47}$$

where k_{ij} is called the rigidity matrix. Similarly, the external work by prescribed volume and surface forces is represented in a discrete form

$$\Phi(q_i) = Q_i q_i \tag{6.48}$$

where Q_i are nodal forces representing volume forces and surface tractions. The theorem of minimum potential energy here implies that the approximation q_i of the solution minimizes the potential

$$\min_{q_i = q_i^d} \mathcal{E}_{\text{pot}}(q_i) = \min \left[W(q_i, q_j) - \Phi(q_i)\right] \tag{6.49}$$

Thus

$$\frac{\partial \mathcal{E}_{\text{pot}}(q_i)}{\partial q_i} = k_{ij} q_j - Q_i = 0 \tag{6.50}$$

We will see an application of this discretized theorem of minimum potential energy in the Problem Set in Section 6.7.

6.3 VARIATIONAL METHOD II: THEOREM OF MINIMUM COMPLEMENTARY ENERGY

The theorem of minimum potential energy defines an upper (energy) bound of the elastic solution. In order to determine the relevance of this upper bound solution, it is useful to develop a lower bound solution. This lower bound solution is based on the theorem of minimum complementary energy.

6.3.1 1D Think Model

We consider again a one-parameter structural system shown in Figure 6.5a. However, our focus is now on the complementary energy of this system . At the material level, this complementary energy is defined by (6.18):

$$\psi^*(\sigma) + \psi(\varepsilon) = \sigma\varepsilon \tag{6.51}$$

where $\psi^*(\sigma)$ is the complementary part to the elastic potential $\psi(\varepsilon)$. This is shown in Figure 6.5b. In the linear case,

$$\psi(\varepsilon) = \frac{1}{2} E \varepsilon^2 \Leftrightarrow \psi^*(\sigma) = \frac{\sigma^2}{2E}$$

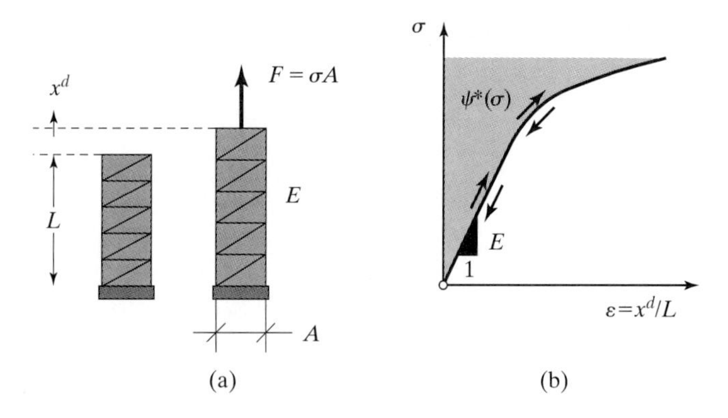

FIGURE 6.5: 1D elastic material system: (a) one-parameter think model; (b) complementary energy at the material level.

Exercise 54. Determine the state equation corresponding to the complementary elastic potential $\psi^*(\sigma)$.

We recall (5.5):

$$\varphi dt = \sigma d\varepsilon - d\psi = 0$$

Use of (6.51) in (5.5) yields

$$\sigma d\varepsilon - d[\sigma\varepsilon - \psi^*(\sigma)] = 0$$

Thus,

$$\varphi dt = -\varepsilon d\sigma + d\psi^*(\sigma) = \left[\frac{\partial\psi^*(\sigma)}{\partial\sigma} - \varepsilon\right] d\sigma = 0$$

This leads to the state equation:

$$\varepsilon = \frac{\partial\psi^*(\sigma)}{\partial\sigma}$$

which is the inverted elastic state equation (5.7). Energy $\psi^*(\sigma) = \sigma\varepsilon - \psi(\varepsilon)$ is also called the Legendre–Fenchel transform of the potential $\psi(\varepsilon)$. We could have used this Legendre–Fenchel transform to invert the elastic state equation in Section 5.3.4. ■

At the structural level, the complementary energy $\mathcal{E}_{\text{com}}$ is the difference between the internal complementary energy in the system, noted $W^*(\sigma)$, and the external complementary energy, $\Phi^*(F = \sigma A)$, supplied to the system by a prescribed displacement x^d:

$$\mathcal{E}_{\text{com}}(\sigma = F/A) = W^*(\sigma) - \Phi^*(F = \sigma A) \tag{6.52}$$

For the one-parameter structural system displayed in Figure 6.5, we have

$$W^*(\sigma) = \psi^*(\sigma)AL = \frac{\sigma^2 AL}{2E}; \quad \Phi^*(F = \sigma A) = Fx^d \tag{6.53}$$

We should note that the condition $\sigma = F/A$ can be seen as the condition for stress σ to be statically admissible.

For an additional prescribed displacement dx^d, the reaction force realizes the following additional external work increment:

$$d\mathcal{W}_{\text{ext}} = F dx^d \tag{6.54}$$

According to the first and second laws, this work increment is stored into the system as free energy $dW = d\psi AL$, which can be expressed by the Legendre–Fenchel transform (see Exercise 54):

$$dW = d(\sigma\varepsilon - \psi^*) AL = dF x^d + F dx^d - dW^* \tag{6.55}$$

Use of (6.54) and (6.55) in (6.26) yields

$$d\mathcal{D} = d\mathcal{W}_{\text{ext}} - dW = -dF x^d + dW^* \equiv 0 \tag{6.56}$$

or, equivalently, in terms of the complimentary energy (6.52) [i.e., $d\mathcal{E}_{\text{com}} = dW^* - (F dx^d + dF x^d)$],

$$d\mathcal{D} = d\mathcal{E}_{\text{com}} + F dx^d \equiv 0 \tag{6.57}$$

Hence, at constant prescribed displacement ($dx^d = 0$), the change in complimentary energy is zero. Since the principal unknown in this complementary energy formulation is stress $\sigma = F/A$, it follows that

$$d\mathcal{D} = \frac{\partial \mathcal{E}_{\text{com}}(\sigma)}{\partial \sigma} d\sigma \equiv 0 \rightarrow \forall d\sigma; \quad \frac{\partial \mathcal{E}_{\text{com}}(\sigma)}{\partial \sigma} = 0 \tag{6.58}$$

Hence, for the one-parameter system,

$$d\mathcal{D} = \frac{\partial}{\partial \sigma}\left(\frac{\sigma^2 AL}{2E} - A\sigma x^d\right) d\sigma \equiv 0 \rightarrow \forall d\sigma; \quad \sigma = E\frac{x^d}{L} \tag{6.59}$$

If we note the convexity of the complementary energy function $\mathcal{E}_{\text{com}}(\sigma)$, we realize that the stress solution $\sigma = Ex^d/L$ minimizes, among all statically admissible stresses, the complementary energy $\mathcal{E}_{\text{com}}$. Therefore, for any σ':

$$\mathcal{E}_{\text{com}}(\sigma) = W^*(\sigma) - \Phi^*(F) \leq W^*(\sigma') - \Phi^*(F') = \mathcal{E}_{\text{com}}(\sigma') \tag{6.60}$$

or, equivalently, using (6.53),

$$\begin{aligned} \psi^*(\sigma') - \psi^*(\sigma) &\geq \frac{x^d}{AL}[F' - F] = \frac{\sigma}{E}(\sigma' - \sigma) \\ &\geq \frac{\partial \psi^*(\sigma)}{\partial \sigma}(\sigma' - \sigma) \end{aligned} \tag{6.61}$$

Inequality (6.61) is nothing but the convexity condition of the complementary material energy $\psi^*(\sigma)$.

6.3.2 3D Stress-Based Variational Formulation

We want to generalize the one-parameter think model to the 3D case. The starting point of the 3D stress-based variational formulation is a statically admissible stress field $\boldsymbol{\sigma}'$ as defined in Section 4.1:

$$\text{in } \Omega : \left\{ \begin{array}{l} \mathbf{T}' = \boldsymbol{\sigma}' \cdot \mathbf{n} \\ \operatorname{div} \boldsymbol{\sigma}' + \rho \mathbf{f} = 0 \\ \boldsymbol{\sigma}' = {}^t\boldsymbol{\sigma}' \end{array} \right\}; \quad \text{on } \partial\Omega_{\mathbf{T}^d} : \mathbf{T}^d = \mathbf{T}'(\mathbf{n}) = \boldsymbol{\sigma}' \cdot \mathbf{n} \tag{6.62}$$

The theorem of virtual work applied to $\boldsymbol{\sigma}'$ reads

$$\underbrace{\int_\Omega \boldsymbol{\xi} \cdot (\rho \mathbf{f})\, d\Omega + \int_{\partial\Omega_{\mathbf{T}^d}} \boldsymbol{\xi} \cdot \mathbf{T}^d da}_{\Phi(\boldsymbol{\xi})} + \underbrace{\int_{\partial\Omega_{\boldsymbol{\xi}^d}} \boldsymbol{\xi}^d \cdot \mathbf{T}' da}_{\Phi^*(\boldsymbol{\sigma}')} = \int_\Omega \boldsymbol{\sigma}' : \boldsymbol{\varepsilon} d\Omega \tag{6.63}$$

In the same way, the theorem of virtual work applied to the stress solution $\boldsymbol{\sigma}$ reads

$$\underbrace{\int_\Omega \boldsymbol{\xi} \cdot (\rho \mathbf{f})\, d\Omega + \int_{\partial\Omega_{\mathbf{T}^d}} \boldsymbol{\xi} \cdot \mathbf{T}^d da}_{\Phi(\boldsymbol{\xi})} + \underbrace{\int_{\partial\Omega_{\boldsymbol{\xi}^d}} \boldsymbol{\xi}^d \cdot \mathbf{T} da}_{\Phi^*(\boldsymbol{\sigma})} = \int_\Omega \boldsymbol{\sigma} : \boldsymbol{\varepsilon} d\Omega \tag{6.64}$$

Subtracting (6.64) from (6.63), while using the very definition of the stress vector ($\mathbf{T}' = \boldsymbol{\sigma}' \cdot \mathbf{n}$ and $\mathbf{T} = \boldsymbol{\sigma} \cdot \mathbf{n}$), yields

$$\underbrace{\int_{\partial\Omega_{\boldsymbol{\xi}^d}} \boldsymbol{\xi}^d \cdot (\boldsymbol{\sigma}' - \boldsymbol{\sigma}) \cdot \mathbf{n} da}_{\Phi^*(\boldsymbol{\sigma}') - \Phi^*(\boldsymbol{\sigma})} = \int_\Omega (\boldsymbol{\sigma}' - \boldsymbol{\sigma}) : \boldsymbol{\varepsilon} d\Omega \tag{6.65}$$

Note here that the external work term $\Phi(\boldsymbol{\xi})$ related to the body forces and prescribed surface force on $\partial\Omega_{\mathbf{T}^d}$ skip out of the stress-based variational formulation. Indeed, both are taken into account through the restriction to statically admissible stress fields, $\boldsymbol{\sigma}$ and $\boldsymbol{\sigma}'$ (i.e., to stress fields that satisfy the momentum balance equation and the boundary condition $\mathbf{T}' = \mathbf{T}^d$ on $\partial\Omega_{\mathbf{T}^d}$).

Furthermore, the state equation in the complementary energy formulation reads

$$\boldsymbol{\varepsilon} = \frac{\partial \psi^*(\boldsymbol{\sigma})}{\partial \boldsymbol{\sigma}}; \quad \psi^*(\boldsymbol{\sigma}) = \frac{1}{2} \boldsymbol{\sigma} : \boldsymbol{\Lambda} : \boldsymbol{\sigma} \tag{6.66}$$

where $\boldsymbol{\Lambda} = \mathbf{C}^{-1}$ is the elastic compliance tensor (have a look back at Section 5.3.4). In addition, the convexity condition (6.21) of the material complementary energy $\psi^*(\boldsymbol{\sigma})$ reads

$$\frac{\partial \psi^*(\boldsymbol{\sigma})}{\partial \boldsymbol{\sigma}} : (\boldsymbol{\sigma}' - \boldsymbol{\sigma}) \leq \psi^*(\boldsymbol{\sigma}') - \psi^*(\boldsymbol{\sigma}) \tag{6.67}$$

Finally, use of (6.66) and (6.67) in (6.65) yields

$$\underbrace{\int_{\partial\Omega_{\boldsymbol{\xi}^d}} \boldsymbol{\xi}^d \cdot (\boldsymbol{\sigma}' - \boldsymbol{\sigma}) \cdot \mathbf{n} da}_{\Phi^*(\boldsymbol{\sigma}') - \Phi^*(\boldsymbol{\sigma})} = \int_\Omega (\boldsymbol{\sigma}' - \boldsymbol{\sigma}) : \frac{\partial \psi^*(\boldsymbol{\sigma})}{\partial \boldsymbol{\sigma}} d\Omega \leq \underbrace{\int_\Omega \left[\psi^*(\boldsymbol{\sigma}') - \psi^*(\boldsymbol{\sigma}) \right] d\Omega}_{W^*(\boldsymbol{\sigma}') - W^*(\boldsymbol{\sigma})}$$

or, equivalently, after rearrangement,

$$\underbrace{W^*(\boldsymbol{\sigma}) - \Phi^*(\boldsymbol{\sigma})}_{\mathcal{E}_{\text{com}}(\boldsymbol{\sigma})} \leq \underbrace{W^*(\boldsymbol{\sigma}') - \Phi^*(\boldsymbol{\sigma}')}_{\mathcal{E}_{\text{com}}(\boldsymbol{\sigma}')} \tag{6.68}$$

This is the theorem of minimum complementary energy.

THEOREM 5 of minimum complementary energy: Among all statically admissible stress fields $\boldsymbol{\sigma}'$, the stress solution $\boldsymbol{\sigma}$ minimizes the (convex) elastic complementary energy:

$$\mathcal{E}_{\text{com}}(\boldsymbol{\sigma}) = \min_{\substack{\mathbf{T}'=\mathbf{T}^d \text{ on } \partial\Omega_{\mathbf{T}^d} \\ \operatorname{div}\boldsymbol{\sigma}'+\rho\mathbf{f}=0 \\ \boldsymbol{\sigma}'={}^t\boldsymbol{\sigma}'}} [W^*(\boldsymbol{\sigma}') - \Phi^*(\boldsymbol{\sigma}')] \tag{6.69}$$

where

$$\mathcal{E}_{\text{com}}(\boldsymbol{\sigma}') = \underbrace{\int_\Omega \psi^*(\boldsymbol{\sigma}')\,d\Omega}_{W^*(\boldsymbol{\sigma}')} - \underbrace{\int_{\partial\Omega_{\boldsymbol{\xi}^d}} \boldsymbol{\xi}^d \cdot (\boldsymbol{\sigma}' \cdot \mathbf{n})\,da}_{\Phi^*(\boldsymbol{\sigma}')} \tag{6.70}$$

This theorem holds for any elastic behavior, whether linear or nonlinear. The displacement $\boldsymbol{\xi}' = \boldsymbol{\xi}'(\boldsymbol{\sigma}')$ that can be derived by means of an elastic constitutive relation from stress $\boldsymbol{\sigma}'$ is not required to be kinematically admissible.

The theorem of minimum complementary energy sets out a second optimization problem: To seek statically admissible stress fields $\boldsymbol{\sigma}'$ that come the closest to the minimum of the complementary energy (i.e., to the stress solution of the elasticity problem).

6.3.3 Stress Method: Application to Linear Isotropic Elastic Material Systems

Consider an isotropic elastic material system. For a given stress field $\boldsymbol{\sigma}'$, the local elastic material behavior is defined by the complementary elastic potential:

$$\psi^*(\boldsymbol{\sigma}') = -\frac{\nu}{2E}(\operatorname{tr}\boldsymbol{\sigma}')^2 + \frac{1+\nu}{2E}\operatorname{tr}(\boldsymbol{\sigma}' \cdot \boldsymbol{\sigma}') \tag{6.71}$$

where E is the Young's modulus and ν the Poisson ratio (have a look back at Section 5.3.4).[7] The internal complementary energy in an isotropic elastic material system therefore reads

$$W^*(\boldsymbol{\sigma}') = \int_\Omega \frac{1}{2E}\left[-\nu(\operatorname{tr}\boldsymbol{\sigma}')^2 + (1+\nu)\operatorname{tr}(\boldsymbol{\sigma}' \cdot \boldsymbol{\sigma}')\right] d\Omega \tag{6.72}$$

[7]With the Legendre–Fenchel transform $\psi^*(\boldsymbol{\sigma})$, the isotropic strain equation of state can be derived:

$$\boldsymbol{\varepsilon} = \frac{\partial\psi^*(\boldsymbol{\sigma})}{\partial\boldsymbol{\sigma}} = -\frac{\nu}{E}(\operatorname{tr}\boldsymbol{\sigma})\mathbf{1} + \frac{1+\nu}{2E}\boldsymbol{\sigma}$$

The external complementary energy supplied to the system in form of prescribed displacement $\boldsymbol{\xi}^d$ reads

$$\Phi^*(\boldsymbol{\sigma}') = \int_{\partial\Omega_{\boldsymbol{\xi}^d}} \boldsymbol{\xi}^d \cdot (\boldsymbol{\sigma}' \cdot \mathbf{n})\, da \tag{6.73}$$

where $\boldsymbol{\sigma}'$ is any statically admissible stress field. Equations (6.72) and (6.73) enter the theorem of minimum complementary energy (6.68) or (6.69). We should note that the isotropic elastic material parameters E and ν in (6.72) need not be constant (i.e., homogeneous) in the structural domain Ω.

Exercise 55. We consider again the heterogeneous material system shown in Figure 6.4, but we will now consider a stress field of the form

$$\boldsymbol{\sigma}' = -p\mathbf{1}$$

Determine a second energy bound for the material system.

We verify that $\boldsymbol{\sigma}'$ is statically admissible, satisfying the equilibrium condition ($\operatorname{div}\boldsymbol{\sigma}' = 0$), the symmetry of the stress tensor ($\boldsymbol{\sigma}' = {}^t\boldsymbol{\sigma}'$), and the stress boundary condition on $\partial\Omega_{\mathbf{T}^d}$:

$$\text{on } \partial\Omega_{\mathbf{T}^d} : \mathbf{T}^d \equiv \mathbf{T}' = \boldsymbol{\sigma}' \cdot \mathbf{n} = -p\mathbf{n}$$

The stress invariants read

$$\operatorname{tr}\boldsymbol{\sigma}' = -3p; \quad \operatorname{tr}(\boldsymbol{\sigma}' \cdot \boldsymbol{\sigma}') = 3p^2$$

The internal complementary energy (6.72) reads

$$\begin{aligned} W^*(\boldsymbol{\sigma}') &= \int_\Omega \left[-\frac{\nu(\mathbf{x})}{2E(\mathbf{x})} (\operatorname{tr}\boldsymbol{\sigma}')^2 + \frac{1+\nu(\mathbf{x})}{2E(\mathbf{x})} \operatorname{tr}(\boldsymbol{\sigma}' \cdot \boldsymbol{\sigma}') \right] d\Omega \\ &= \int_\Omega \left[-\frac{9\nu(\mathbf{x})}{2E(\mathbf{x})} p^2 + \frac{1+\nu(\mathbf{x})}{2E(\mathbf{x})} 3p^2 \right] d\Omega \\ &= \frac{p^2}{2} \int_\Omega \frac{d\Omega}{K(\mathbf{x})} = \frac{p^2\Omega}{2} \left\langle \frac{1}{K(\mathbf{x})} \right\rangle \end{aligned}$$

where we used (5.52), and the volume average:

$$K(\mathbf{x}) = \frac{E(\mathbf{x})}{3(1-2\nu(\mathbf{x}))}; \quad \left\langle \frac{1}{K(\mathbf{x})} \right\rangle = \frac{1}{\Omega} \int_\Omega \frac{d\Omega}{K(\mathbf{x})}$$

Furthermore, since $\partial\Omega_{\mathbf{T}^d} = \partial\Omega$, the work provided by prescribed displacement [i.e., Eq. (6.73)] is zero:

$$\Phi^*(\boldsymbol{\sigma}') = \int_{\partial\Omega_{\boldsymbol{\xi}^d}} \boldsymbol{\xi}^d \cdot (\boldsymbol{\sigma}' \cdot \mathbf{n})\, da = 0$$

The complementary energy of the system for the chosen stress field $\boldsymbol{\sigma}'$ reads

$$\mathcal{E}_{\text{com}}(\boldsymbol{\sigma}') = W^*(\boldsymbol{\sigma}') - \Phi^*(\boldsymbol{\sigma}') = \frac{p^2\Omega}{2} \left\langle \frac{1}{K(\mathbf{x})} \right\rangle$$

According to (6.68), this complementary energy $\mathcal{E}_{\text{com}}(\boldsymbol{\sigma}')$ is an upper bound of the exact complementary energy $\mathcal{E}_{\text{com}}(\boldsymbol{\sigma})$; but according to (6.20) also a lower bound to the exact potential energy $\mathcal{E}_{\text{pot}}(\boldsymbol{\xi})$:

$$-\frac{p^2\Omega}{2}\left\langle\frac{1}{K(\mathbf{x})}\right\rangle = -\mathcal{E}_{\text{com}}(\boldsymbol{\sigma}') \leq -\mathcal{E}_{\text{com}}(\boldsymbol{\sigma}) = \mathcal{E}_{\text{pot}}(\boldsymbol{\xi}) \qquad \blacksquare$$

6.3.4 Stress Method in Linear Elastic Structural Mechanics

The underlying idea of structural mechanics of beam-type problems is to reduce the three-dimensional information of a 3D material system to a 1D structural system oriented along the beam axis. As for the variational method based on statically admissible stress fields, this consists of compressing, by means of the reduction formulas (see Section 3.3.3), the information contained in stress tensor $\boldsymbol{\sigma}'$ to global force and moment quantities. The principle of the stress variational method remains the same. It suffices to consider as internal complementary energy

$$W^*(\boldsymbol{\sigma}') = \int_\Omega \psi^*(\boldsymbol{\sigma}')\,d\Omega = \frac{1}{2}\int_L \left[\frac{N'^2}{EA} + \frac{M_y'^2}{EI_y} + \frac{M_z'^2}{EI_z} + \frac{M_x'^2}{GJ}\right] dx \tag{6.74}$$

where N' is the axial force; M_x', M_y', M_z' are the components of the moments defined by $(3.23)_2$; and A, I_y, I_z, J are section parameters (i.e., area, bending inertia moment around y- and z-axes, and torsional inertia moment).

6.4 UPPER AND LOWER BOUNDS: CLAPEYRON'S FORMULA

We recall the theorem of virtual work (6.20) applied to an elastic material system:

$$\underbrace{-\left(W^*(\boldsymbol{\sigma}) - \Phi^*(\boldsymbol{\sigma})\right)}_{-\mathcal{E}_{\text{com}}(\boldsymbol{\sigma})} = \underbrace{W(\boldsymbol{\varepsilon}) - \Phi(\boldsymbol{\xi})}_{\mathcal{E}_{\text{pot}}(\boldsymbol{\xi})} \tag{6.75}$$

Provided that $\boldsymbol{\xi}$ and $\boldsymbol{\sigma}$ are solutions of the elasticity problem, Eq. (6.75) states that the potential energy $\mathcal{E}_{\text{pot}}(\boldsymbol{\xi})$ of the displacement solution $\boldsymbol{\xi}$ is equal with opposite sign to the complementary energy $\mathcal{E}_{\text{com}}(\boldsymbol{\sigma})$ of the stress solution $\boldsymbol{\sigma}$. Furthermore, the potential energy $\mathcal{E}_{\text{pot}}(\boldsymbol{\xi})$ is defined by the theorem of minimum potential energy (6.37) and the complementary energy $\mathcal{E}_{\text{com}}(\boldsymbol{\sigma})$ by the theorem of minimum complementary energy (6.68). Therefore, substituting these two theorems of variational methods into (6.75) yields

$$\underbrace{-W^*(\boldsymbol{\sigma}') + \Phi^*(\boldsymbol{\sigma}') \leq -W^*(\boldsymbol{\sigma}) + \Phi^*(\boldsymbol{\sigma})}_{\substack{-\mathcal{E}_{\text{com}}(\boldsymbol{\sigma}') \leq -\mathcal{E}_{\text{com}}(\boldsymbol{\sigma}) \\ \text{Statically Admissible Stress}}} = \underbrace{W(\boldsymbol{\varepsilon}) - \Phi(\boldsymbol{\xi}) \leq W(\boldsymbol{\varepsilon}') - \Phi(\boldsymbol{\xi}')}_{\substack{\mathcal{E}_{\text{pot}}(\boldsymbol{\varepsilon}) \leq \mathcal{E}_{\text{pot}}(\boldsymbol{\varepsilon}') \\ \text{Kinematically Admissible Displacement}}} \tag{6.76}$$

Hence, in terms of energy, the stress method based on the minimum of the complementary energy, $\mathcal{E}_{\text{com}}(\boldsymbol{\sigma}')$, provides a lower bound, and the displacement method based on the minimum of the potential energy, $\mathcal{E}_{\text{pot}}(\boldsymbol{\varepsilon}')$, an upper bound of the elastic solution for which lower and upper bound coincide [i.e., (6.75)]. We therefore have obtained a formidable means of framing the exact elasticity solution by two independent bounds.

Finally, for linear elastic material systems, the internal work at a material level $\psi^*(\boldsymbol{\sigma}) + \psi(\boldsymbol{\varepsilon}) = \boldsymbol{\sigma} : \boldsymbol{\varepsilon}$ simplifies to

$$\psi^*(\boldsymbol{\sigma}) = \psi(\boldsymbol{\varepsilon}) = \frac{1}{2}\boldsymbol{\sigma} : \boldsymbol{\varepsilon} \tag{6.77}$$

Hence, provided that $\boldsymbol{\sigma}$ and $\boldsymbol{\varepsilon}$ are the solutions of the problem, it is

$$W^*(\boldsymbol{\sigma}) = W(\boldsymbol{\varepsilon}) \tag{6.78}$$

Note clearly that Eq. (6.78) is restricted to linear elastic material systems. It states that the internal potential energy is equal to the internal complementary energy. In addition, provided that $\boldsymbol{\sigma}$,$\boldsymbol{\varepsilon}$ and $\boldsymbol{\xi}$ are the solutions of the problem, a combination of (6.75) and (6.78) yields

$$W(\boldsymbol{\varepsilon}) = W^*(\boldsymbol{\sigma}) = \frac{1}{2}\left[\Phi(\boldsymbol{\xi}) + \Phi^*(\boldsymbol{\sigma})\right] \tag{6.79}$$

Equation (6.79) is known as *Clapeyron's formula*: The internal potential energy (respectively, the internal complementary energy) can be evaluated from the externally supplied work.

Exercise 56. In the exercises of Sections 6.2.3 and 6.3.3 we determined, respectively, an upper and a lower energy bound of the exact potential of the heterogeneous material system displayed in Figure 6.4:

$$-\frac{p^2\Omega}{2}\left\langle\frac{1}{K(\mathbf{x})}\right\rangle = -\mathcal{E}_{\text{com}}(\boldsymbol{\sigma}') \leq -\mathcal{E}_{\text{com}}(\boldsymbol{\sigma}) = \mathcal{E}_{\text{pot}}(\boldsymbol{\xi}) \leq \mathcal{E}_{\text{pot}}(\boldsymbol{\xi}') = -\frac{p^2\Omega}{2\left\langle K(\mathbf{x})\right\rangle}$$

By means of Clapeyron's formula, determine the exact potential energy of an equivalent homogenized material sample, and give an upper and a lower bound for the apparent bulk modulus K_{app} of the heterogeneous material.

For the solutions ($\boldsymbol{\sigma}$ and $\boldsymbol{\xi}$) of the linear elastic material system, the exact energy stored in the system is given by (6.79). The potential energy $\mathcal{E}_{\text{pot}}(\boldsymbol{\xi})$, therefore, reads

$$\begin{aligned}\mathcal{E}_{\text{pot}}(\boldsymbol{\xi}) &= \frac{1}{2}\left[\Phi(\boldsymbol{\xi}) + \Phi^*(\boldsymbol{\sigma})\right] - \Phi(\boldsymbol{\xi}) \\ &= \frac{1}{2}\left[\Phi^*(\boldsymbol{\sigma}) - \Phi(\boldsymbol{\xi})\right] = -\mathcal{E}_{\text{com}}(\boldsymbol{\sigma})\end{aligned}$$

For the problem at hand—with $\partial\Omega = \partial\Omega_{\mathbf{T}^d} \rightarrow \Phi^*(\boldsymbol{\sigma}) = 0$ (see Exercise 55 in Section 6.3.3)—the target (solution) potential energy reduces to (see Exercise 53 in Section 6.2.3)

$$\mathcal{E}_{\text{pot}}(\boldsymbol{\xi}) = -\frac{1}{2}\Phi(\boldsymbol{\xi}) = -\frac{1}{2}\int_{\partial\Omega_{\mathbf{T}^d}} \boldsymbol{\xi}\cdot\mathbf{T}^d da = \frac{p}{2}\int_{\partial\Omega_{\mathbf{T}^d}} \boldsymbol{\xi}\cdot\mathbf{n} da = \frac{p\Delta\Omega}{2}$$

where we made use of the fact that the closed surface integral of the normal displacement over the surface represents the overall volume variation (see Exercise 53 Section 6.2.3):

$$\Delta\Omega = \int_{\partial\Omega_{\mathbf{T}^d}=\partial\Omega} \boldsymbol{\xi}\cdot\mathbf{n}da = \oint \boldsymbol{\xi}\cdot\mathbf{n}da$$

Furthermore, following Section 5.3.1, the overall volume variation $\Delta\Omega$ resulting from this pressure application can be associated with an apparent bulk modulus of the equivalent homogenized material system:

$$\frac{\Delta\Omega}{\Omega} = -\frac{p}{K^{\text{app}}}$$

Finally, using the previous results in the energy bound inequality, we obtain

$$-\frac{p^2\Omega}{2}\left\langle\frac{1}{K(\mathbf{x})}\right\rangle \le -\frac{p^2\Omega}{2K^{\text{app}}} \le -\frac{p^2\Omega}{2\left\langle K(\mathbf{x})\right\rangle}$$

This leads to the following bounds for the apparent bulk modulus:

$$\left\langle\frac{1}{K(\mathbf{x})}\right\rangle^{-1} \le K^{\text{app}} \le \left\langle K(\mathbf{x})\right\rangle$$

Note that the upper bound of K^{app} corresponds to the upper energy bound obtained with a kinematically admissible displacement field $\boldsymbol{\xi}'$ (Exercise 53 in Section 6.2.3), while the lower bound corresponds to the lower energy bound using the theorem of minimum complementary energy with the statically admissible stress field $\boldsymbol{\sigma}'$ (Exercise 55 in Section 6.3.3). Note clearly that

$$\left\langle\frac{1}{K(\mathbf{x})}\right\rangle^{-1} = \left[\frac{1}{\Omega}\int_\Omega \frac{d\Omega}{K(\mathbf{x})}\right]^{-1} \ne \left\langle K(\mathbf{x})\right\rangle = \frac{1}{\Omega}\int_\Omega K(\mathbf{x})\,d\Omega$$ ■

6.5 TRAINING SET: EFFECTIVE MODULUS OF A HETEROGENEOUS MATERIAL SYSTEM

Variational methods based on energy minima are powerful tools in mechanics of heterogeneous materials, for which material properties are not homogeneously distributed in a sample. We have already seen one example throughout this chapter, dealing with the bulk modulus of a heterogeneous material system (i.e., exercises in Sections 6.2.3, 6.3.3, and 6.4). Another example concerned with the apparent Young's modulus of a heterogeneous material system is the focus of this training set. We consider a direct tension test on a cylinder specimen (section A, length L) as shown in Figure 6.6 (on the next page). At $z = 0$, the sample is in contact without friction with a fixed rigid plate. At $z = L$, a displacement δ in the z-direction is prescribed. The cylinder surface is stress free. The sample is composed of a *heterogeneous* linear isotropic material. The elastic properties λ and G depend on the position vector. We want to give an upper and a lower bound of the apparent (or effective) elasticity modulus E_{app} of the heterogeneous linear isotropic elastic material.

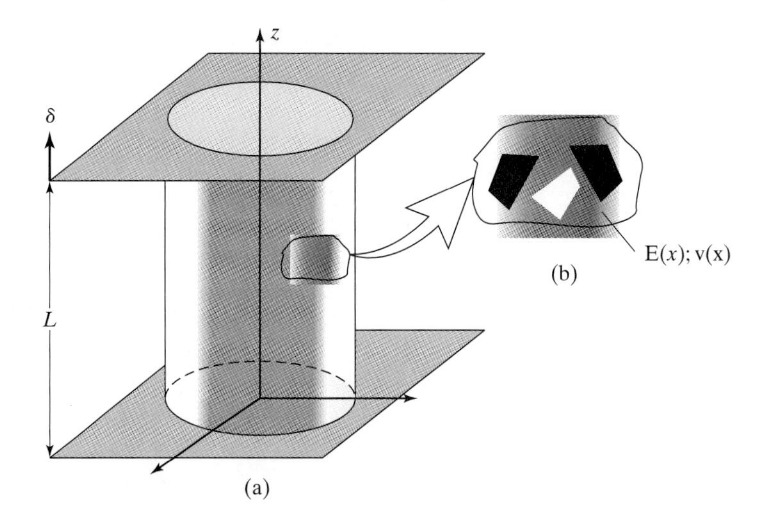

FIGURE 6.6: (a) Direct tension test on (b) a heterogeneous material system.

6.5.1 Clapeyron's Formula

The starting point of the energy approach is the study of the target solution of the problem, the apparent elasticity modulus E_{app}. This effective stiffness is the Young's modulus of an equivalent homogeneous material sample, or homogenized sample. There are different ways of determining the corresponding energy solution for the homogeneous sample. For a linear isotropic material, Clapeyron's formula can be applied.

Exercise 57. Determine the potential energy for the equivalent homogeneous sample.

We use Clapeyron's formula to determine the potential energy of the exact (equivalent homogeneous) solution:

$$\begin{aligned} W(\boldsymbol{\varepsilon}) &= W^*(\boldsymbol{\sigma}) = \frac{1}{2}[\Phi(\boldsymbol{\xi}) + \Phi^*(\boldsymbol{\sigma})] \\ &= \frac{1}{2}\left[\int_\Omega \boldsymbol{\xi}\cdot(\rho\mathbf{f})d\Omega + \int_{\partial\Omega_{\mathbf{T}^d}} \boldsymbol{\xi}\cdot\mathbf{T}^d da + \int_{\partial\Omega_{\boldsymbol{\xi}^d}} \boldsymbol{\xi}^d\cdot\mathbf{T}da\right] \end{aligned}$$

Disregarding body forces, that is,

$$\int_\Omega \boldsymbol{\xi}\cdot(\rho\mathbf{f})\,d\Omega = 0$$

and for $\mathbf{T}^d = 0$ on $\partial\Omega_{\mathbf{T}^d}$,

$$\int_{\partial\Omega_{\mathbf{T}^d}} \boldsymbol{\xi}\cdot\mathbf{T}^d da = 0$$

the potential energy of the exact solution reduces here to

$$W(\varepsilon) = W^*(\boldsymbol{\sigma}) = \frac{1}{2}\int_{\partial\Omega_{\boldsymbol{\xi}^d}} \boldsymbol{\xi}^d \cdot \mathbf{T} da$$

The only nonzero contribution results from the prescribed displacement at $z = L$:

$$\boldsymbol{\xi}^d = \delta \mathbf{e}_z$$

and

$$\mathbf{T} = \mathbf{T}(\mathbf{n} = \mathbf{e}_z)$$

Hence,

$$W(\varepsilon) = W^*(\boldsymbol{\sigma}) = \frac{1}{2}\delta \int_{\partial\Omega_{\boldsymbol{\xi}^d}} \mathbf{T}(\mathbf{e}_z) \cdot \mathbf{e}_z da = \frac{1}{2}\delta \int_{\partial\Omega_{\boldsymbol{\xi}^d}} \sigma_{zz} da = \frac{1}{2}\delta N$$

Here, N is the resultant normal force at $z = L$, as introduced in Section 3.3.3. This normal force is related to the strain δ/L by

$$N = E_{\text{app}} A \frac{\delta}{L}$$

with E_{app} the apparent E-modulus at the level of the (equivalent homogeneous) structure. The energy of the exact homogenized solution reads

$$\mathcal{E}_{\text{pot}}(\varepsilon) = -\mathcal{E}_{\text{com}}(\boldsymbol{\sigma}) = \frac{1}{2} E_{\text{app}} A \frac{\delta^2}{L} \qquad \blacksquare$$

This energy is the target solution of the upper and lower bound variational approach.

6.5.2 Upper Bound: Minimum Potential Energy

We are interested in exploring the displacement based variational formulation for the heterogeneous material sample. The displacement method is based on the choice of a kinematically admissible displacement field $\boldsymbol{\xi}'$.

Exercise 58. We consider the case of a heterogeneous material sample, for which $\lambda = \lambda(\mathbf{x})$ and $G = G(\mathbf{x})$. Inspired by the homogenous solution,[8] we consider in a first approach the following displacement field:

$$\boldsymbol{\xi}' = \frac{\delta}{L}(z\mathbf{e}_z - \alpha r \mathbf{e}_r)$$

Determine a first bound of the apparent elasticity modulus of the heterogeneous material system. (Remark: In the homogeneous case, $\alpha = \nu$, the Poisson ratio.) In return, for

[8]The homogeneous linear elastic solution of a direct tension test reads

$$\boldsymbol{\xi} = \frac{\delta}{L}[z\mathbf{e}_z - \nu r \mathbf{e}_r]$$

where ν is the Poisson ratio.

the variational problem of a heterogeneous material system at hand, $\alpha \neq \nu$ is the unknown degree of freedom.

We first need to check that $\boldsymbol{\xi}'$ is kinematically admissible: It satisfies the displacement boundary condition,

$$\boldsymbol{\xi}^d \cdot \mathbf{e}_z = \boldsymbol{\xi}'(z = L) \cdot \mathbf{e}_z = \delta$$

The strain tensor reads[9]

$$\boldsymbol{\varepsilon}' = \frac{\delta}{L}(\mathbf{e}_z \otimes \mathbf{e}_z - \alpha(\mathbf{e}_r \otimes \mathbf{e}_r + \mathbf{e}_\theta \otimes \mathbf{e}_\theta)$$

and the invariants[10]

$$\operatorname{tr} \boldsymbol{\varepsilon}' = \frac{\delta}{L}(1 - 2\alpha); \quad \operatorname{tr}(\boldsymbol{\varepsilon}' \cdot \boldsymbol{\varepsilon}') = \left(\frac{\delta}{L}\right)^2 (1 + 2\alpha^2)$$

The potential energy reads

$$\mathcal{E}_{\text{pot}}(\boldsymbol{\varepsilon}') = W(\boldsymbol{\varepsilon}') - \Phi(\boldsymbol{\xi}')$$

Since $\mathbf{T}^d = 0$, and volume forces disregarded,

$$\Phi(\boldsymbol{\xi}') = \int_\Omega \boldsymbol{\xi}' \cdot (\rho \mathbf{f})\, d\Omega + \int_{\partial\Omega_{\mathbf{T}^d}} \boldsymbol{\xi}' \cdot \mathbf{T}^d da = 0$$

Hence, with $\Phi(\boldsymbol{\xi}') = 0$, the potential energy is given by

$$\begin{aligned} \mathcal{E}_{\text{pot}}(\boldsymbol{\varepsilon}') &= \int_\Omega \left[\frac{\lambda(\mathbf{x})}{2}(\operatorname{tr} \boldsymbol{\varepsilon}')^2 + G(\mathbf{x}) \operatorname{tr}(\boldsymbol{\varepsilon}' \cdot \boldsymbol{\varepsilon}')\right] d\Omega \\ &= \left(\frac{\delta}{L}\right)^2 \int_\Omega \left[\frac{\lambda(\mathbf{x})}{2}(1 - 2\alpha)^2 + G(\mathbf{x})(1 + 2\alpha^2)\right] d\Omega \end{aligned}$$

We introduce the volume average of the Lamé constants $\lambda(\mathbf{x})$ and $G(\mathbf{x})$:

$$\begin{aligned} \langle \lambda \rangle &= \frac{1}{\Omega} \int_\Omega \lambda(\mathbf{x})\, d\Omega = \frac{1}{AL} \int_\Omega \lambda(\mathbf{x})\, d\Omega \\ \langle G \rangle &= \frac{1}{\Omega} \int_\Omega G(\mathbf{x})\, d\Omega = \frac{1}{AL} \int_\Omega G(\mathbf{x})\, d\Omega \end{aligned}$$

---

[9] We recall for the considered cylinder coordinate setting

$$\varepsilon'_{zz} = \frac{\partial \xi'_z}{\partial z}; \quad \varepsilon'_{rr} = \frac{\partial \xi'_r}{\partial r}; \quad \varepsilon'_{\theta\theta} = \frac{\xi'_r}{r}$$

[10] The second invariant can be determined in matrix form from

$$(\boldsymbol{\varepsilon}' \cdot \boldsymbol{\varepsilon}') = \left(\frac{\delta}{l}\right)^2 \begin{bmatrix} 1 & & \\ & -\alpha & \\ & & -\alpha \end{bmatrix} \begin{bmatrix} 1 & & \\ & -\alpha & \\ & & -\alpha \end{bmatrix} = \left(\frac{\delta}{l}\right)^2 \begin{bmatrix} 1 & & \\ & \alpha^2 & \\ & & \alpha^2 \end{bmatrix}$$

Thus,

$$\operatorname{tr}(\boldsymbol{\varepsilon}' \cdot \boldsymbol{\varepsilon}') = [1 + 2\alpha^2] \left(\frac{\delta}{l}\right)^2$$

The potential energy $\mathcal{E}_{\text{pot}}(\varepsilon')$ can be rewritten in the form

$$\mathcal{E}_{\text{pot}}(\varepsilon') = \frac{1}{2}\left(\frac{\delta}{L}\right)^2 AL\left[(1-2\alpha)^2\langle\lambda\rangle + 2(1+2\alpha^2)\langle G\rangle\right]$$

This potential energy has a minimum for

$$\delta = \text{const} \to d\mathcal{E}_{\text{pot}} = 0 : \frac{\partial \mathcal{E}_{\text{pot}}}{\partial \alpha} = 0$$

Thus,

$$-4(1-2\alpha_m)\langle\lambda\rangle + 4\alpha_m\langle G\rangle = 0 \to \alpha_m = \frac{\langle\lambda\rangle}{2\left[\langle\lambda\rangle + \langle G\rangle\right]}$$

where α_m is the value of α minimizing $\mathcal{E}_{\text{pot}}(\varepsilon')$.[11]
Finally, applying the theorem of minimum potential energy to the tension test on the heterogeneous material sample yields

$$\begin{aligned} \mathcal{E}_{\text{pot}}(\varepsilon) &\leq \mathcal{E}_{\text{pot}}(\varepsilon') \\ \frac{1}{2}\left(\frac{\delta}{L}\right)^2 E_{\text{app}}AL &\leq \frac{1}{2}\left(\frac{\delta}{L}\right)^2 AL\left[(1-2\alpha_m)^2\langle\lambda\rangle + 2(1+2\alpha_m^2)\langle G\rangle\right] \end{aligned}$$

where $\mathcal{E}_{\text{pot}}(\varepsilon)$ is the target potential energy determined in Section 6.5.1. Thus,

$$\begin{aligned} E_{\text{app}} &\leq (1-2\alpha_m)^2\langle\lambda\rangle + 2(1+2\alpha_m^2)\langle G\rangle \\ E_{\text{app}} &\leq \langle G\rangle\frac{3\langle\lambda\rangle + 2\langle G\rangle}{\langle\lambda\rangle + \langle G\rangle} \end{aligned}$$

The displacement based variational method provides an upper bound of the apparent elasticity modulus of the heterogeneous sample (i.e., a displacement based solution overestimates the stiffness of the sample). ■

6.5.3 Lower Bound: Minimum Complementary Energy

Exercise 59. Determine another bound of the apparent E-modulus, E_{app}, of the heterogeneous sample in a tension test (see Figure 6.6), using the theorem of minimum complementary energy.

Inspired by the homogeneous stress solution in a direct tension test, a first approach consists of considering a constant uniaxial stress field in the heterogeneous sample:

$$\boldsymbol{\sigma}' = \sigma'\mathbf{e}_z \otimes \mathbf{e}_z$$

[11]Note that

$$\alpha_m = \frac{\langle\lambda\rangle}{2\left[\langle\lambda\rangle + \langle G\rangle\right]} \neq \frac{1}{\Omega}\int_\Omega \nu(\mathbf{x})\,d\Omega = \frac{1}{\Omega}\int_\Omega \frac{\lambda(\mathbf{x})}{2[\lambda(\mathbf{x}) + G(\mathbf{x})]}d\Omega$$

In the homogeneous case, $\sigma' = \sigma = EA\delta/L$ (see Section 6.3.1). This is not the case when considering the heterogeneous nature of the material system. From the standpoint of variational methods, σ' is the stress unknown degree of freedom of the stress problem.

The only restriction on $\boldsymbol{\sigma}'$ is that it is a statically admissible stress field, satisfying

1. The momentum balance equation:

$$\text{in } \Omega : \frac{\partial \sigma'_{zz}}{\partial z} = 0 \leftrightarrow \sigma'_{zz} = \sigma'$$

2. The stress boundary condition:

$$\text{on } \partial\Omega_{\mathbf{T}^d} : \mathbf{T}^d(\mathbf{e}_r) = 0$$

3. The (frictionless) contact condition at $z = 0$ and $z = L$:

$$\mathbf{T}^d(\mathbf{e}_z) \cdot \mathbf{e}_r = \sigma'_{zr} = 0; \quad \mathbf{T}^d(\mathbf{e}_z) \cdot \mathbf{e}_\theta = \sigma'_{z\theta} = 0$$

Clearly, $\boldsymbol{\sigma}'$ is statically admissible.

The complementary energy reads

$$\mathcal{E}_{\text{com}}(\boldsymbol{\sigma}') = W^*(\boldsymbol{\sigma}') - \Phi^*(\boldsymbol{\sigma}')$$

where

$$W^*(\boldsymbol{\sigma}') = \int_\Omega \left[-\frac{\nu}{2E(\mathbf{x})} \sigma'^2 + \frac{1+\nu}{2E(\mathbf{x})} \sigma'^2 \right] d\Omega = \frac{\sigma'^2}{2} \int_\Omega \frac{1}{E(\mathbf{x})} d\Omega$$

and

$$\Phi^*(\boldsymbol{\sigma}') = \int_{\partial\Omega_{\boldsymbol{\xi}^d}} \boldsymbol{\xi}^d \cdot (\boldsymbol{\sigma}' \cdot \mathbf{n})\, da = \int_A \delta \mathbf{e}_z \cdot (\boldsymbol{\sigma}' \cdot \mathbf{e}_z)\, da = \delta A \sigma'$$

Using the volume average $\langle 1/E \rangle$,

$$\left\langle \frac{1}{E} \right\rangle = \frac{1}{\Omega} \int_\Omega \frac{1}{E(\mathbf{x})} d\Omega = \frac{1}{AL} \int_\Omega \frac{1}{E(\mathbf{x})} d\Omega$$

the complementary energy reads

$$\mathcal{E}_{\text{com}}(\boldsymbol{\sigma}') = \frac{\sigma'^2}{2} AL \left\langle \frac{1}{E} \right\rangle - \delta A \sigma'$$

Minimizing this expression of $\mathcal{E}_{\text{com}}(\boldsymbol{\sigma}')$ with regard to the stress unknown σ' gives

$$d\mathcal{E}_{\text{com}} = 0 : \frac{\partial \mathcal{E}_{\text{com}}(\boldsymbol{\sigma}')}{\partial \sigma'} = 0 \rightarrow \sigma'_m = \frac{1}{\left\langle \frac{1}{E} \right\rangle} \frac{\delta}{L}$$

where σ'_m is the value of σ' minimizing $\mathcal{E}_{\text{com}}(\boldsymbol{\sigma}')$.

Finally, application of the theorem of minimum complementary energy gives:

$$\begin{aligned} \mathcal{E}_{\text{com}}(\boldsymbol{\sigma}') &\geq \mathcal{E}_{\text{com}}(\boldsymbol{\sigma}) \\ \frac{\sigma_m'^2}{2} AL \left\langle \frac{1}{E} \right\rangle - \delta A \sigma_m' &\geq -\frac{1}{2} E_{\text{app}} A \frac{\delta^2}{L} \end{aligned}$$

where $\mathcal{E}_{\text{com}}(\boldsymbol{\sigma}) = -\mathcal{E}_{\text{pot}}(\boldsymbol{\varepsilon})$ is the target complementary energy determined in Section 6.5.1. Last, use of the homogenized stress solution $\sigma_m' = \frac{1}{\langle 1/E \rangle} \frac{\delta}{L}$ yields

$$\frac{1}{\left\langle \frac{1}{E} \right\rangle} \leq E_{\text{app}}$$

The stress-based variational method provides a lower bound of the apparent elasticity modulus of the heterogeneous sample. ■

6.5.4 The Voigt–Reuss Elasticity Bounds

The variational methods provide an upper and lower bound of the effective modulus of a heterogeneous sample:

$$\frac{1}{\left\langle \frac{1}{E} \right\rangle} \leq E_{\text{app}} \leq \langle G \rangle \frac{3\langle \lambda \rangle + 2\langle G \rangle}{\langle \lambda \rangle + \langle G \rangle}$$

For a two-phase heterogeneous material system with the same Poisson ratio in both phases, the lower bound is known as Reuss bound, the upper one as the Voigt bound.

Exercise 60. Consider a two-phase isotropic elastic material system characterized by the elasticity properties $E_1 > E_2$ and $\nu_1 = \nu_2$, and the volume concentrations

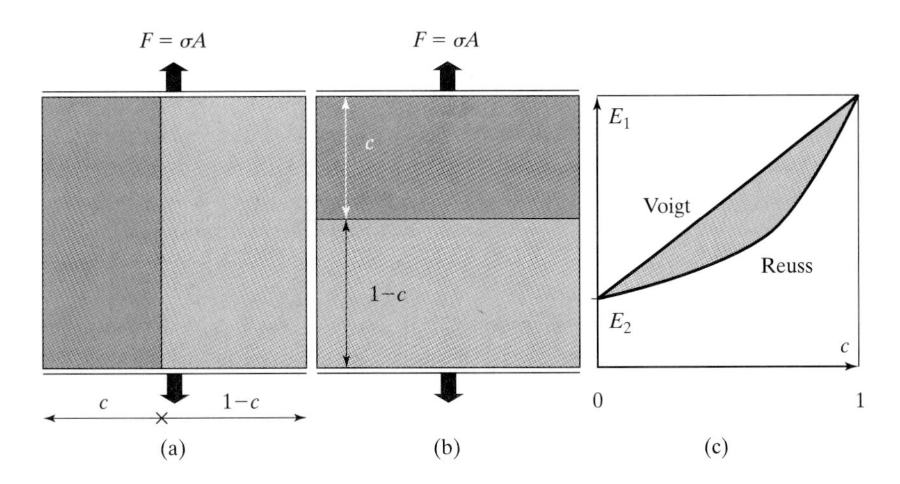

FIGURE 6.7: Illustration of upper and lower bound of a two-phase material system, having the same Poisson ratio: (a) Voigt composite model; (b) Reuss composite model; (c) effective modulus domain.

$c_1 = c$ and $c_2 = 1 - c$ with $c \in [0, 1]$. Determine the effective E-modulus for this two-phase material system as a function of the volume concentration c. Give a 1D graphical representation of the upper and lower bounds, corresponding, respectively, to the Voigt bound and the Reuss bound (See Figure 6.7.
For the lower bound,

$$\left\langle \frac{1}{E} \right\rangle = \frac{1}{\Omega} \int_\Omega \frac{1}{E(\mathbf{x})} d\Omega = \frac{c}{E_1} + \frac{1-c}{E_2}$$

For the upper bound, provided $\nu_1 = \nu_2$,

$$\langle G \rangle \frac{3\langle \lambda \rangle + 2\langle G \rangle}{\langle \lambda \rangle + \langle G \rangle} = \langle E \rangle = E_1 c + E_2(1-c)$$

Thus

$$\frac{E_1 E_2}{E_1(1-c) + E_2 c} \leq E_{\text{app}} \leq E_1 c + E_2(1-c)$$

The lower and upper bounds correspond, respectively, to a constant stress and a constant strain in both phases. This is sketched in Figure 6.7a (Voigt model), and Figure 6.7b (Reuss model). The real effective Young's modulus lies within the Voigt–Reuss bounds, sketched in Figure 6.7c. ∎

6.6 APPENDIX CHAPTER 6: A THERMODYNAMIC ARGUMENT

We want to give a thermodynamic argument for the minimum energy theorems by studying the energy transformation in the elastic material system. To this end, we derive the theorems from the first law and the second law of thermodynamics.

Proceeding as in Section 5.2.2, the first law reads:

$$d\mathbb{U} = d\mathcal{W}_{\text{ext}} + d\mathcal{Q} \tag{6.80}$$

$\mathbb{U}$ is the internal energy of the entire system, and according to the energy balance (i.e., the first law) its variation results from the external energy supply in the form of work (the term $d\mathcal{W}_{\text{ext}}$) and heat (the term $d\mathcal{Q}$). The first term is obtained by applying the theorem of virtual work (6.17) with a real displacement increment $d\boldsymbol{\xi}$:

$$d\mathcal{W}_{\text{ext}}(d\boldsymbol{\xi}, \boldsymbol{\sigma}) = \underbrace{\int_\Omega \rho \mathbf{f} \cdot d\boldsymbol{\xi}\, d\Omega + \int_{\partial\Omega_{\mathbf{T}^d}} \mathbf{T}^d \cdot d\boldsymbol{\xi}\, da}_{d\Phi(d\boldsymbol{\xi})} + \underbrace{\int_{\partial\Omega_{\boldsymbol{\xi}^d}} \mathbf{T} \cdot d\boldsymbol{\xi}^d da}_{d\Phi^*(\boldsymbol{\sigma})} \tag{6.81}$$

For simplicity, we will assume isothermal evolutions ($T = T_0$). In this case, the entropy balance of the system expressed by the second law reads

$$d\mathcal{D} = T_0 dS - d\mathcal{Q} \geq 0 \tag{6.82}$$

where dS is the total entropy variation in the system and $d\mathcal{D}$ is the overall dissipation (i.e., the change of useful mechanical work into heat form). Due to the assumed elastic behavior $d\mathcal{D} = 0$, and therefore

$$d\mathcal{D} = 0 = d\mathcal{W}_{\text{ext}} + T_0 dS - d\mathbb{U} = 0 \tag{6.83}$$

Finally, if we introduce the global free energy of the system, that is,

$$W = \int_{\Omega_t} \psi d\Omega = \mathbb{U} - T_0 \mathcal{S} \tag{6.84}$$

we obtain the Clausius–Duhem equality for the entire system in elastic evolutions:

$$d\mathcal{D} = 0 : d\mathcal{W}_{\text{ext}} - dW = 0 \tag{6.85}$$

It states that the amount of externally supplied work, $d\mathcal{W}_{\text{ext}}$, is stored in the system as free energy, dW. It is instructive to introduce the potential energy:

$$\mathcal{E}_{\text{pot}} = \underbrace{\int_{\Omega} \psi d\Omega}_{W(\varepsilon)} - \underbrace{\left(\int_{\Omega} \rho \mathbf{f} \cdot \boldsymbol{\xi} d\Omega + \int_{\partial_0 \Omega_{\mathbf{T}^d}} \mathbf{T}^d \cdot \boldsymbol{\xi} da \right)}_{\Phi(\boldsymbol{\xi})} \tag{6.86}$$

Note that the increment of the potential energy $\mathcal{E}_{\text{pot}}(\boldsymbol{\xi})$ reads

$$d\mathcal{E}_{\text{pot}} = \underbrace{\int_{\Omega} d\psi \, d\Omega}_{dW} - \left[\underbrace{\left(\int_{\Omega} \rho \mathbf{f} \cdot d\boldsymbol{\xi} \, d\Omega + \int_{\partial \Omega_{\mathbf{T}^d}} \mathbf{T}^d \cdot d\boldsymbol{\xi} \, da \right)}_{d\Phi(d\boldsymbol{\xi})} + \int_{\partial \Omega_{\mathbf{T}^d}} d\mathbf{T}^d \cdot \boldsymbol{\xi} da \right] \tag{6.87}$$

Here, we assumed that the externally imposed volume force density $\rho \mathbf{f}$ is constant (as is the case for gravity forces), while $d\mathbf{T}^d$ is the stress vector increment prescribed at $\partial \Omega_{\mathbf{T}^d}$.[12] Use of $d\mathcal{E}_{\text{pot}}$ in the Clausius–Duhem equality yields

$$\begin{aligned} d\mathcal{D} &= \underbrace{\int_{\Omega} \rho \mathbf{f} \cdot d\boldsymbol{\xi} \, d\Omega + \int_{\partial \Omega_{\mathbf{T}^d}} \mathbf{T}^d \cdot d\boldsymbol{\xi} \, da}_{d\Phi(d\boldsymbol{\xi})} + \underbrace{\int_{\partial \Omega_{\boldsymbol{\xi}^d}} \mathbf{T} \cdot d\boldsymbol{\xi}^d da}_{d\Phi^*(\boldsymbol{\sigma})} - dW \\ &= -d\mathcal{E}_{\text{pot}} + \underbrace{\int_{\partial \Omega_{\boldsymbol{\xi}^d}} \mathbf{T} \cdot d\boldsymbol{\xi}^d da}_{d\Phi^*(\boldsymbol{\sigma})} + \int_{\partial \Omega_{\mathbf{T}^d}} d\mathbf{T}^d \cdot \boldsymbol{\xi} da = 0 \end{aligned} \tag{6.88}$$

Therefore, provided the convexity of the potential energy, the displacement solution $\boldsymbol{\xi}$ realizes, at constant prescribed surface forces (i.e., $d\mathbf{T}^d = 0$) and constant prescribed displacements (i.e., $d\boldsymbol{\xi}^d = 0$), a minimum:

$$d\mathbf{T}^d = d\boldsymbol{\xi}^d = 0 : d\mathcal{D} = -d\mathcal{E}_{\text{pot}} = -\frac{\partial}{\partial \boldsymbol{\xi}} \Big[W(\varepsilon) - \Phi(\boldsymbol{\xi}) \Big] \cdot d\boldsymbol{\xi} = 0 \tag{6.89}$$

[12] The last term in the incremental formulation of the potential energy (the term $d\mathcal{E}_{\text{pot}} = \ldots - \int_{\partial \Omega_{\mathbf{T}^d}} d\mathbf{T}^d \cdot \boldsymbol{\xi} da$) does not figure in the expression of the incrementally supplied external work $d\mathcal{W}_{\text{ext}} = d\mathcal{W}_{\text{ext}}(d\boldsymbol{\xi}, \boldsymbol{\sigma})$, which—we repeat—is an application of the virtual work theorem (6.17) with an incremental displacement field $d\boldsymbol{\xi}$. In return, $d\mathcal{E}_{\text{pot}}$ is the total differential of $\mathcal{E}_{\text{pot}}$ with regard to all its arguments. We highlight this difference through the expressions $d\mathcal{W}_{\text{ext}} = d\mathcal{W}_{\text{ext}}(d\boldsymbol{\xi}, \boldsymbol{\sigma})$ and $d\Phi = d\Phi(d\boldsymbol{\xi})$.

$$\forall d\boldsymbol{\xi};\quad d\mathcal{D} = 0 \Leftrightarrow \frac{\partial}{\partial \boldsymbol{\xi}}\left[W(\boldsymbol{\varepsilon}) - \Phi(\boldsymbol{\xi})\right] = 0 \tag{6.90}$$

We have used this line of arguments in Section 6.2.1.

Exercise 61. Following the line of arguments developed in Section 6.3.1, show that the stress solution $\boldsymbol{\sigma}$ realizes, at constant prescribed surface forces (i.e., $d\mathbf{T}^d = 0$) and constant prescribed displacements (i.e., $d\boldsymbol{\xi}^d = 0$), a minimum of the complementary energy:

$$d\mathbf{T}^d = d\boldsymbol{\xi}^d = 0 : d\mathcal{D} = d\mathcal{E}_{\text{com}} = \frac{\partial}{\partial \boldsymbol{\sigma}}\left[W^*(\boldsymbol{\sigma}) - \Phi^*(\boldsymbol{\sigma})\right] : d\boldsymbol{\sigma} = 0 \tag{6.91}$$

$$\forall d\boldsymbol{\sigma};\quad d\mathcal{D} = 0 \Leftrightarrow \frac{\partial}{\partial \boldsymbol{\sigma}}\left[W^*(\boldsymbol{\sigma}) - \Phi^*(\boldsymbol{\sigma})\right] = 0 \tag{6.92}$$

■

6.7 PROBLEM SET: WATER FILLING OF A GRAVITY DAM

We consider a gravity dam of triangular shape (height H, top angle α) clamped on its base OA (zero displacement prescribed); see Figure 6.8. The downstream surface AB is stress free. The upstream wall OB is subjected to the water pressure (body force $\rho_w g$) over a height $OB' = mH$, where $m \in [0,1]$. The remaining surface $B'B$ is stress free. For the purpose of analysis, we assume that the gravity dam is composed of a linear isotropic elastic material (E, ν or λ, G) with volume mass ρ_d. With regard to the dimension of the gravity dam in the Oz-direction, the problem can be treated as a plane strain problem with regard to the plane parallel to Oxy.

We want to evaluate the displacement of the dam along the upstream wall OB with water filling $m \in [0,1]$.

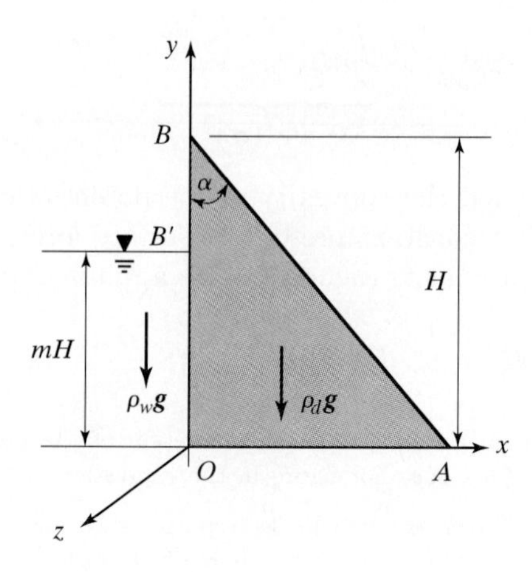

FIGURE 6.8: Problem set: simplified geometry of a gravity dam.

1. **First Approximation**: We consider a displacement field of the form

$$\boldsymbol{\xi}' = a_0 \frac{y}{H} \mathbf{e}_x + b \frac{y}{H} \mathbf{e}_y$$

Using the displacement based variational method, give a first approximation of the displacement along OB. Discuss the corresponding stress solution. Show that the displacement-based variational approach can be recast in the format of the finite element method:

$$k_{ij} q_j = Q_j$$

where k_{ij} are the components of the stiffness matrix, q_j are the unknown nodal displacements, and Q_j are nodal forces. Determine k_{ij} and Q_j for the given approximation of the displacement $\boldsymbol{\xi}'$ in the gravity dam.

2. **Second Approximation**: To improve the solution, we consider a second displacement field:

$$\boldsymbol{\xi}' = a(y) \mathbf{e}_x + b \frac{y}{H} \mathbf{e}_y$$

Based on the shortcomings of the previous approximation, propose a better solution for $a(y)$, and show that this solution is a better approximation for the displacement along OB. (Energy argument required) Discuss the corresponding stress solution.

3. **Finite Element Approximation**: Consider $m = 1$ and $\rho_d = 0$. By means of the finite element method, show that the horizontal displacement solution converges with increasing number of elements. To this end, normalize the obtained displacement at $y = H$ by

$$\Xi = \frac{\xi_x (x = 0, y = H)}{\left(\frac{\rho_w g H^2}{3G \tan \alpha} \right)}$$

with G the shear modulus. Represent your results in the form of a graph giving Ξ as a function of the number of elements over OB.

6.7.1 First Approximation: Linear Triangle Element

Displacement-based variational method (see Section 6.2.1):

$$\mathcal{E}_{\text{pot}}(\boldsymbol{\xi}) = \min_{\boldsymbol{\xi}' = \boldsymbol{\xi}^d \text{ on } \partial\Omega_{\boldsymbol{\xi}^d}} \left[W(\boldsymbol{\varepsilon}') - \Phi(\boldsymbol{\xi}') \right]$$

For an isotropic material (see Section 6.2.3)

$$W(\boldsymbol{\varepsilon}') = \int_\Omega \left[\frac{\lambda}{2} (\operatorname{tr} \boldsymbol{\varepsilon}')^2 + G \operatorname{tr} (\boldsymbol{\varepsilon}' \cdot \boldsymbol{\varepsilon}') \right] d\Omega$$

$$\Phi(\boldsymbol{\xi}') = \int_\Omega \boldsymbol{\xi}' \cdot (\rho \mathbf{f}) \, d\Omega + \int_{\partial\Omega_{\mathbf{T}^d}} \boldsymbol{\xi}' \cdot \mathbf{T}^d da$$

The steps of the displacement-based variational method are as follows:

1. Choose a kinematically admissible displacement field $\boldsymbol{\xi}' = \boldsymbol{\xi}^d$ on $\partial_{\boldsymbol{\xi}^d}\Omega$.
2. Determine the potential energy $\mathcal{E}_{\text{pot}}(\boldsymbol{\xi}')$ that is,
   - (a) The internal elastic energy $W(\boldsymbol{\varepsilon}')$
   - (b) The external work of prescribed volume and surface forces $\Phi(\boldsymbol{\xi}')$.
3. Minimize the solution with regard to displacement unknowns.

Kinematically Admissible Displacement Field

For the application of the displacement-based variational method, we first need to check whether the given displacement field $\boldsymbol{\xi}'$ is kinematically admissible:

$$\boldsymbol{\xi}' = a_0 \frac{y}{H}\mathbf{e}_x + b\frac{y}{H}\mathbf{e}_y$$

where a_0 and b represent the horizontal and vertical displacement of the dam crown (Point B). This displacement field is kinematically admissible, since

$$\text{on } \partial_{\boldsymbol{\xi}^d}\Omega = OA : \boldsymbol{\xi}'(y=0) = \boldsymbol{\xi}^d = \mathbf{0}$$

Potential Energy Component I: Elastic Energy

Strain:

$$(\boldsymbol{\varepsilon}') = \frac{1}{2H}\begin{bmatrix} 0 & a_0 & 0 \\ & 2b & 0 \\ \text{sym} & & 0 \end{bmatrix}$$

Invariants of strain tensor (third invariant not required since isotropic behavior):

$$\operatorname{tr}\boldsymbol{\varepsilon}' = b/H$$

$$\operatorname{tr}(\boldsymbol{\varepsilon}'\cdot\boldsymbol{\varepsilon}') = \frac{1}{2}\left(\frac{a_0}{H}\right)^2 + \left(\frac{b}{H}\right)^2$$

Elastic energy on material level:

$$\begin{aligned} \text{in } d\Omega : \psi(\boldsymbol{\varepsilon}') &= \frac{\lambda}{2}(b/H)^2 + G\left(\frac{1}{2}(a_0/H)^2 + (b/H)^2\right) \\ &= \frac{1}{2}G\left(\frac{a_0}{H}\right)^2 + \left(\frac{\lambda}{2}+G\right)\left(\frac{b}{H}\right)^2 = \text{const} \end{aligned}$$

Elastic energy on structural level:

$$\text{in } \Omega = A1_z = \frac{H^2\tan\alpha}{2}1_z : W(\boldsymbol{\varepsilon}'_{(0)}) = \int_\Omega \psi(\boldsymbol{\varepsilon}'_{(0)})d\Omega = \psi(a_0,b)A1_z$$

where A is the section of the triangle, and 1_z is the unit length in the z-direction.

Potential Energy Component II: External Work by Prescribed Forces

External work by prescribed volume forces (gravity $\mathbf{f} = -g\mathbf{e}_y$):

$$\Phi^v(\boldsymbol{\xi}') = \int_\Omega \boldsymbol{\xi}' \cdot (\rho \mathbf{f})\, d\Omega = -\int_\Omega \left(a_0 \frac{y}{H}\mathbf{e}_x + b\frac{y}{H}\mathbf{e}_y\right) \cdot (\rho_d g \mathbf{e}_y) d\Omega = -\frac{gb}{H}\int_\Omega \rho_d y\, d\Omega$$

Note that $\int_\Omega \rho_d y\, d\Omega$ represents close to a multiplied constant the center of gravity of in the y-direction of the triangle:

$$y_G = \frac{1}{M}\int_\Omega \rho_d y\, d\Omega = \frac{H}{3}$$

with $M = \rho_d \Omega$ the mass of the dam. Thus,

$$\Phi^v(\boldsymbol{\xi}') = -\frac{bgM}{H} y_G = -\frac{b\rho_d g}{3} A 1_z$$

Prescribed surface forces along OB' (hydrostatic water pressure):

$$\text{on } OB' : \mathbf{T}^d = p(y)\mathbf{e}_x = \rho_w g(mH - y)\mathbf{e}_x$$

Thus,

$$\begin{aligned} \Phi^T(\boldsymbol{\xi}') &= \int_{\partial\Omega_{\mathbf{T}^d}} \boldsymbol{\xi}' \cdot \mathbf{T}^d dS = \int_{\partial\Omega_{\mathbf{T}^d}} \left(a_0 \frac{y}{H}\mathbf{e}_x + b\frac{y}{H}\mathbf{e}_y\right) \cdot \rho_w g(mH - y)\mathbf{e}_x dS \\ &= \int_{\partial\Omega_{\mathbf{T}^d}} a_0 \frac{y}{H}\rho_w g(mH - y)\, dS \end{aligned}$$

The surface force allows for a straightforward geometric interpretation: It represents the work done by the force resultant F on OB' along the displacement $\xi_x^{'}(x_{G(OB')})$ at the center of gravity of the triangle pressure block:

$$\Phi^T(\boldsymbol{\xi}') = F\xi_x^{'}(y_{G(OB')})$$

The force resultant is

$$F = \frac{1}{2}\rho_w g(mH)^2 1_z$$

The displacement is

$$\xi_x^{'}(y_{G(OB')} = mH/3) = a_0 \frac{m}{3}$$

Thus

$$\Phi^T(\boldsymbol{\xi}') = \frac{a_0}{6}\rho_w g m^3 H^2 1_z$$

In summary,

$$\Phi(\boldsymbol{\xi}') = \left(-\frac{b\rho_d g}{3} A + \frac{a_0}{6}\rho_w g m^3 H^2\right) 1_z$$

Minimization Problem

Potential energy:

$$\begin{aligned}\mathcal{E}_{\text{pot}}(\boldsymbol{\xi}') &= W(\boldsymbol{\varepsilon}') - \Phi(\boldsymbol{\xi}') \\ &= \left[\frac{1}{2}G\left(\frac{a_0}{H}\right)^2 A + \left(\frac{\lambda}{2}+G\right)\left(\frac{b}{H}\right)^2 A + b\frac{\rho_d g}{3}A - \frac{a_0}{6}\rho_w g m^3 H^2\right] 1_z\end{aligned}$$

At constant prescribed volume and surface forces,

$$d\mathcal{E}_{\text{pot}}(a,b) = 0 : \frac{\partial \mathcal{E}_{\text{pot}}}{\partial a_0} da_0 + \frac{\partial \mathcal{E}_{\text{pot}}}{\partial b} db = 0$$

Since da_0 and db can take any value, it follows that

$$\forall da_0, \forall db : \frac{\partial \mathcal{E}_{\text{pot}}}{\partial a_0} = 0; \quad \frac{\partial \mathcal{E}_{\text{pot}}}{\partial b} = 0$$

Therefore, here (with $A = \frac{1}{2}H^2 \tan\alpha$)

$$\frac{\partial \mathcal{E}_{\text{pot}}}{\partial a} = \left[\frac{G}{H}\left(\frac{a_0}{H}\right)A - \frac{1}{6}\rho_w g m^3 H^2\right] 1_z = 0 \rightarrow \frac{a_0}{H} = \frac{\rho_w g m^3 H}{3G\tan\alpha}$$

$$\frac{\partial \mathcal{E}_{\text{pot}}}{\partial b} = \left[\frac{1}{H}(\lambda + 2G)\left(\frac{b}{H}\right) + \frac{\rho_d g}{3}\right] A 1_z = 0 \rightarrow \frac{b}{H} = -\frac{\rho_d g H}{3(\lambda + 2G)}$$

Note that the horizontal displacement a_0 scales with m^3 and that it does not depend on the vertical displacement, which results only from the dams deadweight.

Stress Solution

The elastic stresses associated with the displacement field $\boldsymbol{\xi}'$ read

$$\sigma_{ij} = 2G\varepsilon'_{ij} + \lambda\varepsilon'_{kk}\delta_{ij}$$

$$\sigma_{xx} = \sigma_{zz} = -\frac{\rho_d g H}{3(1+2G/\lambda)}; \quad \sigma_{xy} = \frac{\rho_w g m^3 H}{3\tan\alpha}; \quad \sigma_{yy} = -\frac{\rho_d g H}{3}$$

This stress field is not statically admissible. It does *not* satisfy the following:

1. The momentum balance equation in the y-direction:

$$\frac{\partial \sigma_{xy}}{\partial x} + \frac{\partial \sigma_{yy}}{\partial y} + \frac{\partial \sigma_{yz}}{\partial z} - \rho_d g = 0$$

2. The boundary conditions:

$$\text{on } OB' : \mathbf{T}^d = \rho_w g(mH - y)\mathbf{e}_x \neq \boldsymbol{\sigma}\cdot\mathbf{n} = -\boldsymbol{\sigma}\cdot\mathbf{e}_x = -\sigma_{xx}\mathbf{e}_x - \sigma_{xy}\mathbf{e}_y$$

$$\text{on } B'B : \mathbf{T}^d = \mathbf{0} \neq -\boldsymbol{\sigma}\cdot\mathbf{e}_x$$

$$\begin{aligned}\text{on } AB \quad &: \quad \mathbf{T}^d = \mathbf{0} \neq \boldsymbol{\sigma}\cdot\mathbf{n} = \boldsymbol{\sigma}\cdot(\sin\alpha\mathbf{e}_x + \cos\alpha\mathbf{e}_y) \\ &= \sin\alpha(\sigma_{xx}\mathbf{e}_x + \sigma_{xy}\mathbf{e}_y) + \cos\alpha(\sigma_{xy}\mathbf{e}_x + \sigma_{yy}\mathbf{e}_y) \\ &= (\sigma_{xx}\sin\alpha + \sigma_{xy}\cos\alpha)\mathbf{e}_x + (\sigma_{xy}\sin\alpha + \sigma_{yy}\cos\alpha)\mathbf{e}_y\end{aligned}$$

Finite Element Method

The solution we derived is nothing but the development of a linear elastic triangle finite element, for which the displacement is approximated by linear basis function, and $q_1 = a_0$ and $q_2 = b$ as nodal displacements. The elastic energy in discretized form reads

$$W(q_i, q_j) = \frac{1}{2} k_{ij} q_i q_j = \frac{1}{2} \begin{pmatrix} a_0 & b \end{pmatrix} \begin{bmatrix} k_{aa} & k_{ab} \\ \text{sym} & k_{bb} \end{bmatrix} \begin{pmatrix} a_0 \\ b \end{pmatrix}$$

where k_{ij} are the components of the stiffness matrix

$$k_{aa} = \frac{G}{H^2} A 1_z; \quad k_{ab} = k_{ba} = 0; \quad k_{bb} = \frac{\lambda + 2G}{H^2} A 1_z$$

The external work reads

$$\Phi(q_i) = Q_i q_i = \begin{pmatrix} Q_a & Q_b \end{pmatrix} \begin{pmatrix} a_0 \\ b \end{pmatrix}$$

where Q_i are the nodal forces

$$Q_a = \frac{\rho_w g m^3 H^2}{6} 1_z; \quad Q_b = -\frac{\rho_d g}{3} A 1_z$$

The solution $q_i = (a_0, b)$ minimizes the potential energy

$$\mathcal{E}_{\text{pot}}(q_i, q_j) = W(q_i, q_j) - \Phi(q_i)$$

Thus,

$$d\mathcal{E}_{\text{pot}} = 0 \rightarrow \frac{\partial [W(q_i, q_j) - \Phi(q_i)]}{\partial q_i} = k_{ij} q_j - Q_i = 0$$

$$\begin{bmatrix} \frac{G}{H^2} A 1_z & 0 \\ 0 & \frac{\lambda+2G}{H^2} A 1_z \end{bmatrix} \begin{pmatrix} a_0 \\ b \end{pmatrix} - \begin{pmatrix} \frac{1}{6} \rho_w g m^3 H^2 1_z \\ -\frac{1}{3} \rho_d g A 1_z \end{pmatrix} = 0$$

6.7.2 Second Approximation: Higher-Order Polynomials

Zero Distortion Condition.

The main shortcoming of the first solution is that it does not respect the rectangular angle at O. The strain $\varepsilon'_{xy} = a_0/H$ represents the distortion. Therefore, an improvement can be achieved by considering a displacement field, which respects, in addition to the clamped boundary condition along OA,

$$\varepsilon'_{xy}(y = 0) = 0$$

For instance, consider a polynomial development of the displacement field

$$\xi'_x = a(y) = a_0 \frac{y}{H} + a_1 \left(\frac{y}{H} \right)^2 + \ldots$$

This displacement is still kinematically admissible but obeys, in addition, the zero distortion condition for

$$\varepsilon'_{xy}(y=0) = 0 \leftrightarrow a_0 = 0$$

$$\varepsilon'_{xy}(y) = \frac{1}{2H}\left(2a_1\left(\frac{y}{H}\right) + ...\right)$$

Potential Energy.

In what follows, we restrict ourselves to a linear distortion variation along y, thus

$$(\varepsilon') = \frac{1}{H}\begin{bmatrix} 0 & a_1\left(\frac{y}{H}\right) & 0 \\ & b & 0 \\ \text{sym} & & 0 \end{bmatrix}$$

$$\operatorname{tr}\varepsilon' = b/H; \quad \operatorname{tr}(\varepsilon'\cdot\varepsilon') = 2\left(\frac{a_1}{H}\right)^2\left(\frac{y}{H}\right)^2 + \left(\frac{b}{H}\right)^2$$

Elastic energy:

$$\begin{aligned} W(\varepsilon') &= \int_\Omega \left[\left(\frac{\lambda}{2}+G\right)(b/H)^2 + 2G\left(\frac{a_1}{H}\right)^2\left(\frac{y}{H}\right)^2\right] d\Omega \\ &= \left[\frac{G}{6}\left(\frac{a_1}{H}\right)^2 + \left(\frac{\lambda}{2}+G\right)\left(\frac{b}{H}\right)^2\right] A1_z \end{aligned}$$

External energy of body forces (same as before):

$$\Phi^v(\boldsymbol{\xi}') = -\frac{bgM}{H}y_G = -\frac{b\rho_d g}{3}A1_z$$

External energy of surface forces (see Section 6.7.1):

$$\Phi^T(\boldsymbol{\xi}') = \int_{OB'} \boldsymbol{\xi}'\cdot\mathbf{T}^d dS = F\xi'_x(y_{G(OB')})$$

with

$$F = \frac{1}{2}\rho_w g(mH)^2 1_z$$

$$\xi'_x(y_{G(OB')} = mH/3) = a_1\left(\frac{m}{3}\right)^2$$

Thus

$$\Phi^T(\boldsymbol{\xi}') = \frac{a_1}{18}\rho_w g m^4 H^2 1_z$$

Total external work:

$$\Phi(\boldsymbol{\xi}') = \left[-\frac{b\rho_d g}{3}A + \frac{a_1}{18}\rho_w g m^4 H^2\right]1_z$$

Potential energy:

$$\begin{aligned} \mathcal{E}_{\text{pot}}(\boldsymbol{\xi}') &= W(\varepsilon') - \Phi(\boldsymbol{\xi}') \\ &= \left[\frac{G}{6}\left(\frac{a_1}{H}\right)^2 A + \left(\frac{\lambda}{2}+G\right)\left(\frac{b}{H}\right)^2 A + b\frac{\rho_d g}{3}A - \frac{a_1}{18}\rho_w g m^4 H^2\right]1_z \end{aligned}$$

This potential energy is not smaller than the one determined with a linear displacement along OB. Therefore, from an energy point of view, it is not closer to the exact solution:

$$\mathcal{E}_{\text{pot}}(\boldsymbol{\xi}) = \min_{\boldsymbol{\xi}'=\boldsymbol{\xi}^d \text{ on } \partial\Omega_{\boldsymbol{\xi}^d}} [W(\boldsymbol{\varepsilon}') - \Phi(\boldsymbol{\xi}')]$$

Displacement Solution.

$$\frac{\partial \mathcal{E}_{\text{pot}}}{\partial a_1} = \left[\frac{G}{3H}\left(\frac{a_1}{H}\right)A - \frac{1}{18}\rho_w g m^4 H^2\right] 1_z = 0 \rightarrow \frac{a_1}{H} = \frac{\rho_w g m^4 H}{3G \tan\alpha}$$

$$\frac{\partial \mathcal{E}_{\text{pot}}}{\partial a_1} = 0 \rightarrow \frac{b}{H} = -\frac{\rho_d g H}{3(\lambda + 2G)}$$

Stress Solution.

The only stress that changes is the shear stress:

$$\sigma_{xy} = \frac{2\rho_w g m^4 H}{3\tan\alpha}\left(\frac{y}{H}\right)$$

The stress field is not statically admissible.

6.7.3 Finite Element Approximation

From the previous analysis, it turns out that for $m = 1$ the horizontal displacement at the dam crown a_1 coincides with the one obtained with a linear base function (i.e., $a_1 = a_0$), and the solutions differ only in shape in between $0 < y/H < 1$. This motivates us to explore the convergence of the finite element analysis by the dimensionless parameter:

$$\Xi = \frac{\xi_x(x = 0, y = H)}{\left(\frac{\rho_w g H^2}{3G \tan\alpha}\right)}$$

6.8 PROBLEM SET: ELASTIC SETTLEMENT BOUNDS BELOW A CIRCULAR FOUNDATION

We consider a soil layer of height h on a rigid substrate (see Figure 6.9). The soil layer is subjected at its surface to a pressure load of intensity p of a circular foundation of radius R. The soil layer is assumed to behave as an isotropic linear elastic. We want to give an approximation of the settlement at $z = 0$ below the circular foundation.

1. **Statically Admissible Stress Field**: For the purpose of analysis, we separate the domain Ω in two subdomains, denoted, respectively, Ω_1 and Ω_2, as shown in Figure 6.9. In these domains, we consider the following stress fields:

   (a) In Ω_1 defined by $0 \le z \le h$ and $r \le R$:

$$\sigma'_{rr} = q'; \quad \sigma'_{\theta\theta} = q''; \quad \sigma'_{zz} = \sigma \text{ (other } \sigma'_{ij} = 0)$$

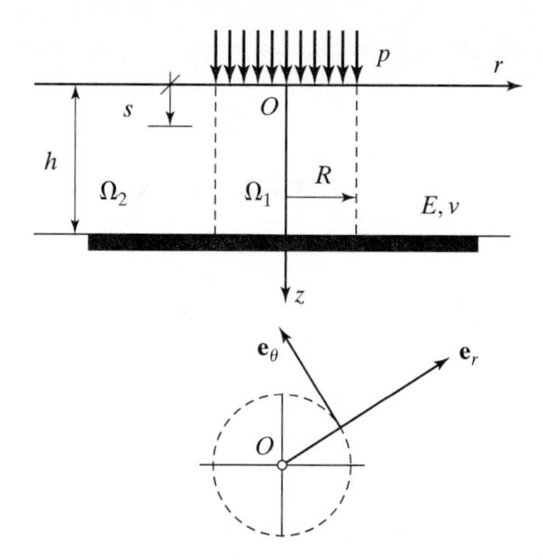

FIGURE 6.9: Problem set: settlement below a circular foundation.

(b) In Ω_2 defined by $0 \leq z \leq h$ and $r \geq R$:

$$\sigma'_{rr} = -q(R/r)^2; \quad \sigma'_{\theta\theta} = q(R/r)^2 \text{ (other } \sigma'_{ij} = 0)$$

This stress field is the elastic solution of a cylinder tube with an infinite outer radius subjected to an internal pressure q at $r = R$ (see Training Section 5.5, Chapter 5).

Determine the constants q', q'' and σ, so that the stress field $\boldsymbol{\sigma}'$ is statically admissible in $\Omega = \Omega_1 \cup \Omega_2$. Is the stress field $\boldsymbol{\sigma}'$ the solution of the linear elastic problem?

2. **Lower Energy Bound–Upper Displacement Bound**: We consider the previously found statically admissible stress solution.

(a) For this stress field, show that the complementary energy associated with the stress field $\boldsymbol{\sigma}'$ reads per unit height

$$\frac{\mathcal{E}_{\text{com}}(\boldsymbol{\sigma}')}{h} = \frac{W^*(\boldsymbol{\sigma}') - \Phi^*(\boldsymbol{\sigma}')}{h} = \frac{\pi R^2}{2E}\left[4q^2 + p^2 - 4\nu pq\right]$$

where E and ν denote, respectively, the Young's modulus and the Poisson ratio. (*Hint*: Carefully apply Clapeyron's formula.) In this regard, we recall that the elastic displacement field, say $\boldsymbol{\xi}^{*\prime}$, which is related to the considered stress solution $\boldsymbol{\sigma}'$ in Ω_2 through a linear elastic constitutive law, reads (see Section 5.5.3)

$$\boldsymbol{\xi}^{*\prime} = \xi_r \mathbf{e}_r = q\frac{R^2}{2Gr}\mathbf{e}_r$$

(b) Let s be the elastic settlement at $z = 0$, which represents the component of the displacement field in the z-direction. Let $\langle s \rangle$ be the average settlement below the circular foundation, that is,

$$\langle s \rangle = \frac{1}{\pi R^2} \int_{A=\pi R^2} s(z=0,r,\theta)\, da$$

By using the stress-based variational method, determine an upper bound for $\langle s \rangle$.

3. **Upper Energy Bound–Lower Displacement Bound**: Can the displacement field employed in Part 2 be employed to determine, by means of a displacement-based variational method, a lower bound of $\langle s \rangle$? (Determination not required)

6.8.1 Statically Admissible Stress Field

$\boldsymbol{\sigma}'$ is statically admissible provided that it satisfies the following:

- The equilibrium condition ($\operatorname{div} \boldsymbol{\sigma}' = 0$); thus in Ω_1 for $\boldsymbol{\sigma}' = \text{const}$

$$\sigma_{rr} - \sigma_{\theta\theta} = 0 \rightarrow q' = q''$$

- The symmetry of the stress tensor ($\boldsymbol{\sigma}' = {}^t\boldsymbol{\sigma}'$)
- The force boundary condition on $\partial\Omega_{\mathbf{T}^d}$:

$$z = 0, r \leq R : \mathbf{T}^d = +p\mathbf{e}_z \equiv \mathbf{T}(\mathbf{n} = -\mathbf{e}_z) = -\boldsymbol{\sigma}' \cdot \mathbf{e}_z = -\sigma\mathbf{e}_z \rightarrow \sigma = -p$$

$$z = 0, r \geq R : \mathbf{T}^d = \mathbf{0} \equiv \mathbf{T}(\mathbf{n} = -\mathbf{e}_z) = -\boldsymbol{\sigma}' \cdot \mathbf{e}_z = \mathbf{0}$$

- The stress vector continuity at the interface between Ω_1 and Ω_2:

$$0 \leq z \leq h, r = R : \mathbf{T}(\mathbf{n} = \mathbf{e}_r) + \mathbf{T}(-\mathbf{n} = -\mathbf{e}_r) = \mathbf{0} \leftrightarrow q'\mathbf{e}_r + q\mathbf{e}_r = 0 \rightarrow q' = -q$$

In summary,

$$q' = q'' = -q; \quad \sigma = -p$$

This statically admissible stress field $\boldsymbol{\sigma}'$, however, is not the solution of the problem, since the nonzero stress $\sigma_{\theta\theta}$ is not compatible with the prescribed zero displacement at the rigid substrate (i.e., $\boldsymbol{\xi}^{*'} \neq \boldsymbol{\xi}^d$ on $\partial\Omega_{\boldsymbol{\xi}^d}$). Therefore, the associated displacement field is not kinematically admissible.

6.8.2 Lower Energy Bound–Upper Displacement Bound

A prerequisite for the use of the stress-based variational method is that the stress field $\boldsymbol{\sigma}'$ is statically admissible. We have shown this in Section 6.8.1.

a. Complementary Energy

$$\mathcal{E}_{\text{com}}(\boldsymbol{\sigma}') = W^*(\boldsymbol{\sigma}') - \Phi^*(\boldsymbol{\sigma}')$$

For the problem at hand, $\Phi^*(\boldsymbol{\sigma}') = 0$, since the only prescribed displacement is the zero displacement at $z = h$, thus

$$\Phi^*(\boldsymbol{\sigma}') = \Phi^*(\boldsymbol{\sigma}) = \int_{\partial\Omega_{\boldsymbol{\xi}^d}} \boldsymbol{\xi}^d \cdot (\boldsymbol{\sigma}' \cdot \mathbf{n})\, da = 0$$

To determine the internal complementary energy, we make use of the subdomain separation:

$$W^*(\boldsymbol{\sigma}') = W^*_{\Omega_1}(\boldsymbol{\sigma}') + W^*_{\Omega_2}(\boldsymbol{\sigma}')$$

1. The internal complementary energy stored in Ω_1 is readily determined from the standard formula:

$$W^*_{\Omega_1}(\boldsymbol{\sigma}') = \int_{\Omega_1} \frac{1}{2E} \left[-\nu(\operatorname{tr}\boldsymbol{\sigma}')^2 + (1+\nu)\operatorname{tr}(\boldsymbol{\sigma}' \cdot \boldsymbol{\sigma}')\right] d\Omega_1$$

Noting that

$$\text{in } \Omega_1 : \operatorname{tr}\boldsymbol{\sigma}' = -(2q + p); \quad \operatorname{tr}(\boldsymbol{\sigma}' \cdot \boldsymbol{\sigma}') = 2q^2 + p^2$$

we obtain

$$\begin{aligned} W^*_{\Omega_1}(\boldsymbol{\sigma}') &= \int_{\Omega_1} \frac{1}{2E}\left[-\nu(2q+p)^2 + (1+\nu)(2q^2+p^2)\right] d\Omega_1 \\ &= \frac{\pi R^2 h}{2E}\left[p^2 - 4\nu q p + 2q^2(1-\nu)\right] \end{aligned}$$

2. For the internal complementary energy stored in the infinite domain Ω_2, we make use of the fact that $(\boldsymbol{\sigma}', \boldsymbol{\xi}^{*\prime})$ constitutes the solution of the elastic problem of an internal pressure q prescribed at the inside of a cylinder in an infinite solid. Therefore, the internal complementary energy of Ω_2 can be obtained using Clapeyron's formula, which applies whenever a stress field (here $\boldsymbol{\sigma}'$) is associated through a linear elastic material law to a displacement field (here $\boldsymbol{\xi}^{*\prime}$):

$$W^*_{\Omega_2}(\boldsymbol{\sigma}') = \frac{1}{2}\left[\Phi_{\Omega_2}(\boldsymbol{\xi}^{*\prime}) + \Phi^*_{\Omega_2}(\boldsymbol{\sigma}')\right]$$

Here,

$$\begin{aligned} \Phi_{\Omega_2}(\boldsymbol{\xi}^{*\prime}) &= \int_{\partial\Omega_2} \mathbf{T}^d \cdot \boldsymbol{\xi}^{*\prime} da_2 \\ &= \int_{\partial\Omega_2 = 2\pi R h} q\mathbf{e}_r \cdot \boldsymbol{\xi}^{*\prime}(r = R) da_2 = \frac{q^2 \pi R^2 h}{G} = \frac{2q^2 \pi R^2 h(1+\nu)}{E} \end{aligned}$$

$$\Phi^*_{\Omega_2}(\boldsymbol{\sigma}') = \int_{\partial\Omega_2} \mathbf{T}' \cdot \boldsymbol{\xi}^d da_2 = 0$$

Therefore,

$$W^*_{\Omega_2}(\boldsymbol{\sigma}') = \frac{q^2 \pi R^2 h(1+\nu)}{E}$$

In summary,

$$\begin{aligned}\mathcal{E}_{\text{com}}(\boldsymbol{\sigma}') &= W^*_{\Omega_1}(\boldsymbol{\sigma}') + W^*_{\Omega_2}(\boldsymbol{\sigma}') \\ &= \frac{\pi R^2 h}{2E}\left[p^2 - 4\nu qp + 2q^2(1-\nu) + 2q^2(1+\nu)\right] \\ &= \frac{\pi R^2 h}{2E}\left[p^2 - 4\nu qp + 4q^2\right]\end{aligned}$$

b. Upper Settlement Bound The theorem of minimum complementary energy reads

$$\mathcal{E}_{\text{com}}(\boldsymbol{\sigma}) \leq \mathcal{E}_{\text{com}}(\boldsymbol{\sigma}')$$

1. Target solution: For $(\boldsymbol{\sigma}, \boldsymbol{\xi})$ solution of the problem, $\mathcal{E}_{\text{com}}(\boldsymbol{\sigma})$ can be determined from Clapeyron's formula:

$$\begin{aligned}\mathcal{E}_{\text{com}}(\boldsymbol{\sigma}) &= W^*(\boldsymbol{\sigma}) - \Phi^*(\boldsymbol{\sigma}) \\ &= \frac{1}{2}\left[\Phi(\boldsymbol{\xi}) + \Phi^*(\boldsymbol{\sigma})\right] - \Phi^*(\boldsymbol{\sigma}) \\ &= \frac{1}{2}\left[\Phi(\boldsymbol{\xi}) - \Phi^*(\boldsymbol{\sigma})\right] = -\mathcal{E}_{\text{pot}}(\boldsymbol{\xi})\end{aligned}$$

Noting $s = \xi_z(z=0) = \boldsymbol{\xi}(z=0)\cdot\mathbf{e}_z$, we obtain

$$\begin{aligned}\Phi(\boldsymbol{\xi}) &= \int_{\partial\Omega_{\mathbf{T}^d}} \mathbf{T}^d \cdot \boldsymbol{\xi}\, da \\ &= \int_{\partial\Omega_{\mathbf{T}^d}=\pi R^2} p\mathbf{e}_z \cdot \boldsymbol{\xi}(z=0)\, da = p\int_{\pi R^2} \xi_z(0)\, da = p\pi R^2 \langle s\rangle\end{aligned}$$

$$\Phi^*(\boldsymbol{\sigma}) = \int_{\partial\Omega_{\boldsymbol{\xi}^d}} \mathbf{T}\cdot\boldsymbol{\xi}^d da = 0$$

Thus

$$\mathcal{E}_{\text{com}}(\boldsymbol{\sigma}) = \frac{p\pi R^2}{2}\langle s\rangle$$

2. Minimum of $\mathcal{E}_{\text{com}}(\boldsymbol{\sigma}')$:

$$\frac{\partial\mathcal{E}_{\text{com}}(q)}{\partial q}|_{p=\text{const}} = 0 : \frac{\pi R^2 h}{2E}\left[-4\nu p + 8q\right] = 0$$

It follows that

$$q = \frac{\nu p}{2}$$

$$\mathcal{E}_{\text{com}}(\boldsymbol{\sigma}') = \frac{\pi R^2 h}{2E}p^2\left[1-\nu^2\right]$$

3. Finally, application of the theorem of minimum complementary energy reads

$$\mathcal{E}_{\mathrm{com}}(\boldsymbol{\sigma}) = \frac{p\pi R^2}{2}\langle s\rangle \leq \frac{\pi R^2 h}{2E}p^2\left[1-\nu^2\right] = \mathcal{E}_{\mathrm{com}}(\boldsymbol{\sigma}')$$

and delivers an upper bound for the average surface settlement:

$$\langle s\rangle \leq \frac{ph\left[1-\nu^2\right]}{E} = \frac{ph}{3K}$$

6.8.3 Upper Energy Bound–Lower Displacement Bound

The theorem of minimum potential energy reads

$$\mathcal{E}_{\mathrm{pot}}(\boldsymbol{\xi}) \leq \mathcal{E}_{\mathrm{pot}}(\boldsymbol{\xi}'); \quad \boldsymbol{\xi}' = \boldsymbol{\xi}^d \text{ on } \partial\Omega_{\boldsymbol{\xi}^d}$$

Target solution:

$$\mathcal{E}_{\mathrm{pot}}(\boldsymbol{\xi}) = -\mathcal{E}_{\mathrm{com}}(\boldsymbol{\sigma}) = -\frac{p\pi R^2}{2}\langle s\rangle$$

Upper energy bound:

$$\mathcal{E}_{\mathrm{pot}}(\boldsymbol{\xi}') = W(\boldsymbol{\xi}') - \Phi^*(\boldsymbol{\xi}')$$

The displacement field $\boldsymbol{\xi}^{*\prime}$ cannot been used for $\boldsymbol{\xi}'$, since it is not kinematically admissible:

$$\boldsymbol{\xi}^{*\prime} = \frac{qR^2}{2\mu r}\mathbf{e}_r \neq \boldsymbol{\xi}^d(z=h) = \mathbf{0}$$

The lower bound for the settlement can be determined from

$$\langle s\rangle \geq \frac{2}{p\pi R^2}\min[W(\boldsymbol{\xi}') - \Phi^*(\boldsymbol{\xi}')]$$

where $\boldsymbol{\xi}'$ is a kinematically admissible displacement field.

In summary,

$$\frac{2}{p\pi R^2}\min[W(\boldsymbol{\xi}') - \Phi^*(\boldsymbol{\xi}')] \leq \langle s\rangle \leq \frac{ph}{3K}$$

6.9 PROBLEM SET: TORSION OF AN ELASTIC HETEROGENEOUS CYLINDER

Consider the cylinder sample of height h, and circular section of radius R, shown in Figure 6.10. The vertical fibers of the cylinder are initially parallel to the Oz-axis. The points O and M are the centers of the sections at $z=0$, and $z=h$. The cylinder sample is composed of a heterogeneous linear isotropic material. The sample is subjected to torsion in the following way:

- The cylinder walls are stress free.
- The lower surface at $z=0$ is attached to a rigid fixed plate (no movement).

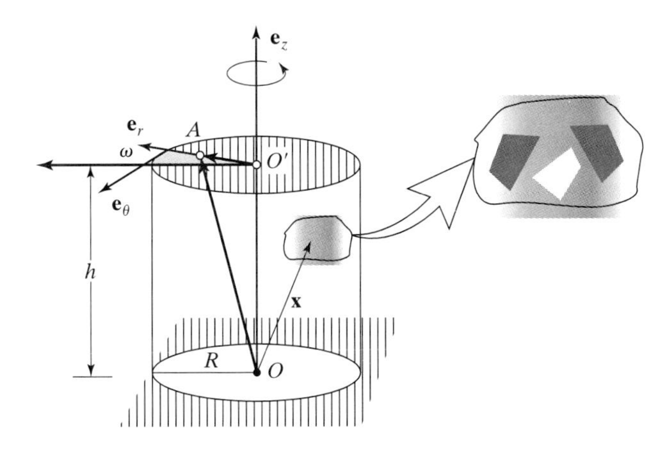

FIGURE 6.10: Problem set: torsion of a heterogeneous cylinder sample.

- The upper surface at $z = h$ is attached to a rotating rigid plate. An infinitesimal rotation ω around the Oz-axis is prescribed. Any point A at this upper surface undergoes an infinitesimal displacement:

$$z = h : \boldsymbol{\xi}^d = \omega \mathbf{e}_z \times \mathbf{MA} = \omega r \mathbf{e}_\theta$$

where $\mathbf{MA} = r\mathbf{e}_r$, and $r \in [0, R]$ is the radial coordinate (see Figure 6.10).

We want to give an upper and a lower bound of the apparent (or effective) shear modulus G of the heterogeneous sample.

1. **Equivalent Homogeneous Sample:** We start with the equivalent homogeneous sample of constant elasticity properties.

   (a) We consider the displacement field:

   $$\text{in } \Omega : \boldsymbol{\xi} = \omega r \frac{z}{h} \mathbf{e}_\theta$$

   Determine the strain tensor and the stress tensor, $\boldsymbol{\sigma}$, using the isotropic linear elastic material law ($G = G_{\text{app}} = \text{const}$). Specify (briefly) why ($\boldsymbol{\sigma}$ and $\boldsymbol{\xi}$) is the solution of the homogeneous problem.

   (b) The $\mathbf{e}_z$-component of the moment vector $\mathcal{M}$ at point M is the torsion moment, denoted M_z. By using the found solution, show that the resultant torsion moment $M_z(z = h)$ reads as

   $$M_z = G_{\text{app}} J \frac{\omega}{h}$$

   where J is the torsion inertia moment.

2. **Heterogeneous Sample**: We now consider the heterogeneous sample [i.e., $G = G(\mathbf{x})$].

(a) The target: Let ($\boldsymbol{\xi}$ and $\boldsymbol{\sigma}$) be the (exact) displacement and stress solution in the heterogeneous sample. For this (unknown) linear elastic solution, show that the elastic energy per unit of height reads as

$$\frac{W(\boldsymbol{\xi})}{h} = \frac{W^*(\boldsymbol{\sigma})}{h} = \frac{1}{2}M_z(z=h)\alpha$$

with $\alpha = \omega/h$.

(b) Upper bound: Inspired by the elastic homogeneous solution, we consider for the heterogeneous sample a displacement field of the form

$$\text{in } \Omega : \boldsymbol{\xi}' = \omega r \frac{z}{h}\mathbf{e}_\theta$$

Determine an upper bound of the apparent shear modulus.

(c) Lower bound: We consider the elastic homogeneous solution developed previously as a first approximation for the stress field in the heterogeneous sample. It is of the form

$$\text{in } \Omega : \boldsymbol{\sigma}' = \sigma_{\theta z}(r)[\mathbf{e}_\theta \otimes \mathbf{e}_z + \mathbf{e}_z \otimes \mathbf{e}_\theta]$$

where $\sigma_{\theta z}(r)$ is assumed to have the same form as the shear stress of the homogeneous sample. Determine a lower bound of the apparent shear modulus.

3. **Composite Cylinder Model**: We want to apply the general solution to the determination of the effective shear modulus of fiber-reinforced materials. One common model is the composite cylinder model. The fiber is considered as one phase, the matrix as the other. The fiber phase of shear modulus G_1 is modeled as a cylinder of radius R_1, embedded within a cylinder of external radius R_2 modeling the matrix phase of shear modulus G_2. Determine the upper and the lower bound. Display the results as a function of the fiber volume concentration c for a typical value of $G_1/G_2 = 10$ (e.g., corresponding to carbon fibers embedded in a polymer matrix). Show that one of the two bounds corresponds precisely to the considered morphology. (*Hint*: Use an energy argument.)

6.9.1 Equivalent Homogeneous Sample.

Solution ($\boldsymbol{\xi}, \boldsymbol{\sigma}$)

The torsion deformation has been studied in Section 2.5. For the displacement field

$$\boldsymbol{\xi} = \omega r \frac{z}{h}\mathbf{e}_\theta$$

the linearized strain tensor reads

$$\boldsymbol{\varepsilon} = \frac{\omega r}{2h}(\mathbf{e}_z \otimes \mathbf{e}_\theta + \mathbf{e}_\theta \otimes \mathbf{e}_z)$$

and the elastic stress tensor

$$\boldsymbol{\sigma} = G\frac{\omega r}{h}(\mathbf{e}_z \otimes \mathbf{e}_\theta + \mathbf{e}_\theta \otimes \mathbf{e}_z)$$

The solution ($\boldsymbol{\xi}, \boldsymbol{\sigma}$) is the solution of the problem, since

1. $\boldsymbol{\xi}$ is kinematically admissible (see Section 2.5.1):

$$\boldsymbol{\xi} = \boldsymbol{\xi}^d \text{ on } \partial\Omega_{\boldsymbol{\xi}^d} : \left\{ \begin{array}{c} z = h : \boldsymbol{\xi}^d = \omega r \mathbf{e}_\theta \\ z = 0 : \boldsymbol{\xi}^d = 0 \end{array} \right\}$$

2. $\boldsymbol{\sigma}$ is statically admissible (cylinder coordinates):

$$\sigma_{\theta z} = \sigma_{z\theta} = G\frac{\omega r}{h}$$

$$\operatorname{div}\boldsymbol{\sigma} = 0 : \frac{\partial \sigma_{\theta z}}{\partial z} = 0; \quad \frac{\partial \sigma_{z\theta}}{\partial \theta} = 0$$

$$\text{on } \partial\Omega_{\mathbf{T}^d} \Leftrightarrow r = R : \mathbf{T} = \mathbf{T}^d \quad \Leftrightarrow \mathbf{T}(\mathbf{e}_r) = 0$$

3. $(\boldsymbol{\xi}, \boldsymbol{\sigma})$ are related through the isotropic linear elastic material law.

Torsion Moment.

Application of the reduction elements (see Section 3.3.3):

$$\begin{aligned} z = h; \ \mathbf{n} = \mathbf{e}_z : \mathcal{M} &= \int_{\partial\Omega} \left(\mathbf{MA} \times \boldsymbol{\sigma} \cdot \mathbf{n} \right) da \\ &= \int_{\pi R^2} \left(r\mathbf{e}_r \times \sigma_{z\theta}\mathbf{e}_\theta \right) da = \left(\int_{\pi R^2} r\sigma_{z\theta}\, da \right) \mathbf{e}_z \\ &= \left(G_{\text{app}} \frac{\omega}{h} \int_{\pi R^2} r^2 da \right) \mathbf{e}_z = \left(G_{\text{app}} J \frac{\omega}{h} \right) \mathbf{e}_z = M_z\, \mathbf{e}_z \end{aligned}$$

with J the torsion inertia moment:

$$J = \int_{\pi R^2} r^2 da$$

6.9.2 Heterogeneous Sample

Target Energy Solution.

The exact solution $(\boldsymbol{\xi}, \boldsymbol{\sigma})$ for the torsion of the heterogeneous material sample is not known. However, the linearity of the elastic behavior allows us to determine the internal energy from the external work through the use of Clapeyron's formula (see Section 6.4):

$$W(\boldsymbol{\varepsilon}) = W^*(\boldsymbol{\sigma}) = \frac{1}{2}\left[\Phi(\boldsymbol{\xi}) + \Phi^*(\boldsymbol{\sigma}) \right]$$

Neglecting deadweight and due to the stress boundary conditions, the external work of prescribed volume and surface forces is zero:

$$\Phi(\boldsymbol{\xi}) = \int_\Omega \boldsymbol{\xi} \cdot (\rho \mathbf{f})\, d\Omega + \int_{\partial\Omega_{\mathbf{T}^d}} \boldsymbol{\xi} \cdot \mathbf{T}^d da = 0$$

In turn, the external work associated with prescribed displacements reduces to

$$\begin{aligned} z = h;\ \mathbf{n} = \mathbf{e}_z : \Phi^*(\boldsymbol{\sigma}) &= \int_{\partial\Omega_{\boldsymbol{\xi}^d}} \boldsymbol{\xi}^d \cdot (\boldsymbol{\sigma}\cdot\mathbf{n})\, da \\ &= \int_{\pi R^2} \omega r \mathbf{e}_\theta \cdot (\sigma_{z\theta}\mathbf{e}_\theta)\, da \\ &= \omega \int_{\pi R^2} r\sigma_{z\theta} da = \omega M_z \end{aligned}$$

Therefore,

$$\frac{W(\boldsymbol{\xi})}{h} = \frac{W^*(\boldsymbol{\sigma})}{h} = \frac{\Phi^*(\boldsymbol{\sigma})}{2h} = \frac{1}{2}M_z(h)\frac{\omega}{h}$$

Finally, replacing $M_z(h)$ by the expression derived for the equivalent homogeneous sample ($M_z = G_{\text{app}}J\frac{\omega}{h}$), we obtain the potential energy of the sought solution in the form

$$\mathcal{E}_{\text{pot}}(\boldsymbol{\xi}) = W(\boldsymbol{\varepsilon}) - \Phi(\boldsymbol{\xi}) = \frac{1}{2}G_{\text{app}}J\left(\frac{\omega}{h}\right)^2 h$$

Upper Bound.

The displacement field $\boldsymbol{\xi}'$ is kinematically admissible. The invariants of the strain tensors are

$$\text{tr}\,\boldsymbol{\varepsilon}' = 0;\quad \text{tr}\,(\boldsymbol{\varepsilon}'\cdot\boldsymbol{\varepsilon}') = 2\varepsilon_{zr}^{'2} = \frac{1}{2}\left(\frac{\omega r}{h}\right)^2$$

Internal energy:

$$\begin{aligned} W(\boldsymbol{\varepsilon}') &= \int_\Omega \left[\frac{\lambda(\mathbf{x})}{2}(\text{tr}\,\boldsymbol{\varepsilon}')^2 + G(\mathbf{x})\text{tr}\,(\boldsymbol{\varepsilon}'\cdot\boldsymbol{\varepsilon}')\right] d\Omega \\ &= \int_\Omega \left[G(\mathbf{x})\frac{1}{2}\left(\frac{\omega r}{h}\right)^2\right] d\Omega = \frac{1}{2}\Omega\left\langle Gr^2\right\rangle\left(\frac{\omega}{h}\right)^2 \end{aligned}$$

where $\langle Gr^2\rangle$ denotes the volume average of the torsion stiffness:

$$\left\langle Gr^2\right\rangle = \frac{1}{\Omega}\int_\Omega G(\mathbf{x})r^2 d\Omega$$

External work of volume and surface forces:

$$\Phi(\boldsymbol{\xi}') = \int_\Omega \boldsymbol{\xi}'\cdot(\rho\mathbf{f})\, d\Omega + \int_{\partial\Omega_{\mathbf{T}^d}} \boldsymbol{\xi}'\cdot\mathbf{T}^d da = 0$$

Potential energy:

$$\mathcal{E}_{\text{pot}}(\boldsymbol{\xi}') = W(\boldsymbol{\varepsilon}') - \Phi(\boldsymbol{\xi}') = \frac{1}{2}\Omega\left\langle Gr^2\right\rangle\left(\frac{\omega}{h}\right)^2$$

Application of the theorem of minimum potential theory:

$$\mathcal{E}_{\text{pot}}(\boldsymbol{\xi}) = \frac{1}{2}G_{\text{app}}J\left(\frac{\omega}{h}\right)^2 h \le \frac{1}{2}\left(\frac{\omega}{h}\right)^2\Omega\left\langle Gr^2\right\rangle = \mathcal{E}_{\text{pot}}(\boldsymbol{\xi}')$$

leads to the upper bound of the apparent shear modulus:

$$G_{\text{app}} \leq \frac{\Omega}{Jh} \langle Gr^2 \rangle = \frac{1}{Jh} \int_{\Omega} G(\mathbf{x}) r^2 d\Omega$$

Lower Bound.

We consider the following statically admissible stress field:

$$\boldsymbol{\sigma}' = \tau \frac{r}{h} \left[\mathbf{e}_\theta \otimes \mathbf{e}_z + \mathbf{e}_z \otimes \mathbf{e}_\theta \right]$$

Invariants:

$$\operatorname{tr} \boldsymbol{\sigma}' = 0; \quad \operatorname{tr}(\boldsymbol{\sigma}' \cdot \boldsymbol{\sigma}') = 2\sigma_{zr}'^2 = 2\left(\tau \frac{r}{h}\right)^2$$

Internal complementary energy:

$$\begin{aligned} W^*(\boldsymbol{\sigma}') &= \int_{\Omega} \frac{1}{2E(\mathbf{x})} \left[-\nu(\mathbf{x})(\operatorname{tr} \boldsymbol{\sigma}')^2 + (1 + \nu(\mathbf{x})) \operatorname{tr}(\boldsymbol{\sigma}' \cdot \boldsymbol{\sigma}') \right] d\Omega \\ &= \int_{\Omega} \left[\frac{1}{4G(\mathbf{x})} \operatorname{tr}(\boldsymbol{\sigma}' \cdot \boldsymbol{\sigma}') \right] d\Omega = \int_{\Omega} \left[\frac{\tau^2}{2G(\mathbf{x})} \left(\frac{r}{h}\right)^2 \right] d\Omega \\ &= \frac{\tau^2}{2h^2} \int_{\Omega} \left[\frac{r^2}{G(\mathbf{x})} \right] d\Omega = \frac{\tau^2}{2h^2} \Omega \left\langle \frac{r^2}{G} \right\rangle \end{aligned}$$

where $\langle r^2/G \rangle$ denotes the volume average:

$$\left\langle \frac{r^2}{G} \right\rangle = \frac{1}{\Omega} \int_{\Omega} \left[\frac{r^2}{G(\mathbf{x})} \right] d\Omega$$

External work (prescribed displacements):

$$\Phi^*(\boldsymbol{\sigma}') = \omega \int_{\pi R^2} r\sigma_{z\theta}' da = \frac{\omega \tau}{h} \int_{\pi R^2} r^2 da = \tau J \frac{\omega}{h}$$

Complementary energy:

$$\begin{aligned} \mathcal{E}_{\text{com}}(\boldsymbol{\sigma}') &= W^*(\boldsymbol{\sigma}') - \Phi^*(\boldsymbol{\sigma}') \\ &= \frac{\tau^2}{2h^2} \Omega \left\langle \frac{r^2}{G} \right\rangle - \tau J \frac{\omega}{h} \end{aligned}$$

Minimization:

$$d\mathcal{E}_{\text{com}}(\boldsymbol{\sigma}') = 0 : \frac{\partial \mathcal{E}_{\text{com}}(\boldsymbol{\sigma}')}{\partial \tau} = \frac{\tau}{h^2} \Omega \left\langle \frac{r^2}{G} \right\rangle - \frac{\omega}{h} J = 0$$

$$\tau = \omega \frac{Jh}{\Omega} \left\langle \frac{r^2}{G} \right\rangle^{-1}$$

Therefore,

$$\mathcal{E}_{\text{com}}(\boldsymbol{\sigma}') = -\frac{1}{2} \left\langle \frac{r^2}{G} \right\rangle^{-1} J^2 \frac{\omega^2}{\Omega}$$

Application of the theorem of minimum complementary energy:

$$-\mathcal{E}_{\text{com}}(\boldsymbol{\sigma}') \leq -\mathcal{E}_{\text{com}}(\boldsymbol{\sigma}) = \mathcal{E}_{\text{pot}}(\boldsymbol{\xi})$$

that is,

$$-\frac{1}{2}\left\langle\frac{r^2}{G}\right\rangle^{-1} J^2\frac{\omega^2}{\Omega} \leq -\frac{1}{2}G_{\text{app}}J\left(\frac{\omega}{h}\right)^2 h$$

leads to the lower bound:

$$\frac{Jh}{\Omega}\left\langle\frac{r^2}{G}\right\rangle^{-1} = \frac{Jh}{\int_\Omega\left[\frac{r^2}{G(\mathbf{X})}\right]d\Omega} \leq G_{\text{app}}$$

6.9.3 Composite Cylinder Model

We summarize the found results. Application of the two minimum theorems

$$-\mathcal{E}_{\text{com}}(\boldsymbol{\sigma}') \leq -\mathcal{E}_{\text{com}}(\boldsymbol{\sigma}) = \mathcal{E}_{\text{pot}}(\boldsymbol{\xi}) \leq \mathcal{E}_{\text{pot}}(\boldsymbol{\xi}')$$

reads here

$$\frac{Jh}{\int_\Omega\left[\frac{r^2}{G(\mathbf{X})}\right]d\Omega} = \frac{Jh}{\Omega}\left\langle\frac{r^2}{G}\right\rangle^{-1} \leq G_{\text{app}} \leq \frac{\Omega}{Jh}\langle Gr^2\rangle = \frac{1}{Jh}\int_\Omega G(\mathbf{x})r^2 d\Omega$$

The composite model is characterized by two subdomains:

$$\begin{aligned} 0 < r < R_1 \quad &: \quad G(\mathbf{x}) = G_1 \\ R_1 < r \leq R_2 \quad &: \quad G(\mathbf{x}) = G_2 \end{aligned}$$

For the geometric configuration

$$d\Omega = dh\, dA = r\, dr\, d\theta\, dh$$

Upper Bound.

$$\frac{1}{Jh}\int_\Omega G(\mathbf{x})r^2 d\Omega = \frac{J_1G_1}{J} + \frac{J_2G_2}{J}$$

where

$$J_1 = 2\pi\int_{r=0}^{r=R_1} r^3 dr = \pi\frac{R_1^4}{2} = \frac{A_1^2}{2\pi}; \quad J_2 = \pi\frac{R_2^4 - R_1^4}{2} = \frac{A^2}{2\pi} - \frac{A_1^2}{2\pi}; \quad J = \frac{A^2}{2\pi}$$

Therefore,

$$G_{\text{app}} \leq \frac{\Omega}{Jh}\langle Gr^2\rangle = c^2G_1 + (1-c^2)G_2$$

with c the fiber volume concentration:

$$c = \frac{A_1h}{Ah} \in [0,1]$$

Lower Bound.

$$\frac{J}{\int_{\pi R_2^2}\left[\frac{r^2}{G(\mathbf{x})}\right] dA} = \frac{J}{\frac{J_1}{G_1}+\frac{J_2}{G_2}} = \frac{1}{\frac{c^2}{G_1}+\frac{(1-c^2)}{G_2}}$$

Thus,

$$\frac{1}{\frac{c^2}{G_1}+\frac{(1-c^2)}{G_2}} \leq G_{\text{app}}$$

Morphology and Energy.

The results for $G_1/G_2 = 10$ are displayed in Figure 6.11. We note quite a difference between the upper and lower bounds with increasing fiber volume concentration. To check whether one of the two bounds corresponds to the geometric model of a fiber embedded in a cylinder, it suffices to check the energies. According to Clapeyron's formula, we have with $\Phi(\boldsymbol{\xi}) = 0$

$$\mathcal{E}_{\text{pot}}(\boldsymbol{\xi}) = W(\boldsymbol{\varepsilon}) = \frac{1}{2}\Phi^*(\boldsymbol{\sigma}) = \frac{1}{2}M_z\omega = \frac{\omega}{2}\int_{\pi R^2} r\sigma_{z\theta}(r)\, da$$

Noting $\sigma_{z\theta}(r) = G(r)\omega r/h$ in the considered morphology, the minimum potential energy corresponding to the solution reads

$$\mathcal{E}_{\text{pot}}(\boldsymbol{\xi}) = \frac{\omega^2}{2h}\int_{\pi R^2} G(r)r^2 da$$

Therefore, we need to give proof that the potential energy $\mathcal{E}_{\text{pot}}(\boldsymbol{\xi}')$ associated with the given configuration coincides with this minimum. The potential energy for the considered configuration for $G(\mathbf{x}) = G(r)$ has been determined in the form

$$\mathcal{E}_{\text{pot}}(\boldsymbol{\xi}') = \frac{1}{2}\left(\frac{\omega}{h}\right)^2 \int_{\Omega} G(r)r^2 d\Omega = \frac{\omega^2}{2h}\int_{\pi R^2} G(r)r^2 da$$

Therefore,

$$\mathcal{E}_{\text{pot}}(\boldsymbol{\xi}) = \mathcal{E}_{\text{pot}}(\boldsymbol{\xi}') \Leftrightarrow G_{\text{app}} = \frac{\Omega}{Jh}\left\langle Gr^2 \right\rangle = c^2 G_1 + (1-c^2)G_2$$

This proves that the upper bound corresponds to the exact solution for the considered configuration.

Note that the lower bound could be improved by considering two cylinders on top of each other, the first representing the fibers the second the matrix. In this case,

$$\frac{1}{\frac{c}{G_1}+\frac{1-c}{G_2}} = \frac{Jh}{\Omega}\left\langle\frac{r^2}{G}\right\rangle^{-1} \leq G_{\text{app}}$$

where $c = h_1/h$. This improved lower bound is also displayed in Figure 6.11.

6.10 PROBLEM SET: ELASTICITY BOUNDS OF MICROFLEXURAL STRUCTURES (MEMS TYPE)

Microelectromechanical systems (MEMS) are miniature electromechanical sensors and actuators that are used as accelerometers, micropumps, microturbines, and optical switches. The majority of MEMS devices are based on the movement of cantilever beams or thin membranes. MEMS are often made of thin layers of polysilicon, amorphous silicon, silica, oxides, or polyimides, and the material is deposited in layers on some substratum (see Figure 6.12a). We want to give a lower bound and an upper bound for the apparent elasticity modulus of such microflexural structures.

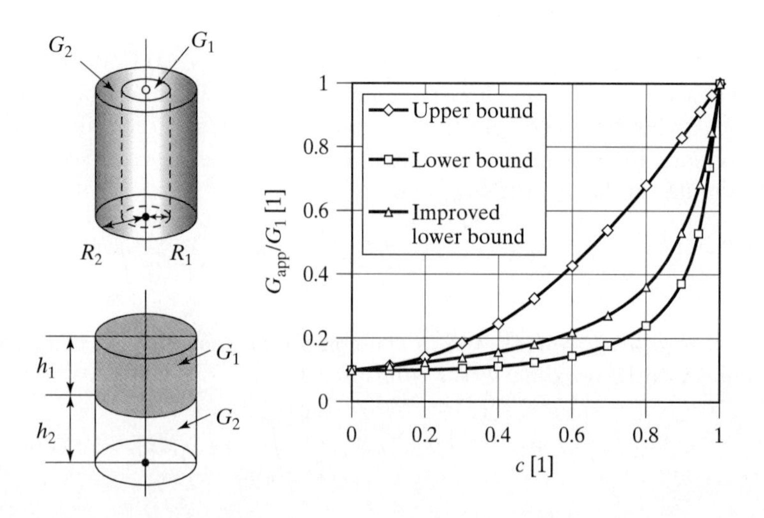

FIGURE 6.11: Cylinder composite models for fiber reinforced composite materials ($G_1/G_2 = 10$).

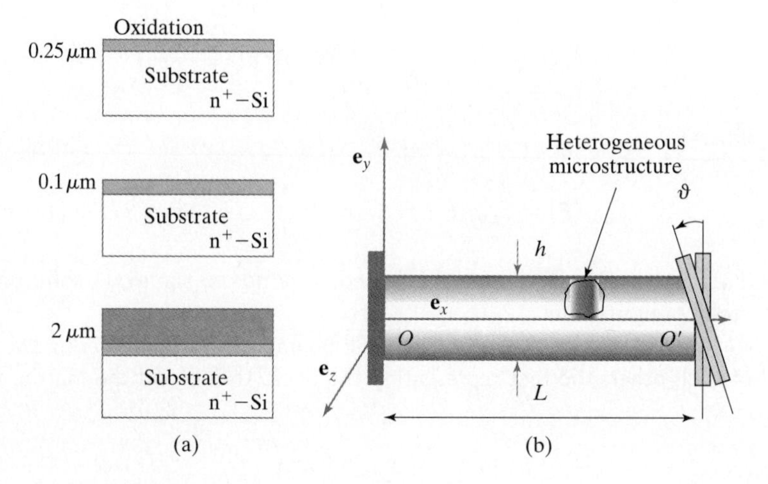

FIGURE 6.12: Problem set: microflexural system: (a) layered deposition manufacturing; (b) cantilever structure.

To this end, we consider a cantilever beam of length L along the x-axis and of height h and width b, as shown in Figure 6.12b. At $x = 0$, the section (centered around O) is in contact without friction with a fixed rigid plate. At the other end, at $x = L$, the section (centered around O') fixed on a frictionless rigid plate is subjected to a rotation of intensity $\vartheta \ll 1$ around the z-axis, such that

$$\text{at } x = L : \xi_x^d = [\mathbf{O'M} \times \vartheta \mathbf{e}_z] \cdot \mathbf{e}_x = -\vartheta y$$

The surface of the beam is stress free. In a very first approximation, we consider that the heterogeneous matter constituting the beam is a linear isotropic elastic material defined by Young's modulus $E(\mathbf{x})$ and Poisson's ratio $\nu(\mathbf{x})$.

1. **Equivalent Homogeneous Sample:** Let E_{app} be the apparent elastic modulus of the equivalent homogeneous sample. By considering a uniaxial stress field of the form

$$\text{in } \Omega : \boldsymbol{\sigma} = \sigma(y) \mathbf{e}_x \otimes \mathbf{e}_x$$

determine $\sigma(y)$, using the isotropic linear elastic material law ($E = E_{\text{app}} = \text{const}$), and show that the bending moment at $x = L$ obeys to the section-type constitutive law:

$$M_z = E_{\text{app}} I_z \frac{\vartheta}{L}$$

where I_z is the bending inertia moment around the z-axis. Specify briefly why $(\boldsymbol{\xi}, \boldsymbol{\sigma})$ is the solution of the equivalent homogeneous bending problem (determination of $\boldsymbol{\xi}$ not required).

2. **Heterogeneous Sample**: We now consider the heterogeneous bending sample [i.e., $E = E(\mathbf{x}), \nu = \nu(\mathbf{x})$].

   (a) **Target Solution**: Let $(\boldsymbol{\xi}, \boldsymbol{\sigma})$ be the (exact) displacement and stress solution in the heterogeneous sample subjected to the rotation at $x = L$. For this solution, show that the elastic energy stored per unit length in the beam reads

$$\frac{W(\boldsymbol{\xi})}{L} = \frac{W^*(\boldsymbol{\sigma})}{L} = \frac{1}{2} M_z(x = L) \frac{\vartheta}{L}$$

   (b) **Lower Bound**: Inspired by the elastic homogeneous solution just developed, we consider for the heterogeneous sample, an approximated stress field of the form

$$\text{in } \Omega : \boldsymbol{\sigma}' = \sigma' \frac{y}{L} \mathbf{e}_x \otimes \mathbf{e}_x$$

   Determine a lower bound of the apparent E-modulus of the heterogeneous sample.

   (c) **Upper Bound**: Inspired by the elastic homogeneous solution (see Problem Section 5.7), we consider for the heterogeneous sample a displacement field of the form

$$\text{in } \Omega : \boldsymbol{\xi}' = -\vartheta y \frac{x}{L} \mathbf{e}_x + \frac{\vartheta}{2L} \left(x^2 + \alpha (y^2 - z^2) \right) \mathbf{e}_y + \alpha \frac{\vartheta}{L} y z \mathbf{e}_z$$

   Determine an upper bound of the apparent E-modulus of the heterogeneous sample.

3. **Application to a Two-Layer Beam**: We consider that the beam is composed of two layers of thickness h_1 and $h_2 = h - h_1$. E_1 and E_2 are the elasticity moduli of layer 1 and layer 2, respectively. The Poisson's ratio in both layers is assumed to be the same. Determine the lower and upper bounds of the apparent elasticity modulus E_{app} of the bending beam. Display the results as a function of the relative thickness $c = h_1/h$ for a value of $E_1/E_2 = 2$ (e.g., corresponding to a quartz-layer SiO_2 of $E_1 = 380$ GPa obtained by thermal oxidation of a silicon substratum of $E_2 = 190$ GPa). Show that one of the two bounds corresponds precisely to the considered geometric situation.

6.10.1 The Equivalent Homogeneous Sample

The equivalent homogeneous sample was solved in Problem Section 5.7.

6.10.2 Heterogeneous Sample

Target Solution.

$(\boldsymbol{\xi}, \boldsymbol{\sigma})$ is the (exact) displacement and stress solution. Therefore, Clapeyron's formula applies:

$$W(\boldsymbol{\xi}) = W(\boldsymbol{\sigma}) = \frac{1}{2}\Big[\Phi(\boldsymbol{\xi}) + \Phi^*(\boldsymbol{\sigma})\Big]$$

Here, $\Phi(\boldsymbol{\xi}) = 0$. Thus,

$$\Phi^*(\boldsymbol{\sigma}) = \int_{\partial\Omega_{\boldsymbol{\xi}^d}} \mathbf{T} \cdot \boldsymbol{\xi}^d da = -\vartheta \int_{A, x=L} \sigma_{xx} y \, da = \vartheta M_z(x = L)$$

It follows that

$$W(\boldsymbol{\xi}) = W(\boldsymbol{\sigma}) = \frac{1}{2}\vartheta M_z$$

and

$$\mathcal{E}_{\text{pot}}(\boldsymbol{\xi}) = \frac{1}{2}\vartheta M_z = -\mathcal{E}_{\text{com}}(\boldsymbol{\sigma})$$

Finally, if we use the (given) section type constitutive law of the bending beam, we obtain the energy of the solution in the form

$$\mathcal{E}_{\text{pot}}(\boldsymbol{\xi}) = E_{\text{app}} I_z \frac{\vartheta^2}{2L} = -\mathcal{E}_{\text{com}}(\boldsymbol{\sigma})$$

Lower Energy Bound.

$$\mathcal{E}_{\text{com}}(\boldsymbol{\sigma}) \le \mathcal{E}_{\text{com}}(\boldsymbol{\sigma}') = W^*(\boldsymbol{\sigma}') - \Phi^*(\boldsymbol{\sigma}')$$

The stress field $\boldsymbol{\sigma}'$ is statically admissible, since

$$\text{in } \Omega : \operatorname{div} \boldsymbol{\sigma}' = 0 \leftrightarrow \frac{\partial}{\partial x}\left(\sigma' \frac{y}{L}\right) = 0; \quad \boldsymbol{\sigma}' = {}^t\boldsymbol{\sigma}'$$

$$\text{on } \partial\Omega_{\mathbf{T}^d}: \quad \begin{array}{l} \text{on surface: } \mathbf{T}^d = 0 \equiv \mathbf{T}'(\mathbf{n} = \pm\mathbf{e}_y; \mathbf{n} = \pm\mathbf{e}_z) \\ \text{on } x = 0, x = L : \mathbf{T}'(\mathbf{n} = \pm\mathbf{e}_x)\cdot\mathbf{e}_y = \mathbf{T}'(\mathbf{n} = \pm\mathbf{e}_x)\cdot\mathbf{e}_z = 0 \end{array}$$

Internal complementary energy:

$$\begin{aligned} W^*(\boldsymbol{\sigma}') &= \int_\Omega \frac{1}{2E(\mathbf{x})}\Big[-\nu(\mathbf{x})(\operatorname{tr}\boldsymbol{\sigma}')^2 + (1+\nu(\mathbf{x}))\operatorname{tr}(\boldsymbol{\sigma}'\cdot\boldsymbol{\sigma}')\Big]d\Omega \\ &= \int_\Omega \frac{1}{2E(\mathbf{x})}\left[-\nu(\mathbf{x})\left(\sigma'\frac{y}{L}\right)^2 + (1+\nu(\mathbf{x}))\left(\sigma'\frac{y}{L}\right)^2\right]d\Omega \\ &= \left(\frac{\sigma'}{L}\right)^2\int_\Omega \frac{y^2}{2E(\mathbf{x})}d\Omega = \frac{1}{2}\Omega\left\langle\frac{y^2}{E}\right\rangle\left(\frac{\sigma'}{L}\right)^2 \end{aligned}$$

where $\langle y^2/E\rangle$ is the volume average:

$$\left\langle\frac{y^2}{E}\right\rangle = \frac{1}{\Omega}\int_\Omega \frac{y^2}{E(\mathbf{x})}d\Omega$$

External work by prescribed displacements on $\partial\Omega_{\boldsymbol{\xi}^d}$ ($\mathbf{T}' = \sigma' y/L\mathbf{e}_x; \boldsymbol{\xi}^d(x = L) = -\vartheta y\mathbf{e}_x$):

$$\Phi^*(\boldsymbol{\sigma}') = \int_{\partial\Omega_{\boldsymbol{\xi}^d}} \mathbf{T}'\cdot\boldsymbol{\xi}^d da = -\frac{\vartheta\sigma'}{L}\int_{A,x=L} y^2 da = -\frac{\sigma'}{L}I_z\vartheta$$

Complementary energy:

$$\mathcal{E}_{\text{com}}(\boldsymbol{\sigma}') = \frac{1}{2}\Omega\left\langle\frac{y^2}{E}\right\rangle\left(\frac{\sigma'}{L}\right)^2 + \frac{\sigma'}{L}I_z\vartheta$$

Minimization delivers

$$\frac{\partial\mathcal{E}_{\text{com}}(\boldsymbol{\sigma}')}{\partial(\sigma'/L)} = \Omega\left\langle\frac{y^2}{E}\right\rangle\frac{\sigma'}{L} + \vartheta I_z = 0 \rightarrow \frac{\sigma'}{L} = -\frac{\vartheta I_z}{\Omega}\left\langle\frac{y^2}{E}\right\rangle^{-1}$$

Therefore,

$$\mathcal{E}_{\text{com}}(\boldsymbol{\sigma}') = -\frac{1}{2}\frac{(\vartheta I_z)^2}{\Omega}\left\langle\frac{y^2}{E}\right\rangle^{-1}$$

The stress-based variational method, therefore, delivers the following lower bound for the elasticity modulus:

$$\begin{aligned} \mathcal{E}_{\text{com}}(\boldsymbol{\sigma}) &\le \mathcal{E}_{\text{com}}(\boldsymbol{\sigma}') \Leftrightarrow -E_{\text{app}}I_z\frac{\vartheta^2}{2L} \le -\frac{1}{2}\frac{(\vartheta I_z)^2}{\Omega}\left\langle\frac{y^2}{E}\right\rangle^{-1} \\ &\rightarrow E_{\text{app}} \ge \frac{I_z L}{\Omega}\left\langle\frac{y^2}{E}\right\rangle^{-1} = \frac{I_z L}{\int_\Omega \frac{y^2}{E(\mathbf{x})}d\Omega} \end{aligned}$$

Upper Energy Bound.

$$\mathcal{E}_{\text{pot}}(\boldsymbol{\xi}) \leq \mathcal{E}_{\text{pot}}(\boldsymbol{\xi}') = W(\boldsymbol{\varepsilon}') - \Phi(\boldsymbol{\xi}')$$

The displacement field $\boldsymbol{\xi}'$ is kinematically admissible, since

$$\text{on } \partial\Omega_{\boldsymbol{\xi}^d} : \begin{array}{l} \text{at } x = 0 : \xi_x^d = 0 \equiv \boldsymbol{\xi}'(x=0) \cdot \mathbf{e}_x \\ \text{at } x = L : \xi_x^d = -\vartheta y \equiv \boldsymbol{\xi}'(x=L) \cdot \mathbf{e}_x \end{array}$$

The components of the strain tensor read (see Problem Section 5.7.1)

$$\varepsilon'_{xx} = \frac{\partial \xi'_x}{\partial x} = -\frac{\vartheta y}{L};\quad \varepsilon'_{yy} = \frac{\partial \xi'_y}{\partial y} = \alpha\frac{\vartheta y}{L};\quad \varepsilon'_{zz} = \frac{\partial \xi'_z}{\partial z} = \alpha\frac{\vartheta y}{L}$$

$$2\varepsilon'_{xy} = \frac{\partial \xi'_x}{\partial y} + \frac{\partial \xi'_y}{\partial x} = 0;\quad 2\varepsilon'_{yz} = \frac{\partial \xi'_y}{\partial z} + \frac{\partial \xi'_z}{\partial y} = 0;\quad 2\varepsilon'_{xz} = \frac{\partial \xi'_x}{\partial z} + \frac{\partial \xi'_z}{\partial x} = 0$$

and the invariants are

$$\operatorname{tr}\boldsymbol{\varepsilon}' = -\frac{\vartheta y}{L}(1-2\alpha)$$

$$\operatorname{tr}(\boldsymbol{\varepsilon}' \cdot \boldsymbol{\varepsilon}') = \left(\frac{\vartheta y}{L}\right)^2 (1+2\alpha^2)$$

Internal potential energy (structural free energy):

$$\begin{aligned} W(\boldsymbol{\varepsilon}') &= \int_\Omega \left[\frac{\lambda(\mathbf{x})}{2}(\operatorname{tr}\boldsymbol{\varepsilon}')^2 + G(\mathbf{x})\operatorname{tr}(\boldsymbol{\varepsilon}' \cdot \boldsymbol{\varepsilon}')\right] d\Omega \\ &= \left(\frac{\vartheta}{L}\right)^2 \int_\Omega \left[\frac{\lambda(\mathbf{x})}{2} y^2(1-2\alpha)^2 + G(\mathbf{x}) y^2 (1+2\alpha^2)\right] d\Omega \\ &= \left(\frac{\vartheta}{L}\right)^2 \int_\Omega \left[\frac{\lambda(\mathbf{x})}{2} y^2(1-4\alpha+4\alpha^2) + G(\mathbf{x}) y^2 (1+2\alpha^2)\right] d\Omega \end{aligned}$$

External work provided by prescribed forces:

$$\Phi(\boldsymbol{\xi}') = 0$$

Potential energy:

$$\mathcal{E}_{\text{pot}}(\boldsymbol{\xi}') = \left(\frac{\vartheta}{L}\right)^2 \int_\Omega \left[\frac{\lambda(\mathbf{x})}{2} y^2(1-4\alpha+4\alpha^2) + G(\mathbf{x}) y^2 (1+2\alpha^2)\right] d\Omega$$

Minimization delivers

$$\frac{\partial \mathcal{E}_{\text{pot}}(\boldsymbol{\xi}')}{\partial \alpha} = \left(\frac{\vartheta}{L}\right)^2 \int_\Omega \left[\frac{\lambda(\mathbf{x})}{2} y^2(-4+8\alpha) + G(\mathbf{x}) y^2 (4\alpha)\right] d\Omega = 0$$

$$\rightarrow \alpha = \frac{\int_\Omega y^2 \lambda(\mathbf{x})\, d\Omega}{2\int_\Omega y^2\left[G(\mathbf{x}) + \lambda(\mathbf{x})\right] d\Omega} = \frac{\langle y^2 \lambda\rangle}{2\left[\langle y^2 G\rangle + \langle y^2\lambda\rangle\right]}$$

where $\langle y^2\lambda\rangle$ and $\langle y^2 G\rangle$ are the volume averages:

$$\langle y^2\lambda\rangle = \frac{1}{\Omega}\int_\Omega y^2\lambda(\mathbf{x})\,d\Omega; \quad \langle y^2 G\rangle = \frac{1}{\Omega}\int_\Omega y^2 G(\mathbf{x})\,d\Omega$$

Therefore,

$$\begin{aligned}\mathcal{E}_{\text{pot}}(\boldsymbol{\xi}') &= \left(\frac{\vartheta}{L}\right)^2\left[\frac{1}{2}(1-2\alpha)^2\int_\Omega y^2\lambda(\mathbf{x})\,d\Omega + (1+2\alpha^2)\int_\Omega G(\mathbf{x})y^2 d\Omega\right]\\ &= \left(\frac{\vartheta}{L}\right)^2\Omega\left[\frac{1}{2}(1-2\alpha)^2\langle y^2\lambda\rangle + (1+2\alpha^2)\langle y^2 G\rangle\right]\end{aligned}$$

The displacement-based variational method, therefore, delivers the following upper bound:

$$\begin{aligned}\mathcal{E}_{\text{pot}}(\boldsymbol{\xi}) &\leq \mathcal{E}_{\text{pot}}(\boldsymbol{\xi}')\\ E_{\text{app}} &\leq \frac{\Omega}{I_z L}\left[(1-2\alpha)^2\langle y^2\lambda\rangle + 2(1+2\alpha^2)\langle y^2 G\rangle\right]\\ &\leq \frac{\Omega}{I_z L}\langle y^2 G\rangle\frac{3\langle y^2\lambda\rangle + 2\langle y^2 G\rangle}{\langle y^2\lambda\rangle + \langle y^2 G\rangle}\end{aligned}$$

For the same Poisson ratio in all phases, the previous expression simplifies to

$$\forall \mathbf{x}, \nu = \text{const} : E_{\text{app}} \leq \frac{\Omega}{I_z L}\langle y^2 E\rangle = \frac{1}{I_z L}\int_\Omega y^2 E(\mathbf{x})\,d\Omega$$

6.10.3 Application to Two-Layer Beam

$$\frac{I_z L}{\Omega}\left\langle\frac{y^2}{E}\right\rangle^{-1} \leq E_{\text{app}} \leq \frac{\Omega}{I_z L}\langle y^2 G\rangle\frac{3\langle y^2\lambda\rangle + 2\langle y^2 G\rangle}{\langle y^2\lambda\rangle + \langle y^2 G\rangle}$$

and for $\forall \mathbf{x}, \nu = \text{const}$:

$$\frac{I_z L}{\int_\Omega \frac{y^2}{E(\mathbf{X})}d\Omega} \leq E_{\text{app}} \leq \frac{1}{I_z L}\int_\Omega y^2 E(\mathbf{x})\,d\Omega$$

The beam is composed of two subdomains:

$$\begin{aligned} h/2 - h_1 < y \leq h/2 &\ : \ E(\mathbf{x}) = E_1\\ -h/2 \leq y < h/2 - h_1 &\ : \ E(\mathbf{x}) = E_2\end{aligned}$$

Lower bound:

$$\begin{aligned}E_{\text{app}} &\geq \frac{I_z L}{\int_\Omega \frac{y^2}{E(\mathbf{X})}d\Omega} = \frac{I_z}{b\left[\frac{1}{3E_1}y^3\Big|_{h/2-h_1}^{h/2} + \frac{1}{3E_2}y^3\Big|_{-h/2}^{h/2-h_1}\right]}\\ E_{\text{app}} &\geq \frac{1}{4\left\{\frac{1}{E_1}\left[(1/2)^3 - (1/2-c)^3\right] + \frac{1}{E_2}\left[(1/2-c)^3 + (1/2)^3\right]\right\}}\end{aligned}$$

where $c = h_1/h$ is the volume concentration of material 1. Thus,

$$\frac{E_{\text{app}}}{E_1} \geq \frac{0.25}{\left[(1/2)^3 - (1/2-c)^3\right] + \frac{E_1}{E_2}\left[(1/2-c)^3 + (1/2)^3\right]}$$

Upper bound:

$$\begin{aligned} E_{\text{app}} &\leq \frac{1}{I_z L}\int_\Omega y^2 E(\mathbf{x})\, d\Omega = \frac{1}{I_z} b\left[\frac{E_1}{3} y^3\Big|_{h/2-h_1}^{h/2} + \frac{E_2}{3} y^3\Big|_{-h/2}^{h/2-h_1}\right] \\ E_{\text{app}} &\leq 4\left\{E_1\left[(1/2)^3 - (1/2-c)^3\right] + E_2\left[(1/2-c)^3 + (1/2)^3\right]\right\} \end{aligned}$$

Thus,

$$\frac{E_{\text{app}}}{E_1} \leq 4\left\{\left[(1/2)^3 - (1/2-c)^3\right] + \frac{E_2}{E_1}\left[(1/2-c)^3 + (1/2)^3\right]\right\}$$

Figure 6.13 illustrates the upper and lower bound for a stiffness ratio $E_1/E_2 = 2$. The figure also gives the Voigt–Reuss bounds, which correspond to the elasticity bounds of a heterogeneous material sample in direct tension. We note that for low volume ratios the Voigt–Reuss bounds underestimate the bending elasticity modulus, and for large volume ratios it is the inverse.

The upper bound corresponds to the geometric configuration of a layered beam element, since

$$\mathcal{E}_{\text{pot}}(\boldsymbol{\xi}) = \mathcal{E}_{\text{pot}}(\boldsymbol{\xi}')$$

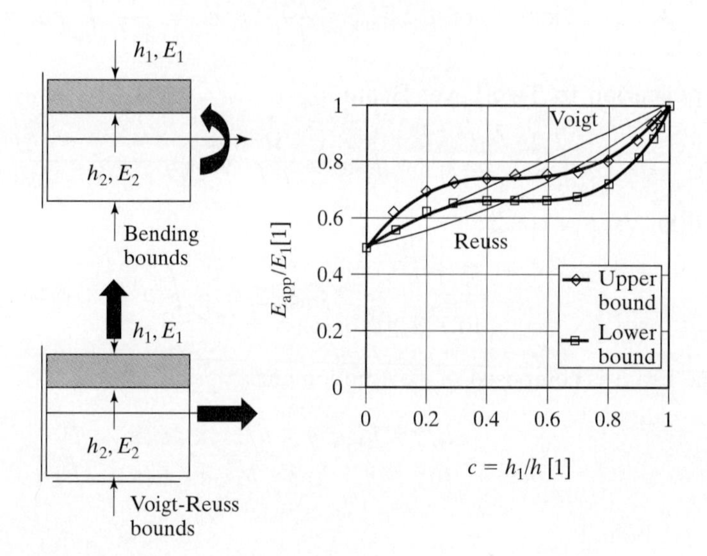

FIGURE 6.13: Upper and lower bounds of the bending elasticity modulus of a two-layered microflexural structure.

PART FOUR

PLASTICITY AND YIELD DESIGN

CHAPTER 7

1D Plasticity: An Energy Approach

Plasticity is the propriety of materials to undergo irreversible deformation beyond a certain stress threshold. For metals, the plastic behavior results from stress-induced irreversible sliding between the crystalline planes of the microstructure of the material. In the case of geomaterials, plasticity can be seen as the macroscopic representation of the irreversible sliding of particles, grains, and so on that compose the heterogeneous matter at a microscopic scale. Plasticity models describe this irreversible material behavior. This chapter develops the steps of constitutive modeling of 1D plasticity based on 1D think models of ideal plasticity and hardening plasticity. Through a study of the energy transformation and dissipation, the basic steps of constitutive modeling using an energy approach are developed. This approach will be applied in the next chapter to extend the 1D formulation to 3D hardening plasticity. Finally, we finish this first chapter on plasticity with a study of viscoplastic evolutions. Set within the same energy framework as plasticity, viscoplasticity is the propriety of materials to undergo—beyond a certain threshold—permanent deformation in time: Viscoplasticity is a delayed plastic behavior.

7.1 IDEAL PLASTICITY

7.1.1 Friction Element

Plastic deformation occurs when the stress reaches a threshold. In a 1D think model this threshold can be represented by a friction element as shown in Figure 7.1: The friction strength k restricts the admissible stresses to a domain D_E expressed in

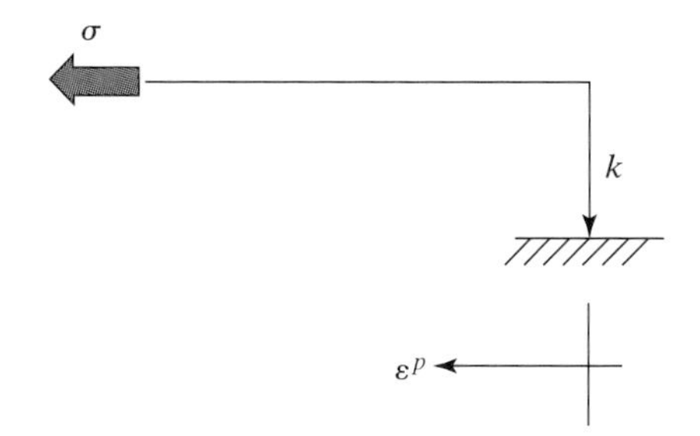

FIGURE 7.1: Friction element.

the form of a scalar loading function $f(\sigma)$:

$$\sigma \in D_E : f(\sigma) \leq 0; \quad f(\sigma) = |\sigma| - k \leq 0 \tag{7.1}$$

When the applied stress intensity $|\sigma|$ is below the friction strength k, the friction element is at rest, and its displacement or plastic sliding ε^p does not develop. In turn, when the stress reaches the stress threshold k, the friction element slides at an undetermined plastic rate $\dot{\varepsilon}^p$ in the direction of the applied stress (i.e., $\sigma\dot{\varepsilon}^p \geq 0$). The plastic (or permanent) sliding ε^p is irreversible: When the applied stress decreases below the friction strength, the friction element stops and the displacement ε^p remains. During plastic sliding, $|\sigma| = k$, it follows that

$$f(\sigma + d\sigma) = f(\sigma) + \frac{\partial f}{\partial \sigma} d\sigma = 0 \Leftrightarrow df = 0 \tag{7.2}$$

Equation (7.2) is called the consistency condition: A load σ that is at the loading surface [i.e., $f(\sigma) = 0$] remains on it (i.e., $df = 0$). Furthermore, the occurrence of the plastic deformation is not related to a time scale associated with the plastic deformation, but only to the stress history (chronology).[1] With regard to the absence of an intrinsic time scale (or a characteristic physical rate) associated with plastic evolution, plastic deformation is considered to occur instantaneously. In the 1D think model, the plastic sliding occurs in the direction of the applied load [i.e., sign (σ)] or, more generally, normal to the loading surface [i.e., sign $(\sigma) = \partial f / \partial \sigma$]. Finally, if we denote by $d\lambda$ the plastic multiplier, expressing the intensity of the plastic flow, (i.e., $d\lambda = |d\varepsilon^p| \geq 0$), the constitutive equations of the friction element can be summarized by

$$d\varepsilon^p = d\lambda \, \text{sign}(\sigma) = d\lambda \frac{\partial f}{\partial \sigma}; \quad f \leq 0; \quad d\lambda \geq 0; \quad d\lambda f = 0 \tag{7.3}$$

[1] The consistency condition is often written in rate form:

$$\dot{f} = \frac{df}{dt} = \frac{\partial f}{\partial \sigma} \frac{d\sigma}{dt} = 0$$

which shows that the time scale of plastic evolution defined by the rate $\dot{f}$ is determined by the stress rate $\dot{\sigma}$ applied to the system. For instance, if a sudden stress is applied to the system [in form of a Heaviside $\sigma(t) = \sigma H(t - t_0)$], the stress rate at $t = t_0$ is infinite, as is the time scale of plastic deformation. This is why we prefer reasoning in terms of infinitesimal increments ($df = 0$) rather than in rates.

The first equation is known as the plastic flow rule, and the three other conditions, which represent the plastic loading–unloading conditions of the friction element, are often referred to as Kuhn–Tucker conditions.

7.1.2 1D Think Model of Ideal Elastoplasticity

The friction element defines a strength domain and the occurrence of plastic deformation. Below the strength the material may behave elastically. In this case we speak of elastoplasticity, and the loading function (7.1) defines the elasticity domain D_E. If, in addition, the stress threshold remains constant during plastic deformation, we speak of an ideal elastoplastic behavior. The corresponding 1D think model is sketched in Figure 7.2a: a spring (stiffness E, initial unit length $l = 1$ $[L]$), in series with a friction element (strength k). The first captures the elasticity, the second the plastic behavior defined by (7.3). Application of an external stress σ to the system represented in the 1D think model (e.g., a force per unit surface) leads to a relative length increase $\varepsilon = \Delta l/[L]$. When the stress reaches the strength k, the sliding of the friction element ensures that the stress in the spring remains constant. After a complete unloading to $\sigma = 0$, an irreversible relative displacement ε is recorded. This irreversible displacement corresponds to the total sliding ε^p undergone by the friction element during loading. A novel stress application leads to the sliding of the friction element once the stress σ reaches the same stress threshold k. This stress-strain response is given in Figure 7.2b.

The study of this loading–unloading-reloading cycle reveals the essential ingredients of ideal elastoplasticity:

1. As long as the stress is below the friction element strength, the system response is elastic. For a linear spring behavior, the stress in the spring reads

$$\sigma = E(\varepsilon - \varepsilon^p) \tag{7.4}$$

The domain D_E defined by Eq. (7.1) contains all *plastically admissible* stress states: It is the elasticity domain of the spring-friction device.

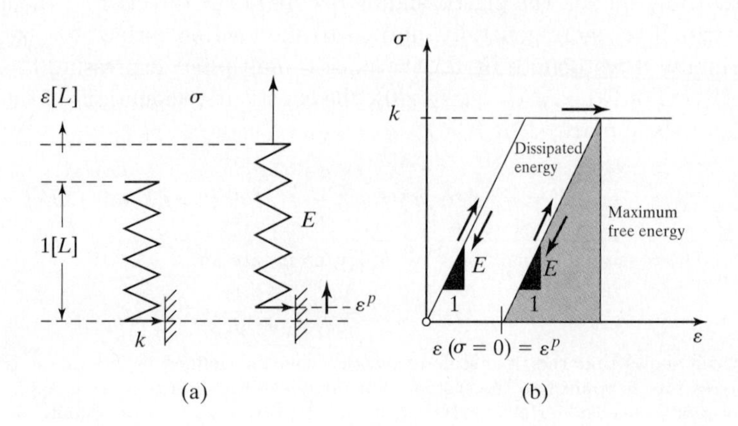

FIGURE 7.2: 1D ideal plasticity: (a) think model; (b) material response.

2. When the stress reaches the friction element strength, plastic deformation occurs at constant stress, with $d\varepsilon = d\varepsilon^p$, following the plastic flow rule (7.3).

The only measurable variable that can be controlled from the outside is the total strain ε. The plastic strain, ε^p, can only be accessed by means of measurements of the strain under specific well-defined (stress-strain) unloading conditions. Neither its magnitude, ε^p, nor its increment (or rate), $d\varepsilon^p = \dot{\varepsilon}^p dt$, can be controlled from the outside. This is due to the elasticity component of the behavior, which causes the stress in the spring. The plastic variable, therefore, appears as an internal or hidden variable. For the spring-friction element the plastic strain derives no more from a (geometrically measurable) displacement field (i.e., $\varepsilon^p \neq u^p/[L]$ in Figure 7.2a). Without any further specification, the only available information on $d\varepsilon^p$ is its direction sign (σ), which corresponds to the orientation of the slippage plane in the 1D think model.

7.1.3 Energy Considerations: The Principle of Maximum Plastic Work

We have a look on the energy side of the 1D material problem. For $f(\sigma) = |\sigma| - k < 0$, the system's response is elastic: The energy provided during time dt in the form of work to the system (i.e., the strain energy $\sigma d\varepsilon$) is entirely stored in the spring in the form of free energy ψ as mechanical work recoverable later:

$$f(\sigma) < 0 : \sigma d\varepsilon - d\psi = 0 \tag{7.5}$$

The free energy of the 1D system is the energy stored in the spring. From Figure 7.2a, it reads

$$\psi = \psi(\varepsilon, \varepsilon^p) = \frac{1}{2} E(\varepsilon - \varepsilon^p)^2 \tag{7.6}$$

Use of (7.6) in (7.5) yields for $d\varepsilon^p = 0$

$$\sigma = \frac{\partial \psi(\varepsilon, \varepsilon^p)}{\partial \varepsilon} = E(\varepsilon - \varepsilon^p) \tag{7.7}$$

Equation (7.7) designates stress σ as the driving force of the energy variation associated with the total relative length variation $\varepsilon = \Delta l/[L]$ prescribed from the outside. It has been derived for $d\varepsilon^p = 0$, which holds for any elastic response of the system, including the onset of any unloading process (i.e., for $|\sigma| - k = 0$ and $df < 0$) irrespective of the present value of ε^p. In short, Eq. (7.7) holds at any time: It is the equation of the state of the spring considered alone.

On the other hand, for $|\sigma| - k = 0$ and $df = 0$, the strain energy $\sigma d\varepsilon$ is not entirely stored in the system in the form of free energy $d\psi$. A part of this energy, say $d\mathcal{D}$, is dissipated through the irreversible sliding of the friction element along $d\varepsilon^p$:

$$d\mathcal{D} = \sigma d\varepsilon - d\psi \geq 0 \tag{7.8}$$

Use of (7.6) and (7.7) in (7.8) yields

$$d\mathcal{D} = -\frac{\partial \psi}{\partial \varepsilon^p} d\varepsilon^p = \sigma d\varepsilon^p \geq 0 \tag{7.9}$$

or, equivalently,

$$dD = k\,|d\varepsilon^p| = d\lambda k \geq 0 \tag{7.10}$$

Hence, set within an energy approach, stress $\sigma = -\partial\psi/\partial\varepsilon^p$ is (also) the driving force of the plastic deformation; and the quantity

$$dW^p = \sigma d\varepsilon^p \tag{7.11}$$

is the plastic work, which—in ideal plasticity—is entirely dissipated into heat form: dW^p equals the total dissipation $d\mathcal{D}$. Hence, compared with any plastically admissible stress σ' within or at the boundary of the elasticity domain [i.e., $f(\sigma') \leq 0$], the stress σ associated with the plastic strain increment $d\varepsilon^p$ provokes the maximum dissipation, that is,

$$\forall\sigma' : \; f(\sigma') \leq 0; \quad (\sigma - \sigma')\, d\varepsilon^p \geq 0 \tag{7.12}$$

Inequality (7.12) is known as the *principle of maximum plastic work*. It can be intuitively understood as a principle of noneconomy of matter: Among all possible plastically admissible stresses σ', stress σ associated with the plastic strain $d\varepsilon^p$ is the one that produces the maximum plastic work (i.e., the maximum dissipation of supplied mechanical energy $\sigma d\varepsilon$). While somewhat obvious in the 1D case, we will see later that this principle is a suitable concept to extend the 1D flow rule to three dimensions. In fact, the principle of maximum plastic work (7.12) combines the two key elements of the plasticity model: the normality of the flow rule $(7.3)_1$ and the convexity of the elasticity domain D_E. To illustrate this, consider for a moment that σ was within D_E (i.e., $|\sigma| < k$). Since $\sigma - \sigma'$ may be positive or negative, the only possibility to satisfy (7.12) in this case is $d\varepsilon^p = 0$. Hence, in this case (7.12) substitutes for the plastic loading–unloading conditions in (7.3). Next, let σ be at the boundary of D_E, $|\sigma| = k$, and $\sigma' = 0$. In this case, (7.12) requires that $\sigma d\varepsilon^p \geq 0$. Since σ can be positive or negative, it follows that $d\varepsilon^p$ must take the sign of stress σ and follow the normality rule of the flow rule $(7.3)_1$ [i.e., $d\varepsilon^p = d\lambda \operatorname{sign}(\sigma) = d\lambda \frac{\partial f}{\partial \sigma}$, with $d\lambda \geq 0$]. Use of this result in (7.12) yields

$$\forall\sigma' : \; f(\sigma') \leq 0; \quad f(\sigma) = 0; \quad (\sigma - \sigma')\frac{\partial f}{\partial \sigma}\, d\lambda \geq 0 \tag{7.13}$$

Finally, consider an unloading process, for which $f = 0$ and $df = \frac{\partial f}{\partial \sigma} d\sigma \leq 0$. Then, let $\sigma - \sigma' = d\sigma$ in (7.13). To satisfy simultaneously (7.13) and $df \leq 0$ requires $d\lambda = 0$ for the unloading process. Hence, (7.13) fully substitutes for the loading–unloading conditions in (7.3). In addition, (7.13) implies the convexity of D_E. In the 1D case, the convexity of D_E means that if two stresses, say σ_1 and σ_2, lie within the domain D_E, any stress σ' on the segment $\sigma_1 \leq \sigma' \leq \sigma_2$ is within D_E. In the 1D case the convex domain D_E must be a segment $\sigma^- \leq \sigma \leq \sigma^+$ containing the stress $\sigma = 0$, $\sigma^- \leq 0$ and $\sigma^+ \geq 0$. In fact, the zero stress state belongs to the plastically admissible stress states. Since $d\lambda \geq 0$, the convexity of D_E is included in the principle of maximum plastic work (7.13):

$$\forall\sigma' : \; f(\sigma') \leq 0; \quad f(\sigma) = 0; \quad (\sigma - \sigma')\frac{\partial f}{\partial \sigma} \geq 0 \tag{7.14}$$

Furthermore, since $f(\sigma' = 0) \leq 0$, (7.14) yields $\sigma \frac{\partial f}{\partial \sigma} \geq 0$. It follows that $f(\sigma') \leq 0$ defines the semi-infinite line $\sigma' \leq \sigma^+$ for $\sigma = \sigma^+ \geq 0$, and the semi-infinite line

$\sigma' \geq \sigma^-$ for $\sigma = \sigma^- \leq 0$. The intersection of these two semi-infinite lines constitutes the segment $\sigma^- \leq \sigma \leq \sigma^+$.[2]

In summary, the principle of maximum plastic work that stipulates the maximum dissipation into heat form of supplied mechanical energy $\sigma d\varepsilon$ is a founding principle of plasticity models, combining the normality rule [here sign (σ)] and the convexity of the elasticity domain D_E. We will see later that (7.14) extends to more complicated situations, such as hardening plasticity and the 3D case.

7.1.4 1D Thermodynamics of Ideal Plasticity

We take a step back to first principles, to set the previous energy considerations within the larger thermodynamic framework developed in Section 5.2.2. We need the following:

1. The first law (5.10),

$$dU = dW + dQ = \sigma d\varepsilon + dQ \tag{7.15}$$

2. The second law (5.11),

$$\theta dS \geq dQ \tag{7.16}$$

3. The definition of the Helmholtz free energy (5.14),

$$\psi = U - \theta S \tag{7.17}$$

A combination of (7.15) and (7.17) gives

$$\theta dS = \underbrace{\sigma d\varepsilon - S d\theta - d\psi}_{dD \equiv \varphi dt \geq 0} + dQ \tag{7.18}$$

A comparison of the second law (7.16) with (7.18) shows that the total entropy variation dS is composed of two terms of well distinct origin: dQ/θ is the externally supplied (i.e., controllable) entropy variation; $\varphi dt/\theta$ is the spontaneous entropy production, which cannot be controlled from the outside. Therefore, φdt is referred to as *intrinsic* dissipation. It represents the part of the strain energy $\sigma d\varepsilon$ (corrected by the thermal term $Sd\theta$) that is not stored in the form of free energy $d\psi$ but that is irreversibly dissipated into heat form. The inequality $\varphi dt \geq 0$ is known as Clausius–Duhem inequality.

---

[2]Inequality (7.14) extends to the 2D case in the form

$$\forall \sigma', \zeta' : f\left(\sigma', \zeta'\right) \leq 0; \quad f(\sigma, \zeta) = 0; \quad \left(\sigma - \sigma'\right) \frac{\partial f}{\partial \sigma} + \left(\zeta - \zeta'\right) \frac{\partial f}{\partial \zeta} \geq 0$$

If we add the condition $f(\sigma = 0, \zeta = 0) \leq 0$, the previous equation states the convexity of domain $f(\sigma, \zeta) \leq 0$ in the $(\sigma \times \zeta)$-space. Indeed, according to (7.14), $f(\sigma', \zeta') \leq 0$ results in a convex domain formed by the intersection of all the convex semi-infinite half planes that contain the origin, which are bounded by lines normal to the direction $\left(\frac{\partial f}{\partial \sigma}, \frac{\partial f}{\partial \zeta}\right)$ and which pass through the point (σ, ζ). The vector $\left(\frac{\partial f}{\partial \sigma}, \frac{\partial f}{\partial \zeta}\right)$ is oriented in the direction normal to the domain $f(\sigma', \zeta') \leq 0$ at point (σ, ζ), so that the domain is the envelop of the previous lines. These considerations can be extended to the n-D case. We will make use of this extension in subsequent chapters.

Equation (7.18) defines a general framework for macroscopic modeling of materials. The starting point is the choice of the appropriate state variables that describe the energy states of the material system. For the 1D think model of ideal thermoelastoplasticity, these are the strain ε and the temperature θ as external (i.e., controllable) state variables. In addition, the plastic strain ε^p is an internal (i.e., hidden) state variable describing the irreversible sliding of the friction element. Thus formally, in terms of the free energy ψ,

$$\psi = \psi(\varepsilon, \theta, \varepsilon^p) \tag{7.19}$$

The second step consists of determining the state equations. To this end, we use (7.19) in the expression of the intrinsic dissipation φdt:

$$\varphi dt = \left(\sigma - \frac{\partial \psi}{\partial \varepsilon}\right) d\varepsilon - \left(S + \frac{\partial \psi}{\partial \theta}\right) d\theta - \frac{\partial \psi}{\partial \varepsilon^p} d\varepsilon^p \geq 0 \tag{7.20}$$

For elastic (reversible) evolutions, $f(\sigma) < 0 \Leftrightarrow d\varepsilon^p = 0$, the dissipation is zero. This leads to the state equations (5.15) (have a quick look back at Section 5.2.2)

$$\sigma = \left(\frac{\partial \psi}{\partial \varepsilon}\right)_{\theta, \varepsilon^p} ; \quad S = -\left(\frac{\partial \psi}{\partial \theta}\right)_{\varepsilon, \varepsilon^p} \tag{7.21}$$

Since we can always define evolutions in which ε and θ develop independent of each other, these state equations still hold when plastic deformations occur. Thus, with (7.21), inequality (7.20) becomes

$$\varphi dt = \Sigma d\varepsilon^p \geq 0; \quad \Sigma = -\frac{\partial \psi}{\partial \varepsilon^p} \tag{7.22}$$

From (7.22), we formally identify $\Sigma = -\partial\psi/\partial\varepsilon^p$ as the driving force of the irreversible deformation $d\varepsilon^p$: It is the thermodynamic force that provokes the dissipation (i.e., the transformation of useful mechanical work—stored as free energy ψ—into heat form). In the particular case of the 1D ideal plasticity think model, this driving force is the stress (i.e., $\Sigma = \sigma$).

The evolution of the external state variables, ε and θ, can be controlled from the outside. This is not the case of the internal state variables. The evolution of internal variables is spontaneous and cannot be prescribed from the outside. Hence, the last step in the constitutive modeling consists of developing the appropriate *complementary* evolution law of the internal state variable, $d\varepsilon^p$. Since force Σ is the driving force of the plastic strain $d\varepsilon^p$, it is natural that this complementary evolution law relates $d\varepsilon^p$ and $\Sigma = -\partial\psi/\partial\varepsilon^p$. The only restriction for this complementary evolution law is to satisfy the inequality (7.22). In the case of the 1D ideal plasticity model, this complementary evolution law is the flow rule (7.3), and application of the principle of maximum plastic work (7.12) satisfies automatically this thermodynamic requirement. From an energy point of view, the constitutive model is then complete (closed set of equations).

Exercise 62. Develop the 1D constitutive model of a linear thermoelastic-ideal-plastic material (*cf.* Figure 7.2). For adiabatic conditions, determine the temperature change in a strain-driven experiment.

In Section 5.2.2, we have developed the free energy of a thermoelastic material within the framework of physical linearization.[3] For the variable set (7.19), the free energy of the linear thermoelastic-ideal-plastic material is obtained from (5.19) by replacing ε by $\varepsilon - \varepsilon^p$:

$$\psi = \psi(\varepsilon, \theta = T - T_0) = \frac{1}{2}\left(-\frac{\mathcal{C}}{T_0}\theta^2 - 2\alpha E\theta(\varepsilon - \varepsilon^p) + E(\varepsilon - \varepsilon^p)^2\right)$$

$\mathcal{C}$ is the heat capacity, and α the thermal dilatation coefficient. Use in (7.21) yields the state equations

$$\sigma = \left(\frac{\partial\psi}{\partial\varepsilon}\right)_{T,\varepsilon^p} = E(\varepsilon - \varepsilon^p - \alpha\theta)$$

$$S = -\left(\frac{\partial\psi}{\partial T}\right)_{\varepsilon,\varepsilon^p} = \frac{\mathcal{C}\theta}{T_0} + \alpha E(\varepsilon - \varepsilon^p)$$

In addition, (7.22) reads here

$$\varphi dt = -\frac{\partial\psi}{\partial\varepsilon^p}d\varepsilon^p = E(\varepsilon - \varepsilon^p - \alpha\theta)d\varepsilon^p \geq 0$$

Hence, as expected from the expression of the free energy,

$$\Sigma = -\frac{\partial\psi}{\partial\varepsilon^p} = \sigma$$

Therefore, the complementary evolution law for $d\varepsilon^p$ reads

$$d\varepsilon^p = d\lambda\mathcal{F}\left(\Sigma = \sigma\right)$$

where function $\mathcal{F}(\sigma)$ must satisfy

$$\varphi dt = d\lambda\,\sigma\mathcal{F}\left(\sigma\right) \geq 0$$

This is (almost) all thermodynamics can give us for designing the appropriate flow rule of plastic sliding. The energy approach needs to be completed by the description of the physical process of plastic sliding (i.e., for the 1D model in hand), by the flow rule (7.3) (i.e., by the intensity of plastic sliding $d\lambda \geq 0$), and by the orientation $\mathcal{F}(\sigma) = \text{sign}(\sigma)$. Finally, use of the flow rule in the fundamental inequality $\varphi dt \geq 0$ yields

$$\varphi dt = d\lambda\sigma\text{sign}\left(\sigma\right) \geq 0$$

If we note that $\sigma\text{sign}(\sigma) = |\sigma|$, we can replace σ by its value on the loading surface [i.e., Eq. (7.1)]. We obtain

$$f(\sigma) = 0 : \varphi dt = \sigma d\varepsilon^p = d\lambda k \geq 0$$

[3]We should note here that the framework of physical linearization only applies to external state variables, for which the relevance of the assumed linearity can be controlled. In other words, the linear thermoelastoplastic behavior is based on an additional assumption, the hypothesis of infinitesimal elastic strain $\varepsilon - \varepsilon^p$.

Since $d\lambda \geq 0$, the nonnegativity of the dissipation implies here the nonnegativity of the strength threshold, $k \geq 0$.
Last, in a strain-driven experiment, in which $d\varepsilon^p = d\varepsilon$ [i.e., Eq. (7.7) for $f = 0$], use of the previous expression of the dissipation in (7.18), together with the entropy equation of the state of the linear thermoelastic ideal plastic material, gives

$$f(\sigma) = 0 : TdS \approx T_0 dS = \mathcal{C}dT = dQ + k\mathrm{sign}\,(\sigma)\,d\varepsilon$$

In an isothermal experiment ($dT = 0$), the amount of energy dissipated into heat form corresponds in the stress-strain diagram to the area enclosed by a complete load-unloading cycle. This is shown in Figure 7.2b. In turn, in an adiabatic experiment, for which $dQ = 0$, the plastic dissipation φdt leads to a temperature rise $dT \geq 0$. In other words, a part of the useful mechanical work is irreversibly dissipated into heat form due to plastic sliding. Adding to this plastic dissipation, the elastic latent heat effects (see the exercise in Section 5.2.2), the total temperature rise in the strain-driven experiment under adiabatic conditions reads

$$dQ = 0 : \theta = T - T_0 = \frac{1}{\mathcal{C}/T_0}\left[-\alpha E \varepsilon_0 + k(\varepsilon - \varepsilon_0)\mathrm{sign}\,(\sigma)\right]$$

where ε_0 is the strain at which plastic deformation starts (i.e., in the adiabatic test set-up)

$$\varepsilon_0 = \left(\frac{\mathcal{C}/T_0}{\mathcal{C}/T_0 + \alpha^2 E}\right)\frac{k\mathrm{sign}\,(\sigma)}{E} = \frac{k\mathrm{sign}\,(\sigma)}{\bar{E}}$$

with $\bar{E} \geq E$ the adiabatic modulus (see the exercise in Section 5.2.2). We should note that the sign of the latent deformation heat depends on the sign of the prescribed strain (temperature increase in compression, temperature decrease in tension). In addition, this heat source is reversible. By contrast, the heat term associated with plastic dissipation is always positive due to the second law—irrespective of the sign of the prescribed strain increment $d\varepsilon$. ■

Remark It could have been appealing to define the onset of plastic deformation in the plasticity criterion by a strain threshold (ε_0) instead of a stress threshold (strength). From the study of the response of a 1D material system under adiabatic test conditions, it appears that such strain-space formulations of plasticity have an enormous drawback, because the threshold ε_0 varies in a reversible manner with thermal conditions (latent heat of elastic deformation). In other words, a strain threshold is not intrinsic to the matter but depends on the boundary conditions.

7.2 HARDENING PLASTICITY

7.2.1 1D Think Model of Hardening Plasticity

The second example we consider is 1D hardening plasticity. In contrast to ideal plasticity, hardening plasticity has a nonconstant elasticity domain: The threshold of plastic deformation increases (hardening) or decreases (softening). We speak of plastic hardening (or plastic softening) if this change of the threshold is caused by the plastic deformation undergone by the material. A simple 1D model of hardening plasticity (with plastic hardening) is displayed in Figure 7.3a. In addition to the ideal plasticity model (see Figure 7.2a), another spring of stiffness H in series with the elastic spring (stiffness E) is activated during plastic loading. The force in the elastic spring still reads

$$\sigma = E(\varepsilon - \varepsilon^p) \tag{7.23}$$

Let χ be the relative length increase of the hardening spring, and ζ the associated force resisting this length increase (i.e., negative in tension, positive in compression):

$$\zeta = -H\chi \tag{7.24}$$

Force ζ is called hardening force. Both stress σ and hardening force ζ are required to describe the elasticity domain of the plastic hardening material:

$$\sigma \in D_E : f(\sigma, \zeta) = |\sigma + \zeta| - k \leq 0 \tag{7.25}$$

In addition, the (kinematical) compatibility between the plastic deformation ε^p and the hardening deformation χ in this 1D system implies

$$\chi \equiv \varepsilon^p \tag{7.26}$$

From an energy point of view, ε^p and χ are two independent variables: ε^p defines the dissipation induced by the irreversible sliding of the friction element, while χ is associated with the elastic strain energy in the hardening spring. In return, Eq. (7.26) is an additional condition that defines the compatibility between the irreversible sliding of the friction element and the elastic hardening spring elongation. In the framework of plasticity, this compatibility is ensured by the plastic multiplier $d\lambda \geq 0$, expressing the intensity of plastic loading according to the Kuhn–Tucker conditions:

$$d\lambda \geq 0; \quad f \leq 0; \quad d\lambda f = 0 \tag{7.27}$$

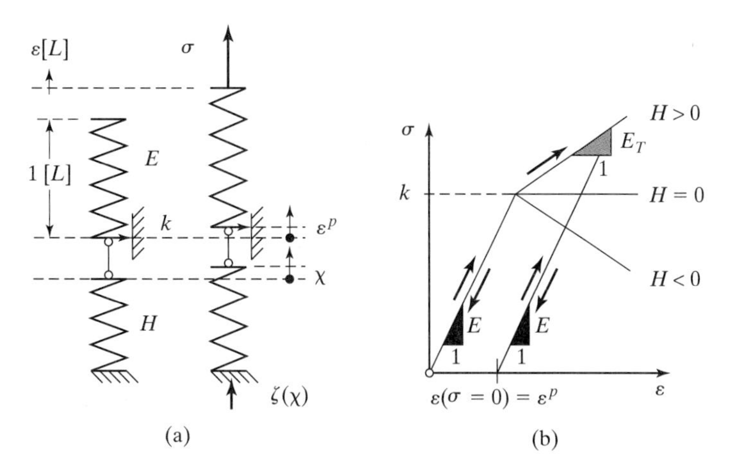

FIGURE 7.3: 1D hardening plasticity: (a) think model; (b) material response.

In addition, we need to specify the directions of both the plastic sliding $d\varepsilon^p$ and the direction of the elastic hardening spring elongation $d\chi$. In the 1D think model given in Figure 7.3a, these directions coincide and are defined by sign $(\sigma + \zeta)$; thus

$$d\varepsilon^p = d\lambda \text{sign}\,(\sigma + \zeta); \quad d\chi = d\lambda \text{sign}\,(\sigma + \zeta) \tag{7.28}$$

or, more generally,

$$d\varepsilon^p = d\lambda \frac{\partial f}{\partial \sigma}; \quad d\chi = d\lambda \frac{\partial f}{\partial \zeta} \tag{7.29}$$

where $\partial f / \partial \sigma$ represents the direction of the plastic sliding, and $\partial f / \partial \zeta$ the direction of the hardening spring elongation. Equations (7.29) extend the normality rule of the ideal plasticity model to plastic hardening.

Furthermore, during plastic loading, a loading point σ that is at the loading surface remains on the loading surface. Hence, any infinitesimal change of stress $\sigma + d\sigma$ implies a change in the hardening force $\zeta + d\zeta$:

$$f(\sigma + d\sigma, \zeta + d\zeta) = f(\sigma, \zeta) + \frac{\partial f}{\partial \sigma} d\sigma + \frac{\partial f}{\partial \zeta} d\zeta = 0 \tag{7.30}$$

Since $f(\sigma, \zeta) = 0$, the consistency condition of hardening plasticity reads

$$df = \frac{\partial f}{\partial \sigma} d\sigma + \frac{\partial f}{\partial \zeta} d\zeta = 0 \tag{7.31}$$

Use of (7.24) and $(7.28)_2$ in (7.31) yields

$$df = \frac{\partial f}{\partial \sigma} d\sigma - H d\lambda = 0 \tag{7.32}$$

Equation (7.32) leads to the plastic multiplier and the hardening modulus:

$$d\lambda = \frac{1}{H}\left(\frac{\partial f}{\partial \sigma} d\sigma\right) = \frac{d\sigma \text{sign}\,(\sigma + \zeta)}{H}; \quad H = -\left(\frac{\partial f}{\partial \zeta}\right)^2 \frac{\partial \zeta}{\partial \chi} \tag{7.33}$$

Finally, use of (7.33) in $(7.28)_1$ yields the plastic strain increment,

$$d\varepsilon^p = \frac{d\sigma}{H} = \frac{E d\varepsilon}{E + H} \tag{7.34}$$

and the tangential stress-strain relation

$$d\sigma = \frac{EH}{E + H} d\varepsilon \tag{7.35}$$

The tangent modulus $E_T = EH/(E + H)$ can be determined from a material test (see Figure 7.3b). We note that plastic hardening is defined by $H > 0$, which corresponds to an increase of the elasticity domain. In turn, for $H < 0$ the elasticity domain shrinks. This is the case of plastic softening. For $H = 0$, the ideal plastic case is recovered.

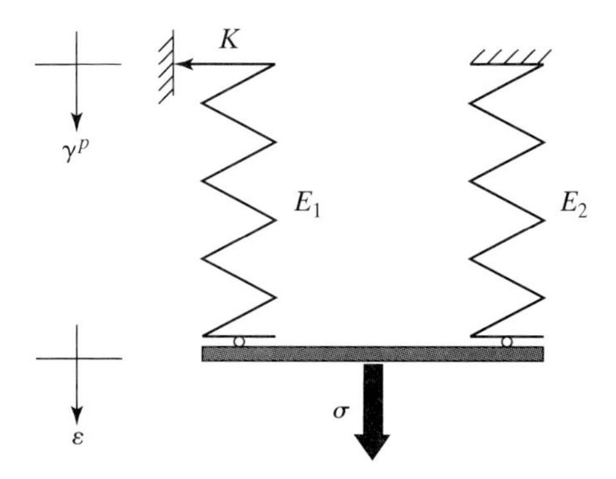

FIGURE 7.4: Plastic hardening think model.

7.2.2 An Instructive Exercise: An Alternative 1D Think Model of Hardening Plasticity

The choice of two distinct variables, plastic strain ε^p and hardening strain χ, to represent a plastic hardening behavior may appear, on first sight, surprising. The aim of this exercise is to work out the underlying reason for this separation, which is related to elastic energy stored in the microstructure of the material and associated with eigenstresses at a scale below the macrolevel of material description.

We consider the 1D hardening plasticity model shown in Figure 7.4. It is composed of an ideal plasticity component (spring of stiffness E_1 in series with a friction element of strength κ), in parallel with another spring of stiffness E_2. We note γ^p the plastic sliding of the friction element.

1. Determine the initial strength k of the model.
2. Consider a loading process that goes beyond this stress threshold and that is followed by a complete unloading process to a zero stress $\sigma = 0$. At the end of the unloading process, determine as a function of γ^p
   - (a) The overall irreversible plastic strain ε^p;
   - (b) The residual stresses σ_1^{res} and σ_2^{res} in the two springs;
   - (c) The elastic energy that remains trapped in the two springs.
3. Finally, determine the loading function and show that this 1D hardening plasticity model is equivalent to the think model shown in Figure 7.3a.

Onset of Plastic Deformation.

The total stress σ is the sum of the two stresses in the springs:

$$\sigma = \sigma_1 + \sigma_2 \tag{7.36}$$

Prior to plastic sliding of the friction element, the stress σ_1 in spring E_1 and the stress in spring E_2 relate to the strain ε by

$$\gamma^p = 0 : \quad \sigma_1 = E_1\varepsilon; \quad \sigma_2 = E_2\varepsilon \tag{7.37}$$

and to the total stress σ by

$$\gamma^p = 0: \ \sigma_1 = \frac{E_1}{E_1 + E_2}\sigma; \quad \sigma_2 = \frac{E_2}{E_1 + E_2}\sigma \tag{7.38}$$

The constant $E_i/(E_1 + E_2)$ in (7.38) may be seen as a stress localization or stress concentration factor: It localizes the macroscopic stress σ in the spring E_i. The stress on the friction element is $\sigma_1 = \sigma - \sigma_2$. It follows from (7.38) that the macroscopic strength corresponding to the onset of plastic deformations is related to the friction strength κ by

$$k = \frac{E_1 + E_2}{E_1}\kappa \tag{7.39}$$

Plastic Hardening Deformation.

When the stress intensity $|\sigma|$ is greater than the initial stress threshold k, the friction element irreversibly slides, and the elongation of spring E_1 becomes $\varepsilon - \gamma^p$. The partial stresses and total stresses read

$$\sigma_1 = E_1(\varepsilon - \gamma^p); \quad \sigma_2 = E_2\varepsilon; \quad \sigma = E_1(\varepsilon - \gamma^p) + E_2\varepsilon \tag{7.40}$$

After a total unloading to $\sigma = 0$, the observable strain ε is the irreversible part, that is, the plastic strain ε^p:

$$\sigma = 0 : \varepsilon = \varepsilon^p = \frac{E_1}{E_1 + E_2}\gamma^p \tag{7.41}$$

While the total stress is zero after unloading, the partial stresses are not. In fact, substitution of (7.41) in (7.40) yields the residual stresses in the springs:

$$\sigma_1^{\text{res}} = -\frac{E_1 E_2}{E_1 + E_2}\gamma^p = -E_2\varepsilon^p; \quad \sigma_2^{\text{res}} = \frac{E_1 E_2}{E_1 + E_2}\gamma^p = E_2\varepsilon^p \tag{7.42}$$

These stresses are associated with a residual elastic energy, say $\mathcal{U}$, in the springs:

$$\mathcal{U} = \frac{1}{2}\frac{E_1 E_2}{E_1 + E_2}(\gamma^p)^2 = \frac{1}{2}\frac{E_2}{E_1}(E_1 + E_2)(\varepsilon^p)^2 \tag{7.43}$$

The deformation of the springs, after a complete unloading, ensures the kinematical compatibility of the microscopic plastic sliding γ^p. In fact, the overall elongation of the ideal plasticity component (spring E_1+ friction element κ) must be equal to the elongation of spring E_2. In other words, the microscopic plastic sliding is not kinematically compatible alone, without the microscopic elastic elongation of spring E_2. This induces a self-balanced microscopic stress field, $\sigma_1^{\text{res}} = -\sigma_2^{\text{res}}$, associated with some residual elastic energy $\mathcal{U}$ trapped in the two springs. This is why energy $\mathcal{U}$ is called frozen energy, in the self-explanatory sense that it is not recovered during the unloading process. It is precisely the incompatibility of the plastic sliding γ^p at the microscopic level of the components of the system, which is at the origin of the macroscopic plastic hardening phenomenon and which leads to an appreciable change of macroscopic strength during plastic evolution.

Link with Macroscopic Plastic Hardening Think Model.

At any time the elasticity domain D_E is defined by

$$\sigma \in D_E : \ f = |\sigma - E_2\varepsilon| - \kappa \leq 0 \tag{7.44}$$

In order to express the loading function in a form that outlines that the change in the threshold stress is associated with plastic deformation, we decompose the strain ε in the elastic recoverable part and the irreversible or plastic part:

$$\varepsilon = \frac{\sigma}{E_1 + E_2} + \varepsilon^p \tag{7.45}$$

Substitution of (7.45) in (7.44) delivers the loading function f in the form

$$f = \left| \sigma - \frac{E_2}{E_1} (E_1 + E_2)\, \varepsilon^p \right| - \frac{E_1 + E_2}{E_1} \kappa \leq 0 \tag{7.46}$$

The found expression of the loading function is of the same format as (7.25), and the 1D think model displayed in Figure 7.4 is equivalent to the macroscopic plastic hardening model in Figure 7.3a, if we note that

$$E = E_1 + E_2; \quad H = \alpha E_2; \quad k = \alpha \kappa; \quad \alpha = \frac{E}{E_1} \tag{7.47}$$

On the other hand, the 1D think model in Figure 7.4 suggests that the appropriate hardening variable χ is not the macroscopic plastic strain but the microscopic one, that is,

$$\chi \equiv \gamma^p \tag{7.48}$$

This choice implies the following form of the macroscopic plastic hardening strength domain:

$$f(\sigma, \zeta) = |\sigma + \alpha\zeta| - k \leq 0 \tag{7.49}$$

where ζ is the hardening force now defined by

$$\zeta = -\frac{E_2}{\alpha}\chi = -h\gamma^p \tag{7.50}$$

Indeed, the amplifying factor α of the hardening force in (7.49)[4] ensures the compatibility between the macroscopic irreversible strain increment $d\varepsilon^p$ and the microplastic sliding $d\gamma^p$, derived from the flow and hardening rule (7.29):

$$d\varepsilon^p = d\lambda \operatorname{sign}(\sigma + \alpha\zeta); \quad d\chi = \alpha d\lambda \operatorname{sign}(\sigma + \alpha\zeta) = \alpha d\varepsilon^p = d\gamma^p \tag{7.51}$$

The plastic strain increment $d\varepsilon^p$ and the hardening strain increment $d\chi$ are related to the microscopic plastic sliding, and the intensity of this plastic sliding is expressed macroscopically by the plastic multiplier $d\lambda = |d\varepsilon^p| = |d\gamma^p|/\alpha$; from (7.33),

$$d\lambda = \frac{d\sigma \operatorname{sign}(\sigma + \alpha\zeta)}{H}; \quad H = -\alpha^2 \frac{\partial \zeta}{\partial \chi} = \alpha E_2 \tag{7.52}$$

[4]It is instructive to note that factor $\alpha = E/E_1$ is the inverse of the stress concentration factor introduced in (7.38) to relate the microscopic spring stress σ_1 to the macroscopic stress σ. In fact, the amplifying factor α scales the elastic force in spring E_1 to the macroscale, where it appears as hardening associated with the frozen energy.

This leads to the plastic strain increments,

$$d\varepsilon^p = \frac{d\sigma}{\alpha E_2}; \quad d\gamma^p = \frac{d\sigma}{E_2} \tag{7.53}$$

and the tangential stress-strain relation:

$$d\sigma = \left[\frac{\alpha E_2 (E_1 + E_2)}{E_1 + (1+\alpha)E_2}\right] d\varepsilon \tag{7.54}$$

Remark Plastic hardening is of a different physical origin than the irreversible plastic strain, which requires a specific hardening variable. Our choice $\chi = \gamma^p$ as hardening variable refers to the incompatibility of the microplastic sliding, which is at the origin of the frozen energy, $\mathcal{U} = \mathcal{U}(\chi = \gamma^p)$. But we should stress on the fact that the intrinsic character of the law governing the hardening is not based on the choice of a particular hardening variable, but rather on the intrinsic values of the frozen energy $\mathcal{U}$ and the material strength. In fact, we may well have chosen, for convenience, $\chi = \varepsilon^p$ as the hardening variable, because ε^p can be directly measured at the macrolevel. In this case, however, ε^p plays two distinct roles that must be distinguished: ε^p stands for the irreversible part of the strain after a complete unloading; but, at the same time, it is also a good candidate for measuring the degree of energy frozen at a microlevel [i.e., $\mathcal{U} = \mathcal{U}(\chi = \varepsilon^p)$]. This choice corresponds then strictly to the macroscopic plastic hardening think model in Figure 7.3a.

7.2.3 Energy Considerations: The Frozen Energy

We have just seen that the difference between ideal plasticity and hardening plasticity is the existence of an elastic energy reserve trapped somewhere at a microlevel. The response in the elastic range $f(\sigma,\zeta) < 0$ is the same as the one of the ideal elastoplasticity model: The strain energy, $\sigma d\varepsilon$, is stored in the elastic spring of stiffness E in the form of free energy ψ; and this stored energy can be recovered later. With a similar reasoning as employed in Section 7.1.3, we have

$$f(\sigma,\zeta) < 0 : \sigma d\varepsilon - d\psi = 0 \Leftrightarrow \sigma = \frac{\partial \psi}{\partial \varepsilon} = E(\varepsilon - \varepsilon^p) \tag{7.55}$$

On the other hand, the free energy of the 1D hardening system displayed in Figure 7.3a is the sum of the macroscopic elasticity, $\psi^{\mathrm{el}}(\varepsilon,\varepsilon^p) = \frac{1}{2}E(\varepsilon - \varepsilon^p)^2$, and the frozen energy, $\mathcal{U}(\chi) = \frac{1}{2}H\chi^2$, and both change their value during plastic loading [i.e., for $f(\sigma,\zeta) = 0$ and $df = 0$]. In this case, the strain energy $\sigma d\varepsilon$ is not entirely stored in the system in the form of free energy $d\psi = d(\psi^{\mathrm{el}} + \mathcal{U})$, but partly dissipated into heat form:

$$dD = \sigma d\varepsilon - d(\psi^{\mathrm{el}} + \mathcal{U}) = \sigma d\varepsilon^p - d\mathcal{U} \geq 0; \quad \sigma = -\frac{\partial \psi^{\mathrm{el}}}{\partial \varepsilon^p} = E(\varepsilon - \varepsilon^p) \tag{7.56}$$

We here made use of (7.55) and (7.9): As in the ideal plasticity model, stress σ is identified as the driving force of both the total strain and the plastic strain.

However, in contrast to ideal plasticity, the plastic work $dW^p = \sigma d\varepsilon^p$ in the plastic hardening model is no more equal to the dissipation $d\mathcal{D}$:

$$d\mathcal{D} = dW^p - d\mathcal{U} \geq 0 \tag{7.57}$$

In fact, during plastic deformation, a part of the plastic work dW^p is converted into frozen energy $d\mathcal{U} = H\chi \times d\chi$, and therefore not dissipated into heat form:

$$d\mathcal{D} = \sigma d\varepsilon^p + \zeta d\chi \geq 0; \quad \zeta = -\frac{\partial \mathcal{U}}{\partial \chi} = -H\chi \tag{7.58}$$

In (7.58), ζ is the hardening force (i.e., the force in the hardening spring resisting the movement of the friction element). During plastic deformation, the hardening force changes its values, and with it the elastic energy in the hardening spring. After a complete unloading process (i.e., restoring a zero stress state), the total strain reduces to its plastic component, $\varepsilon = \varepsilon^p$, and $\psi^{\text{el}} = 0$. The free energy related to this unloaded state is the energy $\psi = \mathcal{U}(\chi = \varepsilon^p) = \frac{1}{2}H(\varepsilon^p)^2$. This energy, which was not recovered during the unloading process, remains trapped in the hardening spring due to the irreversible sliding ε^p of the friction element: It is frozen by plastic hardening.

Finally, we should note that the total amount of work dissipated by plastic deformation and hardening deformation is identical to the dissipated work of the ideal plastic model (i.e., $d\mathcal{D} = k\,|d\varepsilon^p|$). Indeed, $d\mathcal{D}$ is the part of the plastic strain work converted into heat form due to the sliding of the friction element in Figure 7.3a. To show this, it suffices to substitute the flow and hardening rule (7.28) in (7.58):

$$d\mathcal{D} = d\lambda\,[\sigma + \zeta]\,\text{sign}\,(\sigma + \zeta) \geq 0 \tag{7.59}$$

If we note that $d\lambda = |d\varepsilon^p|$, and that $[\sigma + \zeta]\,\text{sign}\,(\sigma + \zeta) = |\sigma + \zeta| = k$ for $f(\sigma, \zeta) = 0$, we verify

$$d\mathcal{D} = \sigma d\varepsilon^p - d\mathcal{U} = d\lambda k \geq 0 \tag{7.60}$$

In the case of plastic hardening ($H > 0$), there is an increase of the frozen hardening energy ($d\mathcal{U} > 0$), thus an increase of the microelasticity reserve, which can be recovered later on to do work. In turn, in the case of plastic softening ($H < 0$), a part of the stored microelastic energy is released ($d\mathcal{U} < 0$) and dissipated into heat form. Figure 7.5 summarizes these results. The sign of the hardening modulus H determines the sign of the variation of the frozen energy $d\mathcal{U}$ during plastic loading. This physical significance of the hardening modulus can be captured from (7.32) and (7.58):

$$H = -\left(\frac{\partial f}{\partial \zeta}\right)^2 \frac{\partial \zeta}{\partial \chi} = \left(\frac{\partial f}{\partial \zeta}\right)^2 \frac{\partial^2 \mathcal{U}}{\partial \chi^2} \tag{7.61}$$

Equation (7.61) gives an energy interpretation of the hardening modulus.

Last, it is worthwhile to note that the principle of maximum plastic work [that is, Equation (7.12)] we developed for ideal plasticity extends to the hardening case in the form

$$\forall\ (\sigma', \zeta'):\ f(\sigma', \zeta') \leq 0;\ \ (\sigma - \sigma')\,d\varepsilon^p + (\zeta - \zeta')\,d\chi \geq 0 \tag{7.62}$$

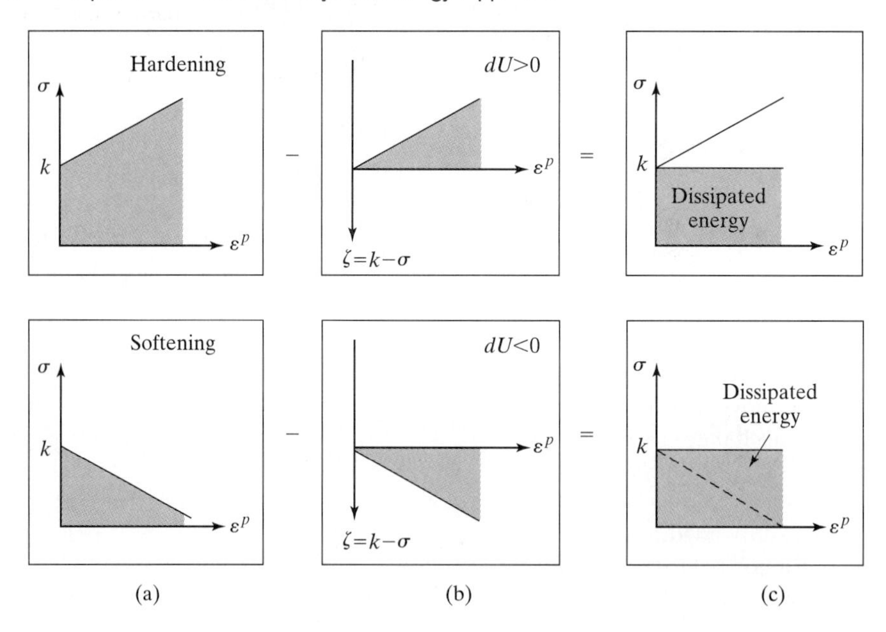

FIGURE 7.5: Dissipation of mechanical work into heat form due to (a) plastic deformation; (b) storage (hardening) or release (softening) of the frozen energy; (c) total dissipation.

or, equivalently,

$$(\sigma - \sigma') \frac{\partial f}{\partial \sigma} + (\zeta - \zeta') \frac{\partial f}{\partial \zeta} \geq 0 \tag{7.63}$$

Expression (7.62) combines the normality rule of both plastic deformation and hardening deformation, according to the flow and hardening rule (7.29), with the convexity of the elasticity domain D_E with regard to the two arguments σ and ζ, defining the elasticity domain in the $(\sigma \times \zeta)$-plane. This convexity, which is expressed by inequality (7.63), ensures the maximum plastic dissipation. We will refer to this principle when extending the 1D hardening plasticity model to 3D.

7.2.4 1D Thermodynamics of Hardening Plasticity

We summarize the key elements of the 1D hardening plasticity model within the framework of thermodynamics: In addition to the state variables of ideal plasticity (strain ε, temperature θ, plastic strain ε^p), hardening plasticity requires the consideration of another internal state variable χ modeling the hardening deformation; thus formally

$$\psi = \psi(\varepsilon, \theta, \varepsilon^p, \chi) \tag{7.64}$$

The energy associated with plastic hardening phenomena is the *frozen* energy. On a microlevel of material description, this energy refers to an elastic energy stored in the microstructure. The irreversible sliding of particles at the microlevel during plastic loading involves a microelasticity contribution. After a complete unloading of the matter, a part of this microelastic deformation remains. Indeed, it is required to ensure the kinematical compatibility of the irreversible sliding (since the plastic deformation does not derive from a displacement field). The energy associated with these microelastic deformation and residual eigenstresses at a microlevel is not released during unloading, but frozen at the microlevel of the heterogeneous matter. This stored energy constitutes a microelastic reserve that can be released in form of apparent hardening or softening at a macroscopic level, affecting the evolution of the macroscopic strength domain of the material. At the macroscopic level of constitutive modeling, this microelastic deformation is taken into account by the hardening variable χ and the associated frozen energy $\mathcal{U}(\chi)$:

$$\psi = \psi^{\text{el}}(\varepsilon, \theta, \varepsilon^p) + \mathcal{U}(\chi) \tag{7.65}$$

Here, we assumed that the elastic strain energy, $\psi^{\text{el}}(\varepsilon, \theta, \varepsilon^p)$, and the frozen energy, $\mathcal{U}(\chi)$, do not affect each other.[5] The first term in (7.65) corresponds to the thermoelastoplastic energy defined by (7.19). It includes the elastic strain energy, the latent heat of elastic deformation, and the heat capacity energy term. In turn, the frozen energy $\mathcal{U}(\chi)$ corresponds to the microelastic energy associated with hardening. For the 1D think model in Figure 7.3a, $\mathcal{U}(\chi)$ represents the elastic strain hardening energy stored in the hardening spring (i.e., in the case of linear hardening behavior),

$$\mathcal{U}(\chi) = \frac{1}{2} H \chi^2 \tag{7.66}$$

Both ψ^{el} and $\mathcal{U}$ compose the free energy $\psi = U - \theta S$ of the material system. Hence, analogously to (7.18), the first law and the second law lead to the Clausius–Duhem inequality of the hardening plasticity model:

$$TdS = \underbrace{\sigma d\varepsilon - Sd\theta - d(\psi^{\text{el}} + \mathcal{U})}_{d\mathcal{D} \equiv \varphi dt \geq 0} + dQ \tag{7.67}$$

Equation (7.67) expresses the nonnegativity of the intrinsic dissipation φdt associated with both plastic deformation and hardening deformation:

$$\varphi dt = \sigma d\varepsilon - Sd\theta - d(\psi^{\text{el}} + \mathcal{U}) \geq 0 \tag{7.68}$$

Use of the free energy expression (7.65) in (7.68) yields

$$\varphi dt = \left(\sigma - \frac{\partial \psi^{\text{el}}}{\partial \varepsilon}\right) d\varepsilon - \left(S + \frac{\partial \psi^{\text{el}}}{\partial \theta}\right) d\theta - \frac{\partial \psi^{\text{el}}}{\partial \varepsilon^p} d\varepsilon^p - \frac{\partial \mathcal{U}}{\partial \chi} d\chi \geq 0 \tag{7.69}$$

---

[5]This assumption is often referred to as the hypothesis of separation of energies. It is consistent with the existence of an isothermal infinitesimal elasticity,

$$d\sigma = E(d\varepsilon - d\varepsilon^p)$$

and requires the linearity of the stress state equation with respect to the elastic strain $\varepsilon - \varepsilon^p$. Furthermore, this assumption also implies that the frozen energy does not depend on the temperature. This comes to assume that latent heat effects related to the deformation of the hardening spring are negligible. A more comprehensive approach would need to consider this effect, which results in a loading function that depends on temperature.

A comparison with (7.20) shows that the state equations (7.21) still hold in the case of hardening plasticity. Using (7.21) in (7.69) it follows that

$$\varphi dt = \sigma d\varepsilon^p + \zeta d\chi \geq 0 \tag{7.70}$$

with

$$\sigma = -\frac{\partial \psi}{\partial \varepsilon^p}; \quad \zeta = -\frac{\partial \mathcal{U}}{\partial \chi} \tag{7.71}$$

We have already seen that the identification of force σ as the driving force of the plastic dissipation [i.e., (7.22)] implies a specific form of the complementary plastic evolution law. In addition, expression (7.70) formally designates the hardening force ζ as the driving force of the dissipation associated with hardening deformation $d\chi$. Hence, due to its hidden nature as an internal state variable, the complementary evolution law for $d\chi$ is a law that relates the hardening deformation $d\chi$ to the hardening force $\zeta = -\partial \mathcal{U}/\partial \chi$. The only requirement on the complementary evolution laws is the nonnegativity of the dissipation (7.70). This formal requirement needs to be fulfilled by the specific evolution laws of hardening plasticity, describing the kinematics of the physical processes involved. In the case of 1D hardening plasticity, these are the plastic flow rule $(7.29)_1$, and the hardening rule $(7.29)_2$, while the compatibility between plastic and hardening evolution is ensured by the plastic multiplier $d\lambda \geq 0$. Use of (7.29) in (7.70) gives

$$\varphi dt = d\lambda \left[\sigma \frac{\partial f(\sigma, \zeta)}{\partial \sigma} + \zeta \frac{\partial f(\sigma, \zeta)}{\partial \zeta} \right] \geq 0 \tag{7.72}$$

Since $d\lambda \geq 0$, the term in brackets of (7.72) must be greater than or equal to zero. Hence, the second law imposes certain restrictions on the directions, $\partial f/\partial \sigma$ and $\partial f/\partial \zeta$, taken, respectively, by the plastic strain increment, $d\varepsilon^p$, and the hardening strain, $d\chi$, and thus on function $f(\sigma, \zeta)$. It is the convexity of the loading function $f(\sigma, \zeta)$ with respect to its arguments, the thermodynamic forces σ and ζ, that ensures the nonnegativity of the intrinsic dissipation associated with plastic and hardening deformation. This convexity of the elasticity domain, expressed by (7.63), ensures the maximum plastic dissipation.

7.2.5 Kinematic versus Isotropic Hardening

The elasticity domain of the 1D hardening plasticity model undergoes a translation of the center $\sigma = 0$, due to plastic hardening, while the maximum segment, $\max \sigma - \min \sigma = 2k$, remains constant irrespective of the value of the hardening force. The response of the system in the $(\sigma \times \varepsilon)$-plane and in the $(\sigma \times \zeta)$-plane is shown in Figure 7.6. The plastic admissible stress states in the $(\sigma \times \zeta)$-plane are situated between two parallel lines, representing the loading function $f = 0$, inclined by $d\sigma/d\zeta = -1$.[6] The $(\sigma \times \zeta)$-plane is the plane of generalized (thermodynamic) forces. It is on the boundary of the elasticity domain that dissipative mechanisms

[6] In the case of the 1D think model of Figure 7.3, with $f = |\sigma + \alpha\zeta| - k = 0$, the slope in Figure 7.6b would be $d\sigma/d\zeta = -\alpha$; and the center of the elasticity domain would translate along a straight line $\sigma = -\alpha\zeta$ in the $(\sigma \times \zeta)$-plane, while maintaining the maximum stress segment constant, $\max \sigma - \min \sigma = 2k$.

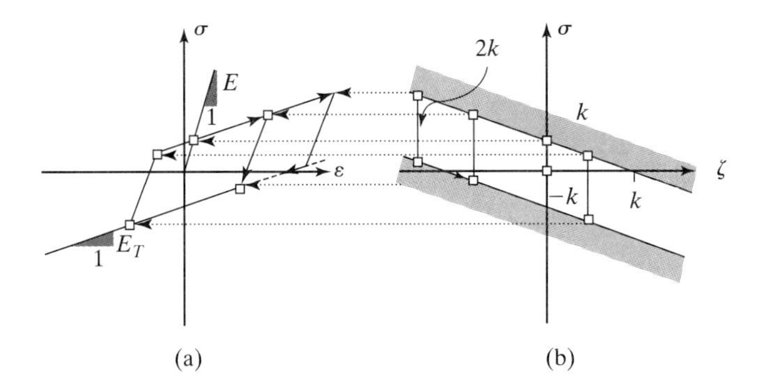

FIGURE 7.6: Kinematic hardening: (a) in the stress-strain plane; (b) in the generalized force plane.

are active within the material. The response shows that the hardening model translates the center of the elasticity domain, in the $(\sigma \times \zeta)$-plane, along a straight line $\sigma = -\zeta$. The consequence of this translation is that when the load is reversed, plastic evolutions will start at a lower stress intensity than during the original loading from a virgin state of the material, for which $\sigma = \zeta = 0$. This effect is known as Bauschinger effect, and it characterizes a large number of steel and steel-type materials. Such a hardening behavior, which translates the center of the elasticity domain, is called *kinematic* hardening.

By contrast, some materials undergo a plastic hardening (or softening) that expands (or shrinks) simultaneously the elasticity domain in all directions (i.e., in the 1D case in both tension and compression). The evolution of the hardening, therefore, is independent of the sign of the stress applied. Such a plastic hardening behavior is referred to as *isotropic* hardening: The boundary of the elasticity domain dilates (hardening) or shrinks (softening) in an isotropic way around the center, $\sigma = 0$. In the 1D case, this isotropic hardening behavior can be captured by the following loading function:

$$f(\sigma, \zeta) = |\sigma| + \zeta - k \leq 0 \tag{7.73}$$

Use in the flow and hardening rule (7.29) yields

$$d\varepsilon^p = d\lambda \frac{\partial f}{\partial \sigma} = d\lambda \, \mathrm{sign}\,(\sigma); \quad d\chi = d\lambda = \sqrt{d\varepsilon^p \times d\varepsilon^p} \tag{7.74}$$

From (7.74), it turns out that the plastic hardening variable in the isotropic hardening plasticity model is not the plastic strain, but the equivalent plastic strain intensity:

$$\chi \equiv \int \sqrt{d\varepsilon^p \times d\varepsilon^p} \tag{7.75}$$

This hardening variable enters the expression of the frozen energy of the isotropic hardening plasticity model and of the associated hardening force:

$$\mathcal{U} = \mathcal{U}\left(\chi \equiv \int \sqrt{d\varepsilon^p \times d\varepsilon^p}\right); \quad \zeta = -\frac{\partial \mathcal{U}}{\partial \chi} \tag{7.76}$$

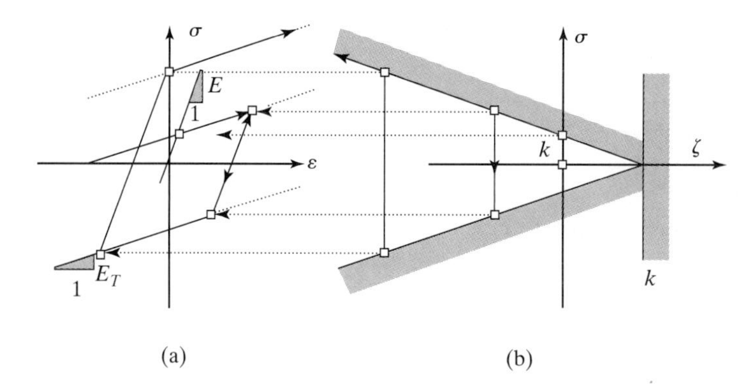

FIGURE 7.7: Isotropic hardening: (a) stress-strain response; (b) generalized force plane.

The consistency condition (7.31) applied to the loading function (7.73) and the energy definition of the hardening modulus (7.61) deliver the plastic multiplier:

$$d\lambda = \frac{d\sigma}{H}\operatorname{sign}(\sigma); \quad H = \frac{\partial^2 \mathcal{U}}{\partial \chi^2} \tag{7.77}$$

For a linear plastic hardening for which $\zeta = -H\chi$, the isotropic plastic hardening model yields the same tangential stress-strain relation as the kinematic hardening model; and the only difference is the isotropic expansion ($H > 0$), or isotropic shrinkage ($H < 0$) of the elasticity domain around the center $\sigma = 0$. The isotropic hardening case is illustrated in Figure 7.7 in the $(\sigma \times \varepsilon)$-plane and in the $(\sigma \times \zeta)$-plane.

7.3 VISCOPLASTICITY

Some engineering materials have pronounced rate effects: Strength and stiffness of the material depend on the rate of the stress applied. In geomaterials, these rate effects result often from viscous contacts between grains and adsorbed water films on particles that compose the heterogeneous matter on a microscopic level. When a sudden load is applied, viscous effects oppose the matter to move apart uncontrolled. This leads to a delay in the plastic response of the material system[7]:

[7]A similar delayed plastic behavior occurs in materials with crystalline microstructure: Beyond the shear threshold, the stress induced irreversible movement of some atoms on the surface of a slippage plane is governed by a certain probability, proportional to $\exp(-U/kT)$, where U is an activation energy, k the Boltzmann constant ($k = 1.38 \times 10^{-23}$ J/K/atom), and T the absolute temperature. The different energy levels at which the movement of the atoms occur are associated with different rate coefficients, which lead to a noninstantaneous plastic sliding. The plastic behavior is delayed in time. In addition, as the temperature increases, the probability of atoms to move increases, and consequently the rate of sliding. Viscous coefficients of some materials are often found to be proportional to $\exp(U/kT)$.

The apparent strength and stiffness depend on the applied stress rate. The time-independent plasticity model cannot capture these viscous effects. Still, as we will discuss later, the plasticity model extends straightforwardly to viscoplasticity, if we permit—during viscoplastic deformation—stress states outside the elasticity domain. This will also turn out as an engineering means to capture rate effects. The objective of this last section on 1D plasticity is to construct a macroscopic material model that captures roughly this time-dependent plastic phenomenon at the macroscopic scale of engineering material description.

7.3.1 1D Viscoplasticity Model

The starting point is the analysis of the 1D material behavior. We want to model the following behavior:

1. Below a stress threshold, say $|\sigma| < k$, the response of the material is instantaneous and linear elastic.
2. For $|\sigma| > k$, the material undergoes a time-dependent deformation: The material *creeps*. After unloading, a permanent strain is recorded. The intensity of this strain increases with the time of load application.
3. The apparent strength of the material is rate dependent: The higher the loading rate, $\dot{\sigma}$, the higher the plastically admissible stress. For an infinitely slow loading process, $\dot{\sigma} \to 0$, the material behaves like an ideal plastic material, with $|\sigma| \leq k$.

There are several models that can capture this material behavior. We focus here on a plastic model with time-dependent or viscous features (i.e., a viscoplastic material model). A 1D think model that captures the model requirements is given in Figure 7.8. It is composed of three elements in series: an elastic spring of stiffness E, a plastic slider of stiffness k, and a dashpot of viscosity η. We note ε the total strain, and ε^{vp} the displacement of the friction element.

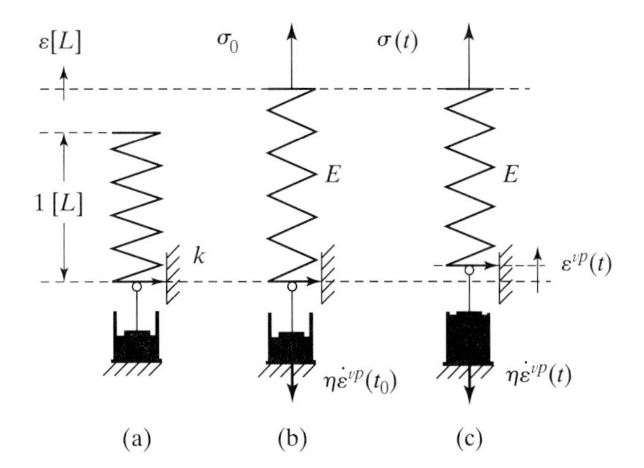

FIGURE 7.8: 1D viscoplasticity model: (a) relaxed state; (b) instantaneous response; (c) time-dependent stress relaxation.

Exercise 63. Determine the model's response to

1. A stress step loading:

$$\sigma = \sigma_0 H(t - t_0); \quad |\sigma_0| > k$$

where $H(x)$ is the Heaviside step function, $H(x < 0) = 0$; $H(x \geq 0) = 1$.

2. A strain step loading:

$$\varepsilon = \varepsilon_0 H(t - t_0); \quad |\varepsilon_0| > k/E$$

where t_0 is the time of load/strain application.

We note σ the stress in the elastic spring reading:

$$\sigma = E(\varepsilon - \varepsilon^{vp}) \tag{7.78}$$

For $|\sigma| < k$, the system's response is elastic. As in plasticity, this threshold defines when plastic deformation occurs. However, in contrast to plasticity, a loading point σ is not confined within an elasticity domain. In fact, if we note $\eta\dot{\varepsilon}^{vp}$ the viscous force in the dashpot, equilibrium around the friction element yields

$$|\sigma| \geq k : \sigma - k \operatorname{sign}(\sigma) = \eta\dot{\varepsilon}^{vp}$$

or, equivalently,

$$\dot{\varepsilon}^{vp} = \frac{|\sigma| - k}{\eta} \operatorname{sign}(\sigma) = \frac{\langle f(\sigma)\rangle}{\eta}\frac{\partial f}{\partial \sigma} \tag{7.79}$$

where $\langle x\rangle = \frac{1}{2}(x + |x|)$ denotes the positive part of x. Hence, in contrast to time-independent plasticity, a loading point σ can be situated outside of the elasticity domain of the material. If $f = |\sigma| - k > 0$, irreversible deformation occurs. Equation (7.79) sets out how this irreversible deformation occurs. It is the viscoplastic flow rule.

1. Plastic Creep: We consider the stress step loading $\sigma = \sigma_0 H(t - t_0)$. For $t > t_0$, the material deforms in time: It creeps at a constant rate proportional to the overstress $|\sigma_0| - k$:

$$\dot{\varepsilon}^{vp} = \frac{\langle|\sigma_0| - k\rangle}{\eta} \operatorname{sign}(\sigma_0) \rightarrow \varepsilon^{vp} = \int_{s=t_0}^{s=t} \dot{\varepsilon}^{vp} ds = \langle|\sigma_0| - k\rangle \frac{\langle t - t_0\rangle}{\eta} \operatorname{sign}(\sigma_0)$$

In (7.78), we obtain

$$\varepsilon(t) = \frac{\sigma_0}{E} + \langle f(\sigma_0)\rangle \frac{\langle t - t_0\rangle}{\eta}\frac{\partial f}{\partial \sigma}$$

The time-dependent deformation (plastic creep) is shown in Figure 7.9.

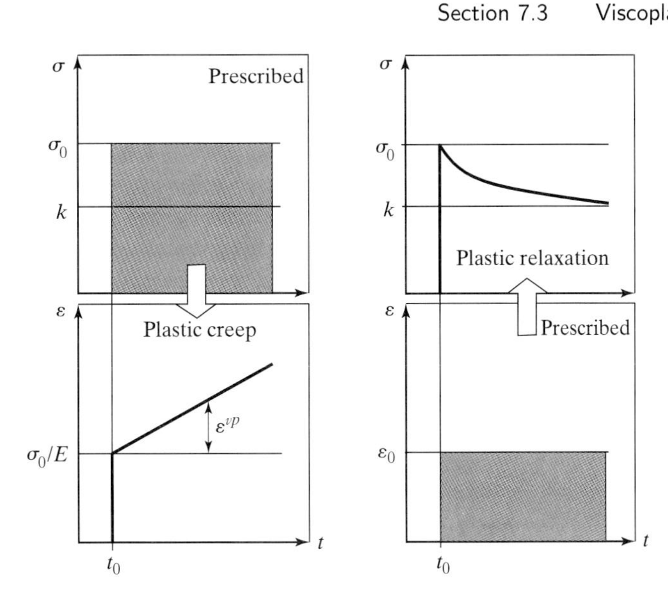

FIGURE 7.9: Viscoplastic model response: (left) plastic creep; (right) plastic stress relaxation.

2. Plastic Stress Relaxation: We consider the strain step loading $\varepsilon = \varepsilon_0 H(t - t_0)$. Substitution in (7.114) yields

$$\varepsilon_0 H(t - t_0) = \frac{\sigma(t)}{E} + \varepsilon^{vp}(t)$$

Deriving this equation with respect to time $t > t_0$ and substituting the result into the viscoplastic flow rule yields

$$|\sigma| \geq k : \sigma - k \operatorname{sign}(\sigma) = -\frac{\eta}{E}\dot{\sigma}$$

and, after integration,

$$\sigma = k \operatorname{sign}(\sigma) + [\sigma_0 - k \operatorname{sign}(\sigma)] \exp\left[-\frac{t - t_0}{\tau}\right]$$

or, equivalently,

$$\langle f(\sigma(t))\rangle = \langle f(\sigma_0)\rangle \exp\left[-\frac{t - t_0}{\tau_r}\right]$$

where $\sigma_0 = \sigma(t = t_0) = E\varepsilon_0$ is the instantaneous stress at the moment of load application, and $\tau_r = \eta/E$ is a characteristic relaxation time. The previous equations describe the plastic stress relaxation of the viscoplastic model. Due to viscoplastic deformation, a loading point σ_0 that is outside the elasticity domain returns asymptotically back to the loading function. The time scale of this relaxation process is defined by the characteristic relaxation time τ_r. The time-dependent stress relaxation is illustrated in Figure 7.9, and the Figures 7.8b and 7.8c illustrate the viscoplastic relaxation process in the 1D think model. ■

Note: in the case of a zero-strength threshold, the viscoplastic model reduces to the viscoelastic Maxwell model composed of an elastic spring in series with a dashpot.

7.3.2 Apparent Rate Effects

We have seen that the viscoplastic loading function permits stress states beyond the intrinsic material strength k (i.e., outside the elasticity domain, to which the stress returns in time). This is one possible key to capture rate effects on the apparent material strength and stiffness.[8]

Exercise 64. We consider a force-driven experience, characterized by a prescribed constant stress rate $\mathring{\sigma} > 0$, such that $\sigma(t) = \mathring{\sigma}t$. Show that the apparent stiffness of a viscoplastic material, which can be measured in such an experiment, depends on the loading rate.

From (7.83) and (7.79), we obtain

$$\mathring{\sigma} = E(\dot{\varepsilon} - \dot{\varepsilon}^{vp}) = E\dot{\varepsilon} - \frac{\langle \mathring{\sigma}t - k \rangle}{\tau_r}$$

or, equivalently, for $\sigma(t) > k$,

$$\mathring{\sigma} = \frac{E}{1 + \frac{t}{\tau_r}}\dot{\varepsilon} + \frac{k}{\tau_r + t}$$

where $\tau_r = \eta/E$ is the relaxation time. For $t \ll \tau_c$, we obtain

$$d\varepsilon = \frac{1}{E}\left(1 - \frac{k}{\tau_r \mathring{\sigma}}\right)\mathring{\sigma}dt$$

Therefore, in such an experience, the apparent stiffness $E_{\text{app}} = d\sigma/d\varepsilon$ will be found to be stress rate dependent [i.e., $E_{\text{app}}/E = F(\mathring{\sigma})$]: The higher the stress rate, the higher the apparent elastic stiffness. This stress rate dependency per se is not intrinsic to the matter, but results from the activation of viscous mechanisms during the time of loading $t \ll \tau_r = \eta/E$. ■

Note that rate effects on strength and stiffness are temporary and depend on the time scale of observation: The loading time must be much smaller than the relaxation time in order to observe the phenomenon. This is why we refer to these rate effects as apparent.

7.3.3 1D Thermodynamics

We have a look on the energy side of the viscoplastic model. The purpose is twofold: (1) To appreciate the difference of the viscoplastic model with the time-independent plastic model; and (2) to develop a framework for the possible extension of the model to more complicated situations, such as viscoplastic hardening and three dimensions.

[8] We will see in Volume II of this textbook a refined model of rate effects, derived within the context of damage and fracture mechanics.

Exercise 65. Develop the 1D viscoplastic model within the framework of thermodynamics of irreversible processes.

The starting point is the expression of the intrinsic dissipation rate. For isothermal evolutions, it reads

$$\varphi = \sigma\dot{\varepsilon} - \dot{\psi} \geq 0 \tag{7.80}$$

where a dot denotes time derivation, $\dot{x} = dx/dt$. The state variables required to describe the energy states of the elementary system are the total strain ε and the visco-plastic strain ε^{vp}. From Figure 7.8, the free energy corresponds to the elastic energy of the spring of stiffness E,

$$\psi = \frac{1}{2}E(\varepsilon - \varepsilon^{vp})^2 \tag{7.81}$$

Use in (7.80) leads to

$$\varphi = \left(\sigma - \frac{\partial\psi}{\partial\varepsilon}\right)\dot{\varepsilon} - \frac{\partial\psi}{\partial\varepsilon^{vp}}\dot{\varepsilon}^{vp} = \sigma\dot{\varepsilon}^{vp} \geq 0 \tag{7.82}$$

From this equation, it turns out that the state equations of the elasto-viscoplastic model are of the same form as the one of the plastic model:

$$\sigma = \frac{\partial\psi}{\partial\varepsilon} = -\frac{\partial\psi}{\partial\varepsilon^{vp}} = E(\varepsilon - \varepsilon^{vp}) \tag{7.83}$$

On first sight, the same can be said about the dissipation. The amount of energy that is dissipated during time interval dt is the viscoplastic work $d\mathcal{D} = dW^{vp} = \varphi dt = \sigma d\varepsilon^{vp}$. On a closer look (on the 1D think model of Figure 7.8), it turns out that the dissipation has two sources: The plastic energy dissipation due to the sliding of the friction element along $d\varepsilon^{vp}$, [i.e., k sign$(\sigma)d\varepsilon^{vp}$], that is, Eq. (7.10); and the dissipation associated with the movement of the dashpot of viscosity η, that is,

$$\varphi = \sigma\dot{\varepsilon}^{vp} = k\ \text{sign}\,(\sigma)\dot{\varepsilon}^{vp} + (\eta\dot{\varepsilon}^{vp})\dot{\varepsilon}^{vp} \geq 0 \tag{7.84}$$

where $\eta\dot{\varepsilon}^{vp} = \sigma - k\ \text{sign}\,(\sigma)$ is the viscous force in the dashpot. An immediate thermodynamic requirement that results from (7.84) is the nonnegativity of the viscosity η. ■

7.3.4 Normal Dissipative Mechanism

From (7.84) we identify stress σ as the driving force of the viscoplastic strain rate $\dot{\varepsilon}^{vp}$. From a strict thermodynamic standpoint, this identification implies that the complementary evolution for the viscoplastic strain relates the rate to its associated force:

$$\varphi = \sigma\dot{\varepsilon}^{vp} \geq 0 : \dot{\varepsilon}^{vp} = \mathcal{F}(\sigma) \rightarrow \varphi = \sigma\mathcal{F}(\sigma) \geq 0 \tag{7.85}$$

The Clausius–Duhem inequality restricts possible choices for function $\mathcal{F}(\sigma)$. One mathematical trick to ensure the nonnegativity of the dissipation is to define function $\mathcal{F}(\sigma)$ as the partial derivative of a function, say $\mathcal{D}$, which is convex with regard to its argument, here σ. Thus,

$$\dot{\varepsilon}^{vp} = \frac{\partial\mathcal{D}}{\partial\sigma};\quad \varphi = \sigma\frac{\partial\mathcal{D}}{\partial\sigma} \geq 0 \tag{7.86}$$

The postulated convexity of $\mathcal{D}$, therefore, ensures the nonnegativity of the dissipation. Function $\mathcal{D}$ is often referred to as dissipation potential. Complementary evolution laws that derive from a dissipation potential are called *normal dissipative mechanisms*. The normality rule is not really a (proven) physical principle, but rather a mathematical result, in the same way as the principle of maximum plastic work. In fact, it may well be seen as an extension of the principle of plastic work to viscoplastic evolution: The viscoplastic flow obeys the normality rule (i.e., normal to the dissipation potential $\mathcal{D}$), and this dissipation potential is convex. The corresponding expression of $\mathcal{D}$ for the viscoplastic flow rule (7.79) is

$$\mathcal{D} = \frac{1}{2\eta}\langle f(\sigma)\rangle^2; \quad \dot{\varepsilon}^{vp} = \frac{\partial \mathcal{D}}{\partial \sigma} = \frac{\langle f(\sigma)\rangle}{\eta}\frac{\partial f}{\partial \sigma} \tag{7.87}$$

We verify that $\mathcal{D}$ is convex. We also note that if the viscosity tends to zero (i.e., $\eta \to 0$), also $\langle f(\boldsymbol{\sigma})\rangle$ must go to zero, since the viscoplastic rate cannot be infinite. Then the factor $\langle f\rangle/\eta$ is undetermined as the plastic multiplier $d\lambda$ in the ideal plastic case, which is then recovered.

The framework of normal dissipative mechanisms is a safe orientation for a thermodynamic sound formulation of constitutive equations. For instance, imagine that experimental observation brought about a different exponent than 2 in expression (7.87) [i.e., a nonlinear dependence of the strain rate with regard to the overstress $\langle f(\sigma)\rangle = |\sigma| - k$]. In this case, a generalization of (7.87) is required; for example, in the form

$$\mathcal{D} = \frac{1}{(1+n)\eta}\langle f(\sigma)\rangle^{1+n}; \quad \dot{\varepsilon}^{vp} = \frac{\partial \mathcal{D}}{\partial \sigma} = \frac{\langle f(\sigma)\rangle^n}{\eta}\frac{\partial f}{\partial \sigma} \tag{7.88}$$

In this case, the convexity of the dissipation potential requires $n > 0$. In this sense, the concept of normal dissipative mechanisms provides some safe guidelines for constitutive modeling—not more, not less: It is a sufficient condition that ensures the thermodynamic stability of the material. This represents an argument in favor of its physical justification, but it cannot be considered as a general physical principle.

7.4 TRAINING SET: 1D CYCLIC PLASTICITY

Many materials respond to moderate cyclic loading by an accumulation of permanent deformation. This is known for different steel types used (e.g., for railroads). But also bituminous materials employed for pavements are known to accumulate permanent deformation under heavy cyclic traffic loading. For instance, this accumulation leads to the formation of ruts, which is one main cause of damage in bituminous pavements. The hardening models, whether kinematic or isotropic, developed in this chapter are insensitive to cyclic loading (with constant stress amplitude). The modeled material response remains the same irrespective of the cycles. We want to model this cyclic plasticity phenomenon within the framework of plasticity. This challenging training set reviews the basic steps of constitutive modeling using an energy approach: 1D rheological model, choice of state variables, free energy formulation, state equations, complementary evolution laws, and thermodynamic restrictions.

7.4.1 The Stéfani Model

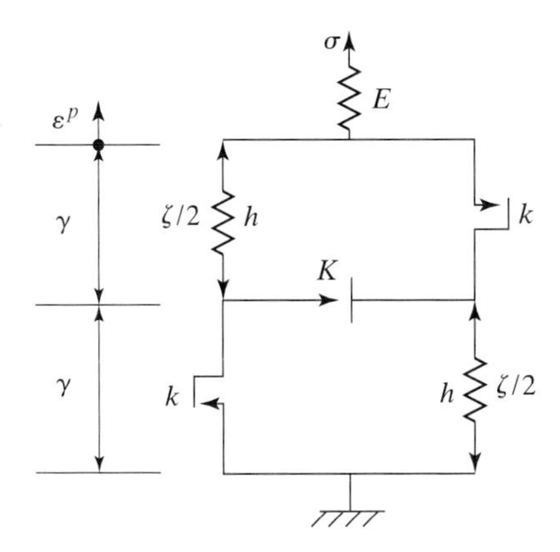

FIGURE 7.10: Stéfani's model.

The starting point of the constitutive modeling is (often) a 1D think model that captures the main features of the phenomena at stake. Figure 7.10 shows a 1D rheological device developed by Stéfani[9] for rutting in bituminous pavements. The Stéfani model is composed of an elastic spring (stiffness E) in series with two parallel head-to-foot spring-friction elements (spring modulus h, friction strength k). These subsystems are linked by a further friction element of strength K. We want to analyze the response of this material model.

Exercise 66. The Stéfani model is a plasticity model with loading functions that define the plastically admissible stress states of the model. Determine the loading functions of the Stéfani model.

The loading function defines when plastic deformation occurs. In the Stéfani model, there are two friction mechanisms, the sliding of the friction elements of strength k, and the sliding of the friction element of strength K. Hence, there are two loading functions that define the plastically admissible stress states of the Stéfani model.[10] These loading functions can be determined from elementary equilibrium considerations, shown in Figure 7.11. If we note σ the applied stress, and ζ the sum of the resisting forces in the hardening springs of stiffness h (negative in tension, positive in compression), the stress states are constrained by the following two loading functions:

$$f_k(\sigma, \zeta) = |\sigma + \zeta/2| - k \leq 0 \tag{7.89}$$

$$f_K(\sigma, \zeta) = |\sigma + \zeta| - K \leq 0 \tag{7.90}$$

Loading functions $f_k(\sigma, \zeta)$ and $f_K(\sigma, \zeta)$ define when sliding of the friction elements

[9] Stéfani, C. (1996). "Modèle Analogique pour l'Orniérage. Révision de la Note du 28 Novembre 1995." Internal Note, December 20, 1996, Laboratoire Central des Ponts et Chaussées, unpublished.

[10] Plasticity models with more than one loading function defining the plastically admissible stress states are (often) referred to as multisurface plasticity models.

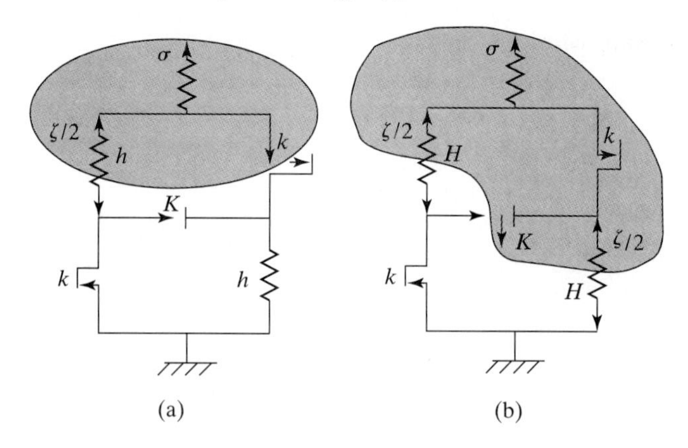

FIGURE 7.11: Loading functions: (a) $f_k \leq 0$; (b) $f_K \leq 0$.

of strength k and K occur. Whether the one or the other frictional mechanisms is activated is determined by

$$\sigma \in D_E : f(\sigma, \zeta) = \max(f_k, f_K) \leq 0 \tag{7.91}$$

Equation (7.91), with (7.89) and (7.90), defines the plastically admissible stress states of the Stéfani model. Finally, the onset of plastic evolutions is determined by the strength ratio K/k. In what follows, we assume

$$K \geq 2k \tag{7.92}$$

The elasticity domain defined by the loading functions is shown in Figure 7.12. In the $(\sigma \times \zeta)$-plane, functions (7.89) and (7.90) define a closed strength domain of plastically admissible stress states.

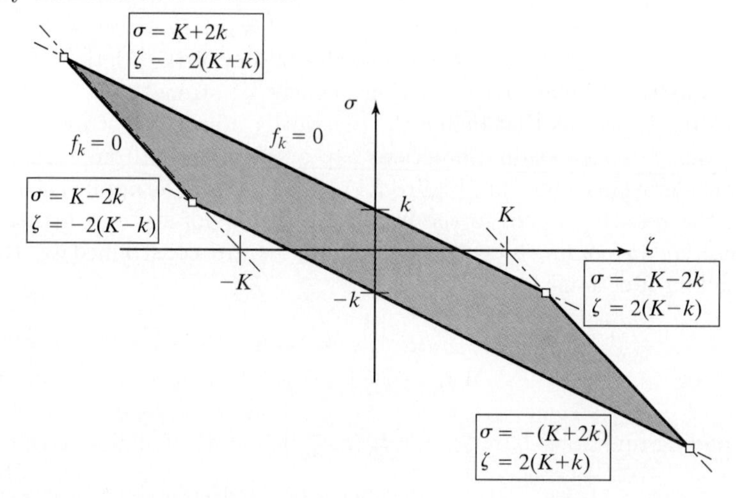

FIGURE 7.12: Elasticity domain of Stéfani's model in the generalized $(\sigma \times \zeta)$-force plane: two-surface plasticity model. ■

7.4.2 Energy Approach: The Stéfani Model

Choice of State Variables and State Equations.

We have a look at the energy side of the Stéfani model. The starting point is an appropriate choice of the state variables and the corresponding equations of state, which define the energy states of the system. There are two types of state variables: external and internal state variables.

Exercise 67. We assume isothermal evolutions. Choose the appropriate state variables of the Stéfani model, and determine the corresponding state equations.

The spring-friction device has three degrees of freedom (*cf.* Figure 7.10): the total strain (change in length) of the device, $\varepsilon = \Delta l/[L]$, the total change in length of the parallel spring-friction device, ε^p, and the relative length variation γ of the springs of stiffness, h (for symmetry reason, the two springs have the same stiffness h). Thus, formally in terms of the free energy,

$$\psi = \psi(\varepsilon, \varepsilon^p, \gamma) \tag{7.93}$$

From Figure 7.10 it appears that the only controllable state variable is the total strain $\varepsilon = \Delta l/[L]$, while ε^p and γ are internal state variables that are required to describe the frictional mechanisms in the parallel spring-friction device. The evolution of ε^p and γ cannot be prescribed from the outside.
The free energy can be explicitly expressed from Figure 7.10 (sum of the elastic spring energies):

$$\psi = \psi^{\text{el}}(\varepsilon, \varepsilon^p) + \mathcal{U}(\gamma) = \frac{1}{2}E(\varepsilon - \varepsilon^p)^2 + h\gamma^2 \tag{7.94}$$

Use of (7.94) in the Clausius–Duhem inequality (7.68) yields (for isothermal evolutions)

$$\begin{aligned} \varphi dt &= \sigma d\varepsilon - d(\psi^{\text{el}} + \mathcal{U}) \\ &= \left(\sigma - \frac{\partial \psi^{\text{el}}}{\partial \varepsilon}\right) d\varepsilon - \frac{\partial \psi^{\text{el}}}{\partial \varepsilon^p} d\varepsilon^p - \frac{\partial \mathcal{U}}{\partial \gamma} d\gamma \\ &= \sigma d\varepsilon^p + \zeta d\gamma \geq 0 \end{aligned} \tag{7.95}$$

Thus, the state equations are

$$\sigma = \frac{\partial \psi^{\text{el}}}{\partial \varepsilon} = -\frac{\partial \psi^{\text{el}}}{\partial \varepsilon^p} = E(\varepsilon - \varepsilon^p) \tag{7.96}$$

$$\zeta = -\frac{\partial \mathcal{U}}{\partial \gamma} = -2h\gamma \tag{7.97}$$

■

Complementary Evolution Laws: Flow and Hardening Rules of the Two-Surface Plasticity Model.

From (7.94) to (7.97), it appears that the Stéfani model is an (almost) standard elastoplastic model with plastic hardening: Stress σ is the driving force of irreversible strains $d\varepsilon^p$, and hardening force ζ is the driving force of hardening strain $d\gamma$. However, the evolution of these internal variables depends on the active friction mechanism. For convenience, we call the friction mechanism k and K the mechanism involving the sliding of the friction elements, respectively, of strength k and K. The two frictional mechanisms are displayed in Figure 7.13.

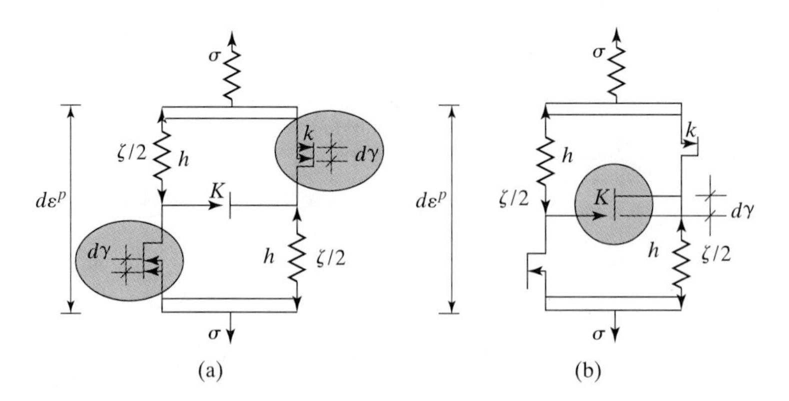

FIGURE 7.13: The two frictional mechanisms of the Stéfani model.

Exercise 68. By using the flow and hardening rule, determine the tangential stress-strain relation of the Stéfani model.

We first consider friction mechanism k (see Figure 7.13a). The flow and hardening rules (7.29) read

$$\text{if } (f_k = 0; f_K < 0) : \left\{ \begin{array}{c} d\varepsilon^p = d\lambda \partial_\sigma f_k = d\lambda \, \text{sign}\,(\sigma + \tfrac{1}{2}\zeta) \\ d\gamma = d\lambda \partial_\zeta f_k = \tfrac{1}{2} d\lambda \, \text{sign}\,(\sigma + \zeta/2) = \tfrac{1}{2} d\varepsilon^p \end{array} \right\} \tag{7.98}$$

where $\partial_\sigma f_k = \partial f_k / \partial \sigma$ and $\partial_\zeta f_k = \partial f_k / \partial \zeta$ denote the directions taken by the plastic strain increments, $d\varepsilon^p$ and $d\gamma$, respectively. The plastic multiplier is obtained from the consistency condition:

$$df_k = 0 : d\lambda = \frac{1}{H_k} \frac{\partial f_k}{\partial \sigma} d\sigma = \frac{d\sigma}{H_k} \text{sign}\,(\sigma + \zeta/2) \tag{7.99}$$

H_k is the hardening modulus associated with the frictional mechanism k. It is obtained by using (7.89) and (7.94) in (7.61):

$$H_k = \left(\frac{\partial f_k}{\partial \zeta} \right)^2 \frac{\partial^2 \mathcal{U}}{\partial \chi^2} = \left(\frac{1}{2} \right)^2 2h = \frac{h}{2} \tag{7.100}$$

Finally, using (7.99) and (7.100) in (7.98), the flow and hardening rules associated with the friction mechanism k become

$$\text{if}(f_k = 0; f_K < 0) : d\varepsilon^p = \frac{2d\sigma}{h}; \quad d\gamma = \frac{d\sigma}{h} \tag{7.101}$$

Last, from (7.96) and (7.101), we obtain the tangential stress-strain relation:

$$\text{if } (f_k = 0; f_K < 0) : d\sigma = E(d\varepsilon - d\varepsilon^p) = \frac{Eh}{h+2E} d\varepsilon; \quad d\zeta = -2d\sigma \tag{7.102}$$

In a similar way, we proceed with the friction mechanism K (see Figure 7.13b). The flow and hardening rules (7.29) read

$$\text{if } (f_k < 0; f_K = 0) : \left\{ \begin{array}{c} d\varepsilon^p = d\lambda \partial_\sigma f_K = d\lambda \operatorname{sign}(\sigma + \zeta) \\ d\gamma = d\lambda \partial_\zeta f_K = d\lambda \operatorname{sign}(\sigma + \zeta) = d\varepsilon^p \end{array} \right\} \tag{7.103}$$

The plastic multiplier,

$$df_K = 0 : d\lambda = \frac{1}{H_K} \frac{\partial f_K}{\partial \sigma} d\sigma = \frac{d\sigma}{H_K} \operatorname{sign}(\sigma + \zeta) \tag{7.104}$$

and the hardening modulus,

$$H_K = \left(\frac{\partial f_K}{\partial \zeta} \right)^2 \frac{\partial^2 \mathcal{U}}{\partial \chi^2} = (1)^2 \, 2h = 2h \tag{7.105}$$

lead to

$$\text{if } (f_k < 0; f_K = 0; df_K = 0) : d\varepsilon^p = d\gamma = \frac{d\sigma}{2h} \tag{7.106}$$

The tangential stress-strain relation reads

$$\text{if } (f_k < 0; f_K = 0) : d\sigma = E(d\varepsilon - d\varepsilon^p) = \frac{2Eh}{2h+E} d\varepsilon; \quad d\zeta = -d\sigma \tag{7.107}$$

A comparison of (7.106) and (7.101) reveals that the evolution of the plastic strain and the hardening strain depends on the activation of the particular friction mechanism, with different tangential moduli [i.e., (7.102) and (7.107)] during plastic loading. We will see later that this is the key to the modeling of the ratchetting phenomenon under cyclic loading.

Finally, the case when both mechanisms are active corresponds to loading points $(\sigma \times \zeta)$ where the loading surfaces f_k and f_K intersect (*cf.* Figure 7.12). In these corners, both the stress σ and the hardening force ζ have fixed values. In other words, the behavior in this corner-regime is ideal plastic:

$$\text{if } (f_k = 0; f_K = 0) : \left\{ \begin{array}{c} d\sigma = 0 \leftrightarrow d\varepsilon^p = d\varepsilon \\ d\zeta = 0 \leftrightarrow d\gamma = 0 \end{array} \right\} \tag{7.108}$$

■

Dissipation.

Last, we want to determine the amount of energy which is dissipated during the plastic sliding of friction elements k and K, respectively.

Exercise 69. Determine the dissipation of the Stéfani model.

For the friction mechanism k, the work dissipated by the sliding of the two friction elements along $d\gamma$ (see Figure 7.13a) reads, with (7.101),

$$\text{if } (f_k = df_k = 0; f_K < 0) : \varphi_k dt = 2k|d\gamma| = \frac{2k|d\sigma|}{h} \geq 0 \tag{7.109}$$

Analogously, for the friction mechanism K (see Figure 7.13b) with (7.106),

$$\text{if } (f_k < 0; f_K = df_K = 0) : \varphi_K dt = K|d\gamma| = \frac{K|d\sigma|}{2h} \geq 0 \tag{7.110}$$

In addition, in (7.92), we assumed $K \geq 2k$. This assumption ensures a lower effective strength of the friction mechanism k. It is the first friction mechanism that is activated by plastic loading. In turn, from an energy point of view [i.e., (7.109) and (7.110)], this assumption implies

$$K \geq 2k \leftrightarrow \varphi_k \leq 2\varphi_K \tag{7.111}$$

■

Note that k and K are strength quantities, which can be accessed by mechanical tests (e.g., data fitting of stress-strain behavior). In turn, the energy approach provides us with a second independent access to the constitutive model. Indeed, $\varphi_k dt$ and $\varphi_K dt$ are heat quantities that can be accessed (at least on the paper) by calorimetric measurements.

7.4.3 Model Response to 1D Cyclic Loading

We consider a cyclic loading:

$$\sigma(t) = \sigma_{\max} F(t); \quad F(t) \in [0; 1] \tag{7.112}$$

where $\sigma_{\max}$ is the stress amplitude. For $\sigma_{\max} < K$ only the friction mechanism k is activated, and the model response corresponds to the one of a single surface hardening plasticity model, shown in Figure 7.6. In addition, from Figure 7.12, $\sigma_{\max} = K + 2k$ corresponds to the maximum plastically admissible stress in the (tension) test. Hence, we are interested here in stress amplitudes:

$$K < \sigma_{\max} < K + 2k \tag{7.113}$$

Exercise 70. Determine the residual strain after the first cycle. Illustrate the model response in the $(\sigma \times \varepsilon)$- and $(\sigma \times \zeta)$-planes.

We start with a first cycle from the initial conditions $\sigma = \zeta = 0$. The response is shown in Figure 7.14 in the stress-strain diagram and the stress-hardening force diagram.

In the first cycle, the response is purely elastic for $\sigma < k$. In Figure 7.14, this elastic response corresponds to the load path 01:

$$\text{Phase } 01\colon 0 \leq \sigma < k \rightarrow d\sigma = E d\varepsilon$$

Beyond $\sigma = k$, the response becomes elastoplastic. During the load path 12 the friction mechanism k (i.e., $f_k = 0; f_K < 0$) is activated. With (7.102),

$$\text{Phase } 12\colon k \leq \sigma < \sigma_{\max} \rightarrow d\sigma = \frac{Eh}{h + 2E} d\varepsilon$$

Consider now the unloading case from $\sigma_{\max}$ to $\sigma = 0$. Depending on stress amplitude $\sigma_{\max}$, we can distinguish the following:

- $\sigma_{\max} \leq 2k$: The unloading is purely elastic.
- $2k < \sigma_{\max} \leq K$: The unloading path comprises an elastic phase for $\sigma \in [\sigma_{\max} - k; \sigma_{\max}]$, followed by an elastoplastic phase activating the frictional mechanism k: The model response corresponds to the one of a single surface hardening plasticity model, shown in Figure 7.6.
- $K < \sigma_{\max} < K + 2k$: The unloading path comprises an elastic phase followed by an elastoplastic one, during which the two frictional mechanisms are successively activated. This is the case we consider [i.e., Eq. (7.113)].

During loading (path $01 - 12$), the hardening force is altered due to the activation of the friction mechanism k. According to (7.102), its value at the onset of unloading is

$$\zeta = -2\Delta\sigma_{12} = -2(\sigma_{\max} - k)$$

Upon unloading the friction mechanism, K is activated when

$$f_K = 0 \leftrightarrow \sigma = 2\sigma_{\max} - (K + 2k)$$

Therefore, the elastic part of the unloading path is defined by:

$$\text{Phase } 23\colon \sigma_{\max} > \sigma > 2\sigma_{\max} - (K + 2k) \rightarrow d\sigma = E d\varepsilon$$

This elastic phase is followed by the activation of friction mechanism K:

$$\text{Phase } 34\colon 2\sigma_{\max} - (K + 2k) > \sigma > K - 2k \rightarrow d\sigma = \frac{2Eh}{2h + E} d\varepsilon$$

followed by the activation of the friction mechanism k:

$$\text{Phase } 45\colon K - 2k > \sigma \geq 0 \rightarrow d\sigma = \frac{Eh}{h + 2E} d\varepsilon$$

At the end of the first cycle, the residual strain is the sum of the plastic strains accumulated during plastic sliding in the loading phase 12 (k), and the plastic unloading phases 34 (K) and 45 (k). The finite stress increments in these plastic phases read

$$\Delta\sigma_{12} = \sigma_{\max} - k > 0; \quad \Delta\sigma_{34} = 2(K - \sigma_{\max}) < 0; \quad \Delta\sigma_{45} = -K + 2k < 0$$

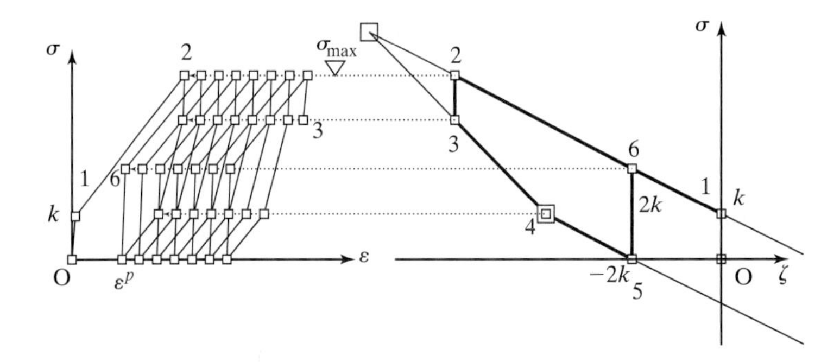

FIGURE 7.14: Model response to cyclic loading: stress-strain diagram and stress-hardening force diagram.

Use of the flow rules (7.101) and (7.106) delivers the residual strain after the first cycle:

$$\varepsilon^p_{(1)} = \frac{2(\Delta\sigma_{12} + \Delta\sigma_{45})}{h} + \frac{\Delta\sigma_{34}}{2h} = \frac{2k}{h} + \frac{\sigma_{\max} - K}{h} \tag{7.114}$$

■

In a similar way, we can determine the energy dissipated during the first cycle: It is the sum of the dissipations accumulated during plastic sliding in the loading phase 12 (k), and the plastic unloading phases 34 (K) and 45 (k). Use of the finite stress increments in these plastic phases in (7.109) and (7.110) yields

$$\begin{aligned} \mathcal{D}_{(1)} &= \int_{t=0}^{t=t_5} \varphi dt = \underbrace{\varphi_k(t_2 - t_1)}_{\frac{2k|\Delta\sigma_{12}|}{h}} + \underbrace{\varphi_K(t_4 - t_3)}_{\frac{K|\Delta\sigma_{34}|}{2h}} + \underbrace{\varphi_k(t_5 - t_4)}_{\frac{2k|\Delta\sigma_{45}|}{h}} \\ &= \frac{2k^2}{h} + \frac{2k(\sigma_{\max} + K - 4k) + K(\sigma_{\max} - K)}{h} \end{aligned} \tag{7.115}$$

where times $t_1 \ldots t_5$ define the chronology (history) of the stress cycle. For the stress amplitude bounds defined by (7.113), the minimum and maximum energy dissipations read

$$\min \mathcal{D}_{(1)} = \frac{2k^2}{h} + \frac{2k(K - 6k)}{h} \tag{7.116}$$

$$\max \mathcal{D}_{(1)} = \frac{2k^2}{h} + \frac{2k(3K - 2k)}{h} \tag{7.117}$$

Figure 7.14 shows the model's response in the $(\sigma \times \varepsilon)$- and $(\sigma \times \zeta)$-plane. At the end of the first cycle, the residual hardening force is equal to $\zeta_{(1)} = -2k$. This is the starting point for the second cycle.

Exercise 71. Determine the residual strain after the nth cycle.

With $\zeta_{(1)} = -2k$ as initial condition, the elastic reloading is defined by

$$\text{Phase } 56: 0 < \sigma < 2k \rightarrow d\sigma = Ed\varepsilon$$

When $\sigma = 2k$, the friction mechanism k is reactivated. In the stress-strain diagram, this loading point is situated below the straight line 12 of the first cycle. This is due to the activation of the friction mechanism K during unloading 34. Beyond this loading point, the response is redundant. In the $(\sigma \times \zeta)$-plane, successive unloading from $\sigma_{\max}$ follows the path 23–34–45; but in the stress-strain diagram the response is shifted by

$$\Delta\varepsilon^p_{(i>1)} = \frac{\sigma_{\max} - K}{h} \tag{7.118}$$

where $\Delta\varepsilon^p_{(i>1)}$ corresponds to the increase in residual strain per cycle. Hence, from (7.114) and (7.118), the total residual strain after the nth cycle reads

$$\varepsilon^p_{(n)} = \frac{2k}{h} + n\frac{\sigma_{\max} - K}{h} \tag{7.119}$$

This accumulation of plastic strain is called the ratchetting phenomenon, modeled here by means of a two-surface plasticity model. In the case of the linear model developed here, the ratchetting rate is constant. ■

The Stéfani model has a constant ratcheting rate. From an energy point, this implies that the amount of energy consumed by dissipation per cycle is also constant:

$$\mathcal{D}_{(n)} = \frac{2k^2}{h} + n\Delta\mathcal{D} \tag{7.120}$$

$\Delta\mathcal{D}$ is the energy consumed per loading cycle:

$$\frac{2k(K-6k)}{h} \leq \Delta\mathcal{D} = \frac{2k(\sigma_{\max} + K - 4k) + K(\sigma_{\max} - K)}{h} \leq \frac{2k(3K-2k)}{h} \tag{7.121}$$

Exercise 72. Illustrate the dissipation per cycle in the $(\sigma \times \varepsilon^p)$-diagram. ■

7.5 PROBLEM SET: THE THREE-TRUSS ANALOGY

The macroscopic hardening plasticity model is characterized from an energy point of view by two quantities: the elastic strain energy $\psi^{\mathrm{el}}(\varepsilon - \varepsilon^p)$ and the frozen hardening energy $\mathcal{U}(\chi)$, where χ denote some plastic hardening variables. The frozen energy is a macroscopic representation of an energy related to a microelasticity stored at the level of the heterogeneous matter. The aim of this exercise is to elucidate the microscopic origin of this hardening energy. This microscopic origin is often described by the three-truss analogy, shown in Figure 7.15a. The structure is composed of three deformable trusses of the same initial length L, the same section A, and hinges on both endsides. At $y = -L$, the trusses are fixed on a rigid device. On the other end, the trusses are connected to a rigid (i.e., undeformable) beam. The system is subjected to a load F, which is applied to the rigid beam at midspan between trusses 2 and 3 (*cf.* Figure 7.15a). The trusses are composed of the same ideal elastoplastic material. The individual truss response is given in Figure 7.15b: For $|N| \leq K$, the truss behaves elastically, defined by the elastic section stiffness EA; N is the axial force in the truss, and K the maximum section force.

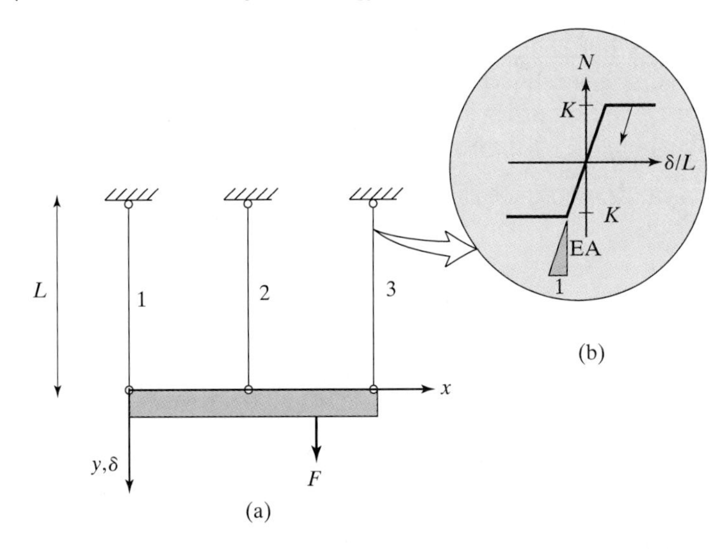

FIGURE 7.15: Problem set: three-truss analogy of a heterogeneous plastic hardening material.

The truss system can be associated with a simplified microstructure of an elasto-plastic material system: The truss forces $N_{i=1,2,3}$ represent the forces at a microscale, while the applied force F represents the macroscopic force. The rigid beam ensures the kinematical compatibility between the deformation on the two scales: The vertical end displacements $\delta_{i=1,2,3}$ of the trusses, and the displacement u at the point of load application.

Part I—Loading: We want to determine the macroscopic load deflection curve of the system, that is,

$$0 \leq F = F(u) \leq F_{\text{ult}}$$

for an increasing load, $dF \geq 0$. F_{ult} is the maximum force the system can support.

1. **Kinematic Compatibility**: Determine the relation that ensures the kinematic compatibility between the displacement $\delta_{i=1,2,3}$ of the trusses. Does this relation depend on the material behavior of the truss element?
2. **Elastic Loading** $0 \leq F < F^I$: We consider a loading from an initial (virgin) material state. Determine the response of the system (i.e., displacements $\delta_{i=1,2,3}$, u and truss forces $N_{i=1,2,3}$) as a function of the applied force in the elastic range. In this regard, use the theorem of minimum of potential energy. Which truss reaches first the section force threshold K? Determine the initial elasticity domain of the overall system, the maximum load F^I associated with it, and the system's response at $F = F^I$.
3. **Elastoplastic Loading** $F^I \leq F < F^{II}$: With $F \geq F^I$ plastic deformation takes place. For this loading range, determine the system's response (i.e., displacements $\delta_{i=1,2,3}$, u and truss forces $N_{i=1,2,3}$) and the permanent strains $\varepsilon^p_{i=1,2,3} = \delta^p_{i,2,3}/L$ in the truss elements. Which truss reaches next

the section force threshold K? How much energy has been dissipated into heat form at the end of this loading phase?

4. **Ultimate Load**: By means of a precise energy reasoning, show that the ultimate load-bearing capacity of the system is

$$F^{II} = F_{\text{ult}}$$

Part II—Unloading: We consider now an unloading from a maximum force $F^I \leq \max F < F^{II}$.

1. **Elastic Unloading**: Show that the complete unloading from $\max F$ to $F = 0$ is elastic. Determine the system's response during this elastic unloading process.
2. **Frozen Elasticity**: Determine the residual displacements $\delta^r_{i=1,2,3}$, u^r, and the residual truss forces $N_{i=1,2,3}$, and determine the elastic energy stored in the system after a complete unloading. Show that this frozen energy can be expressed as a function of only the plastic strain $\max \varepsilon^p_{i=1,2,3} = \max \delta^p_{i,2,3}/L$ developed in the bar during loading to $\max F$. Comment.
3. **Summary**: Discuss briefly the following issues:

    (a) The macroscopic load deflection curve $F = F(u)$ for loading cycles in $-\max F \leq F(t) \leq \max F$. For convenience, sketch the overall system's response for a loading cycle in form of a dimensionless graph $\bar{F} = \bar{F}(\bar{u})$, where $\bar{F} = F/K$, and $\bar{u} = uEA/KL$.

    (b) The loading–unloading of the dimensionless truss forces $\bar{N}_i = N_i/K$ as a function of the dimensionless load $\bar{F} = F/K$.

Part III—The Equivalent Macroscopic Hardening Plasticity Model: We want to give a macroscopic representation of the microscopic response of the material system developed in Parts I and II. The aim is to develop a macroscopic plastic hardening model that captures the same features as the microscopic one concerning the load-deflection curve $F = F(u) \rightarrow \sigma = \sigma(\varepsilon)$ and the associated energy variation.

1. Develop a 1D-hardening plasticity spring-friction device (think model) that captures the macroscopic load deflection curve. Detail the state equations and the plasticity criteria required to describe the behavior.
2. Develop the model within the framework of thermodynamics of irreversible processes:

    (a) Determine the free energy, the state equations, and the complementary evolution laws.

    (b) By comparing the macroscopic energy and its transformation (i.e., dissipation) with the one of the microscopic model, determine the model parameters (stiffness, hardening modulus, strength parameters, etc.).

7.5.1 Part I: Loading

Kinematic Compatibility.

The rigid bar ensures a pure geometric compatibility between the deformable microstructure (truss displacements $\delta_i, i = 1, 2, 3$) and the macroscopic displacement u, irrespective of the material behavior and the loading:

$$\delta = \delta_1 + (\delta_2 - \delta_1)\frac{x}{H}$$

$$\delta_3 = \delta(x = 2H) = -\delta_1 + 2\delta_2; \quad u = -0.5\delta_1 + 1.5\delta_2$$

This reduces the number of displacement unknowns from 3 to 2.

Elastic Loading.

We will use the theorem of minimum potential energy to determine the elastic response of the system:

$$d\mathcal{E}_{\text{pot}} = d[W(\delta_i) - \Phi(u)]_{F^d=\text{const}} = 0$$

with the following:

- $W(\delta_i)$ is the internal elastic energy stored in the system. In the elastic range it reads

$$W(\delta_i) = \frac{EA}{2L}\left[\delta_1^2 + \delta_2^2 + \delta_3^2\right] = \frac{EA}{2L}\left[2\delta_1^2 - 4\delta_1\delta_2 5\delta_2^2 + 4\delta_2^2\right]$$

- $\Phi(u)$ is the external work supplied by the prescribed forces ($F = F^d$):

$$\Phi(u) = F^d u = F[-0.5\delta_1 + 1.5\delta_2]$$

Thus,

$$\mathcal{E}_{\text{pot}} = \frac{EA}{2L}\left[2\delta_1^2 - 4\delta_1\delta_2 + 5\delta_2^2\right] - F\left[-0.5\delta_1 + 1.5\delta_2\right]$$

Finally, $d\mathcal{E}_{\text{pot}}|_{F^d=\text{const}} = 0$ implies

$$d\mathcal{E}_{\text{pot}} = \frac{\partial \mathcal{E}_{\text{pot}}}{\partial \delta_1} d\delta_1 + \frac{\partial \mathcal{E}_{\text{pot}}}{\partial \delta_2} d\delta_2 = 0$$

or, equivalently,

$$\begin{pmatrix} d\delta_1 & d\delta_2 \end{pmatrix}\left[\frac{EA}{L}\begin{bmatrix} 2 & -2 \\ -2 & 5 \end{bmatrix}\begin{pmatrix} \delta_1 \\ \delta_2 \end{pmatrix} - \begin{pmatrix} -0.5F \\ 1.5F \end{pmatrix}\right] = 0$$

Since the previous equation must hold for any value of $d\delta_1, d\delta_2$, it follows that

$$\forall(d\delta_1, d\delta_2) : \begin{pmatrix} \delta_1 \\ \delta_2 \end{pmatrix} = F\begin{pmatrix} -0.5 \\ 1.5 \end{pmatrix}$$

Elastic solution $0 \leq F < F^I$:

$$\begin{pmatrix} \delta_1 \\ \delta_2 \end{pmatrix} = \frac{LF}{12EA} \begin{pmatrix} 1 \\ 4 \end{pmatrix}; \quad \rightarrow \delta_3 = 2\delta_2 - \delta_1 = \frac{7LF}{12EA}$$

$$N_1 = \frac{F}{12} < N_2 = \frac{F}{3} < N_3 = \frac{7F}{12}$$

$$u = \frac{11LF}{24EA} \leftrightarrow F = \left[\frac{24EA}{11L}\right] u$$

Elastic limit load $F = F^I$:

$$f = \max_{i=1,2,3}(|N_i| - K) = |N_3| - K \leq 0 \leftrightarrow F \leq F^I = \frac{12}{7}K$$

$F^d = F^I$ corresponds to the applied force at the onset of plastic yielding of truss 3. The displacements at the elastic limit of the system are

$$u^I = \frac{11L}{14EA}K$$

$$\delta_1^I = \frac{L}{7EA}K; \quad \delta_2^I = \frac{4L}{7EA}K; \quad \delta_3^I = \frac{L}{EA}K$$

Elastoplastic Loading.

With $F^I > \frac{12}{7}K$, truss 3 yields, and the truss force N_3 remains constant during subsequent loading:

$$N_3 = EA(\delta_3 - \delta_3^p) = K \leftrightarrow d\delta_3 = d\delta_3^p$$

where δ_3^p is the permanent displacement in truss 3. The increase of elastic energy stored in the system is therefore limited to the elastic energy stored in truss 1 and 2, thus

$$dW = d\left[\frac{EA}{2L}(\delta_1^2 + \delta_2^2 + (\delta_3 - \delta_3^p)^2)\right] = d\left[\frac{EA}{2L}(\delta_1^2 + \delta_2^2)\right]$$

The external work increase provided from the outside remains the same:

$$d\Phi = F^d du = F^d d[-0.5\delta_1 + 1.5\delta_2]$$

In turn, due to yielding of truss 3, the external work increment $d\Phi$ is not entirely stored in the form of elastic energy dW in the system but is dissipated into heat form. The Clausius–Duhem inequality expressed at the structural level reads

$$d\mathcal{D} = d\Phi - dW > 0$$

or, equivalently,

$$d\mathcal{E}_{\text{pot}} + d\mathcal{D} = 0$$

The dissipation $d\mathcal{D}$ is readily determined:

$$d\mathcal{D} = K d\delta_3^p = K d\delta_3 = K d(-\delta_1 + 2\delta_2)$$

where we made use of the geometric compatibility condition. Summing up the previous terms gives

$$\underbrace{\left[\frac{EA}{L}(\delta_1 d\delta_1 + \delta_2 d\delta_2)\right]}_{dW} - \underbrace{F^d[-0.5d\delta_1 + 1.5d\delta_2]}_{d\Phi} + \underbrace{K(-d\delta_1 + 2d\delta_2)}_{d\mathcal{D}} = 0$$

or, equivalently,

$$\left(d\delta_1 \quad d\delta_2 \right) \left[\frac{EA}{L}\begin{bmatrix} 1 & 0 \\ 0 & 1 \end{bmatrix}\begin{pmatrix} \delta_1 \\ \delta_2 \end{pmatrix} - \begin{pmatrix} -0.5F \\ 1.5F \end{pmatrix} + \begin{pmatrix} -K \\ 2K \end{pmatrix}\right] = 0$$

Provided that this holds for any value of $d\delta_1, d\delta_2$, we obtain the elastic response of trusses 1 and 2 during the elastoplastic evolution:

$$\begin{pmatrix} \delta_1 \\ \delta_2 \end{pmatrix} = -\frac{L}{EA}\begin{pmatrix} -K + 0.5F^d \\ 2K - 1.5F^d \end{pmatrix}$$

Use of the kinematic compatibility condition delivers δ_3:

$$\delta_3 = 2\delta_2 - \delta_1 = (3.5F - 5K)\frac{L}{EA}$$

$$\delta_3^p = \delta_3 - \delta_3^I = (3.5F - 6K)\frac{L}{EA}$$

where $\delta_3^I = \frac{L}{EA}K$ is the displacement of truss 3 at the onset of yielding.

Force solution:

$$N_1 = \frac{EA}{L}\delta_1 = -0.5F + K$$

$$N_2 = \frac{EA}{L}\delta_2 = 1.5F - 2K$$

$$N_3 = \frac{EA}{L}(\delta_3 - \delta_3^p) = \frac{EA}{L}\delta_3^I = K$$

Current elasticity domain:

$$f = \max(|N| - K) = |N_2| - K \leq 0 \leftrightarrow F^I \leq F \leq F^{II} = 2K$$

At a load level of $F^1 = 2K$, truss 2 enters the plastic range:

$$N_1^{II} = 0 < N_2^{II} = N_3^{II} = K$$

The corresponding displacements at this load level are

$$\delta_1^{II} = 0; \quad \delta_2^{II} = \frac{LK}{EA}; \quad \delta_3^{II} = \frac{2LK}{EA}$$

$$u^{II} = 1.5\frac{KL}{EA}$$

Note that this elastoplastic loading phase $F^I \leq F \leq F^{II}$ led to a decrease of the force N_1^{II} in truss 1.

Ultimate Load.

With a similar reasoning as was just employed, the internal elastic energy beyond $F = F^{II}$ reads

$$dW = d\left[\frac{EA}{2L}(\delta_1^2 + (\delta_2 - \delta_2^p)^2 + (\delta_3 - \delta_3^p)^2)\right] = d\left[\frac{EA}{2L}\delta_1^2\right]$$

The external work is still

$$d\Phi = F^d du = F^d[-0.5d\delta_1 + 1.5d\delta_2]$$

And the dissipation reads now

$$d\mathcal{D} = K(d\delta_2^p + d\delta_3^p) = K(d\delta_2 + d\delta_3) = Kd(-d\delta_1 + 3d\delta_2)$$

The Clausius–Duhem inequality reads

$$d\mathcal{D} = K(-d\delta_1 + 3d\delta_2) = -\frac{EA\delta_1}{L}d\delta_1 + F^d[-0.5d\delta_1 + 1.5d\delta_2] \geq 0$$

or, equivalently,

$$\begin{pmatrix} d\delta_1 & d\delta_2 \end{pmatrix}\left[\frac{EA}{L}\begin{bmatrix} 1 & 0 \\ 0 & 0 \end{bmatrix}\begin{pmatrix} \delta_1 \\ \delta_2 \end{pmatrix} - \begin{pmatrix} -0.5F \\ 1.5F \end{pmatrix} + \begin{pmatrix} -K \\ 3K \end{pmatrix}\right] = 0$$

Since this equation must hold for any value of $d\delta_1, d\delta_2$, it turns out that the maximum load supported by the system is

$$F_{\text{ult}} = 2K$$

For $F^d > 2K$, $d\mathcal{D} + d\mathcal{E}_{\text{pot}} < 0$, or, equivalently, $d\mathcal{E}_{\text{pot}} < -d\mathcal{D}$, which violates the second law.[11]

Note: An alternative argument, but still related to the existence of sufficient internal elastic energy, can be based on the stiffness matrix:

$$k_{ij} = \frac{\partial \mathcal{E}_{\text{pot}}}{\partial \delta_i \partial \delta_j}$$

A stability requirement of elasticity is that the stiffness matrix must be positive definite. From a mathematical point of view, this ensures that k_{ij} can be inverted and requires a nonzero determinant. During elastic loading, we have

$$0 \leq F < F^I : \det k_{ij}^{\text{el}} = \left|\frac{EA}{L}\begin{bmatrix} 2 & -2 \\ -2 & 5 \end{bmatrix}\right| = 6\left(\frac{EA}{L}\right)^2$$

and during elastoplastic loading

$$F^I \leq F < F^{II} : \det k_{ij}^{\text{ep}} = \left|\frac{EA}{L}\begin{bmatrix} 1 & 0 \\ 0 & 1 \end{bmatrix}\right| = \left(\frac{EA}{L}\right)^2$$

[11] Indeed, $d\mathcal{E}_{\text{pot}} < -d\mathcal{D}$ would mean that the variation of the potential energy is smaller than the amount of energy dissipated into heat form. Since it is potential energy that is dissipated into heat form, this makes no sense physically.

Finally, at ultimate load,

$$F = F^{II} + dF : \det k_{ij}^p = \left| \frac{EA}{L} \begin{bmatrix} 1 & 0 \\ 0 & 0 \end{bmatrix} \right| = 0$$

The determinant is zero, which indicates that there is not sufficient elasticity reserve to accommodate the external work increase by any load increase dF. In other words, the material system deforms in an uncontrolled fashion.

7.5.2 Part II: Unloading

$$F^I = \frac{12}{7}K < \max F < F^{II} = 2K$$

Initial displacement conditions:

$$\max \delta_1 = (-0.5 \max F + K)\frac{L}{EA}$$

$$\max \delta_2 = (1.5 \max F - 2K)\frac{L}{EA}$$

$$\max \delta_3 = (3.5 \max F - 5K)\frac{L}{EA}; \quad \max \delta_3^p = (3.5 \max F - 6K)\frac{L}{EA}$$

Forces:

$$\max N_1 = \frac{EA}{L} \max \delta_1 = -0.5 \max F + K$$

$$\max N_2 = \frac{EA}{L} \max \delta_2 = 1.5 \max F - 2K$$

$$\max N_3 = \frac{EA}{L}(\max \delta_3 - \max \delta_3^p) = K$$

Elastic Unloading.

During an elastic unloading the dissipation is zero ($d\mathcal{D} = 0$), and the plastic deformation $\max \delta_3^p$ in truss 3 remains constant.

Internal energy:

$$\begin{aligned} W &= W(\delta_1, \delta_2, \delta_3) = \frac{EA}{2L}\left[\delta_1^2 + \delta_2^2 + (\delta_3 - \max \delta_3^p)^2\right] \\ &= \frac{EA}{2L}\left[\delta_1^2 + \delta_2^2 + (2\delta_2 - \delta_1 - \max \delta_3^p)^2\right] \end{aligned}$$

External work by prescribed forces (as before):

$$\Phi = F^d u = F[-0.5\delta_1 + 1.5\delta_2]$$

Theorem of minimum potential energy ($d\mathcal{E}_{\text{pot}}|_{F^d=\text{const}} = 0$):

$$\begin{pmatrix} d\delta_1 & d\delta_2 \end{pmatrix} \left[\frac{EA}{L}\begin{bmatrix} 2 & -2 \\ -2 & 5 \end{bmatrix}\begin{pmatrix} \delta_1 \\ \delta_2 \end{pmatrix} - \begin{pmatrix} -0.5F \\ 1.5F \end{pmatrix} - \max \delta_3^p \frac{EA}{L}\begin{pmatrix} -1 \\ 2 \end{pmatrix}\right] = 0$$

Solution:

$$\begin{pmatrix} \delta_1 \\ \delta_2 \end{pmatrix} = \frac{LF}{12EA}\begin{pmatrix} 1 \\ 4 \end{pmatrix} + \frac{\max \delta_3^p}{6}\begin{pmatrix} -1 \\ 2 \end{pmatrix}$$

From the geometric compatibility condition and the initial condition for $\max \delta_3^p$, it follows that

$$\delta_3 = 2\delta_2 - \delta_1 = \frac{7LF}{12EA} + \frac{5}{6}\max \delta_3^p = \frac{7LF}{12EA} + \frac{5}{6}(3.5\max F - 6K)\frac{L}{EA}$$

$$N_1 = \frac{EA}{L}\delta_1 = \frac{F}{12} - \frac{7\max F}{12} + K$$

$$N_2 = \frac{EA}{L}\delta_2 = \frac{F}{3} + \frac{7}{6}\max F - 2K$$

$$N_3 = \frac{EA}{L}(\delta_3 - \max \delta_3^p) = \frac{7F}{12} - \frac{7\max F}{12} + K$$

or, equivalently, noting $\Delta F = F - \max F \leq 0$ the unloading force increment,

$$\begin{pmatrix} N_1 \\ N_2 \\ N_3 \end{pmatrix} = \frac{\Delta F}{12}\begin{pmatrix} 1 \\ 4 \\ 7 \end{pmatrix} + \frac{\max F}{2}\begin{pmatrix} -1 \\ 3 \\ 0 \end{pmatrix} + K\begin{pmatrix} 1 \\ -2 \\ 1 \end{pmatrix}$$

Finally, we verify the condition of elastic unloading:

$$f = \max_{i=1,2,3}(|N_i| - K) \leq 0 \leftrightarrow |\Delta F| \leq 2K$$

Hence, in the considered range of $\frac{12}{7}K < \max F < F^{II} = 2K$, the total unloading to $F = 0$ with $\Delta F = -\max F$ is elastic.

Frozen Elasticity.

After a complete unloading, the following residual displacements remain in the truss:

$$F = 0: \begin{pmatrix} \delta_1^r \\ \delta_2^r \\ \delta_3^r \end{pmatrix} = \frac{\max \delta_3^p}{6}\begin{pmatrix} -1 \\ 2 \\ 5 \end{pmatrix}$$

These residual displacements satisfy the geometric compatibility condition:

$$\delta_1^r - 2\delta_2^r + \delta_3^r = 0$$

The residual forces in the bars are

$$F = 0: \begin{pmatrix} N_1 \\ N_2 \\ N_3 \end{pmatrix} = \frac{7\max F}{12}\begin{pmatrix} -1 \\ 2 \\ -1 \end{pmatrix} + K\begin{pmatrix} 1 \\ -2 \\ 1 \end{pmatrix}$$

The truss forces are self-balanced. They are associated with a part of the elastic energy put into the system during loading, which was not recovered during unloading

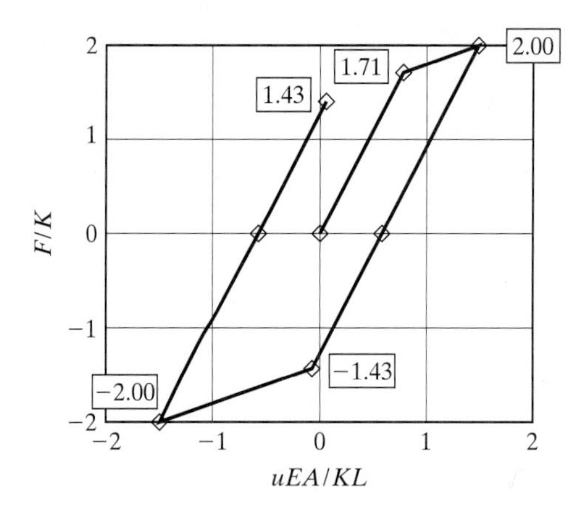

FIGURE 7.16: Macroscopic material response: normalized force displacement relation.

but was stored in the system due to the prescribed geometric compatibility. This elastic energy reads

$$\mathcal{U} = W(\delta_1^r, \delta_2^r, \delta_3^r) = \frac{EA}{2L}[\delta_1^{r2} + \delta_2^{r2} + (\delta_3^r - \max \delta_3^p)^2]$$

or, equivalently, as a function of the plastic deformation undergone during the loading phase,

$$\mathcal{U} = \mathcal{U}(\max \delta_3^p) = \frac{EA}{12L} \max \delta_3^{p2}$$

This elastic energy is frozen at a microlevel during plastic loading. It is therefore called the frozen hardening energy. Finally, the total residual displacement at the point of load application reads

$$u^r = -0.5\delta_1^r + 1.5\delta_2^r = \frac{\max \delta_3^p}{6}[0.5 + 3] = \frac{7}{12} \max \delta_3^p$$

This relation, which results from the kinematic compatibility relation, establishes the link between the micro permanent deformation $\max \delta_3^p$ and the macro permanent deformation u^r, or, in other words, between the hardening variable and the overall permanent deformation.

Summary: Load Cycle and Strength Domain.

The response of the material system is displayed in Figures 7.16 and 7.17. Figure 7.16 shows the load-displacement curve, and Figure 7.17 the truss forces versus the applied load. Note the unloading of truss 1 during elastoplastic loading, which simultaneously leads to an overproportional force increase in truss 2. This redistribution of the elastic energy stored in the microstructure leads to the apparent hardening phenomenon. This stored energy is only partly recovered during unloading: Truss 2 remains in tension, while trusses 1 and 2 are in compression. Therefore, during elastoplastic loading, a part of the external work supplied to the system remains stored in the system. This is where the origin of the plastic hardening phenomenon lies.

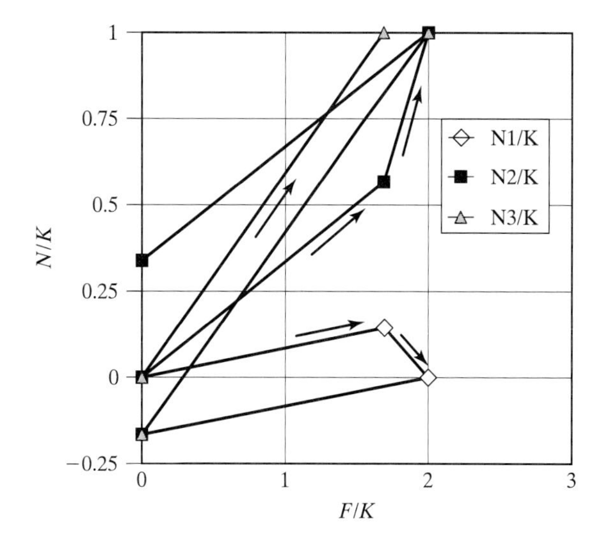

FIGURE 7.17: Microscopic material response: normalized truss forces versus macroscopic loading.

7.5.3 Equivalent Macroscopic Model

1D Hardening Plasticity Spring Friction.

The 1D think model for an equivalent macroscopic model capturing the analyzed material behavior is sketched in Figure 7.18. It is composed of two springs in series, one representing the elasticity of the material (stiffness E^*), the other the hardening phenomenon (stiffness H), and two frictional devices of strength k_1 and k_2, related by

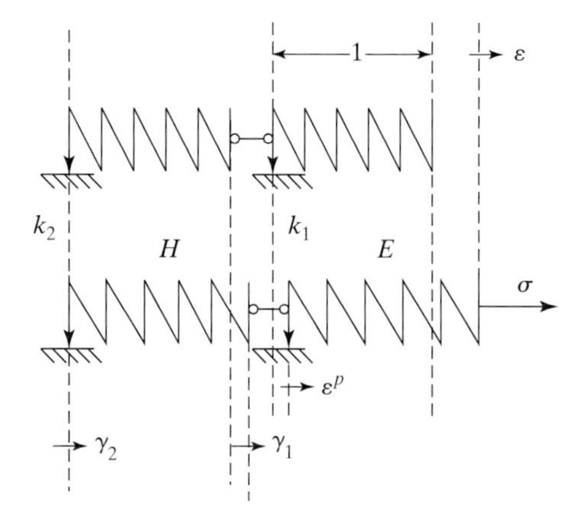

FIGURE 7.18: Equivalent macroscopic think model.

$$\frac{k_2}{k_1} = \frac{F^I}{F^{II}} = \frac{7}{6}$$

The state equations read

$$\sigma = E^*(\varepsilon - \varepsilon^p)$$

$$\Gamma = -H(\gamma_1 - \gamma_2)$$

where Γ is the hardening force (positive in compression) in the hardening spring, and $\gamma_1 - \gamma_2$ the relative change in length of this spring. From elementary equilibrium at the friction elements, the plasticity criteria are obtained in the form

$$f_1(\sigma, \Gamma) = |\sigma + \Gamma| - k_1 \leq 0$$

$$f_2 = |\Gamma| - k_2 \leq 0$$

Macroscopic Constitutive Modeling.

We choose to describe the material behavior through four state variables: the total strain ε, the permanent macroscopic strain ε^p, and two internal plastic variables γ_1 and γ_2.

Free energy:

$$\psi = \psi^{\text{el}}(\varepsilon, \varepsilon^p) + \mathcal{U}(\gamma_1, \gamma_2)$$

where $\psi^{\text{el}}(\varepsilon, \varepsilon^p)$ is the elastic strain energy, and $U(\gamma_1, \gamma_2)$ the frozen energy. From Figure 7.18,

$$\psi^{\text{el}}(\varepsilon, \varepsilon^p) = \frac{1}{2} E^*(\varepsilon - \varepsilon^p)^2$$

$$\mathcal{U}(\gamma_1, \gamma_2) = \frac{1}{2} H(\gamma_1 - \gamma_2)^2$$

Driving forces (from Clausius–Duhem inequality):

$$\varphi dt = \sigma d\varepsilon - d(\psi^{\text{el}} + \mathcal{U}) = \sigma d\varepsilon^p + \Gamma(d\gamma_1 - d\gamma_2) \geq 0$$

Thus,

$$\sigma = \frac{\partial \psi}{\partial \varepsilon} = -\frac{\partial \psi}{\partial \varepsilon^p} = E^*(\varepsilon - \varepsilon^p)$$

$$\Gamma = -\frac{\partial \psi}{\partial \gamma_1} = -H(\gamma_1 - \gamma_2)$$

$$-\Gamma = -\frac{\partial \psi}{\partial \gamma_2} = H(\gamma_1 - \gamma_2)$$

Stress σ is the driving force of total and plastic deformation, and hardening force Γ is the driving force of hardening slidings γ_1 and $-\gamma_2$.

Complementary evolution laws:

$$f_1 = 0 : d\varepsilon^p = d\lambda_1 \frac{\partial f_1}{\partial \sigma} = d\lambda_1 \text{sign}\,(\sigma + \Gamma)$$

$$f_1 = 0 : d\gamma_1 = d\lambda_1 \frac{\partial f_1}{\partial \Gamma} = d\lambda_1 \text{sign}\,(\sigma + \Gamma)$$

$$f_2 = 0 : d\gamma_2 = d\lambda_2 \frac{\partial f_2}{\partial(-\Gamma)} = -d\lambda_2 \text{sign}\,(\Gamma)$$

Consistency conditions:

$$df_1 = 0 : \frac{\partial f_1}{\partial \sigma} d\sigma + \frac{\partial f_1}{\partial \Gamma} d\Gamma = 0$$

1. Elastoplastic loading:

$$f_1 = 0; \quad f_2 < 0 : df_1 = 0 \rightarrow d\sigma = -d\Gamma = H d\gamma_1$$

It follows that

$$d\lambda_1 = \frac{d\sigma}{H} \text{sign}\,(\sigma + \Gamma)$$

$$d\varepsilon^p = d\gamma_1 = \frac{d\sigma}{H}$$

$$d\sigma = E^* \left(d\varepsilon - \frac{d\sigma}{H} \right) \rightarrow d\sigma = \frac{E^* H}{E^* + H} d\varepsilon$$

2. Plastic loading:

$$f_1 = 0; \quad f_2 = 0 : \begin{pmatrix} df_1 = 0 \\ df_2 = 0 \end{pmatrix} \rightarrow \begin{pmatrix} d\sigma = -d\Gamma = H(d\gamma_1 - d\gamma_2) \\ d\Gamma = 0 \end{pmatrix}$$

It follows that

$$d\sigma = d\Gamma = 0 : d\gamma_1 = d\gamma_2 = d\varepsilon^p = d\varepsilon$$

Material parameters: The tangential force-displacement relations read

$$F \leq F^I : dF = k^{\text{el}} du; \quad k^{\text{el}} = \frac{24}{11} \frac{EA}{L}$$

$$F^I < F < F^{II} = dF = k^{\text{ep}} du; \quad k^{\text{ep}} = \frac{2}{5} \frac{EA}{L}$$

The ratio $k^{\text{ep}} / k^{\text{el}}$ characterizes the macroscopic behavior:

$$\frac{k^{\text{ep}}}{k^{\text{el}}} = \frac{H}{E^* + H} = \frac{11}{60} \rightarrow \frac{H}{E^*} = \frac{11}{49}$$

Elastic energy: The energy of the macro- and micromodel must coincide. Thus, in the elastic range,

$$d\mathcal{E}_{\text{pot}} = 0 : dW = d\Phi = F du = \int_\Omega \sigma d\varepsilon \, d\Omega = \sigma d\varepsilon \Omega$$

Integration gives

$$\frac{k^{\text{el}} u^2}{2} = \frac{E^* \varepsilon^2}{2} \Omega$$

If we let $\varepsilon = u/L$, we obtain

$$\frac{E^*}{E} = \frac{24}{11} \frac{AL}{\Omega} = \frac{24}{11} c$$

where $c = AL/\Omega$ is the truss volume concentration.

Dissipation: Similarly, the energy that is dissipated during elastoplastic loading must be the same. In the truss system, we have

$$\mathcal{D} = K \int d\delta^p = \frac{K^2 L}{EA}$$

and in the equivalent continuum system,

$$\mathcal{D} = \int_{t^I}^{t^{II}} \int_\Omega \varphi dt \, d\Omega = k_1 \Omega \int d\varepsilon^p = k_1 \Omega \frac{\sigma}{H} \Big|_{\sigma=k_1}^{\sigma=k_2} = k_1^2 \frac{49}{66 E^*} \Omega$$

Thus,

$$k_1 = K \sqrt{\frac{66 E^* L}{49 E A \Omega}} = \frac{12}{7} \frac{KL}{\Omega} = \frac{12}{7} kc$$

$$k_2 = 2 \frac{KL}{\Omega} = 2kc$$

where $k = K/A$ is the strength of the material composing the truss.

7.6 PROBLEM SET: CREEP HESITANCY

Some metals creep beyond a certain stress threshold k. After a partial unloading from a stress $\sigma_{\max} > k$ to a stress $k < \sigma < \sigma_{\max}$, the creep stops. However, after some time, the material starts again to creep. This phenomenon is called creep hesitancy. We want to study this phenomenon by means of a macroscopic material model. The 1D think model proposed by Mandel[12], is given in Figure 7.19. It is composed of two springs of stiffness E and H (unit length $1[L]$) separated by a friction element of friction strength k. In addition, the hardening spring (stiffness H) is coupled in series with a dashpot of viscosity η. We note σ the applied stress, and $\varepsilon = \Delta L/L$ the measurable strain (relative length variation) at the point of stress application.

1. **Elasticity Domain:** Our first focus is the elasticity domain defined by Mandel's model sketched in Figure 7.19a.

[12] Mandel, J. (1978). *Propriétés Mécaniques des Matériaux*, Eyrolles, Paris, France.

(a) Let Γ be the tension force in the spring of stiffness H. Determine the elasticity domain that defines the plastically admissible stress states $\sigma \in D_E$. Illustrate this elasticity domain in a graph in which the x-axis represents Γ and the y-axis represents σ, called the $(\sigma \times \Gamma)$-plane hereafter.

(b) We consider an initial prestress in the hardening spring, say $\Gamma = \Gamma_0$. How does this force vary in time, when the friction element does not move? Illustrate this force evolution in the $(\sigma \times \Gamma)$-plane.

2. **Constitutive Equations**: We have a quick look at the energy side of the model:

   (a) How many state variables are required to describe the energy states of the system? Express the free energy of the system, and determine the associated state equations.

   (b) Determine the complementary evolution laws. Show that the plastic evolution laws are of the standard format obeying the normality rule

$$d\varepsilon^p = d\lambda \frac{\partial f}{\partial \sigma}; \quad d\chi = d\lambda \frac{\partial f}{\partial \zeta}$$

   and can be developed in the form

$$d\varepsilon^p = \frac{d\sigma}{H} + \Gamma \frac{dt}{\eta} = d\chi$$

   where $d\lambda$ is the plastic multiplier, χ is the hardening variable, and ζ is the hardening force. (*Hint*: Develop carefully the consistency condition $df = 0$.)

3. **Instantaneous and Time-Dependent Deformations**: We want to study the response of the system to a step loading $k < \sigma_0 < 3k$ applied at time $t = t_0$, and partially unloaded to $\sigma_1 > k$ at time $t = t_1$(see Figure 7.19b). Before loading the system is stress free and in thermodynamic equilibrium.

   (a) Determine the instantaneous response (strain, forces) of the model for a applied stress $\sigma_0 > k$. How much energy is dissipated into heat form during this instantaneous loading phase?

   (b) Determine the strain at times $t_0^+ < t \leq t_1$, and the amount of energy dissipated into heat form. Sketch the instantaneous and the time dependent response in the $(\sigma \times \Gamma)$-plane.

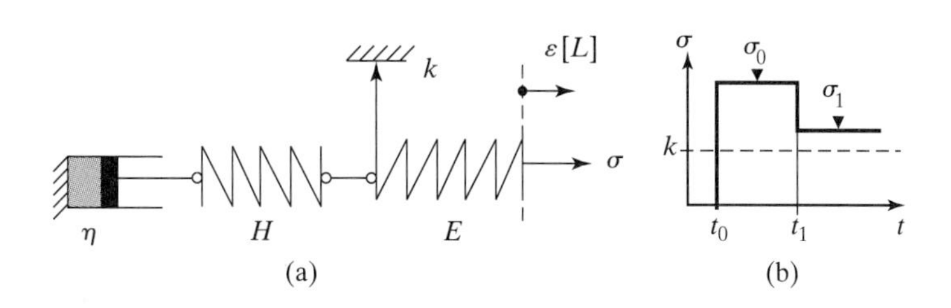

FIGURE 7.19: Problem set: Mandel's model of creep hesitancy.

(c) Determine the instantaneous response after unloading at $t = t_1$, and show the unloading process in the $(\sigma \times \Gamma)$-plane.

(d) By using the result developed in Question 1b, show that after unloading the creep stops only for some time, and that—after a time interval Δt—the creep starts again. Determine Δt and the creep rate, and show this creep hesitancy behavior in the $(\sigma \times \Gamma)$-plane.

7.6.1 Elasticity Domain

Plastically Admissible Stress States.

The elasticity domain defines the plastically admissible stress states:

$$f = |\sigma - \Gamma| - k \leq 0 \leftrightarrow \Gamma - k < \sigma < \Gamma + k$$

In the $(\sigma \times \Gamma)$-plane, this loading function corresponds to two straight lines, within which all admissible stress states are situated. This is sketched in Figure 7.20.

Relaxation of Hardening Force.

When the friction element does not move, an initial prestress in the hardening spring is relaxed in time. Let γ be the displacement of the dashpot. The viscous force is $\eta\dot{\gamma}$ and is equal to the tension force $\Gamma = \Gamma(t)$ in the hardening spring:

$$\eta\dot{\gamma} = \Gamma_0 - H\gamma(t) = H[\gamma(\infty) - \gamma(t)]$$

with $\gamma(\infty) = \Gamma_0/H$. Integration of this differential equation gives

$$\gamma(t) = \gamma(\infty)\left[1 - \exp\left(-\frac{t - t_0}{\tau^r}\right)\right]$$

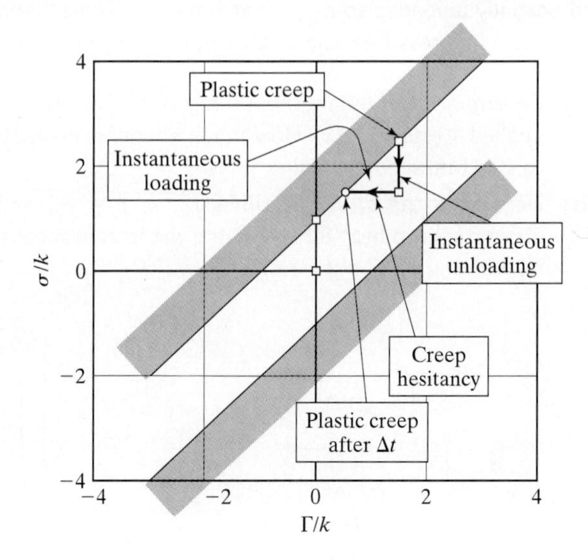

FIGURE 7.20: Generalized force plane $\sigma \times \Gamma$.

where $\tau^r = \eta/H$ is the relaxation time. The tension force relaxes in time:

$$\Gamma(t) = \Gamma_0 \exp\left(-\frac{t-t_0}{\tau^r}\right)$$

This relaxation affects the elasticity domain $f(\sigma, \Gamma) \leq 0$: In the $(\sigma \times \Gamma)$-plane, the relaxation of the hardening force corresponds to a horizontal line describing the change in hardening force at constant stress σ (see Figure 7.20).

7.6.2 Constitutive Equations

State Equations.

The model has *a priori* four state variables: strain ε, permanent deformation ε^p, hardening variable χ, and viscous deformation γ. The free energy reads

$$\psi = \frac{1}{2}\left[E(\varepsilon - \varepsilon^p)^2 + H(\chi - \gamma)^2\right]$$

Use in the Clausius–Duhem inequality yields

$$\begin{aligned}
\varphi dt &= \sigma d\varepsilon - d\psi \\
&= \left[\sigma - \frac{\partial \psi}{\partial \varepsilon}\right] d\varepsilon - \frac{\partial \psi}{\partial \varepsilon^p} d\varepsilon^p - \frac{\partial \psi}{\partial \chi} d\chi - \frac{\partial \psi}{\partial \gamma} d\gamma \\
&= \sigma d\varepsilon^p + \zeta d\chi + \Gamma d\gamma
\end{aligned}$$

and the state equations

$$\sigma = \frac{\partial \psi}{\partial \varepsilon} = -\frac{\partial \psi}{\partial \varepsilon^p} = E(\varepsilon - \varepsilon^p)$$

$$\zeta = -\frac{\partial \psi}{\partial \chi} = -H(\chi - \gamma) = -\Gamma = \frac{\partial \psi}{\partial \gamma}$$

Evolution Laws.

A priori, there are three complementary evolution laws, one for each internal state variable. The plastic flow rule and hardening rule read

$$d\varepsilon^p = d\lambda \frac{\partial f}{\partial \sigma} = d\lambda \operatorname{sign}(\sigma + \zeta) \equiv d\chi = d\lambda \frac{\partial f}{\partial \zeta} = -d\lambda \frac{\partial f}{\partial \Gamma} = d\lambda \operatorname{sign}(\sigma + \zeta)$$

They derive from the loading function $f(\sigma, \zeta = -\Gamma)$ previously defined. The third evolution law is the one of the viscous strain evolution γ, as already used:

$$\eta \dot{\gamma} = \Gamma = H(\chi - \gamma)$$

The second law requires

$$\varphi dt = d\lambda k + \eta \dot{\gamma}^2 dt \geq 0$$

and is satisfied for $k \geq 0$ and $\eta \geq 0$. The plastic multiplier is determined by the consistency condition:

$$df = \frac{\partial f}{\partial \sigma} d\sigma + \frac{\partial f}{\partial \zeta}\frac{\partial \zeta}{\partial \chi} d\chi + \frac{\partial f}{\partial \zeta}\frac{\partial \zeta}{\partial \gamma} d\gamma = 0$$

The plastic hardening modulus is only associated with plastic evolutions (i.e., only with plastic hardening variable $d\chi$):

$$\frac{\partial f}{\partial \zeta}\frac{\partial \zeta}{\partial \chi} d\chi = -H d\lambda$$

Use of the hardening rule gives

$$H = -\left(\frac{\partial f}{\partial \zeta}\right)^2 \frac{\partial \zeta}{\partial \chi}$$

and the plastic multiplier

$$d\lambda = \frac{1}{H}\left[\frac{\partial f}{\partial \sigma} d\sigma + \frac{\partial f}{\partial \zeta}\frac{\partial \zeta}{\partial \gamma} d\gamma\right] = \left(\frac{d\sigma}{H} + d\gamma\right) \operatorname{sign}(\sigma + \zeta)$$

or, in rate form,

$$\dot{\lambda} = \left(\frac{\dot{\sigma}}{H} + \dot{\gamma}\right) \operatorname{sign}(\sigma + \zeta)$$

The plastic strain rate reads

$$\dot{\varepsilon}^p = \frac{\dot{\sigma}}{H} + \dot{\gamma} = \frac{\dot{\sigma}}{H} + \frac{\Gamma}{\eta} = \dot{\chi}$$

The same result could have been obtained from Figure 7.19a, for $f = 0$ (i.e., during plastic loading): The incremental hardening spring elongation is $d\chi - d\gamma = d\varepsilon^p - d\gamma$, and the associated force increment in the spring is $d\sigma$. Note that a pure plastic behavior occurs for $\dot{\gamma} = 0$, while for constant load, $\dot{\sigma} = 0$, it is $\dot{\varepsilon}^p = \dot{\gamma}$ provided that $f = 0$.

7.6.3 Instantaneous and Time-Dependent Deformation

Instantaneous Response.

For $\sigma_0 > k$ the model behaves like a standard hardening plasticity model, since $\dot{\gamma} = 0$. The strain variables after instantaneous loading are

$$t = t_0^+ : \varepsilon_0^p = \frac{\sigma_0 - k}{H} = \chi_0; \quad \gamma_0 = 0$$

$$t = t_0^+ : \varepsilon_0 = \frac{\sigma_0}{E} + \frac{\sigma_0 - k}{H}$$

The forces are

$$t = t_0^+ : \sigma(0) = \sigma_0; \quad \Gamma(0) = -\zeta(0) = \sigma_0 - k$$

The dissipated energy is:

$$\int_{s=t_0^-}^{s=t_0^+} \varphi ds = k \frac{\sigma_0 - k}{H}$$

Plastic Creep.

The stress $\sigma(t - t_0^+ > 0) = \sigma_0$ is constant (i.e., $\dot{\sigma} = 0$). Therefore, the internal variables develop in the same manner:

$$t > t_0^+ : \dot{\varepsilon}^p = \dot{\gamma} = \dot{\chi} = \frac{\Gamma}{\eta}$$

Since

$$t > t_0^+ : \Gamma = -\zeta = H(\chi - \gamma) = H[\chi_0 + \Delta\chi(t) - \gamma(t)] = \sigma_0 - k$$

the strain during this loading phase occurs at a constant rate. The spring forces preserve their lengths. The strains at time t read

$$t > t_0^+ : \varepsilon^p(t) = \varepsilon_0^p + \frac{\sigma_0 - k}{\eta}(t - t_0) = \frac{\sigma_0 - k}{H}\left(1 + \frac{t - t_0}{\tau^r}\right) = \chi(t)$$

$$\gamma(t) = \frac{\sigma_0 - k}{\eta}(t - t_0)$$

Thus the total strain

$$\varepsilon(t) = \frac{\sigma_0}{E} + \frac{(\sigma_0 - k)}{H}\left(1 + \frac{t - t_0}{\tau^r}\right)$$

where $\tau^r = \eta/H$ is the relaxation time. The energy dissipated into heat form is

$$\begin{aligned} \varphi(t - t_0^+) &= k\,(\sigma_0 - k)\,\frac{(t - t_0)}{\eta} + (\sigma_0 - k)^2\,\frac{(t - t_0)}{\eta} \\ &= \frac{\sigma_0\,(\sigma_0 - k)}{H}\frac{(t - t_0)}{\tau^r} \end{aligned}$$

and the total dissipation

$$\int_{s=t_0^-}^{s=t} \varphi ds = k\frac{\sigma_0 - k}{H}\left[1 + \frac{\sigma_0}{k}\frac{(t - t_0)}{\tau^r}\right]$$

Since the forces are unchanged, the loading point (σ, Γ) is fixed in the $(\sigma \times \Gamma)$-plane.

Instantaneous Unloading.

During unloading at $t = t_1$, $\dot{\gamma} = 0$. The tension force in the hardening spring at unloading is

$$\Gamma(t_1) = \sigma_0 - k$$

Use in the loading function gives

$$t = t_1 : f = |\sigma_1 - \Gamma(t_1)| - k = |\sigma_1 - \sigma_0 + k| - k \leq 0$$

Hence, for $\sigma_0 < 3k$ and $\sigma_1 > k$, the point enters the elasticity domain, and $f < 0$. The instantaneous elastic deformation is $\Delta\varepsilon = -\sigma_0/E$, and the plastic strain keeps the value it developed during loading:

$$t \geq t_1^- : \varepsilon^p(t) = \frac{(\sigma_0 - k)}{H}\left(1 + \frac{t_1 - t_0}{\tau^r}\right) = \chi(t)$$

Relaxation of Hardening Force and Restart of Creep.

While there is no plastic evolution after unloading (i.e., $d\varepsilon^p = d\chi = 0$), viscous deformation occurs, triggered by the initial tension force $\Gamma(t_1) = \sigma_0 - k$. In the absence of plastic sliding, the relaxation was determined in Section 7.6.1:

$$t \geq t_1 : \Gamma(t) = \Gamma(t_1) \exp\left(-\frac{t - t_1}{\tau^r}\right)$$

Use in the loading function yields

$$f = 0 : \sigma_1 - \Gamma(t_1) = \sigma_1 - (\sigma_0 - k) \exp\left(-\frac{t - t_1}{\tau^r}\right) - k = 0$$

Thus

$$\Delta t = t_2 - t_1 = \tau^r \ln\left(\frac{\sigma_1 - k}{\sigma_0 - k}\right)$$

During this time, the force Γ relaxes from its initial value $\Gamma(t_1)$ to $\Gamma(t_2) = \sigma_1 - k$, for which $f = 0$. For $t > t_2$ the creep process starts again, on the same lines as developed before. The rate at which this creep occurs is

$$\dot{\varepsilon} = \dot{\varepsilon}^p = \dot{\chi} = \dot{\gamma} = \frac{\Gamma(t_2)}{\eta} = \frac{\sigma_1 - k}{\tau^r H}$$

The total strain reads

$$\varepsilon(t) = \frac{\sigma_1}{E} + \frac{\sigma_0 - k}{H}\left(1 + \frac{t_1 - t_0}{\tau^r}\right) + \frac{\sigma_1 - k}{H}\left(\frac{t - t_1}{\tau^r}\right)$$

The solution in the $(\sigma \times \Gamma)$-plane is shown in Figure 7.20.

CHAPTER 8

Plasticity Models

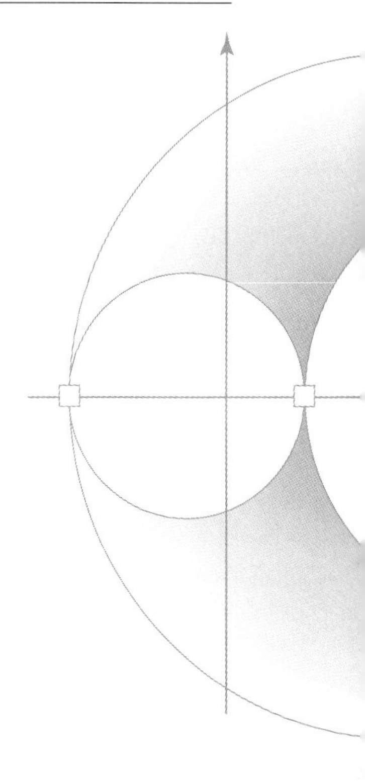

The three elements of plasticity models are (1) the plasticity criterion, (2) the flow and hardening rule, and (3) the hardening (or softening) process. The plasticity criterion defines *when* plastic deformation occurs: *How* this plastic deformation occurs is set forth by the flow and hardening rule. Finally, the (plastic) hardening law provides the link between when and how. This second chapter on plasticity extends the 1D formulation of Chapter 7 to 3D hardening plasticity. Set within the framework of the thermodynamics of irreversible processes, we examine associate and nonassociate plasticity for some common isotropic plasticity models: Von–Mises and Drucker–Prager plasticity. Using the notion of the frozen hardening energy, common plastic hardening models are discussed (i.e., isotropic and kinematic hardening). The training set of this chapter revisits the elements of 3D plasticity through the Cambridge model, a 3D hardening plasticity model, which we examine in the context of plasticity and viscoplasticity.

8.1 ELEMENTS OF 3D IDEAL PLASTICITY MODELS

The plasticity criterion defines *when* plastic deformation occurs; the plastic flow rule defines *how* it occurs. The first is related to the strength domain and defines the elasticity domain of the material. The second deals with the kinematics of plastic deformation. In this section, we restrict ourselves to ideal plasticity. Later we will see how to extend the concepts to hardening plasticity.

8.1.1 Ideal Plasticity Criterion

The plasticity criterion is a generalized strength criterion that defines the elasticity domain of the material. The three-dimensional strength domain accounts for the maximum strength on all possible material planes on which the stress vector $\mathbf{T}(\mathbf{n}) = \boldsymbol{\sigma} \cdot \mathbf{n}$ reaches the material strength; for instance, the shear strength $|\mathbf{t} \cdot \mathbf{T}(\mathbf{n})| \leq \max|\tau|$, the tensile strength $\mathbf{n} \cdot \mathbf{T}(\mathbf{n}) \leq \max \sigma$, or any combination (have a quick look back at Section 4.4). This confinement of the stress components to some maximum strength values suggests a scalar loading function involving the six stress components of the stress tensor:

$$\boldsymbol{\sigma} \in D_E \Leftrightarrow f = f(\boldsymbol{\sigma}) \leq 0 \tag{8.1}$$

Expression (8.1) is the generalization of the 1D plasticity criterion (7.1) to the $\Re^6$-space of the six stress components σ_{ij}. A stress $\boldsymbol{\sigma}$ is called *plastically admissible* if it is within ($f < 0$) or on the boundary ($f = 0$) of the elasticity domain D_E. Furthermore, a loading point on the boundary of D_E remains on it during plastic evolutions. In 1D ideal plasticity, this was expressed by the consistency condition (7.2). The 3D counterparts reads

$$\text{if } f(\boldsymbol{\sigma}) = 0 : f(\boldsymbol{\sigma} + d\boldsymbol{\sigma}) = \; f(\boldsymbol{\sigma}) + \frac{\partial f}{\partial \boldsymbol{\sigma}} : d\boldsymbol{\sigma} = 0 \Leftrightarrow df = 0 \tag{8.2}$$

In (8.2), $\partial f / \partial \boldsymbol{\sigma}$ defines the normal of the tangent plane to the loading surface $f(\boldsymbol{\sigma})$ in the $\Re^6$-space of the six stress components. It is intuitively understood that this normal is an outward normal to the loading function, as sketched in Figure 8.1a. However, without any further specification, this normal may well be oriented to either the outside or inside of the elasticity domain. The orientation of this normal is specified by the convexity of the elasticity domain (have a quick look back at Section 7.1.3). In the 3D case, convexity of the elasticity domain means that the elasticity domain is entirely included on one side of the tangent plane to the loading function oriented by the outward unit normal $\partial f / \partial \boldsymbol{\sigma}$. This is shown in

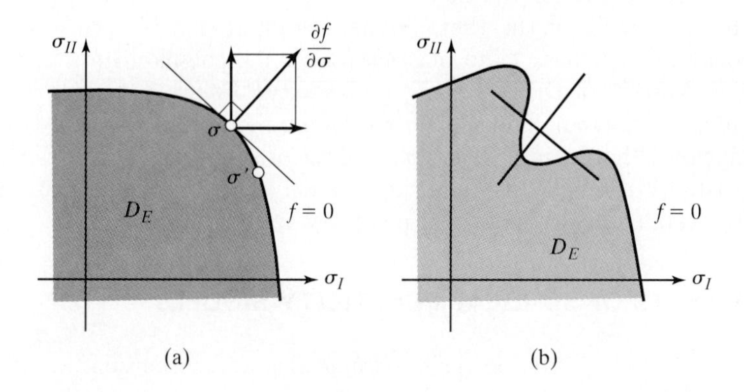

FIGURE 8.1: Elements of ideal plasticity: loading function $f = 0$ and flow rule in stress space: (a) convexixity of loading function and normality rule; (b) not permitted.

Figure 8.1a. A situation like the one displayed in Figure 8.1b is not permitted. The 1D convexity condition (7.14) can be generalized for any regular (i.e., differentiable) loading surface[1]:

$$\frac{\partial f}{\partial \boldsymbol{\sigma}} : (\boldsymbol{\sigma} - \boldsymbol{\sigma}') \geq 0 \tag{8.3}$$

where $\boldsymbol{\sigma}$ and $\boldsymbol{\sigma}'$ are two three-dimensional plastically admissible stress states, one situated on the loading function $f(\boldsymbol{\sigma}) = 0$ associated with $\partial f / \partial \boldsymbol{\sigma}$, the other within or on the boundary $f(\boldsymbol{\sigma}') \leq 0$. We will deliver later the energy argument in favor of this convexity.

Finally, in the case of an isotropic plastic material, the loading function reads as a function of the stress tensor invariants (have a quick look back at Section 4.2.3):

$$\boldsymbol{\sigma} \in D_E \Leftrightarrow f = f(I_1, I_2, I_3) \leq 0 \tag{8.4}$$

where $I_1 = \operatorname{tr} \boldsymbol{\sigma}$, $I_2 = \frac{1}{2}\operatorname{tr}(\boldsymbol{\sigma} \cdot \boldsymbol{\sigma})$ and $I_3 = \frac{1}{3}\operatorname{tr}(\boldsymbol{\sigma} \cdot \boldsymbol{\sigma} \cdot \boldsymbol{\sigma})$ are stress tensor invariants.

8.1.2 Flow Rule of Ideal Plasticity

The flow rule describes the kinematics of plastic deformation. If we denote by $d\boldsymbol{\varepsilon}^p$ the plastic strain increment, the flow rule and the plastic loading–unloading conditions (7.3) read in the 3D case:

$$d\boldsymbol{\varepsilon}^p = d\lambda \frac{\partial f(\boldsymbol{\sigma})}{\partial \boldsymbol{\sigma}}; \quad f \leq 0; \quad d\lambda \geq 0; \quad d\lambda f = 0 \tag{8.5}$$

The (scalar) plastic multiplier $d\lambda \geq 0$ expresses still the intensity of plastic flow, while $\partial f(\boldsymbol{\sigma})/\partial \boldsymbol{\sigma}$ represents the direction taken by the plastic strain increment $d\boldsymbol{\varepsilon}^p$. From a mathematical standpoint, $\partial f(\boldsymbol{\sigma})/\partial \boldsymbol{\sigma}$ is the gradient to the (differentiable) loading function with respect to $\boldsymbol{\sigma}$. Due to the convexity of the loading function expressed by (8.3), this direction corresponds to the outward normal to the boundary of the elasticity domain. This is illustrated in Figure 8.1a.

8.1.3 Von–Mises Plasticity

We have introduced the elements of 3D ideal plasticity model in a rather ad hoc way. This and the next section examine the strength criterion and the plastic flow rule with a closer look on the physical processes involved.

A Shear Strength Criterion.

A shear strength criterion on a slippage plane oriented by unit normal $\mathbf{n}$ (and tangent vector $\mathbf{t}$) reads

$$\boldsymbol{\sigma} \in D_E \leftrightarrow f(\boldsymbol{\sigma}) = |\mathbf{t} \cdot \mathbf{T}(\mathbf{n})| - k \leq 0 \tag{8.6}$$

or, equivalently.

$$\boldsymbol{\sigma} \in D_E \leftrightarrow f(\boldsymbol{\sigma}) = |\boldsymbol{\sigma} : (n \otimes \mathbf{t})| - k \leq 0 \tag{8.7}$$

---

[1] For completeness, note that the convexity condition for any loading surface, whether regular or not, reads $\forall \alpha \in [0,1]$; $\forall \boldsymbol{\sigma}' \neq \boldsymbol{\sigma}$; $f[\alpha \boldsymbol{\sigma} + (1-\alpha)\boldsymbol{\sigma}'] \leq \alpha f(\boldsymbol{\sigma}) + (1-\alpha) f(\boldsymbol{\sigma}')$

where k denotes a shear stress threshold defining the shear strength on a specific (oriented) slippage plane. If the shear strength of the material is independent of the orientation of the slippage plane, on which the stress reaches the threshold, the plasticity criterion is isotropic. In this case, the slippage plane is oriented by the hydrostatic axis. This particular slippage plane is known as *deviator plane*, and the stress vector built on it reads

$$\mathbf{T}(\mathbf{n}) = \sigma_m \mathbf{n} + \tau_{\text{oct}} \mathbf{t} \tag{8.8}$$

where $\mathbf{n}$ is the orientation of the hydrostatic axis:

$$\mathbf{n} = \frac{\sqrt{3}}{3}(\mathbf{u}_I + \mathbf{u}_{II} + \mathbf{u}_{III}) \tag{8.9}$$

$\mathbf{u}_J$ are the principal stress directions. The normal and shear stresses are the octrahedral stresses we derived in Section 4.2.4, reading[2]

$$\sigma_m = \mathbf{n} \cdot \mathbf{T}(\mathbf{n}) = \frac{1}{3}(\sigma_I + \sigma_{II} + \sigma_{III}) \tag{8.10}$$

$$|\tau_{\text{oct}}| = |\mathbf{t} \cdot \mathbf{T}(\mathbf{n})| = \frac{\sqrt{3}}{3}\sqrt{\sum_{J=I,II,III} s_J^2} = \sqrt{\frac{2}{3}}\sqrt{J_2} \tag{8.11}$$

The normal stress $\sigma_m = \frac{1}{3}\text{tr}\,\boldsymbol{\sigma}$ corresponds to the first invariant of the stress tensor; $s_J = \sigma_J - \sigma_m$ are the principal deviator stresses; and $J_2 = \frac{1}{2}\text{tr}\,(\mathbf{s} \cdot \mathbf{s})$ is the second invariant of the stress deviator. J_2 enters the isotropic plasticity criterion.

In addition, we need to have a look at the strength parameter. To this end, consider a shear test for which $\boldsymbol{\sigma} = \tau[\mathbf{n} \otimes \mathbf{t} + \mathbf{t} \otimes \mathbf{n}]$. In this shear test, $\sigma_m = 0$ and $J_2 = \tau^2$, thus $|\tau_{\text{oct}}| = \sqrt{2/3}|\tau|$. Denoting by k the shear strength, it follows that

$$|\max \tau_{\text{oct}}| = \sqrt{2/3}k \Leftrightarrow |\max \tau| = k \tag{8.12}$$

or, equivalently,

$$\boldsymbol{\sigma} \in D_E \Leftrightarrow f(J_2) = \sqrt{J_2} - k \leq 0 \tag{8.13}$$

Plasticity criterion (8.13) is known as *Von–Mises* yield surface. In the principal stress space, it represents a cylinder surface generated around the hydrostatic axis.

---

[2] Have a quick look back at Section 4.2.4 and Figure 4.2. The stress vector built on the deviator plane reads

$$\mathbf{T}(\mathbf{n}) = \frac{\sqrt{3}}{3}\sum_{J=I,II,III} \sigma_J \mathbf{u}_J$$

With (8.8), it follows that

$$\tau_{\text{oct}}\mathbf{t} = \frac{\sqrt{3}}{3}\sum_{J=I,II,III} \sigma_J \mathbf{u}_J - \sigma_m \mathbf{n}$$

Thus, using (8.9),

$$\tau_{\text{oct}}\,\mathbf{t} = \frac{\sqrt{3}}{3}\sum_{J=I,II,III} (\sigma_J - \sigma_m)\,\mathbf{u}_J = \frac{\sqrt{3}}{3}\sum_{J=I,II,III} s_J \mathbf{u}_J$$

This leads to (8.11).

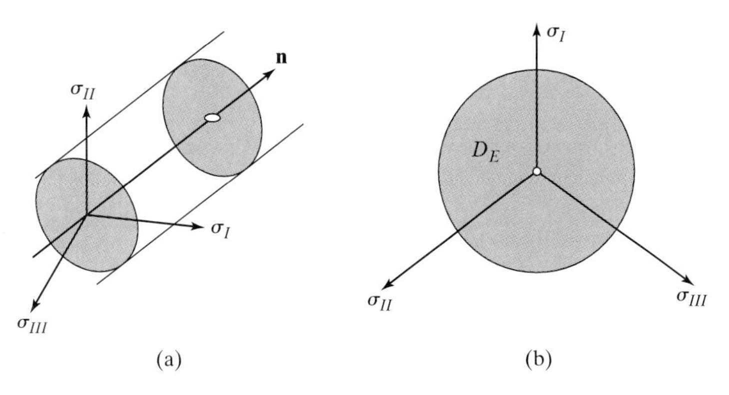

FIGURE 8.2: Von–Mises plasticity criterion: (a) in the principal stress space; (b) in the deviator plane.

This is sketched in Figure 8.2a. In the *deviator plane* (projection normal to the hydrostatic axis), the yield surface represents a circle of radius $\sqrt{2J_2} = \sqrt{2}k$, as displayed in Figure 8.2b.

Plastic Distortion: An Isotropic Shear Flow Rule.

The shear criterion defines the elasticity domain. It is one element of ideal plasticity. The second deals with the kinematics of plastic deformation. To motivate forthcoming developments, we consider a shear test on an isotropic material sample $d\Omega$. A deviator stress state $\boldsymbol{\sigma} = \mathbf{s} = \tau\,(\mathbf{n}\otimes\mathbf{t} + \mathbf{t}\otimes\mathbf{n})$ is prescribed. During loading, a distortion $\theta(\mathbf{n},\mathbf{t}) = 2\varepsilon_{nt}$ occurs (have a quick look back at Section 2.2.1). After unloading, an irreversible distortion, $\theta^p(\mathbf{n},\mathbf{t}) = 2\varepsilon^p_{nt}$, remains, while all other strain components are assumed zero. This is shown in Figure 8.3. Hence, the kinematics of the plastic deformation is given by

$$d\varepsilon^p_{nt} = \frac{d\theta^p(\mathbf{n},\mathbf{t})}{2} = \mathbf{n}\cdot d\boldsymbol{\varepsilon}^p\cdot\mathbf{t} \tag{8.14}$$

If this distortion is the only plastic deformation that occurs during plastic loading, it follows that

$$d\boldsymbol{\varepsilon}^p = \frac{d\theta^p(\mathbf{n},\mathbf{t})}{2}(\mathbf{n}\otimes\mathbf{t} + \mathbf{t}\otimes\mathbf{n}) \tag{8.15}$$

or, equivalently,

$$d\boldsymbol{\varepsilon}^p = d\lambda\frac{\tau}{2|\tau|}(\mathbf{n}\otimes\mathbf{t} + \mathbf{t}\otimes\mathbf{n});\quad (d\varepsilon^p_{ij}) = d\lambda\begin{bmatrix} 0 & \frac{\tau}{2|\tau|} & 0 \\ \frac{\tau}{2|\tau|} & 0 & 0 \\ 0 & 0 & 0 \end{bmatrix} \tag{8.16}$$

where $d\lambda = |d\theta^p(\mathbf{n},\mathbf{t})| = \sqrt{2d\boldsymbol{\varepsilon}^p : d\boldsymbol{\varepsilon}^p}$ quantifies the intensity of the plastic distortion, and $\text{sign}\,(d\theta^p) = \text{sign}\,(\tau) = \tau/|\tau|$ the orientation of the distortion due to shear stress application. Finally, using the expression of the applied stress $\mathbf{s} = \tau[\mathbf{n}\otimes\mathbf{t} + \mathbf{t}\otimes\mathbf{n}]$, and $J_2 = \frac{1}{2}\mathbf{s} : \mathbf{s} = \tau^2$, Eq. (8.16) can be recast in the form

$$d\boldsymbol{\varepsilon}^p = d\lambda\frac{\mathbf{s}}{2\sqrt{J_2}} = d\lambda\frac{\partial f(J_2)}{\partial\boldsymbol{\sigma}} \tag{8.17}$$

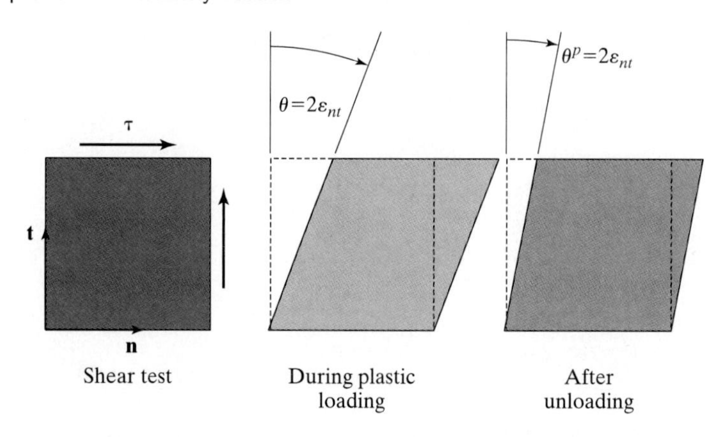

FIGURE 8.3: Plastic distortion in a shear test.

This turns out to be the flow rule (8.5) for a plasticity criterion $f(J_2) = 0$ of the Von-Mises type.

Study of the flow rule of plastic distortion reveals some interesting insight in the structure of 3D plastic evolution laws:

1. The plastic multiplier $d\lambda \geq 0$ expresses the intensity of the plastic deformation; here in the shear test the intensity of the permanent distortion $d\theta^p$. $d\lambda$ is related to plastic shear deformation. The intensity of the plastic distortion is often referred to as *equivalent plastic distortion*:

$$|d\theta^p| = d\lambda \frac{\partial f}{\partial \sqrt{J_2}} = \sqrt{2d\mathbf{e}^p : d\mathbf{e}^p} \tag{8.18}$$

where $d\mathbf{e}^p = d\boldsymbol{\varepsilon}^p - \frac{1}{3}\text{tr}\, d\boldsymbol{\varepsilon}^p \mathbf{1}$ is the plastic strain deviator increment. This has been derived here for a shear test on a Von–Mises material. It holds, however, for all plasticity criteria for which $\partial f / \partial \sqrt{J_2} = 1$:

$$\frac{\partial f}{\partial \sqrt{J_2}} = 1 \rightarrow |d\theta^p| = d\lambda \tag{8.19}$$

(The proof is left to the reader.)

2. The orientation of the plastic flow [i.e., $\partial f(\boldsymbol{\sigma})/\partial \boldsymbol{\sigma}$] in (8.5) has its physical origin in the kinematics of plastic deformation. In the considered shear test, the kinematics refers to the orientation of the plastic distortion and the restriction to a pure elastic and plastic shear deformation.

Exercise 73. For the Von–Mises plasticity model, determine

1. The compressive strength, f'_c, and the tensile strength, f'_t;
2. The plastic strain and the plastic distortion in uniaxial tension and compression. Show that the permanent plastic volume variation is zero.

The stress deviator for a uniaxial stress state $\boldsymbol{\sigma} = \sigma \mathbf{e}_1 \otimes \mathbf{e}_1$ reads

$$(s_{ij}) = \frac{\sigma}{3} \begin{bmatrix} 2 & & \\ & -1 & \\ & & -1 \end{bmatrix}$$

or, equivalently, in tensor notation,

$$\mathbf{s} = \frac{\sigma}{3} (2\mathbf{e}_1 \otimes \mathbf{e}_1 - \mathbf{e}_2 \otimes \mathbf{e}_2 - \mathbf{e}_3 \otimes \mathbf{e}_3)$$

The second deviator invariant reads

$$J_2 = \frac{1}{2} \operatorname{tr}(\mathbf{s} \cdot \mathbf{s}) = \frac{\sigma^2}{3}$$

Substitution in (8.13) yields

$$\boldsymbol{\sigma} = \sigma \mathbf{e}_1 \otimes \mathbf{e}_1 \in D_E \leftrightarrow f(J_2) = \frac{1}{\sqrt{3}} |\sigma| - k \leq 0$$

Thus, the compressive and tensile strengths are

$$f(J_2) = 0 : |\sigma_{\max}| = f_c' = f_t' = \sqrt{3}k$$

As a shear strength criterion, the Von–Mises criterion captures a symmetrical strength behavior of materials: The compressive strength and the tensile strength coincide. Von–Mises materials are pressure-insensitive isotropic materials. Only one material test is required for the determination of the model parameter k.
Use of the stress deviator in (8.17) yields

$$(d\varepsilon_{ij}^p) = d\lambda \left(\frac{\partial f}{\partial \sigma_{ij}} \right) = \frac{d\lambda}{2\sqrt{J_2}} (s_{ij}) = \frac{d\lambda}{2\sqrt{3}} \frac{\sigma}{|\sigma|} \begin{bmatrix} 2 & & \\ & -1 & \\ & & -1 \end{bmatrix}$$

We verify without difficulty that the plastic dilatation is zero:

$$\operatorname{tr} d\boldsymbol{\varepsilon}^p = 0 \leftrightarrow d\boldsymbol{\varepsilon}^p = d\mathbf{e}^p$$

The plastic distortion satisfies (8.19) and reads

$$|d\theta^p| = \sqrt{2 d\mathbf{e}^p : d\mathbf{e}^p} = d\lambda$$

■

8.1.4 Drucker–Prager Plasticity

Many geomaterials deform in an irreversible manner along friction planes, on which normal stresses and shear stresses limit the strength domain of the material. In addition, the kinematics of the associated plastic deformation involves permanent distortion and permanent dilatation.

A Pressure-Dependent Strength Criterion.

We consider again a slippage plane oriented by unit normal $\mathbf{n}$. The critical stress state at which irreversible deformation starts may depend on both stress components of the stress vector, the normal stress component $\sigma = \mathbf{n} \cdot \mathbf{T}(\mathbf{n})$ (or surface traction), and the shear stress $\tau = \mathbf{t} \cdot \mathbf{T}(\mathbf{n})$. A strength criterion depending on both stress vector components is the Mohr–Coulomb criterion, which reads (have a quick look back at Section 4.4.2)

$$\boldsymbol{\sigma} \in D_E \leftrightarrow f(\boldsymbol{\sigma}) = |\tau| + \alpha\sigma - c \leq 0 \tag{8.20}$$

where $\alpha = \tan\varphi$ is the friction coefficient, and c the cohesion of the specific shear-friction plane oriented by unit outward normal $\mathbf{n}$. If these material parameters do not depend on any specific direction, the plasticity criterion is isotropic. Proceeding as in the previous section, we build the stress vector on the deviator plane oriented by the hydrostatic axis [i.e., Eq. (8.9)]. Substitution in (8.20) leads to a strength criterion that depends on the two invariants of the stress tensor:

$$\boldsymbol{\sigma} \in D_E \leftrightarrow f(\sigma_m, J_2) = \sqrt{J_2} + \alpha\sigma_m - c \leq 0 \tag{8.21}$$

This isotropic shear-friction strength criterion is known as *Drucker–Prager* plasticity criterion.[3] For $\alpha = 0$, the Drucker–Prager criterion reduces to the Von–Mises criterion (with $c = k$). In the principal stress space, the Drucker–Prager criterion is a cone around the hydrostatic axis (Figure 8.4a). The vertex of the cone is situated at the hydrostatic axis at $\sigma_I = \sigma_{II} = \sigma_{III} = \sigma_m = c/\alpha \geq 0$. The corresponding hydrostatic strength $c/\alpha \geq 0$ is often referred to as cohesion pressure. Any deviator plane (projection onto the hydrostatic axis, *cf.* Figure 8.4b), defined by a constant mean stress σ_m, intersects this cone along a circle of radius $\sqrt{2J_2} = \sqrt{2}(c - \alpha\sigma_m) \geq 0$. The criterion is open on the negative hydrostatic axis.

Plastic Dilatation: An Isotropic Volumetric Plastic Flow Rule.

We consider again a material test on an isotropic material sample $d\Omega$. In addition to the shear stress, $\mathbf{s} = \tau[\mathbf{n} \otimes \mathbf{t} + \mathbf{t} \otimes \mathbf{n}]$, we consider a hydrostatic stress state:

$$\boldsymbol{\sigma} = \tau[\mathbf{n} \otimes \mathbf{t} + \mathbf{t} \otimes \mathbf{n}] + \sigma_m 1 \tag{8.22}$$

During loading, both a distortion $\theta(\mathbf{n}, \mathbf{t}) = 2\varepsilon_{nt}$ and a volume variation $d\Omega_t \simeq (1 + \operatorname{tr}\boldsymbol{\varepsilon})d\Omega$ occur. After unloading, an irreversible distortion $\theta^p(\mathbf{n}, \mathbf{t}) = 2\varepsilon^p_{nt}$ remains, but also an irreversible change in volume $d\Omega_t \simeq (1 + \operatorname{tr}\boldsymbol{\varepsilon}^p)d\Omega$. Hence, the kinematics of this process can be described by

$$d\varepsilon^p_{nt} = \frac{d\theta^p(\mathbf{n}, \mathbf{t})}{2} = \mathbf{n} \cdot \left(d\boldsymbol{\varepsilon}^p - \tfrac{1}{3}(\operatorname{tr} d\boldsymbol{\varepsilon}^p)1 \right) \cdot \mathbf{t} \tag{8.23}$$

[3]The Drucker–Prager criterion appears as a Mohr–Coulomb criterion defined on the deviator plane:

$$f = |\tau_{\text{oct}}| + \alpha_{\text{oct}}\sigma_m - c_{\text{oct}} \leq 0$$

where α_{oct} and c_{oct} are the friction coefficient and the cohesion in the deviator stress plane. In the material strength plane, $|\tau_{\text{oct}}| = \sqrt{2/3 J_2}$. Hence, α_{oct} and c_{oct} are related to α and c in (8.21) by

$$\alpha_{\text{oct}} = \sqrt{2/3}\alpha; \quad c_{\text{oct}} = \sqrt{2/3}c$$

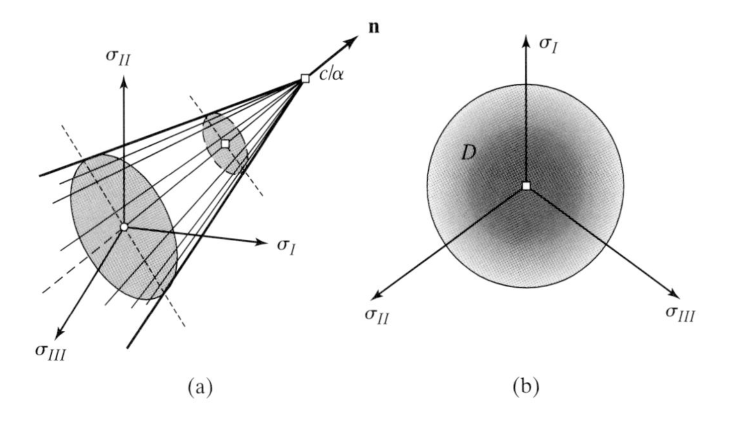

FIGURE 8.4: Drucker–Prager criterion in (a) the principal stress space; (b) the deviator plane.

or, equivalently,

$$d\boldsymbol{\varepsilon}^p = \frac{d\theta^p(\mathbf{n}, \mathbf{t})}{2}(\mathbf{n} \otimes \mathbf{t} + \mathbf{t} \otimes \mathbf{n}) + \frac{1}{3}(\operatorname{tr} d\boldsymbol{\varepsilon}^p)\mathbf{1} \tag{8.24}$$

The first term in (8.24) represents the deviator part $d\mathbf{e}^p = d\boldsymbol{\varepsilon}^p - \frac{1}{3}(\operatorname{tr} d\boldsymbol{\varepsilon}^p)\mathbf{11}$ of the plastic strain increment, and the second the permanent volume variation. Since both are related to the same physical phenomenon (i.e., plastic deformation), it is natural to link both by means of a *plastic dilatation coefficient* δ:

$$\operatorname{tr}(d\boldsymbol{\varepsilon}^p) = \delta |d\theta^p(\mathbf{n}, \mathbf{t})| \tag{8.25}$$

The plastic dilatation coefficient relates the intensity of plastic distortion $|d\theta^p(\mathbf{n}, \mathbf{t})| = \sqrt{2d\mathbf{e}^p : d\mathbf{e}^p}$ to the plastic volume variation. For $\delta > 0$, the material behaves in a plastic dilatant way corresponding to a permanent increase of the volume. If $\delta < 0$, the material is said to be plastically contracting.

Finally, use of (8.25) in (8.24) yields the plastic flow rule:

$$d\boldsymbol{\varepsilon}^p = d\lambda \left(\frac{\tau}{2|\tau|}(\mathbf{n} \otimes \mathbf{t} + \mathbf{t} \otimes \mathbf{n}) + \frac{1}{3}\delta \mathbf{1} \right); \quad (d\varepsilon_{ij}^p) = d\lambda \begin{bmatrix} \frac{1}{3}\delta & \frac{\tau}{2|\tau|} & 0 \\ \frac{\tau}{2|\tau|} & \frac{1}{3}\delta & 0 \\ 0 & 0 & \frac{1}{3}\delta \end{bmatrix} \tag{8.26}$$

or, equivalently,

$$d\boldsymbol{\varepsilon}^p = d\lambda \left(\frac{\mathbf{s}}{2\sqrt{J_2}} + \frac{1}{3}\delta \mathbf{1} \right) \tag{8.27}$$

with $d\lambda = |d\theta^p|$.

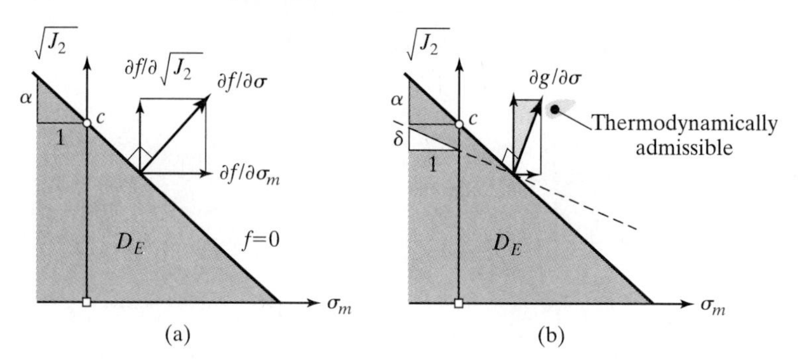

FIGURE 8.5: Plastic flow rules of the Drucker–Prager plasticity model in the stress invariant ($\sigma_m \times \sqrt{J_2}$) half-plane: (a) associated flow rule; (b) nonassociated flow rule.

Last, it is instructive to compare (8.27) with the flow rule (8.5) applied to a loading surface of the Drucker–Prager type (8.21):

$$d\varepsilon^p = d\lambda\frac{\partial f(\sigma_m, J_2)}{\partial \boldsymbol{\sigma}} = d\lambda\left(\frac{\mathrm{s}}{2\sqrt{J_2}} + \frac{1}{3}\alpha 1\right) \tag{8.28}$$

For $\alpha = \delta$, (8.27) and (8.28) coincide: The plastic flow is directed along the outward normal $\partial f/\partial\boldsymbol{\sigma}$ to the boundary of the elasticity domain $f(\sigma_m, J_2) = 0$. However, this normality rule is not a physical law, nor a general principle. In fact, the loading function defines the strength domain, while the flow rule sets out the kinematics of plastic deformation. These two elements of ideal plasticity are independent quantities that describe *when* and *how* plastic deformation occurs. A more general expression of the plastic flow rule than (8.5) is required:

$$d\varepsilon^p = d\lambda\frac{\partial g(\boldsymbol{\sigma})}{\partial \boldsymbol{\sigma}} \tag{8.29}$$

$g(\boldsymbol{\sigma})$ is called the plastic potential. If $g(\boldsymbol{\sigma}) = f(\boldsymbol{\sigma})$, the flow rule (8.29) obeys the normality rule (8.5). In this case, we speak of an *associated* flow rule: The direction of plastic flow represents the outward normal to the loading function. On the contrary, a deviation of this normality rule [i.e., $g(\boldsymbol{\sigma}) \neq f(\boldsymbol{\sigma})$] is referred to as a *nonassociated* flow rule; for instance, in the case of a Drucker–Prager plasticity, $\alpha \neq \delta$. This is illustrated in Figure 8.5.

Exercise 74. For a Drucker–Prager material, determine

1. The compressive strength and the tensile strength;
2. The plastic strain, the plastic distortion, and the plastic volume dilatation in uniaxial tension and compression.

For a uniaxial stress state $\boldsymbol{\sigma} = \sigma\mathbf{e}_1 \otimes \mathbf{e}_1$, the stress deviator reads

$$\mathrm{s} = \frac{1}{3}\sigma[2\mathbf{e}_1 \otimes \mathbf{e}_1 - \mathbf{e}_2 \otimes \mathbf{e}_2 - \mathbf{e}_3 \otimes \mathbf{e}_3]$$

The two stress invariants σ_m and J_2 are

$$\sigma_m = \frac{1}{3}\mathrm{tr}\,\boldsymbol{\sigma} = \frac{\sigma}{3}; \quad J_2 = \frac{1}{2}\mathrm{tr}\,(\mathbf{s}\cdot\mathbf{s}) = \frac{\sigma^2}{3}$$

Substitution in (8.21) yields

$$\boldsymbol{\sigma} = \sigma\mathbf{e}_1 \otimes \mathbf{e}_1 \in D_E \leftrightarrow f(\sigma_m, J_2) = \frac{1}{\sqrt{3}}|\sigma| + \alpha\frac{\sigma}{3} - c \le 0$$

On the boundary of the elasticity domain, we have

$$f(\sigma_m, J_2) = 0 : |\sigma_{\max}| = \frac{3c}{\sqrt{3} + \alpha\mathrm{sign}\,(\sigma)}$$

Thus,

$$f_c' = \frac{3c}{\sqrt{3}-\alpha}; \quad f_t' = \frac{3c}{\sqrt{3}+\alpha}$$

Last, the very existence of the material strengths implies $c \ge 0$ and $0 \le \alpha \le \sqrt{3}$:

$$0 \le c = \frac{2}{\sqrt{3}}\frac{f_c' f_t'}{f_c' + f_t'} \le \frac{1}{\sqrt{3}}f_t'; \quad 0 \le \alpha = \sqrt{3}\frac{f_c' - f_t'}{f_c' + f_t'} \le \sqrt{3}$$

The Drucker–Prager criterion requires two material tests (different homogeneous stress states) for the determination of the model parameters, c and α.
Use of the stress deviator in (8.28) yields

$$(d\varepsilon_{ij}^p) = d\lambda\left(\frac{\partial f}{\partial \sigma_{ij}}\right) = \frac{d\lambda}{3}\begin{bmatrix} \frac{\sqrt{3}\sigma}{|\sigma|} + \alpha & & \\ & -\frac{\sqrt{3}\sigma}{2|\sigma|} + \alpha & \\ & & -\frac{\sqrt{3}\sigma}{2|\sigma|} + \alpha \end{bmatrix}$$

We verify (8.25) with

$$|d\theta^p| = \sqrt{2d\mathbf{e}^p : d\mathbf{e}^p} = d\lambda; \quad \mathrm{tr}\,(d\boldsymbol{\varepsilon}^p) = \alpha d\lambda$$

Recalling the limits of the friction coefficient $0 \le \alpha \le \sqrt{3}$ (imposed by the very existence of the material strengths), the Drucker–Prager plasticity model with an associated flow rule captures the following behavior in uniaxial compression:

$$-\frac{d\lambda}{\sqrt{3}} \le d\varepsilon_{11}^p \le 0; \quad \frac{d\lambda}{2\sqrt{3}} \le d\varepsilon_{22}^p = d\varepsilon_{33}^p \le \frac{\sqrt{3}d\lambda}{2}$$

Analogously, in uniaxial tension,

$$\frac{d\lambda}{\sqrt{3}} \le d\varepsilon_{11}^p \le \frac{2d\lambda}{\sqrt{3}}; \quad -\frac{d\lambda}{2\sqrt{3}} \le d\varepsilon_{22}^p = d\varepsilon_{33}^p \le \frac{d\lambda}{2\sqrt{3}}$$

In both cases, the lower bound corresponds to $\alpha = 0$, and the upper bound to $\alpha = \sqrt{3}$. Within these limits, the material undergoes plastic dilatation [i.e., $\mathrm{tr}\,(d\boldsymbol{\varepsilon}^p) \ge 0$]. We should note that the same restrictions do not apply to a nonassociated flow rule, for which $\alpha \ne \delta$. However, other restrictions may arise related to the nonnegativity of the intrinsic dissipation associated with plastic deformation. ■

8.1.5 Energy Considerations: Plastic Dissipation and Principle of Maximum Plastic Work

During plastic evolutions, a part of the work provided from the outside, $\boldsymbol{\sigma} : d\boldsymbol{\varepsilon}$, is not stored in the system as energy that can be recovered later on to do work. It is dissipated into heat form. In the ideal plasticity case, this dissipation is the plastic work, $d\mathcal{D} = dW^p$. A generalization of (7.11) to three dimension reads

$$d\mathcal{D} = dW^p = \boldsymbol{\sigma} : d\boldsymbol{\varepsilon}^p \geq 0 \tag{8.30}$$

For purposes of illustration, consider the deviator stress state $\boldsymbol{\sigma} = \mathbf{s} = \tau[\mathbf{n} \otimes \mathbf{t} + \mathbf{t} \otimes \mathbf{n}]$ prescribed on a Von–Mises material. In this shear test, the flow rule is given by (8.16). Substitution in (8.30) delivers the plastic work:

$$d\mathcal{D} = \boldsymbol{\sigma} : d\boldsymbol{\varepsilon}^p = d\lambda \frac{\tau}{2|\tau|} \boldsymbol{\sigma} : (\mathbf{n} \otimes \mathbf{t} + \mathbf{t} \otimes \mathbf{n}) = d\lambda\, |\tau| \geq 0 \tag{8.31}$$

or, equivalently, replacing $|\tau|$ by its value on the yield surface, $f = 0 \Leftrightarrow |\tau| = k$,

$$d\mathcal{D} = \boldsymbol{\sigma} : d\boldsymbol{\varepsilon}^p = d\lambda k \geq 0 \tag{8.32}$$

with $d\lambda = |d\theta^p(\mathbf{n}, \mathbf{t})| \geq 0$. This is the same expression as derived in Section 7.1.3. The dissipation in the 3D case remains associated with frictional sliding, and the nonnegativity of the dissipation in the model is satisfied by a nonnegative friction strength $k \geq 0$.

As a second example, consider the mixed shear-mean stress state $\boldsymbol{\sigma} = \tau[\mathbf{n} \otimes \mathbf{t} + \mathbf{t} \otimes \mathbf{n}] + \sigma_m \mathbf{1}$ prescribed on a Drucker–Prager material. Use of the Drucker–Prager flow rule (8.26) in (8.30) yields

$$d\mathcal{D} = \boldsymbol{\sigma} : d\boldsymbol{\varepsilon}^p = d\lambda \boldsymbol{\sigma} : \left(\frac{\tau}{2|\tau|} (\mathbf{n} \otimes \mathbf{t} + \mathbf{t} \otimes \mathbf{n}) + \frac{1}{3} \delta \mathbf{1} \right) = d\lambda \left(|\tau| + \delta \sigma_m \right) \geq 0 \tag{8.33}$$

Consider now the case of associated plasticity, such that the loading function substitutes for the plastic potential (i.e., $f = g \Leftrightarrow \alpha = \delta$). Replacing $|\tau| + \alpha\sigma_m$ by the value on the loading surface, $f = 0 \Leftrightarrow |\tau| + \alpha\sigma_m = c$, the dissipation is due to shear sliding involving cohesion c:

$$d\mathcal{D} = \boldsymbol{\sigma} : d\boldsymbol{\varepsilon}^p = d\lambda c \geq 0 \Leftrightarrow c \geq 0 \tag{8.34}$$

By contrast, in the case of a nonassociated flow rule, $\alpha \neq \delta$, substitution of the yield function $f = 0 \Leftrightarrow \sigma_m = \frac{1}{\alpha} \left(c - |\tau| \right)$ in (8.33) yields

$$d\mathcal{D} = \boldsymbol{\sigma} : d\boldsymbol{\varepsilon}^p = d\lambda \left(|\tau| \left(1 - \frac{\delta}{\alpha} \right) + \frac{\delta}{\alpha} c \right) \geq 0 \tag{8.35}$$

For $\delta/\alpha = 1$, we recover (8.34), (i.e., a pure frictional dissipative mechanism). On the other hand, if $\delta/\alpha \neq 1$, the dissipative mechanism is changed because of the plastic dilating behavior. Since the cohesion c is always greater than or equal to zero, the nonnegativity of the dissipation restricts possible values of the dilatation coefficient; in fact, for a zero shear stress $\delta/\alpha \geq 0$, and for a zero cohesion $\delta/\alpha \leq 1$; thus

$$d\mathcal{D} = \boldsymbol{\sigma} : d\boldsymbol{\varepsilon}^p \geq 0 \Leftrightarrow 0 \leq \delta/\alpha \leq 1 \tag{8.36}$$

In other words, the nonnegativity of the intrinsic dissipation restricts the kinematics of plastic deformation to plastic dilatation [i.e., $0 \leq \mathrm{tr}\,(d\boldsymbol{\varepsilon}^p) = \delta|d\theta^p| \leq \alpha|d\theta^p|$]. This is shown in Figure 8.5b: $\delta = 0$ corresponds to a zero plastic dilating behavior, and $\delta = \alpha$ to a plastic flow following the normality rule. The result shows, from an energy point of view, that the normality rule is not a dogma. The plastic flow may well take any other direction, provided that the plastic work remains greater than or equal to zero. In the case of the Drucker–Prager plasticity model, this is ensured by a dilatancy coefficient that is greater than or equal to zero, and smaller than the friction coefficient of the material.

Remark The required nonnegativity of the dilatancy behavior may well be understood physically if we reset the Drucker–Prager model in the context of a Mohr–Coulomb criterion on a material surface. Dilatation on a surface means that the plastic sliding of the surface leads to an opening of the surface. A plastic contracting behavior is not permitted, since it would correspond to the physically unacceptable situation of an irreversible penetration of one surface into the other.

We want to close these preliminary energy considerations with a precise calculation of the principle of maximum plastic work. A generalization of (7.12) to three dimensions reads

$$\forall f(\boldsymbol{\sigma}) = 0; \quad \forall f(\boldsymbol{\sigma}') \leq 0 : (\boldsymbol{\sigma} - \boldsymbol{\sigma}') : d\boldsymbol{\varepsilon}^p \geq 0 \tag{8.37}$$

Inequality (8.37) states that the plastic work (8.30) of stress $\boldsymbol{\sigma}$ along the plastic strain increment $d\boldsymbol{\varepsilon}^p$ delivers the maximum plastic work in comparison with any other plastically admissible stress state $\boldsymbol{\sigma}'$ situated within or on the boundary of the elasticity domain. This principle of maximum plastic work can be verified (e.g., for the Von–Mises plasticity model). For purposes of illustration, consider the shear test, for which $\boldsymbol{\sigma} : d\boldsymbol{\varepsilon}^p = d\lambda\,|\tau|$ [i.e., 8.32] and $\boldsymbol{\sigma}' : d\boldsymbol{\varepsilon}^p = d\lambda\tau'\mathrm{sign}\,(\tau)$; thus,

$$(\boldsymbol{\sigma} - \boldsymbol{\sigma}') : d\boldsymbol{\varepsilon}^p = d\lambda\,(\tau - \tau')\,\mathrm{sign}\,(\tau) \geq 0 \tag{8.38}$$

which is satisfied for $(\tau - \tau')\,\mathrm{sign}\,(\tau) \geq 0$ (i.e., provided the convexity of the loading function). The principle of maximum plastic work (8.37) combines the two elements of ideal plasticity: convexity of the elasticity domain (8.3), and normality of the flow rule (8.5). While the first condition is a necessary condition in that all plastically admissible stress states are enclosed in the elasticity domain, the second is an additional constraint restricting the directions of the plastic flow. It does not capture the case of a nonassociated flow rule (e.g., the case of the Drucker–Prager plasticity model with $\alpha \neq \delta$).

8.2 THERMODYNAMICS OF 3D HARDENING PLASTICITY

We want to set the preliminary energy considerations in the larger framework of thermodynamics of irreversible processes and to generalize the ideal plasticity model to 3D hardening plasticity.

8.2.1 Elements of 3D Thermodynamics of Irreversible Processes

We have seen that the 1D plasticity model of Chapter 7 generalizes in stress space by replacing the scalar stress σ by the second order stress tensor $\boldsymbol{\sigma}$. In a similar way, we can proceed to generalize the 1D energy approach of hardening plasticity to 3D (have a quick look back at Section 7.2.4). We need the following:

1. The first law,

$$dU = dW + dQ = \boldsymbol{\sigma} : d\boldsymbol{\varepsilon} + dQ \tag{8.39}$$

2. The second law,

$$\theta dS \geq dQ \tag{8.40}$$

3. The expression of the free energy of the hardening material:

$$\psi(\theta, \boldsymbol{\varepsilon}, \boldsymbol{\varepsilon}^p, \boldsymbol{\chi}) = \psi^{\text{el}}(\theta, \boldsymbol{\varepsilon}, \boldsymbol{\varepsilon}^p) + \mathcal{U}(\boldsymbol{\chi}) = U - \theta S \tag{8.41}$$

We recall that the Helmholtz energy $\psi^{\text{el}} = \psi^{\text{el}}(T, \boldsymbol{\varepsilon}, \boldsymbol{\varepsilon}^p, \boldsymbol{\chi})$ expresses the available energy stored in the elementary material volume $d\Omega$ to do work. The energy states are expressed in terms of state variables. In the case of the thermoelasto plastic material with plastic hardening, these are the temperature θ and the strain tensor $\boldsymbol{\varepsilon}$ as external variables; and the plastic variables, the plastic strain tensor $\boldsymbol{\varepsilon}^p$ and the hardening variable $\boldsymbol{\chi}$, as internal state variables. In addition, the total free energy is separated in two terms, the thermoelastic free energy $\psi^{\text{el}}(\theta, \boldsymbol{\varepsilon}, \boldsymbol{\varepsilon}^p)$, and the hardening energy $\mathcal{U}(\boldsymbol{\chi})$. The hardening energy corresponds to the elastic energy stored (or frozen) at a microlevel.

Finally, a combination of the first law (8.39) and the expression of the Helmholtz energy (8.41) leads to the Clausius–Duhem inequality:

$$\theta dS = \underbrace{\boldsymbol{\sigma} : d\boldsymbol{\varepsilon} - S d\theta - d(\psi^{\text{el}} + \mathcal{U})}_{d\mathcal{D} \equiv \varphi dt \geq 0} + dQ \tag{8.42}$$

The second law (8.40) and Eq. (8.42) lead to the expression of the intrinsic dissipation $\varphi dt \geq 0$. It is the amount of energy that is dissipated into heat form:

$$\varphi dt = \boldsymbol{\sigma} : d\boldsymbol{\varepsilon} - S dT - d(\psi^{\text{el}} + \mathcal{U}) \geq 0 \tag{8.43}$$

Thus, with the variable set (8.41),[4]

$$\varphi dt = \left(\boldsymbol{\sigma} - \frac{\partial \psi^{\text{el}}}{\partial \boldsymbol{\varepsilon}}\right) : d\boldsymbol{\varepsilon} - \left(S + \frac{\partial \psi^{\text{el}}}{\partial \theta}\right) d\theta - \frac{\partial \psi^{\text{el}}}{\partial \boldsymbol{\varepsilon}^p} : d\boldsymbol{\varepsilon}^p - \frac{\partial \mathcal{U}}{\partial \boldsymbol{\chi}} \cdot d\boldsymbol{\chi} \geq 0 \tag{8.44}$$

For thermoelastic evolutions, $\varphi dt = 0$ and $d\boldsymbol{\varepsilon}^p = d\boldsymbol{\chi} = 0$; Eq. (8.44) delivers the state equations:

$$\boldsymbol{\sigma} = \left(\frac{\partial \psi^{\text{el}}}{\partial \boldsymbol{\varepsilon}}\right)_{\theta, \boldsymbol{\varepsilon}^p, \boldsymbol{\chi}}; \quad S = -\left(\frac{\partial \psi^{\text{el}}}{\partial \theta}\right)_{\boldsymbol{\varepsilon}, \boldsymbol{\varepsilon}^p, \boldsymbol{\chi}} \tag{8.45}$$

---

[4] At this stage, we do not restrict ourselves to a specific tensorial order of the hardening strain $\boldsymbol{\chi}$. It may be a scalar (*0*-order tensor), a vector (*first*-order tensor), or a *second*-order tensor. As we will see later, the tensorial order of the hardening strain defines different hardening models.

These state equations still hold when plastic deformation occur; thus

$$\varphi dt = dW^p - d\mathcal{U} = \boldsymbol{\sigma} : d\boldsymbol{\varepsilon}^p + \boldsymbol{\zeta} \cdot d\boldsymbol{\chi} \geq 0 \tag{8.46}$$

In (8.46), we designated stress $\boldsymbol{\sigma}$ as the driving force of plastic strain $d\boldsymbol{\varepsilon}^p$, and hardening force $\boldsymbol{\zeta}$ as the driving force of plastic hardening or softening $d\boldsymbol{\chi}$. This is the choice characterizing plasticity models:

$$\boldsymbol{\sigma} = -\frac{\partial \psi^{\text{el}}}{\partial \boldsymbol{\varepsilon}^p}; \quad \boldsymbol{\zeta} = -\frac{\partial \mathcal{U}}{\partial \boldsymbol{\chi}} \tag{8.47}$$

Finally, $(8.45)_1$ and $(8.47)_1$ imply the following remarkable Maxwell symmetries:

$$C_{ijkl} = \frac{\partial \sigma_{ij}}{\partial \varepsilon_{kl}} = -\frac{\partial \sigma_{ij}}{\partial \varepsilon^p_{kl}} = \frac{\partial^2 \psi^{\text{el}}}{\partial \varepsilon_{ij} \partial \varepsilon_{kl}} = -\frac{\partial^2 \psi^{\text{el}}}{\partial \varepsilon^p_{ij} \partial \varepsilon_{kl}} = \frac{\partial^2 \psi^{\text{el}}}{\partial \varepsilon^p_{ij} \partial \varepsilon^p_{kl}} = C_{klij} \tag{8.48}$$

8.2.2 Linear and Isotropic Thermoelastoplasticity

The constitutive equations of a linear thermoelastoplastic material are obtained from the expression of the linear thermoelastic one developed in Section 5.2.3. It suffices to replace the total strain $\boldsymbol{\varepsilon}$ by the elastic strain $\boldsymbol{\varepsilon} - \boldsymbol{\varepsilon}^p$. For instance, the expression of the thermoelastic free energy is

$$\begin{aligned} \psi^{\text{el}} &= \psi^{\text{el}}(\boldsymbol{\varepsilon} - \boldsymbol{\varepsilon}^p, \theta = T - T_0) \\ &= \frac{1}{2}(\boldsymbol{\varepsilon} - \boldsymbol{\varepsilon}^p) : \mathbf{C} : (\boldsymbol{\varepsilon} - \boldsymbol{\varepsilon}^p) - \mathbf{C} : \boldsymbol{\alpha} : (\boldsymbol{\varepsilon} - \boldsymbol{\varepsilon}^p)\theta - \frac{\mathcal{C}\theta^2}{2T_0} \end{aligned} \tag{8.49}$$

Use in (8.45) and (8.46) yields

$$\boldsymbol{\sigma} = \boldsymbol{\sigma}(\boldsymbol{\varepsilon} - \boldsymbol{\varepsilon}^p, \theta) = \frac{\partial \psi^{\text{el}}}{\partial \boldsymbol{\varepsilon}} = -\frac{\partial \psi^{\text{el}}}{\partial \boldsymbol{\varepsilon}^p} = \mathbf{C} : (\boldsymbol{\varepsilon} - \boldsymbol{\varepsilon}^p) - \mathbf{C} : \boldsymbol{\alpha}\theta \tag{8.50}$$

$$S = S(\boldsymbol{\varepsilon} - \boldsymbol{\varepsilon}^p, \theta) = -\frac{\partial \psi^{\text{el}}}{\partial T} = \frac{\mathcal{C}\theta}{T_0} + \mathbf{C} : \boldsymbol{\alpha} : (\boldsymbol{\varepsilon} - \boldsymbol{\varepsilon}^p) \tag{8.51}$$

The material parameters preserve the same physical significance as in the elastic case.

8.2.3 Inversion of State Equations

We want to invert the linear thermoelastoplastic stress equation of state. We have already seen in Section 6.3.1 the dual form (or Legendre–Fenchel transform) of the energy ψ^{el}: the complementary energy ψ^*. Extending this concept to elastoplasticity, we have

$$\psi^{\text{el}}(\boldsymbol{\varepsilon} - \boldsymbol{\varepsilon}^p, \theta) + \psi^*(\boldsymbol{\sigma}, \theta) = \boldsymbol{\sigma} : (\boldsymbol{\varepsilon} - \boldsymbol{\varepsilon}^p) \tag{8.52}$$

Expression (8.52) makes use of the fact that stress $\boldsymbol{\sigma}$ is both the driving force of the total strain $\boldsymbol{\varepsilon}$, and of the plastic strain $\boldsymbol{\varepsilon}^p$. This dual energy role of stress $\boldsymbol{\sigma}$ allows us to reduce the number of state variables in the complementary energy $\psi^*(\boldsymbol{\sigma}, \theta)$ [i.e., 2 instead of 3 in $\psi^{\text{el}} = \psi^{\text{el}}(\boldsymbol{\varepsilon} - \boldsymbol{\varepsilon}^p, \theta)$]. Use of (8.52) in (8.42) yields

$$\theta dS = \underbrace{\boldsymbol{\sigma} : d\boldsymbol{\varepsilon} - Sd\theta - d[\boldsymbol{\sigma} : (\boldsymbol{\varepsilon} - \boldsymbol{\varepsilon}^p) - \psi^*(\boldsymbol{\sigma}, \theta) + \mathcal{U}]}_{d\mathcal{D} \equiv \varphi dt \geq 0} + dQ \tag{8.53}$$

Thus,

$$\begin{aligned} \varphi dt &= \boldsymbol{\sigma} : d\boldsymbol{\varepsilon}^p - (\boldsymbol{\varepsilon} - \boldsymbol{\varepsilon}^p) : d\boldsymbol{\sigma} - S d\theta + d\psi^*(\boldsymbol{\sigma}, \theta) - d\mathcal{U} \\ &= \boldsymbol{\sigma} : d\boldsymbol{\varepsilon}^p - \frac{\partial \mathcal{U}}{\partial \boldsymbol{\chi}} \cdot d\boldsymbol{\chi} - \left((\boldsymbol{\varepsilon} - \boldsymbol{\varepsilon}^p) - \frac{\partial \psi^*}{\partial \boldsymbol{\sigma}} \right) d\boldsymbol{\sigma} - \left(S - \frac{\partial \psi^*}{\partial \theta} \right) d\theta \geq 0 \end{aligned} \tag{8.54}$$

Finally, a comparison of (8.54) with (8.46) yields

$$\boldsymbol{\varepsilon} - \boldsymbol{\varepsilon}^p = \frac{\partial \psi^*}{\partial \boldsymbol{\sigma}}; \quad S = \frac{\partial \psi^*}{\partial \theta} \tag{8.55}$$

In the linear case, the complementary energy for isothermal conditions is given by (6.66). Extending this expression to (linearized) non-isothermal situations, we have

$$\psi^*(\boldsymbol{\sigma}, T) = \frac{1}{2} \boldsymbol{\sigma} : \boldsymbol{\Lambda} : \boldsymbol{\sigma} + \boldsymbol{\alpha} : \boldsymbol{\sigma} \theta + \frac{\mathcal{C}_\sigma \theta^2}{2T_0} \tag{8.56}$$

Use in (8.55) yields

$$\boldsymbol{\varepsilon} - \boldsymbol{\varepsilon}^p = \frac{\partial \psi^*}{\partial \boldsymbol{\sigma}} = \boldsymbol{\Lambda} : \boldsymbol{\sigma} + \boldsymbol{\alpha} \theta \tag{8.57}$$

$$S = \frac{\partial \psi^*}{\partial \theta} = \frac{\mathcal{C}_\sigma \theta}{T_0} + \boldsymbol{\alpha} : \boldsymbol{\sigma} \tag{8.58}$$

where $\boldsymbol{\Lambda} = \mathbf{C}^{-1}$ is the elastic compliance tensor (have a look back at Section 5.3.4):

$$\Lambda_{ijkl} = C^{-1}_{ijkl} = \frac{\partial^2 \psi^*}{\partial \sigma_{ij} \partial \sigma_{kl}} \tag{8.59}$$

$\mathcal{C}_\sigma$ is the volume heat capacity of the material under stress-free conditions; while the last term in (8.58) expresses the (reversible) latent heat related to stress application.

Exercise 75. Derive the state equations of an isotropic thermoelastoplastic material,

1. From the elastic strain energy;
2. From the complimentary energy.

In the isotropic case, the thermoelastic strain energy (or reduced potential, as it excludes the frozen energy) reads as a function of the elastic strain invariants (have a quick look back at Section 5.3). Replacing in (5.41) the total strain by $\boldsymbol{\varepsilon} - \boldsymbol{\varepsilon}^p$ yields

$$\begin{aligned} \psi^{\text{el}} &= \psi^{\text{el}}(\boldsymbol{\varepsilon} - \boldsymbol{\varepsilon}^p, \theta = T - T_0) \\ &= \frac{\lambda}{2} \left[\text{tr}\,(\boldsymbol{\varepsilon} - \boldsymbol{\varepsilon}^p) \right]^2 + G \text{tr} \left[(\boldsymbol{\varepsilon} - \boldsymbol{\varepsilon}^p)^2 \right] - (3\lambda + 2G)\alpha \left[\text{tr}\,(\boldsymbol{\varepsilon} - \boldsymbol{\varepsilon}^p) \right] \theta - \frac{\mathcal{C}\theta^2}{2T_0} \end{aligned}$$

Substitution in (8.45) and (8.46) yields

$$\boldsymbol{\sigma} = \frac{\partial \psi^{\text{el}}}{\partial \boldsymbol{\varepsilon}} = -\frac{\partial \psi^{\text{el}}}{\partial \boldsymbol{\varepsilon}^p} = \lambda \text{tr}\,(\boldsymbol{\varepsilon} - \boldsymbol{\varepsilon}^p)\mathbf{1} + 2G(\boldsymbol{\varepsilon} - \boldsymbol{\varepsilon}^p) - (3\lambda + 2G)\alpha\theta\mathbf{1}$$

$$S = -\frac{\partial \psi^{\mathrm{el}}}{\partial \theta} = \frac{C\theta}{T_0} + (3\lambda + 2G)\alpha \mathrm{tr}\,(\boldsymbol{\varepsilon} - \boldsymbol{\varepsilon}^p)$$

with λ, G the Lamé constants, and α the thermal dilatation coefficient.
In the isotropic case, the complementary energy reads as a function of the stress invariants. For the isothermal case, the expression of the complementary energy was given by (6.71). For (linearized) non-isothermal evolutions, it reads

$$\psi^*(\boldsymbol{\sigma}, \theta) = \psi^*(\boldsymbol{\sigma}) = -\frac{\nu}{2E}(\mathrm{tr}\,\boldsymbol{\sigma})^2 + \frac{1+\nu}{2E}\mathrm{tr}\,(\boldsymbol{\sigma} \cdot \boldsymbol{\sigma}) + \alpha(\mathrm{tr}\,\boldsymbol{\sigma})\theta + \frac{C_\sigma \theta^2}{2T_0}$$

with E the Young's modulus and ν the Poisson ratio. Use in (8.57) gives

$$\boldsymbol{\varepsilon} - \boldsymbol{\varepsilon}^p = \frac{\partial \psi^*}{\partial \boldsymbol{\sigma}} = -\frac{\nu}{E}(\mathrm{tr}\,\boldsymbol{\sigma})\mathbf{1} + \frac{1+\nu}{2E}\boldsymbol{\sigma} + \alpha\theta\mathbf{1}$$

$$S = \frac{\partial \psi^*}{\partial T} = \frac{C_\sigma \theta}{T_0} + \alpha \mathrm{tr}\,\boldsymbol{\sigma}$$

Note that $C_\sigma \neq C$. C is the volume heat capacity measured in a strain free experiment; $C_\sigma = C + T_0\mathbf{C} : \boldsymbol{\alpha} : \boldsymbol{\alpha}$ is the volume heat capacity measured under stress-free conditions. We have encountered this difference in Section 5.2.2, when dealing with 1D thermoelastic properties. ■

8.2.4 Complementary Evolution Laws: Flow Rule and Hardening Rule

The two components of the dissipation of the hardening plasticity model, (8.46), are the plastic work $dW^p = \boldsymbol{\sigma} : d\boldsymbol{\varepsilon}^p$ and the hardening–softening work $-d\mathcal{U} = \boldsymbol{\zeta} \cdot d\boldsymbol{\chi}$. The driving force of the first is stress $\boldsymbol{\sigma}$; the driving force of the second is the hardening force $\boldsymbol{\zeta}$. This formal (thermodynamic) identification is used to design appropriate complementary evolution laws of $d\boldsymbol{\varepsilon}^p$ and $d\boldsymbol{\chi}$ (of which the evolution cannot be controlled from the outside):

$$d\boldsymbol{\varepsilon}^p = d\lambda \mathcal{G}(\boldsymbol{\sigma}); \quad d\boldsymbol{\chi} = d\lambda \mathcal{H}(\boldsymbol{\zeta}) \tag{8.60}$$

In (8.60), $\mathcal{G}(\boldsymbol{\sigma})$ and $\mathcal{H}(\boldsymbol{\zeta})$ are two (still to be specified) functions that define the orientation of the plastic strain and the hardening strain. The plastic evolution laws need to satisfy the fundamental inequality (8.46):

$$\varphi dt = d\lambda \left(\boldsymbol{\sigma} : \mathcal{G}(\boldsymbol{\sigma}) + \boldsymbol{\zeta} \cdot \mathcal{H}(\boldsymbol{\zeta})\right) \geq 0 \tag{8.61}$$

The Hypothesis of Maximum Plastic Work in Hardening Plasticity Models.

We have already seen that the normality rule, $g(\boldsymbol{\sigma}) = f(\boldsymbol{\sigma})$, in ideal plasticity automatically satisfies the second law. This is due to the convexity of the loading function. The normality rule and convexity constitute the principle of maximum plastic work (8.37). This concept extends to hardening plasticity, along the same lines as developed in Section 7.2.3 for the 1D case: Consider two plastically admissible loading points $(\boldsymbol{\sigma}, \boldsymbol{\zeta})$ and $(\boldsymbol{\sigma}', \boldsymbol{\zeta}')$ on a convex loading function $f(\boldsymbol{\sigma}, \boldsymbol{\zeta}) = f(\boldsymbol{\sigma}', \boldsymbol{\zeta}') = 0$. The principle of maximum plastic dissipation needs to be applied to the total dissipation $\varphi dt = dW^p - d\mathcal{U}$ (i.e., to both the plastic work $dW^p = \boldsymbol{\sigma} : d\boldsymbol{\varepsilon}^p$ and the hardening work $-d\mathcal{U} = \boldsymbol{\zeta} \cdot d\boldsymbol{\chi}$):

$$(\boldsymbol{\sigma} - \boldsymbol{\sigma}') : d\boldsymbol{\varepsilon}^p + (\boldsymbol{\zeta} - \boldsymbol{\zeta}') \cdot d\boldsymbol{\chi} \geq 0 \tag{8.62}$$

Two conditions are involved in (8.62):

1. The directions taken by $d\boldsymbol{\varepsilon}^p$ and $d\boldsymbol{\chi}$ follow the normality rule:

$$d\boldsymbol{\varepsilon}^p = d\lambda \mathcal{G}(\boldsymbol{\sigma}, \boldsymbol{\zeta}) = d\lambda \frac{\partial f(\boldsymbol{\sigma}, \boldsymbol{\zeta})}{\partial \boldsymbol{\sigma}} \tag{8.63}$$

$$d\boldsymbol{\chi} = d\lambda \mathcal{H}(\boldsymbol{\sigma}, \boldsymbol{\zeta}) = d\lambda \frac{\partial f(\boldsymbol{\sigma}, \boldsymbol{\zeta})}{\partial \boldsymbol{\zeta}} \tag{8.64}$$

2. The loading function of hardening plasticity, $f(\boldsymbol{\sigma}, \boldsymbol{\zeta}) \leq 0$, is a convex function of its arguments,

$$(\boldsymbol{\sigma} - \boldsymbol{\sigma}') : \frac{\partial f}{\partial \boldsymbol{\sigma}} + (\boldsymbol{\zeta} - \boldsymbol{\zeta}') \cdot \frac{\partial f}{\partial \boldsymbol{\zeta}} \geq 0 \tag{8.65}$$

Provided the general normality rule of the involved plastic plus hardening dissipative mechanisms and the convexity of the loading function, we verify the principle of maximum plastic work (8.62) and satisfy at the same time the second law (8.61). Matter, according to this principle, exhausts its energy reserve at the highest rate. Given, however, the possibility that the plastic strains may take different directions than foreseen by the normality rules (8.63) and (8.64), the principle of maximum plastic work may rather be considered a hypothesis than a principle.

Non-Associated Flow Rule and Non-Associated Hardening Rule.

The only restriction on the plastic evolution laws is the nonnegativity of the intrinsic dissipation (8.61). We have already seen that the non-associated flow rule (8.29) deviates from the normality rule (i.e., $\partial g/\partial \boldsymbol{\sigma} \neq \partial f/\partial \boldsymbol{\sigma}$). Analogously, the hardening rule may as well deviate from the normality rule (8.64); hence

$$d\boldsymbol{\varepsilon}^p = d\lambda \mathcal{G}(\boldsymbol{\sigma}, \boldsymbol{\zeta}) = d\lambda \frac{\partial g(\boldsymbol{\sigma}, \boldsymbol{\zeta})}{\partial \boldsymbol{\sigma}}; \quad d\boldsymbol{\chi} = d\lambda \mathcal{H}(\boldsymbol{\sigma}, \boldsymbol{\zeta}) = d\lambda \frac{\partial h(\boldsymbol{\sigma}, \boldsymbol{\zeta})}{\partial \boldsymbol{\zeta}} \tag{8.66}$$

h is the hardening potential. Functions g and h are chosen to meet the experimental evidence of plastic and hardening deformation. The only thermodynamic restriction they must satisfy is (8.61), which reads here

$$\varphi dt = d\lambda \left(\boldsymbol{\sigma} : \frac{\partial g(\boldsymbol{\sigma}, \boldsymbol{\zeta})}{\partial \boldsymbol{\sigma}} + \boldsymbol{\zeta} \cdot \frac{\partial h(\boldsymbol{\sigma}, \boldsymbol{\zeta})}{\partial \boldsymbol{\zeta}} \right) \geq 0 \tag{8.67}$$

8.3 HARDENING PLASTICITY MODELS

8.3.1 Loading Function and Consistency Condition

The elasticity domain of hardening plastic materials depends on both stress $\boldsymbol{\sigma}$ and hardening force $\boldsymbol{\zeta}$:

$$\boldsymbol{\sigma} \in D_E \leftrightarrow f = f(\boldsymbol{\sigma}, \boldsymbol{\zeta}) \leq 0 \tag{8.68}$$

In addition, during plastic loading, a loading point $\boldsymbol{\sigma}$ that is on the surface remains on it, $f(\boldsymbol{\sigma}, \boldsymbol{\zeta}) = 0$, while carrying it along. This is expressed by the consistency condition:

$$df = \frac{\partial f}{\partial \boldsymbol{\sigma}} : d\boldsymbol{\sigma} + \frac{\partial f}{\partial \boldsymbol{\zeta}} \cdot d\boldsymbol{\zeta} = 0 \tag{8.69}$$

The second term in Eq. (8.69) represents the plastic hardening contribution. It ensures that the stress $\boldsymbol{\sigma} + d\boldsymbol{\sigma}$ remains on the loading surface, by expanding or shrinking the elasticity domain. The first case is referred to as positive hardening, the second as negative hardening or softening. This sign of plastic hardening is incorporated in the plastic hardening modulus H:

$$-\frac{\partial f}{\partial \boldsymbol{\zeta}} \cdot d\boldsymbol{\zeta} = d\lambda H \tag{8.70}$$

From an energy point of view, hardening-softening is associated with the frozen energy $\mathcal{U}(\boldsymbol{\chi})$, attributed to a state of microelastic deformation. Indeed, $-d\mathcal{U} = \boldsymbol{\zeta} \cdot d\boldsymbol{\chi}$ is the part of the plastic work $dW^p = \boldsymbol{\sigma} : d\boldsymbol{\varepsilon}^p$, which is not dissipated into heat form but stored (or frozen) at the microlevel of the heterogeneous material. Use of (8.47) in an incremental form together with the hardening rule $(8.66)_2$ allows us to develop (8.70) in the following form:

$$d\lambda H = -\frac{\partial f}{\partial \boldsymbol{\zeta}} \cdot \left(\frac{d\boldsymbol{\zeta}}{d\boldsymbol{\chi}}\right) \cdot d\boldsymbol{\chi} = \frac{\partial f}{\partial \boldsymbol{\zeta}} \cdot \frac{\partial^2 \mathcal{U}(\boldsymbol{\chi})}{\partial \boldsymbol{\chi}^2} \cdot \left(d\lambda \frac{\partial h}{\partial \boldsymbol{\zeta}}\right) \tag{8.71}$$

We so arrive at an appropriate definition of the hardening modulus H, related to the frozen energy $\mathcal{U}(\boldsymbol{\chi})$:

$$H = \frac{\partial f}{\partial \boldsymbol{\zeta}} \cdot \frac{\partial^2 \mathcal{U}(\boldsymbol{\chi})}{\partial \boldsymbol{\chi}^2} \cdot \frac{\partial h}{\partial \boldsymbol{\zeta}} = \left(\frac{\partial f}{\partial \boldsymbol{\zeta}} \otimes \frac{\partial h}{\partial \boldsymbol{\zeta}}\right) : \frac{\partial^2 \mathcal{U}(\boldsymbol{\chi})}{\partial \boldsymbol{\chi}^2} \tag{8.72}$$

In addition, use of (8.70) in the consistency condition (8.69) delivers the plastic multiplier in the form:

$$d\lambda = \frac{1}{H} \frac{\partial f}{\partial \boldsymbol{\sigma}} : d\boldsymbol{\sigma} = \frac{1}{H} \frac{\partial f}{\partial \sigma_{kl}} d\sigma_{kl} \tag{8.73}$$

Finally, substitution of (8.73) in the plastic flow rule $(8.66)_1$ gives

$$d\boldsymbol{\varepsilon}^p = d\lambda \frac{\partial g}{\partial \boldsymbol{\sigma}} = \frac{1}{H} \left(\frac{\partial g}{\partial \boldsymbol{\sigma}} \otimes \frac{\partial f}{\partial \boldsymbol{\sigma}}\right) : d\boldsymbol{\sigma} \tag{8.74}$$

$$d\varepsilon_{ij}^p = d\lambda \frac{\partial g}{\partial \sigma_{ij}} = \frac{1}{H} \left(\frac{\partial g}{\partial \sigma_{ij}} \frac{\partial f}{\partial \sigma_{kl}}\right) d\sigma_{kl} \tag{8.75}$$

8.3.2 Isotropic and Kinematic Hardening

We are left with setting up the plastic hardening model. This requires two steps:

1. Determination of the hardening force $\boldsymbol{\zeta}$ in the loading function (8.68) from the transformation of the elasticity domain D_E during plastic loading;
2. Determination of the associated hardening variable $\boldsymbol{\chi}$ and its development defined by the hardening rule $(8.66)_2$ during plastic loading.

With this two-step procedure, we experimentally determine the state equation, $\boldsymbol{\zeta} = \boldsymbol{\zeta}(\boldsymbol{\chi})$. Finally, through definition (8.70), we are able to establish the expression of the frozen hardening energy $\mathcal{U}(\boldsymbol{\chi})$ and can determine the plastic hardening modulus H. Last, through calorimetric measurements of the work dissipated into heat during plastic loading [i.e., (8.42) or (8.53)], we can give—at least on the paper—an experimental proof of the relevance of the previously determined state equation $\boldsymbol{\zeta} = \boldsymbol{\zeta}(\boldsymbol{\chi})$.

The hardening force describes the transformation of the strength domain in the stress space during plastic loading. We have already seen, in Section 7.2.5, the 1D version of two types of hardening models. Among all possible transformation modes of the elasticity domain, we distinguish the following:

- *Isotropic hardening*: Through hardening, the elasticity domain in stress space expands (hardening) or shrinks (softening) in an isotropic way around the center, $\boldsymbol{\sigma} = 0$. In this case, the hardening variable reduces to a single scalar ζ. Let $|\boldsymbol{\sigma}^0|$ be the norm of a loading point in the stress space, which reaches—from a virgin material state—for the first time the loading function (see Figure 8.6a)

$$f(\boldsymbol{\sigma}^0, \zeta = 0) = 0 \tag{8.76}$$

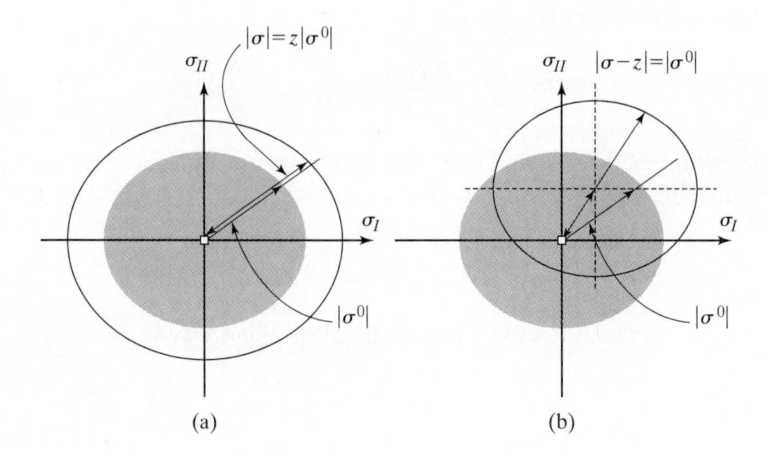

FIGURE 8.6: Common hardening models: (a) isotropic hardening; (b) kinematic hardening.

In the case of isotropic plastic hardening, this norm transforms into $|\boldsymbol{\sigma}| = z\,|\boldsymbol{\sigma}^0|$, where z denotes the experimentally measurable hardening parameter. The new loading point $\boldsymbol{\sigma}$ is still on the loading surface, which has been isotropically transformed:

$$f(\boldsymbol{\sigma} = z\boldsymbol{\sigma}^0, \zeta \neq 0) = 0 \qquad (8.77)$$

Equations (8.76) and (8.77) allow the determination of the scalar hardening force ζ. This isotropic hardening is sketched in Figure 8.6a. It is a generalization of the 1D model displayed in Figure 7.7.

- *Kinematic hardening*: Through hardening, the elasticity domain undergoes a translation in stress space, while maintaining its volume. This translation in the stress space is described by a translation of the center, $\boldsymbol{\sigma} = 0$, and the required hardening force is a second-order tensor, $\boldsymbol{\zeta} = \zeta_{ij}\mathbf{e}_i \otimes \mathbf{e}_j$. Let $|\boldsymbol{\sigma}^0|$ be again the norm of a loading point in the stress space, which reaches—from a virgin material state—for the first time the loading function

$$f(\boldsymbol{\sigma}^0, \boldsymbol{\zeta} = \mathbf{0}) = 0 \qquad (8.78)$$

During plastic loading, the norm preserves its value in the translation,

$$|\boldsymbol{\sigma} - \mathbf{z}| = |\boldsymbol{\sigma}^0| \qquad (8.79)$$

It follows that

$$f(\boldsymbol{\sigma} = \boldsymbol{\sigma}^0 + \mathbf{z}, \boldsymbol{\zeta} = -\mathbf{z}) = 0 \qquad (8.80)$$

where $\mathbf{z}$ corresponds to the measurable translation of the center in stress space. This kinematic hardening is sketched in Figure 8.6b. It is a generalization of the 1D model displayed in Figures 7.4 and 7.6.

- The two previous hardening models can be combined to yield an isotropic and kinematic hardening model.

8.3.3 Hardening Variables

To this point, we have given a mathematical description of the transformation of the elasticity domain during plastic loading. We are left with identifying the associated hardening variable, $\boldsymbol{\chi}$. From an energy point of view, the hardening force, ζ and/or $\boldsymbol{\zeta}$, is the driving force of the change in frozen energy:

$$-d\mathcal{U} = \zeta d\chi; \quad -d\mathcal{U} = \boldsymbol{\zeta} : d\boldsymbol{\chi} \qquad (8.81)$$

In the isotropic hardening case, the associated hardening variable is a scalar, and in the case of kinematic hardening it is a second-order tensor. From an experimental point of view, we need to search for the suitable irreversible strain variable, which matches (best) the development of the hardening force:

$$\zeta(\chi) = -\frac{\partial \mathcal{U}(\chi)}{\partial \chi}; \quad \boldsymbol{\zeta}(\boldsymbol{\chi}) = -\frac{\partial \mathcal{U}(\boldsymbol{\chi})}{\partial \boldsymbol{\chi}} \qquad (8.82)$$

Once the hardening variable is identified, the energy approach offers an independent means of assessing the relevance of the choice of the hardening variable. Indeed, the dissipation in the plastic hardening case is

$$d\mathcal{D} - dW^p = -d\mathcal{U} \tag{8.83}$$

The dissipation can be measured as heat in a calorimetric test, in parallel with the plastic work $dW^p = \boldsymbol{\sigma} : d\boldsymbol{\varepsilon}^p$ assessed by stress-strain measurements in the same test. The two measurements give access to the frozen energy $\mathcal{U}$ and allow for a final energy check of the choice of hardening model, hardening variable, and hardening force.

8.3.4 Example: Drucker–Prager Criterion with Isotropic and Kinematic Hardening

Isotropic Hardening.

Consider a Drucker–Prager material with isotropic hardening:

$$\boldsymbol{\sigma} \in D_E \Leftrightarrow f(\sigma_m = z\sigma_m^0, J_2 = zJ_2^0) = \sqrt{J_2} + \alpha\sigma_m - zc \leq 0 \tag{8.84}$$

where $z \in [0, 1]$ is the experimentally determined hardening parameter. The corresponding hardening force (of stress dimension) reads

$$\boldsymbol{\sigma} \in D_E \Leftrightarrow f(\sigma_m, J_2, \zeta) = \sqrt{J_2} + \alpha\sigma_m - c + \zeta \leq 0 \tag{8.85}$$

with $\zeta = c(1 - z)$.

We consider the case of an associated flow and hardening rule (i.e., $f = g = h$). The flow rule was given by (8.28):

$$d\boldsymbol{\varepsilon}^p = d\lambda \frac{\partial f(\sigma_m, J_2, \zeta)}{\partial \boldsymbol{\sigma}} = d\lambda \left[\frac{\mathbf{s}}{2\sqrt{J_2}} + \frac{1}{3}\alpha \mathbf{1} \right] \tag{8.86}$$

with

$$|d\theta^p| = d\lambda \tag{8.87}$$

In addition, the associated hardening rule (8.64) gives here

$$d\chi = d\lambda \frac{\partial f(\sigma_m, J_2, \zeta)}{\partial \zeta} = d\lambda \tag{8.88}$$

Hence, a comparison of the flow rule (8.87) and the hardening rule (8.88) reveals that the hardening variable χ associated with the isotropic hardening force $\zeta = c(1 - z)$ in (8.85) is the equivalent plastic distortion θ^p:

$$d\chi \equiv |d\theta^p| \to \chi = \theta^p = \int |d\theta^p| \tag{8.89}$$

This is a formal identification, which needs to be completed by the experimental determination of the constitutive relation:

$$\zeta = c(1 - z) = \zeta(\theta^p) = -\frac{\partial \mathcal{U}(\theta^p)}{\partial \theta^p} \tag{8.90}$$

Finally, the hardening modulus reads here

$$H = \left(\frac{\partial f}{\partial \zeta}\right)^2 \frac{\partial^2 \mathcal{U}(\theta^p)}{\partial (\theta^p)^2} = -\frac{\partial \zeta(\theta^p)}{\partial \theta^p} \tag{8.91}$$

The isotropic plastic hardening model is then complete.

Kinematic Hardening.

We consider a Drucker–Prager material with kinematic hardening (i.e., a material for which the elasticity domain undergoes a translation in stress space while maintaining its volume). This model preserves the norm of $\boldsymbol{\sigma} + \boldsymbol{\zeta}$; thus in terms of the invariants that enter the Drucker–Prager criterion,

$$\sigma'_m = \frac{1}{3}\mathrm{tr}\,(\boldsymbol{\sigma} + \boldsymbol{\zeta}) = \sigma_m + \zeta_m \tag{8.92}$$

$$J'_2 = \frac{1}{2}\mathrm{dev}\,[\boldsymbol{\sigma} + \boldsymbol{\zeta}] : \mathrm{dev}\,[\boldsymbol{\sigma} + \boldsymbol{\zeta}] \tag{8.93}$$

where $\mathrm{dev}\,[\mathbf{a}] = \mathbf{a} - \frac{1}{3}(\mathrm{tr}\,\mathbf{a})\mathbf{1}$ denotes the deviator part of the second-order tensor $\mathbf{a}$. Use of these invariants in the Drucker–Prager criterion yields

$$\boldsymbol{\sigma} \in D_E \Leftrightarrow f(\boldsymbol{\sigma}, \boldsymbol{\zeta}) = \sqrt{\frac{1}{2}\mathrm{dev}\,[\boldsymbol{\sigma} + \boldsymbol{\zeta}] : \mathrm{dev}\,[\boldsymbol{\sigma} + \boldsymbol{\zeta}]} + \alpha(\sigma_m + \zeta_m) - c \le 0 \tag{8.94}$$

In the case of an associated flow rule and an associated hardening rule, we can formally identify the hardening variable $\boldsymbol{\chi} \equiv \boldsymbol{\varepsilon}^p$ from a comparison of the flow and hardening rule:

$$d\lambda \frac{\partial f(\boldsymbol{\sigma}, \boldsymbol{\zeta})}{\partial \boldsymbol{\zeta}} = d\boldsymbol{\chi} \equiv d\boldsymbol{\varepsilon}^p = d\lambda \frac{\partial f(\boldsymbol{\sigma}, \boldsymbol{\zeta})}{\partial \boldsymbol{\sigma}} \tag{8.95}$$

Finally, the hardening modulus reads here

$$H = \frac{\partial f}{\partial \boldsymbol{\zeta}} : \frac{\partial^2 \mathcal{U}(\boldsymbol{\varepsilon}^p)}{\partial \boldsymbol{\varepsilon}^{p2}} : \frac{\partial f}{\partial \boldsymbol{\zeta}} = \left(\frac{\partial f}{\partial \boldsymbol{\zeta}} \otimes \frac{\partial f}{\partial \boldsymbol{\zeta}}\right) :: \frac{\partial^2 \mathcal{U}(\boldsymbol{\varepsilon}^p)}{\partial \boldsymbol{\varepsilon}^{p2}} \tag{8.96}$$

where $\mathcal{U}(\boldsymbol{\varepsilon}^p)$ is the frozen energy as a function of the hardening variable $\boldsymbol{\chi} = \boldsymbol{\varepsilon}^p$. This function needs to be determined experimentally. Once function $\mathcal{U}(\boldsymbol{\varepsilon}^p)$ is specified, the model is complete.

8.3.5 Incremental State Equation: Tangent Modulus

Last, we seek for the 3D counterpart of the tangent modulus E_T of the 1D model [i.e., Eq. (7.35)]; that is, the tangent operator that relates the stress increment $d\boldsymbol{\sigma}$ to the increments of external variables $d\boldsymbol{\varepsilon}$ and $d\theta$.

The starting point is the consistency condition (8.69). In a first step, we replace the stress increment $d\boldsymbol{\sigma}$ by the incremental state equation (8.50):

$$\begin{aligned} df &= \frac{\partial f}{\partial \boldsymbol{\sigma}} : d\boldsymbol{\sigma} - H d\lambda \\ &= \frac{\partial f}{\partial \boldsymbol{\sigma}} : \mathbf{C} : (d\boldsymbol{\varepsilon} - d\boldsymbol{\varepsilon}^p - \boldsymbol{\alpha} d\theta) - H d\lambda = 0 \end{aligned} \tag{8.97}$$

Use of the flow rule $(8.66)_1$ in (8.97) yields the plastic multiplier:

$$d\lambda = \frac{\mathbf{f}_{,\sigma} : \mathbf{C}}{H + \mathbf{f}_{,\sigma} : \mathbf{C} : \mathbf{g}_{,\sigma}} : (d\boldsymbol{\varepsilon} - \boldsymbol{\alpha} dT) \tag{8.98}$$

where, for convenience, we let $\mathbf{f}_{,\sigma} = \partial f / \partial \boldsymbol{\sigma}$, $\mathbf{g}_{,\sigma} = \partial g / \partial \boldsymbol{\sigma}$. Substitution of (8.98) in the flow rule and in the incremental state equation gives

$$\begin{aligned} d\boldsymbol{\sigma} &= \mathbf{C} : (d\boldsymbol{\varepsilon} - \boldsymbol{\alpha} dT) - d\lambda \mathbf{C} : \mathbf{g}_{,\sigma} \\ &= \underbrace{\left[\mathbf{C} - \frac{(\mathbf{C} : \mathbf{g}_{,\sigma}) \otimes (\mathbf{f}_{,\sigma} : \mathbf{C})}{H + \mathbf{f}_{,\sigma} : \mathbf{C} : \mathbf{g}_{,\sigma}} \right]}_{\mathbf{C}^T} : (d\boldsymbol{\varepsilon} - \boldsymbol{\alpha} dT) \end{aligned} \tag{8.99}$$

The fourth-order tensor $\mathbf{C}^T$ is called the (isothermal) constitutive elastoplastic tangent stiffness tensor. We will not insist much on this tangent operator. We should, however, note that $\mathbf{C}^T$ does not respect the energy symmetry condition of the elastic stiffness tensor $C_{ijkl} = C_{klij}$ [i.e., Eq. (8.48)]. Indeed, for $\mathbf{g}_{,\sigma} \neq \mathbf{f}_{,\sigma}$ the second term of $\mathbf{C}^T$ does not derive from a potential:

$$\mathbf{g}_{,\sigma} \neq \mathbf{f}_{,\sigma} \rightarrow C^T_{ijkl} \neq C^T_{klij} \tag{8.100}$$

The symmetry is recovered only for an associated flow rule:

$$\mathbf{g}_{,\sigma} = \mathbf{f}_{,\sigma} \rightarrow C^T_{ijkl} = C^T_{klij} \tag{8.101}$$

Exercise 76. Determine the tangent stiffness tensor for an isotropic elastoplastic material of the Von–Mises type (associated flow rule).

What we need is (*1*) the isotropic elastic stiffness tensor (5.53):

$$\mathbf{C} = K \mathbf{1} \otimes \mathbf{1} + 2G \left(\mathbf{I} - \frac{1}{3} \mathbf{1} \otimes \mathbf{1} \right)$$

and (*2*) the gradients of $f = g$ with regard to stress, that is, Eq. (8.17):

$$\mathbf{g}_{,\sigma} = \mathbf{f}_{,\sigma} = \frac{\mathbf{s}}{2\sqrt{J_2}}$$

We note that for any symmetric second-order tensor, say $\mathbf{a} =^t \mathbf{a}$,

$$\mathbf{a} : (\mathbf{1} \otimes \mathbf{1}) = (\mathbf{1} \otimes \mathbf{1}) : \mathbf{a} = \text{tr}\,(\mathbf{a}) \mathbf{1}$$

$$\mathbf{a} : \left(\mathbf{I} - \frac{1}{3} \mathbf{1} \otimes \mathbf{1} \right) = \left(\mathbf{I} - \frac{1}{3} \mathbf{1} \otimes \mathbf{1} \right) : \mathbf{a} = \text{dev}\,[\mathbf{a}] = \mathbf{a} - \frac{1}{3} \text{tr}\,(\mathbf{a}) \mathbf{1}$$

Hence, with $\mathbf{f}_{,\sigma} = \text{dev}\,[\mathbf{f}_{,\sigma}]$, the different terms in (8.99) are obtained in the following form:

$$\mathbf{C} : \mathbf{g}_{,\sigma} = 2G \frac{\mathbf{s}}{2\sqrt{J_2}} = 2G \mathbf{f}_{,\sigma} = \mathbf{f}_{,\sigma} : \mathbf{C}$$

and

$$H + \mathbf{f}_{,\sigma} : \mathbf{C} : \mathbf{g}_{,\sigma} = H + 2G\mathrm{dev}\left[\mathbf{g}_{,\sigma}\right] : \mathrm{dev}\left[\mathbf{f}_{,\sigma}\right] = H + 2G\mathbf{f}_{,\sigma} : \mathbf{f}_{,\sigma} = H + 2G$$

Thus

$$\mathbf{C}^T = K\mathbf{1}\otimes\mathbf{1} + 2G\left[\mathbf{I} - \frac{1}{3}\mathbf{1}\otimes\mathbf{1} - \frac{\mathbf{f}_{,\sigma}\otimes\mathbf{f}_{,\sigma}}{H/2G+1}\right]$$

We verify the symmetry of $C^T_{ijkl} = C^T_{klij}$. ∎

8.4 TRAINING SET: THE CAMBRIDGE (OR CAM–CLAY) MODEL

Many geomaterials are porous materials that undergo substantial volume changes when subjected to external loading. In compression, the material exhibits an irreversible macroscopic contraction. This behavior cannot be captured by the Von-Mises plasticity model ($\mathrm{tr}\, d\boldsymbol{\varepsilon}^p = 0$), nor by the Drucker–Prager plasticity model, restricted to $\mathrm{tr}\, d\boldsymbol{\varepsilon}^p \geq 0$ in both compression and tension. Furthermore, a plastic contraction densifies the matter, and thus leads to an apparent strength increase of the material (i.e., to positive hardening). On the contrary, the loosening of the material, which accompanies a strength decrease (i.e., softening), is associated with an irreversible volume increase. These are in short the features addressed by the Cam–Clay plasticity model, developed by the (UK) Cambridge School (namely, Roscoe, Schofield, Wroth, Parry, and later Burland). Through the Cam–Clay model, we will revisit the different elements of 3D hardening plasticity. For clarity, we will consider the pore space as empty. In Volume II of this textbook, we will study the extension of the model to saturated porous materials.

8.4.1 Cam–Clay Type of Loading Surface

Many porous materials have a negligible tensile strength, compared with their compressive strength, which increases as the material densifies through, for example, pore collapse. A plasticity model of the Cam–Clay type addresses this observation by a closed elasticity domain D_E in stress space. In the $\left(\sqrt{J_2}\times\sigma_m\right)$-plane the elasticity domain is expressed as a quadratic relation between the two stress invariants $\sqrt{J_2}$ and σ_m, closed on both sides of the hydrostatic axis (see Figure 8.7):

$$\boldsymbol{\sigma}\in D_E \Leftrightarrow 2f\left(\boldsymbol{\sigma}, \zeta = p_{\mathrm{cr}}\right) = \frac{3J_2}{m^2} + \left(\sigma_m + p_{\mathrm{cr}}\right)^2 - p_{\mathrm{cr}}^2 \leq 0 \tag{8.102}$$

where m is a material constant. $\zeta = p_{\mathrm{cr}}$ is the hardening force. In the principal stress space $(\sigma_I\times\sigma_{II}\times\sigma_{III})$, loading function (8.102) represents an ellipsoid of revolution around the hydrostatic axis. In the $(\sqrt{J_2}\times\sigma_m)$-plane, this ellipsoid collapses into a half-ellipse having $mp_{\mathrm{cr}}/\sqrt{3}$ and p_{cr} as half-axis, and which is closed on the hydrostatic axis at $\sigma_m = 0$ and $\sigma_m = -2p_{\mathrm{cr}}$. The greater the value of p_{cr}, the larger the elasticity domain. If we let $p_{\mathrm{cr}} = zp^0_{\mathrm{cr}}$, where $z > 0$ is the hardening parameter, and p^0_{cr} be a reference value corresponding to a reference stress state $\boldsymbol{\sigma}^0$ at the boundary, the norm of the stress tensor is transformed according to $|\boldsymbol{\sigma}| = z\left|\boldsymbol{\sigma}^0\right|$; that is,

$$2f\left(\bar{\boldsymbol{\sigma}} = \boldsymbol{\sigma}/p^0_{\mathrm{cr}}, z\right) = \frac{3\bar{J}_2}{m^2} + \left(\bar{\sigma}_m + z\right)^2 - z^2 \leq 0 \tag{8.103}$$

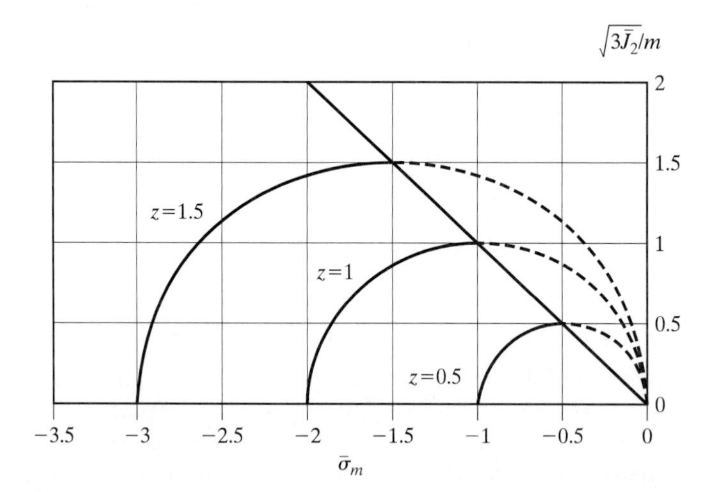

FIGURE 8.7: The Cambridge model: loading function in the normalized stress invariant $\sqrt{3\bar{J}_2} \times \bar{\sigma}_m$ half-plane, with isotropic hardening.

where $\bar{J}_2 = J_2/\left(p_{\text{cr}}^0\right)^2$ and $\bar{\sigma}_m = \sigma_m/p_{\text{cr}}^0$ are the dimensionless stress invariants. This transformation of the elasticity domain corresponds to an isotropic hardening, shown in Figure 8.7 for different values of the hardening parameter $z = p_{\text{cr}}/p_{\text{cr}}^0 > 0$.

Exercise 77. Loading function (8.102) defines the strength domain of a cohesionless material: It requires a confinement pressure, $\sigma_m < 0$, to activate a frictional behavior. Discuss different extensions of the Cambridge model to account for a nonzero cohesion.

There are several possibilities to include a nonzero cohesion. For instance,

$$\boldsymbol{\sigma} \in D_E \Leftrightarrow 2f\left(\boldsymbol{\sigma}, \zeta = p_{\text{cr}}\right) = \frac{3J_2}{m^2} + \left(\sigma_m - \rho + p_{\text{cr}}\right)^2 - p_{\text{cr}}^2 \leq 0$$

where ρ is the cohesion pressure, corresponding to the translation of the elasticity domain along the hydrostatic axis. The ellipse preserves its half-axis, related to only the value of the hardening force $\zeta = p_{\text{cr}}$, but it is now situated between $\max \sigma_m = \rho$ and $\min \sigma_m = \rho - 2p_{\text{cr}}$. In this case, $\sigma_I = \sigma_{II} = \sigma_{III} = \rho/3$ is the fix point of the loading function on the hydrostatic axis.
Another possibility is

$$\boldsymbol{\sigma} \in D_E \Leftrightarrow 2f\left(\boldsymbol{\sigma}, \zeta = p_{\text{cr}}\right) = \frac{3J_2}{m^2} + \left(\sigma_m + 2p_{\text{cr}}\right)\left(\sigma_m - \rho\right) \leq 0$$

for which $\max \sigma_m = \rho$ and $\min \sigma_m = -2p_{\text{cr}}$. In both cases, the consideration of a cohesion pressure requires a second material tests to determine, in addition to m, the cohesion pressure ρ. ■

8.4.2 Cam–Clay Type of Flow Rule

The flow rule defines the kinematics of plastic flow. For an isotropic plasticity model, the kinematics of plastic flow is characterized by the plastic volume change $\operatorname{tr}(d\varepsilon^p)$ and the plastic distortion $|d\theta^p|$. If the flow rule is associated, the plastic flow is directed along the outward normal to the yield surface. The intensity of the plastic flow is such that the loading point remains on the loading surface. This is expressed by the consistency conditions. These are the elements we apply to the Cambridge model.

The flow rule in the Cam–Clay model is assumed to be associated following the normality rule. Letting $f = g$ in $(8.66)_1$ yields

$$d\boldsymbol{\varepsilon}^p = d\lambda \left[\frac{3}{2m^2}\mathbf{s} + \frac{1}{3}(\sigma_m + p_{\text{cr}})\,\mathbf{1} \right] \tag{8.104}$$

For the isotropic plasticity model, the plastic volume change is

$$\operatorname{tr}(d\varepsilon^p) = d\lambda \frac{\partial f}{\partial \sigma_m} = d\lambda\,(\sigma_m + p_{\text{cr}}) \tag{8.105}$$

Since $d\lambda \geq 0$, the sign of the plastic volume change depends on the sign of $\sigma_m + p_{\text{cr}}$:

$$\operatorname{sign}[\operatorname{tr}(d\varepsilon^p)] = \operatorname{sign}(\sigma_m + p_{\text{cr}}) \tag{8.106}$$

More precisely, the plane $\sigma_m + p_{\text{cr}} = 0$ separates a plastic contracting behavior from a plastic dilating behavior. Loading points $\sigma_m + p_{\text{cr}} < 0$ on the loading surface, $f = 0$, provoke plastic contraction $\operatorname{tr}(d\varepsilon^p) < 0$, and for loading points $\sigma_m + p_{\text{cr}} > 0$ it is the inverse: The plastic flow rule predicts plastic dilatation $\operatorname{tr}(d\varepsilon^p) > 0$. Finally, loading points on both the plane $\sigma_m + p_{\text{cr}} = 0$ and the loading surface undergo neither plastic contraction nor plastic dilatation: Plastic

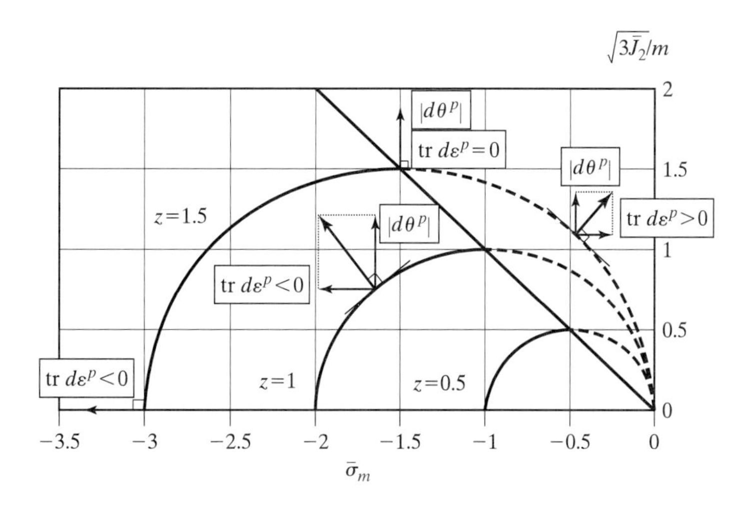

FIGURE 8.8: The Cambridge model: normality rule of plastic deformation.

deformations causes only plastic distortion, as defined by (8.18):

$$|d\theta^p| = d\lambda \frac{\partial f}{\partial \sqrt{J_2}} = d\lambda \frac{3}{m^2}\sqrt{J_2} \tag{8.107}$$

In the $\left(\sqrt{J_2} \times \sigma_m\right)$-plane, $\operatorname{tr}(d\boldsymbol{\varepsilon}^p)$ and $|d\theta^p|$ are the components of the unit outward normal to the loading function. This is shown in Figure 8.8.

Finally, the consistency condition (8.69) and the definition of the hardening modulus (8.70) lead to

$$df = 0 : \frac{3}{2m^2} dJ_2 + (\sigma_m + p_{\text{cr}})\, d\sigma_m = d\lambda H = -\frac{\partial f}{\partial p_{\text{cr}}} dp_{\text{cr}} = -\sigma_m dp_{\text{cr}} \tag{8.108}$$

where $dJ_2 = \mathbf{s} : d\mathbf{s}$. Since $\sigma_m \leq 0$ and $d\lambda \geq 0$, the sign of the hardening modulus H follows the sign of the change of the hardening force, sign (H) = sign (dp_{cr}). This shows that plastic hardening, defined by a positive hardening modulus $H > 0$, corresponds to an expansion of the elasticity domain $dp_{\text{cr}} > 0$; while plastic softening $H < 0$ is associated with a shrinking elasticity domain $dp_{\text{cr}} < 0$.

Last, a combination of the previous equation allows us to express the components of the plastic flow (8.105) and (8.107) in the form of (8.74):

$$\begin{pmatrix} \operatorname{tr}(d\boldsymbol{\varepsilon}^p) \\ |d\theta^p| \end{pmatrix} = \frac{1}{H} \begin{bmatrix} \left(\frac{\partial f}{\partial \sigma_m}\right)^2 & \frac{\partial f}{\partial J_2}\frac{\partial f}{\partial \sigma_m} \\ \frac{\partial f}{\partial J_2}\frac{\partial f}{\partial \sigma_m} & \left(\frac{\partial f}{\partial J_2}\right)^2 \end{bmatrix} \begin{pmatrix} dJ_2 \\ d\sigma_m \end{pmatrix} \tag{8.109}$$

where $\partial f/\partial J_2 = 3/2m^2$ and $\partial f/\partial \sigma_m = \sigma_m + p_{\text{cr}}$.

8.4.3 Cam–Clay Type of Hardening Rule

The hardening law relates the development of the elasticity domain with the plastic deformation. This is achieved by means of a relation between the hardening force, $\zeta = p_{\text{cr}}$, and the hardening variable χ. The relation $\zeta = \zeta(\chi)$ needs to be identified experimentally. Set within an energy approach, this relation stems from the microscopic origin of plastic hardening or softening, the frozen energy $\mathcal{U}(\chi)$. The assumption of the plastic hardening model is that this frozen energy can be expressed at the macrolevel of material description as a function of a plastic hardening variable χ. The evolution of χ is defined by the hardening rule. This is what we apply to the Cambridge model.

Many geomaterials undergo plastic hardening during plastic contraction and plastic softening during plastic dilatation. In fact, the first is associated with a densification of the material at a microlevel, the second with a loosening. On the macrolevel, this densification or loosening is expressed by the plastic volume strain $\operatorname{tr}\boldsymbol{\varepsilon}^p$, which appears as a good candidate for the hardening variable χ:

$$\chi \equiv \operatorname{tr}\boldsymbol{\varepsilon}^p; \quad \zeta \equiv p_{\text{cr}}\left(\chi \equiv \operatorname{tr}\boldsymbol{\varepsilon}^p\right) \tag{8.110}$$

For a particular Cam–Clay type of material, relation (8.110) needs to be identified experimentally by means of triaxial tests, in which the hardening force p_{cr} and the plastic volume strain $\operatorname{tr}\boldsymbol{\varepsilon}^p$ are measured simultaneously. The first is experimentally assessed from the loading function (8.102) by force and pressure measurements, the second by strain measurements after unloading.[5] It is often found that the function $p_{\rm cr} = p_{\rm cr}\left(\operatorname{tr}\boldsymbol{\varepsilon}^p\right)$ follows an exponential relation of the form

$$z = \frac{p_{\rm cr}}{p_{\rm cr}^0} = \exp\left[-k\left(\operatorname{tr}\boldsymbol{\varepsilon}^p - \operatorname{tr}\boldsymbol{\varepsilon}_0^p\right)\right] \tag{8.111}$$

where k is a positive constant, and $p_{\rm cr}^0 = p_{\rm cr}^0\left(\operatorname{tr}\boldsymbol{\varepsilon}_0^p\right)$ a reference value of the hardening force for $\operatorname{tr}\boldsymbol{\varepsilon}^p = \operatorname{tr}\boldsymbol{\varepsilon}_0^p$ and $z = 1$. According to this relation, the hardening force increases when the material plastically contracts, and for plastic dilatation it is the inverse. Once this experimental identification is achieved, we can proceed with setting up the hardening law of the Cam–Clay type of material within the developed energy approach.

We first note that the choice $\chi \equiv \operatorname{tr}\boldsymbol{\varepsilon}^p$ cannot be accommodated by an associated hardening rule, since

$$d\chi = d\lambda\frac{\partial f}{\partial p_{\rm cr}} \neq \operatorname{tr}\left(d\boldsymbol{\varepsilon}^p\right) = d\lambda\frac{\partial f}{\partial \sigma_m} \tag{8.112}$$

It requires a nonassociated hardening rule of the form $(8.66)_2$:

$$d\chi = d\lambda\frac{\partial h}{\partial p_{\rm cr}} \equiv \operatorname{tr}\left(d\boldsymbol{\varepsilon}^p\right) = d\lambda\frac{\partial f}{\partial \sigma_m} = d\lambda\left(\sigma_m + p_{\rm cr}\right) \tag{8.113}$$

Here $h \neq f$ is the nonassociated hardening potential. This hardening potential must be convex with regard to its arguments, and its partial derivative must meet the requirement $d\chi \equiv \operatorname{tr}\left(d\boldsymbol{\varepsilon}^p\right)$. There are several functions that meet this requirement; for instance,

$$2h = \left(\sigma_m + p_{\rm cr}\right)^2 \tag{8.114}$$

[5] It is common practice in soil mechanics to characterize the change of the elasticity domain in the form of void ratio *versus* mean stress curves, determined in triaxial compression, $\boldsymbol{\sigma} = \sigma_m\mathbf{1}$, in which, according to the Cam–Clay loading function, $f = 0 \Leftrightarrow \sigma_m = -2p_{\rm cr}$. The void ratio e is the volume ratio of connected pores over the volume of the solid matrix. For an incompressible solid phase, it is readily understood that the macroscopic relative volume change $\operatorname{tr}\boldsymbol{\varepsilon}$ is only due to a change of the pore space:

$$e = \frac{\operatorname{tr}\boldsymbol{\varepsilon}}{1 - \phi_0}$$

where $1 - \phi_0$ is the constant volume fraction of the solid. If, in addition, the elastic volume change is small in comparison with the total deformation, the volumetric response is dominated by plastic deformation,

$$\operatorname{tr}\boldsymbol{\varepsilon} \approx \operatorname{tr}\boldsymbol{\varepsilon}^p; \quad e \approx e^p = \frac{\operatorname{tr}\boldsymbol{\varepsilon}^p}{1 - \phi_0}$$

This allows us, in triaxial compression, to assess the relation between the hardening force $\zeta = p_{\rm cr} = \sigma_m/2$, and the hardening variable $\chi = \operatorname{tr}\boldsymbol{\varepsilon}^p$ from two independent measurements; the first is the force pressure described from the outside, the second the volume strain recorded during the test. This is what we mean by experimental identification: two independent measurements of the two quantities related by a constitutive relation:

$$\zeta = \zeta(\chi) \Leftrightarrow p_{\rm cr} = p_{\rm cr}(e \approx e^p \propto \operatorname{tr}\boldsymbol{\varepsilon}^p)$$

Next, within the energy approach, $\chi \equiv \operatorname{tr} \varepsilon^p$ means that the plastic volume change is actually the thermodynamic variable associated with the hardening force $\zeta = p_{\text{cr}}$, entering the expression of the dissipation of the plastic hardening model (8.46). It follows from (8.82) that

$$\zeta \equiv p_{\text{cr}} \left(\chi \equiv \operatorname{tr} \varepsilon^p\right) = -\frac{\partial \mathcal{U}\left(\operatorname{tr} \varepsilon^p\right)}{\partial \left(\operatorname{tr} \varepsilon^p\right)} \tag{8.115}$$

where $\mathcal{U}\left(\operatorname{tr} \varepsilon^p\right)$ is the frozen energy. Use of the experimental relation (8.111) in (8.115) yields

$$\mathcal{U}\left(\operatorname{tr} \varepsilon^p\right) = \mathcal{U}\left(\operatorname{tr} \varepsilon_0^p\right) \exp\left[-k\left(\operatorname{tr} \varepsilon^p - \operatorname{tr} \varepsilon_0^p\right)\right] \tag{8.116}$$

where $\mathcal{U}\left(\operatorname{tr} \varepsilon_0^p\right) = p_{\text{cr}}^0/k$ is the frozen energy for $\operatorname{tr} \varepsilon^p = \operatorname{tr} \varepsilon_0^p$. The frozen energy increases for $\operatorname{tr} \varepsilon^p < \operatorname{tr} \varepsilon_0^p$, meaning that the microelasticity reserve stored in the microstructure increases when the material plastically contracts. This corresponds to plastic hardening. By contrast, for $\operatorname{tr} \varepsilon^p > \operatorname{tr} \varepsilon_0^p$, the frozen energy decreases: It is released due to plastic dilatation, which corresponds to plastic softening. We have here an explicit energy interpretation of the plastic hardening and plastic softening phenomenon. In hardening plasticity, the sign of plastic hardening is expressed by the hardening modulus H. It is obtained by using (8.102), (8.114), and (8.116) in (8.72):

$$H = \left(\frac{\partial f}{\partial p_{\text{cr}}} \frac{\partial h}{\partial p_{\text{cr}}}\right) \frac{\partial^2 \mathcal{U}}{\partial \left(\operatorname{tr} \varepsilon^p\right)^2} = \sigma_m \left(\sigma_m + p_{\text{cr}}\right) k p_{\text{cr}} = \sigma_m k p_{\text{cr}} \frac{\operatorname{tr}\left(d\varepsilon^p\right)}{d\lambda} \tag{8.117}$$

Since $\sigma_m \leq 0$ and $d\lambda \geq 0$, the sign of the hardening modulus follows the opposite sign of the plastic dilatation (8.106). It then turns out, from Figure 8.8, that loading points on the loading surface situated in the half $\sigma_m + p_{\text{cr}} < 0$ undergo plastic hardening: $H > 0$, the loading surface expands. In turn, loading points on the loading surface situated in the half $\sigma_m + p_{\text{cr}} > 0$ exhibit plastic softening: $H < 0$, the elasticity domain shrinks. The loading states, which separate the plastic contracting (= hardening) behavior from a plastic dilating (= softening), satisfy $\sigma_m + p_{\text{cr}} = 0$. They exhibit an ideal plastic behavior, $H = 0$. The plastic flow occurs at an undetermined rate. It is for this reason that these loading states are commonly referred to as *critical states*, and the hardening force p_{cr} is called the *critical pressure.* In the $\left(\sqrt{J_2} \times \sigma_m\right)$-plane displayed in Figure 8.8, the critical states are situated on a straight line through the origin, for which

$$\sigma_m + p_{\text{cr}} = 0 : f^{\text{cr}} = \sqrt{J_2} + \alpha \sigma_m = 0 \tag{8.118}$$

where $\alpha = m/\sqrt{3}$. The critical states correspond to loading points situated on a cohesionless Drucker–Prager criterion (8.21).

8.4.4 Thermodynamic Consistency

The Cam–Clay model is a plasticity model with an associated flow rule, but the hardening rule is nonassociated. This requires a (quick) check of the nonnegativity of the dissipation $d\mathcal{D} = dW^p - d\mathcal{U} \geq 0$ associated with plastic and hardening deformation.

We need the following:

1. The plastic work of the Cam–Clay model:

$$dW^p = \boldsymbol{\sigma} : d\boldsymbol{\varepsilon}^p = d\lambda \left[\frac{3}{m^2} J_2 + (\sigma_m + p_{\rm cr})\, \sigma_m \right] = -d\lambda \sigma_m p_{\rm cr} \tag{8.119}$$

where we made use of the flow rule (8.104), and the value taken by $3J_2/m^2$ on the loading function (8.102), $f = 0 \Leftrightarrow 3J_2/m^2 = -\sigma_m (\sigma_m + 2p_{\rm cr})$.

2. The hardening work:

$$-d\mathcal{U} = \zeta d\chi = d\lambda p_{\rm cr} \frac{\partial h}{\partial p_{\rm cr}} = d\lambda p_{\rm cr} (\sigma_m + p_{\rm cr}) \tag{8.120}$$

where h is the hardening potential.

Summing up these two components gives the dissipation (i.e., the transformation of externally supplied energy into heat form)[6]:

$$d\mathcal{D} = dW^p - d\mathcal{U} = d\lambda p_{\rm cr}^2 \geq 0 \tag{8.121}$$

We verify that the second law is satisfied. The Cam–Clay model is then complete.

8.4.5 Cam–Clay Type of Viscoplasticity Model

In the hardening plasticity model, the elasticity domain evolves when a loading point is on the loading surface. By contrast, in the viscoplasticity model the elasticity domain develops when a loading point is situated outside of the elasticity domain, to which it returns asymptotically. The aim of this section is to illustrate the extension of the hardening plasticity model to hardening viscoplasticity, using the case of the Cambridge model.

We have developed in Section 7.3 the elements of the 1D viscoplasticity model. These elements extend to the 3D case, when replacing the scalar quantities by their tensorial counterparts. These are as follows:

1. *The plasticity criterion*: The viscoplasticity model is a plasticity model. Viscoplastic evolutions are defined with respect to the plastic loading function, enclosing the elasticity domain. In the case of the Cambridge model, the loading function is given by (8.102)

$$\boldsymbol{\sigma} \in D_E \Leftrightarrow 2f(\boldsymbol{\sigma}, \zeta = p_{\rm cr}) = \frac{3J_2}{m^2} + (\sigma_m + p_{\rm cr})^2 - p_{\rm cr}^2 \leq 0 \tag{8.122}$$

2. *The viscoplastic flow rule*: In contrast to the plastic model, viscoplastic deformation occurs when a loading point is situated outside of the elasticity

[6] It is instructive to note that the case of an associated hardening rule $h = f$ would correspond to a zero dissipation. In fact, the plastic work $dW^p = -d\lambda\sigma_m p_{\rm cr}$ would be entirely stored into the microstructure in the form of frozen energy $d\mathcal{U} = -d\lambda p_{\rm cr}\sigma_m$, such that $d\mathcal{D} = dW^p - d\mathcal{U} = 0$. While thermodynamically admissible, this case is physically unlikely to occur. It would actually mean that any energy supplied to the system in the form of work would be completely conserved (in the form of hardening energy) to do work later.

domain, that is, for $f > 0$. The 3D counterpart of (7.79) corresponds to an associated viscoplastic flow rule:

$$\frac{d\varepsilon^{vp}}{dt} = \frac{\langle f \rangle}{\eta} \frac{\partial f(\boldsymbol{\sigma})}{\partial \boldsymbol{\sigma}} \tag{8.123}$$

where $\langle f \rangle = \frac{1}{2}(f + |f|)$ denotes the positive part of the loading function, and η the viscosity. Thus, in the case of the Cambridge model,

$$\frac{d\varepsilon^{vp}}{dt} = \frac{\langle f \rangle}{\eta} \left[\frac{3}{2m^2} \mathbf{s} + \frac{1}{3}(\sigma_m + p_{\text{cr}}) \mathbf{1} \right] \tag{8.124}$$

The viscoplastic strain rate is obtained in a straightforward manner from the plastic flow rule, by replacing the plastic multiplier by the viscoplasticity strain rate intensity $\langle f \rangle / \eta$. For instance, the viscoplastic volume strain rate and viscoplastic distortion rate are obtained from (8.105) and (8.107):

$$\text{tr}\left(\frac{d\varepsilon^p}{dt}\right) = \frac{\langle f \rangle}{\eta}(\sigma_m + p_{\text{cr}}) \tag{8.125}$$

$$\left|\frac{d\theta^p}{dt}\right| = \frac{\langle f \rangle}{\eta} \frac{3}{m^2} \sqrt{J_2} \tag{8.126}$$

In the case of an associated viscoplastic flow rule, the viscoplastic flow directions remain normal to the loading function, but the intensity changes: It depends on the overstress $\langle f \rangle$ (i.e., on the distance of the loading point to the yield surface). For completeness, a nonassociated flow rule would read

$$\frac{d\varepsilon^{vp}}{dt} = \frac{\langle f \rangle}{\eta} \frac{\partial g(\boldsymbol{\sigma})}{\partial \boldsymbol{\sigma}} \tag{8.127}$$

where $g \neq f$ is the plastic potential.

3. *The viscoplastic hardening rule*: A viscoplastic hardening rule is required when the elasticity domain evolves during viscoplastic deformation. A straightforward extension of the plastic hardening rule $(8.66)_2$ to viscoplastic hardening is of the form[7]

$$\frac{d\chi}{dt} = \frac{\langle f \rangle}{\eta} \frac{\partial h}{\partial \zeta} \tag{8.128}$$

where h is the hardening potential. Application to the Cambridge model means to let $\zeta = p_{\text{cr}}$ and $\chi \equiv \text{tr}\,\varepsilon^{vp}$ in (8.128), and to substitute for h expression (8.114) developed for the plastic model,

$$\frac{d\chi}{dt} \equiv \text{tr}\left(\frac{d\varepsilon^{vp}}{dt}\right) = \frac{\langle f \rangle}{\eta}(\sigma_m + p_{\text{cr}}) \tag{8.129}$$

[7] We should note, however, that the elasticity domain in viscoplasticity may also develop independent of viscoplastic deformation—in contrast to hardening plasticity. We have encountered such a phenomenon in the Problem Set in Section 7.6 devoted to the creep hesitancy phenomenon. In this case, the viscoplastic flow rule is of the form

$$\frac{d\chi}{dt} = \frac{\langle f \rangle}{\eta} \frac{\partial h}{\partial \zeta} - N\chi$$

where $N \geq 0$. Even in the absence of viscoplastic deformation, $\langle f \rangle = 0$, the hardening variable may change its value, and with it the hardening force $\zeta = \zeta(\chi)$.

4. *The hardening force state equation*: The physical origin of the hardening phenomenon is the same as in the plastic case, associated with an elastic energy stored at the microscale. On the macrolevel, this is taken into account by the frozen energy, $\mathcal{U}\left(\chi \equiv \operatorname{tr} \varepsilon^{vp}\right)$, from which the hardening force derives:

$$\zeta = p_{\mathrm{cr}}\left(\chi \equiv \operatorname{tr} \varepsilon^{vp}\right) = -\frac{\partial \mathcal{U}}{\partial\left(\operatorname{tr} \varepsilon^{vp}\right)} = p_{\mathrm{cr}}^{0} \exp\left[-k\left(\operatorname{tr} \varepsilon^{vp} - \operatorname{tr} \varepsilon_{0}^{vp}\right)\right] \quad (8.130)$$

In summary, viscoplastic flow only occurs for loading points situated outside the current elasticity domain. The material exhibits viscoplastic dilatation if $\sigma_m + p_{\mathrm{cr}} > 0$. This leads to viscoplastic softening: The critical pressure p_{cr} decreases, and the elasticity domain shrinks. For $\sigma_m + p_{\mathrm{cr}} < 0$, it is the inverse: The material undergoes viscoplastic hardening, and the elasticity domain expands. Finally, for loading points outside the elasticity domain, for which $\sigma_m = -p_{\mathrm{cr}}$, the viscoplastic flow occurs at constant volume; the material behaves like an ideal viscoplastic material. But note the difference to the plastic model: In the plastic model, at the critical state $\sigma_m = -p_{\mathrm{cr}}$, plastic flow occurs at an undetermined rate. By contrast, the viscoplastic strain rate is never undetermined, only the loading surface stagnates when $\sigma_m = -p_{\mathrm{cr}}$. In other words, the viscoplastic Cambridge model is free of critical states. Viscous effects regularize the plastic behavior.

8.5 PROBLEM SET: STRENGTH ESTIMATES BY MICROINDENTATION (MICROHARDNESS)

Microindentation tests are frequently used in materials science and engineering to determine the material strength at very fine scales. The test consists of a penetration of a needle-type indenter in a continuous material system. The force required to penetrate is then related to the strength of the material. In this problem set, we propose to develop a simplified triaxial strength model of the microhardness test. We consider that the indenter is a rigid cylinder of radius r_0, situated on the surface of a horizontal half-space. This is sketched in Figure 8.9. A vertical force F is exerted on the cylinder in the direction of the cylinder axis Oz, until it penetrates into the half-space. We suppose that the interface of the cylinder with the half-space is without friction. The value of the force F at this moment is noted $\max F$. We seek to relate this value of $\max F$

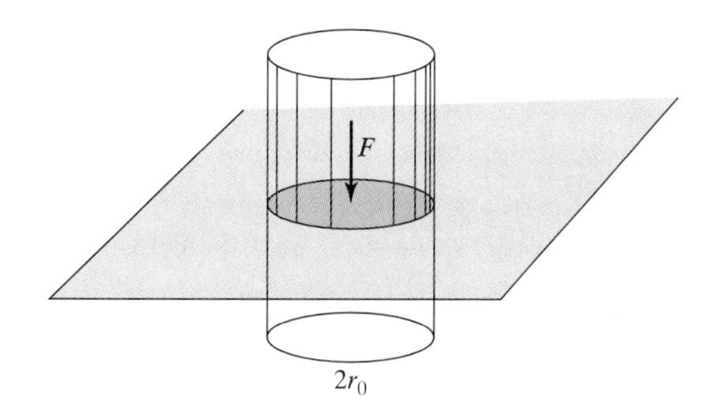

FIGURE 8.9: Microindentation test.

to the strength of the material composing the half-space. This material is considered to follow an isotropic shear strength criterion, of the Von–Mises type, reading

$$f(\boldsymbol{\sigma}) = 0 : \sqrt{J_2} = k$$

where k is the shear strength, and $\sqrt{J_2} = \sqrt{\frac{1}{2}\mathrm{s}:\mathrm{s}}$ is the deviator stress invariant, which reads in principal stresses $\sigma_{J=I,II,III}$

$$J_2 = \frac{1}{6}\left[(\sigma_I - \sigma_{II})^2 + (\sigma_I - \sigma_{III})^2 + (\sigma_{II} - \sigma_{III})^2\right]$$

1. We consider the following stress field in the half-space under the applied load F:

$$\boldsymbol{\sigma} = \boldsymbol{\sigma}(r); \quad (\sigma_{ij}) = \begin{pmatrix} \sigma_{rr}(r) & 0 & 0 \\ & \sigma_{\theta\theta}(r) & 0 \\ \text{sym} & & \sigma_{zz}(r) \end{pmatrix}$$

For the considered loading case, determine the conditions that the stress field σ in the half-space needs to satisfy:

   (a) to be statically admissible,

   (b) to respect the Von–Mises strength criterion.

2. We suppose, in addition, that the stress in the cylinder can be approximated by

$$0 \leq r \leq r_0 : \sigma_{rr} = \sigma_{\theta\theta}; \quad \sigma_{zz} = -p = \text{const}$$

Determine the maximum value of p^+ and the corresponding force F^+, as a function of the strength k, that can be obtained with the stress field from Question 1 of this problem set. Consider first that the stress $\sigma_{\theta\theta}$ is continuous at $r = r_0$. Then show that the admissible value for p^+ increases if the assumption of the continuity of $\sigma_{\theta\theta}$ at $r = r_0$ is omitted.

3. Consider that $\max F$ was determined experimentally. Are the values of F^+ determined in Question 2 of this problem set greater or smaller than the experimentally determined value? Conclude with regard to the strength k that can be assessed with the developed simplified stress-strength model.

8.5.1 Triaxial Stress State and Von–Mises Strength Criterion

Statically Admissible Stress State.

The given triaxial stress field is the same as the one of the circular foundation analyzed in Section 4.6.2. It needs to satisfy the following:

1. Zero surface stress $\sigma_{zz} = 0$ for $r > r_0$ and $z = 0$. It follows that

$$r > r_0 : \sigma_{zz} = 0 \rightarrow (\sigma_{ij}) = \begin{pmatrix} \sigma_{rr}(r) & 0 & 0 \\ & \sigma_{\theta\theta}(r) & 0 \\ \text{sym} & & 0 \end{pmatrix}$$

2. Normal contact without friction between indenter and half-space at $z = 0$:

$$2\pi \int_0^{r_0} \sigma_{zz}(r) r dr = -F \rightarrow r < r_0 : (\sigma_{ij}) = \begin{pmatrix} \sigma_{rr}(r) & 0 & 0 \\ & \sigma_{\theta\theta}(r) & 0 \\ \text{sym} & & \sigma_{zz}(r) \end{pmatrix}$$

3. The radial equilibrium relation,

$$\frac{d\sigma_{rr}}{dr} + \frac{\sigma_{rr} - \sigma_{\theta\theta}}{r} = 0$$

4. The continuity of the radial stress at the interface $r = r_0$:

$$\sigma_{rr}(r_0) = \sigma_{rr}\left(r_0^-\right) = \sigma_{rr}\left(r_0^+\right)$$

Von–Mises Criterion.

The considered stress components correspond to principal stresses. Because of the vertical stress discontinuity at $r = r_0$, we have the following two strength criteria for the given problem:

$$r \leq r_0 : \sqrt{J_2} = \sqrt{\frac{1}{6}\left[(\sigma_{zz} - \sigma_{rr})^2 + (\sigma_{zz} - \sigma_{\theta\theta})^2 + (\sigma_{rr} - \sigma_{\theta\theta})^2\right]} = k$$

$$r \geq r_0 : \sqrt{J_2} = \sqrt{\frac{1}{6}\left[\sigma_{rr}^2 + \sigma_{\theta\theta}^2 + (\sigma_{rr} - \sigma_{\theta\theta})^2\right]} = k$$

8.5.2 Maximum Force

We consider the stress field

$$r \leq r_0 : \sigma_{rr} = \sigma_{\theta\theta}; \quad \sigma_{zz} = -p = -\frac{F}{\pi r_0^2}$$

The first condition implies a constant radial stress in $r \leq r_0$:

$$r \leq r_0 : \frac{d\sigma_{rr}}{dr} = 0$$

The strength criteria read

$$r \leq r_0 : \sqrt{J_2} = \sqrt{\frac{1}{6}\left[2(p + \sigma_{rr})^2\right]} = k$$

$$r \geq r_0 : \sqrt{J_2} = \sqrt{\frac{1}{3}\left[\sigma_{rr}^2 - \sigma_{rr}\sigma_{\theta\theta} + \sigma_{\theta\theta}^2\right]} = k$$

Continuity of the Hoop Stress at the Interface.

Continuity of the hoop stress at the interface means

$$\sigma_{\theta\theta}\left(r_0^+\right) \equiv \sigma_{\theta\theta}\left(r_0^-\right) = \sigma_{rr}\left(r_0\right)$$

Use in the expression of the strength criterion for $r \geq r_0$ (which includes $r = r_0$) gives

$$\sigma_{rr}^2(r_0) = 3k^2$$

In the same way, use in the strength criterion for $r \leq r_0$ (which also includes $r = r_0$) gives

$$\left(p + \sigma_{rr}(r_0)\right)^2 = 3k^2$$

Thus, a combination of the two previous conditions delivers a maximum value for p from

$$p^2 + 2p\sigma_{rr}(r_0) + \sigma_{rr}^2(r_0) = 3k^2$$

$$p^+ = 2\max\left[-\sigma_{rr}(r_0) = \mp\sqrt{3}k\right] = 2\sqrt{3}k$$

and hence a maximum force of

$$F^+ = 2\sqrt{3}\pi r_0^2 k$$

In summary, the stress solution for the considered problem is

$$\sigma_{zz} = -2\sqrt{3}kH\left(r_0 - r\right); \quad \sigma_{rr}\left(r\right) = \sigma_{\theta\theta}\left(r\right) = -\sqrt{3}k$$

where $H\left(r_0 - r\right)$ is the Heaviside function.

Discontinuity of the Hoop Stress at the Interface.

We consider a potential discontinuity of the hoop stress at the interface. The strength criterion must be satisfied on both sides of the interface $r = r_0$ simultaneously; that is:

$$r = r_0^- : \left(p + \sigma_{rr}\right)^2 = 3k^2$$

$$r = r_0^+ : \sigma_{rr}^2 + \sigma_{\theta\theta}^2 - \sigma_{rr}\sigma_{\theta\theta} = 3k^2$$

The maximum value of p depends therefore on the minimum value $\sigma_{rr}(r_0)$ can take. This value is determined by the second strength criterion at $r = r_0^+$. This second strength criterion is shown in Figure 8.10 in a normalized plot of $\sigma_{rr}(r_0)/k$ over $\sigma_{\theta\theta}(r_0)$. From the figure,

$$\sigma_{rr}(r_0) = -2k; \quad \sigma_{\theta\theta}(r_0^+) = -k$$

Finally, use in the strength criterion at $r = r_0^-$ gives a second maximum value for p:

$$p^{++} = k\left(\sqrt{3} + 2\right) > p^+ \rightarrow F^{++} = \left(\sqrt{3} + 2\right)\pi r_0^2 k > F^+$$

However, we still need to give proof that the corresponding stress field in $r \geq r_0^+$ is statically admissible, satisfying the equilibrium equation in the r-direction

$$\frac{d\sigma_{rr}}{dr} + \frac{\sigma_{rr} - \sigma_{\theta\theta}}{r} = 0$$

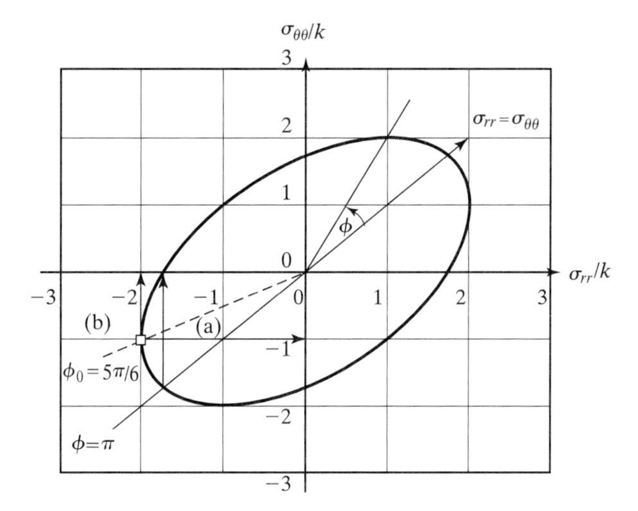

FIGURE 8.10: Von–Mises strength domain for a stress field of the form $\boldsymbol{\sigma} = \sigma_{rr}\mathbf{e}_r \otimes \mathbf{e}_r + \sigma_{\theta\theta}\mathbf{e}_\theta \otimes \mathbf{e}_\theta$.

while satisfying at the same time the strength criterion:

$$r \geq r_0^+ : \sigma_{rr}^2 + \sigma_{\theta\theta}^2 - \sigma_{rr}\sigma_{\theta\theta} = 3k^2$$

We note that the strength domain defined by this strength criterion is a rotated ellipse of parameter form (see Figure 8.10):

$$\begin{aligned}\sigma_{rr} &= k\left(\sqrt{3}\cos\phi - \sin\phi\right) = 2k\cos\left(\phi + \pi/6\right) \\ \sigma_{\theta\theta} &= k\left(\sqrt{3}\cos\phi + \sin\phi\right) = 2k\cos\left(\phi - \pi/6\right)\end{aligned}$$

Use in the equilibrium equation yields

$$-\left(\sqrt{3}\sin\phi + \cos\phi\right)\frac{d\phi}{dr} - \frac{2\sin\phi}{r} = 0$$

which integrates to

$$\ln r + \frac{1}{2}\sqrt{3}\phi\left(r\right) + \frac{1}{2}\ln\left[\sin\left(\phi\left(r\right)\right)\right] + C_1 = 0$$

Use of the boundary condition

$$\sigma_{rr}(r_0) = -2k \Leftrightarrow \cos\left(\phi + \pi/6\right) = -1 \rightarrow \phi_0 = \frac{5}{6}\pi$$

gives

$$\frac{r}{r_0} = \sqrt{\frac{\sin\phi_0}{\sin\phi}}\exp\left(-\frac{\sqrt{3}}{2}\left(\phi - \phi_0\right)\right)$$

The solution interval is $\phi \in \left[\frac{5\pi}{6}, \pi\right]$, where the lower value corresponds to $r = r_0$ and the upper value to $r \rightarrow \infty$. This is also shown in Figure 8.10.

8.5.3 Summary: Model versus Experiment

The chosen statically admissible stress fields are possible solutions of the problem, but the choice is not unique. For instance, σ_{zz} may not be a constant: It may well depend on r, and also on z. Furthermore, shear effects ($\sigma_{rz} \neq 0, \sigma_{r\theta} \neq 0$) at the interface $z = 0, 0 \leq r \leq r_0$ were neglected, which in general increase the strength capacity. Therefore, it is likely that the experimentally determined load $\max F$ may well be superior to the one predicted by the model:

$$F^{++} \leq \max F$$

In other words, the model will rather overestimate the real strength strength of the material:

$$k \leq \frac{\max F}{(\sqrt{3}+2)\,\pi r_0^2}$$

Note that $F^{++} = \max F$ if the plastic behavior follows the principle of maximum plastic work. In this case, the principle relates a stress state to a deformation behavior.

8.6 PROBLEM SET: THIN-WALLED CYLINDER SUBJECTED TO TENSION AND TORSION

We consider a hollow thin-walled cylinder of outer radius R and length L, oriented along the z-axis (see Figure 8.11). The cylinder is made of a Von–Mises material following an associated flow rule. The open end sides of the cylinder are subjected to a tensile stress

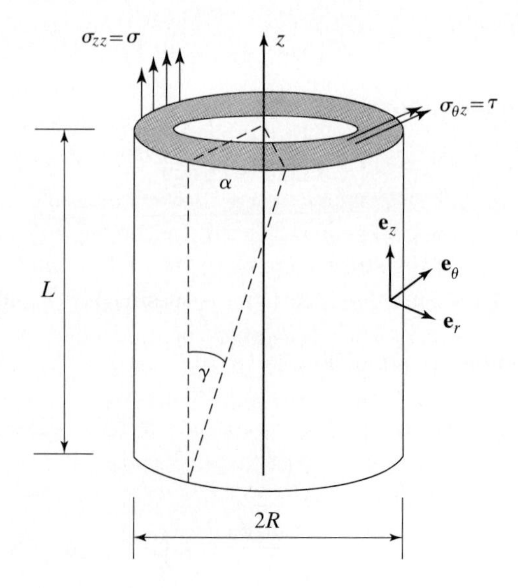

FIGURE 8.11: Problem set: thin-walled cylinder subjected to tension and torsion.

$\sigma_{zz} = \sigma$ in parallel with a shear stress $\sigma_{\theta z} = \tau$ (combined tension–torsion loading). For this problem set, we will adopt the hypothesis of small perturbations.

1. **Loading Surface**: Write the loading function as a function of the applied stresses σ and τ.
2. **Plastic Hardening**: We first consider an isotropic plastic hardening behavior of the material (hardening modulus H).

   (a) We consider the following loading path: The hollow cylinder is loaded in tension until the elastic limit $\sigma = \sigma_0$, which is kept constant hereafter. Then an infinitesimal torsion stress $d\tau$ is applied. For this loading path, determine the plastic strain increments.

   (b) Consider now a proportional tension–torsion loading from a virgin material state, such that $\sigma/\tau = \text{const}$. When the stresses reach the elasticity domain, an additional incremental torsion $d\tau$ is applied. For this loading path, determine the plastic strain increments.

   (c) Display the elasticity domain for the combined tension–torsion loading case in the $(\sigma \times \tau)$-plane. In this plane, show the loading paths considered in parts (a) and (b), and explain the obtained results of plastic strain increments. In particular, display in the $(\sigma \times \tau)$-plane the directions taken by the plastic strain increments.

3. **Elastic Ideal Plastic Loading**: From now on, we suppose that the material behaves ideal plastically.

   (a) How are the previous material responses affected when the material behaves ideal plastically instead of hardening plastically?

   (b) We are interested in the elastoplastic response of the thin cylinder problem:

      i. Elastic Loading: In the elastic range, derive the expression of the strain tensor ε_{ij} as a function of τ and σ. Determine the length increase of the cylinder and the torsion angle α as a function of σ and τ.

      ii. Elastoplastic Loading: In the admissible elastoplastic range, determine the change in radius dR, torsion angle $d\alpha$, and length dL, as a function of the stress increments $d\sigma$ and $d\tau$, and of the plastic multiplier $d\lambda$.

      iii. Plastic Loading: The cylinder tube is loaded in tension to the elastic limit. Then the length of the tube L is fixed, by prescribing an additional increasing torsion angle $\alpha (d\alpha > 0)$. Show that this loading corresponds to a plastic loading. Determine the expressions of the change in radius ΔR during plastic loading and of the applied stresses σ and τ as a function of the torsion angle α. Illustrate this ideal plastic loading in the $(\sigma \times \tau)$-plane. Note that

$$\int \frac{d\phi}{\cos\phi} = \ln\tan(\phi/2 + \pi/4)$$

8.6.1 Loading Surface: Von–Mises Plasticity

For a thin cylinder, we admit that the stress is homogeneous in the cylinder wall. The stress tensor reads

$$\boldsymbol{\sigma} = \sigma \mathbf{e}_z \otimes \mathbf{e}_z + \tau \left(\mathbf{e}_\theta \otimes \mathbf{e}_z + \mathbf{e}_z \otimes \mathbf{e}_\theta \right)$$

The stress deviator $\mathbf{s} = \boldsymbol{\sigma} - \sigma_m \mathbf{1}$ for $\sigma_m = \sigma/3$ is given by

$$(s_{ij}) = \begin{pmatrix} s_{rr} = -\sigma/3 & 0 & 0 \\ 0 & s_{\theta\theta} = -\sigma/3 & s_{\theta z} = \tau \\ 0 & s_{z\theta} = \tau & s_{zz} = 2\sigma/3 \end{pmatrix}$$

The second deviator invariant reads here

$$J_2 = \frac{1}{2} s_{ij} s_{ji} = \frac{1}{2} \left[2 \left(-\sigma/3 \right)^2 + \left(2\sigma/3 \right)^2 + 2\tau^2 \right] = \sigma^2/3 + \tau^2$$

Use in the Von–Mises plasticity criterion gives

$$f = \sqrt{J_2} - k \leq 0 : \sqrt{\sigma^2 + 3\tau^2} - \sqrt{3}k \leq 0$$

For $\tau = 0$, the strength $\sigma_0 = \sqrt{3}k$ turns out to be the uniaxial tensile strength of the Von–Mises material.

8.6.2 Plastic Hardening

Pure Tension Followed by Incremental Torsion.

Stress σ_{zz} reaches the uniaxial strength, and stress $\sigma_{z\theta} = \sigma_{\theta z}$ is zero during elastic loading. Thus at the elastic limit

$$f \left(\boldsymbol{\sigma} = \sigma_0 \mathbf{e}_z \otimes \mathbf{e}_z, \zeta = 0 \right) = 0 : \begin{pmatrix} \sigma_{zz} = \sigma = \sigma_0 = \sqrt{3}k \\ \sigma_{z\theta} = \sigma_{\theta z} = \tau = 0 \end{pmatrix}$$

Then the stress is increased by $d\tau$, and the plastic multiplier reads

$$d\lambda = \frac{1}{H} \frac{\partial f}{\partial \boldsymbol{\sigma}} : d\boldsymbol{\sigma}$$

For the considered loading case,

$$d\boldsymbol{\sigma} = \tau \left(\mathbf{e}_\theta \otimes \mathbf{e}_z + \mathbf{e}_z \otimes \mathbf{e}_\theta \right)$$

and $\frac{\partial f}{\partial \boldsymbol{\sigma}}$ needs to be evaluated for $\boldsymbol{\sigma} = \sigma_0 \mathbf{e}_z \otimes \mathbf{e}_z$:

$$\frac{\partial f}{\partial \boldsymbol{\sigma}} |_{\sigma_0 \mathbf{e}_z \otimes \mathbf{e}_z} = \frac{\mathbf{s}}{2\sqrt{J_2}} = \frac{1}{2\sqrt{3}} \left(-\mathbf{e}_r \otimes \mathbf{e}_r - \mathbf{e}_\theta \otimes \mathbf{e}_\theta + 2\mathbf{e}_z \otimes \mathbf{e}_z \right)$$

Thus,

$$d\lambda = 0 \rightarrow d\varepsilon^p = 0$$

Proportional Tension–Torsion Loading Followed by Incremental Torsion.

The stress state during elastic loading is

$$\boldsymbol{\sigma} = \sigma \left[\mathbf{e}_z \otimes \mathbf{e}_z + \text{const} \left(\mathbf{e}_\theta \otimes \mathbf{e}_z + \mathbf{e}_z \otimes \mathbf{e}_\theta\right)\right]; \; \sigma/\tau = \text{const}$$

The yield surface is reached for

$$f\left(\boldsymbol{\sigma}, \zeta = 0\right) = 0 : |\sigma| \sqrt{1 + 3\text{const}^{\,2}} - \sqrt{3}k \leq 0$$

$$\sigma_{zz} = \sigma = \frac{\sqrt{3}k}{\sqrt{1 + 3\text{const}^{\,2}}}; \quad \sigma_{z\theta} = \sigma_{\theta z} = \text{const} \times \frac{\sqrt{3}k}{\sqrt{1 + 3\text{const}^{\,2}}}$$

The unit outward normal to the yield surface for this elastic limit case is

$$\frac{\partial f}{\partial \boldsymbol{\sigma}} = \frac{\mathbf{s}}{2k}; \quad (s_{ij}) = \begin{pmatrix} -\sigma/3 & 0 & 0 \\ 0 & -\sigma/3 & \tau \\ 0 & \tau & 2\sigma/3 \end{pmatrix}$$

Application of an additional infinitesimal torsion $d\boldsymbol{\sigma} = \tau \left(\mathbf{e}_\theta \otimes \mathbf{e}_z + \mathbf{e}_z \otimes \mathbf{e}_\theta\right)$ yields the plastic multiplier:

$$d\lambda = \frac{1}{H}\frac{1}{2k} \begin{pmatrix} -\sigma/3 & 0 & 0 \\ 0 & -\sigma/3 & \tau \\ 0 & \tau & 2\sigma/3 \end{pmatrix} : \begin{pmatrix} 0 & 0 & 0 \\ 0 & 0 & d\tau \\ 0 & d\tau & 0 \end{pmatrix} = \frac{\tau d\tau}{Hk}$$

The plastic strain increments are obtained from the flow rule:

$$\left(d\varepsilon_{ij}^{p}\right) = \frac{\tau d\tau}{2Hk^2} \begin{pmatrix} -\sigma/3 & 0 & 0 \\ 0 & -\sigma/3 & \tau \\ 0 & \tau & 2\sigma/3 \end{pmatrix}$$

Discussion in the $(\bar{\sigma} \times \bar{\tau})$-plane.

The elasticity domain for the combined tension–torsion stress state in the $(\sigma \times \tau)$-plane is an ellipse centered around the origin, having $\sqrt{3}k$ and k as half-axis. This is displayed in Figure 8.12, in the $(\bar{\sigma} \times \bar{\tau})$-plane, where $\bar{\sigma} = \sigma/k$ and $\bar{\tau} = \tau/k$. In this plane,

- In loading case (a), the stress vector at the onset of plastic deformation is oriented along the σ-axis, while the subsequent loading is oriented in the τ-direction (i.e., normal to the σ-direction). The directions taken by $\partial f/\partial \sigma_{zz}$ and $\partial f/\partial \sigma_{z\theta}$ are

$$\frac{\partial f}{\partial \sigma_{zz}} = \frac{\partial}{\partial \sigma}\sqrt{\sigma^2/3} = \frac{1}{3}\bar{\sigma}; \quad \frac{\partial f}{\partial \sigma_{z\theta}} = \frac{\partial}{\partial \tau}\sqrt{\sigma^2/3} = 0$$

The zero plastic multiplier can be explained in the $(\bar{\sigma} \times \bar{\tau})$-plane, from the normality of vector $(\partial f/\partial \sigma, \partial f/\partial \tau) = (\bar{\sigma}/3, 0)$ and the incremental stress vector $(d\sigma, d\tau) = (0, d\tau)$. It suffices to carry out the scalar product,

$$d\lambda = \frac{1}{H}\left(\bar{\sigma}/3, 0\right)\begin{pmatrix} 0 \\ d\tau \end{pmatrix} = 0$$

This is referred to as neutral plastic loading.

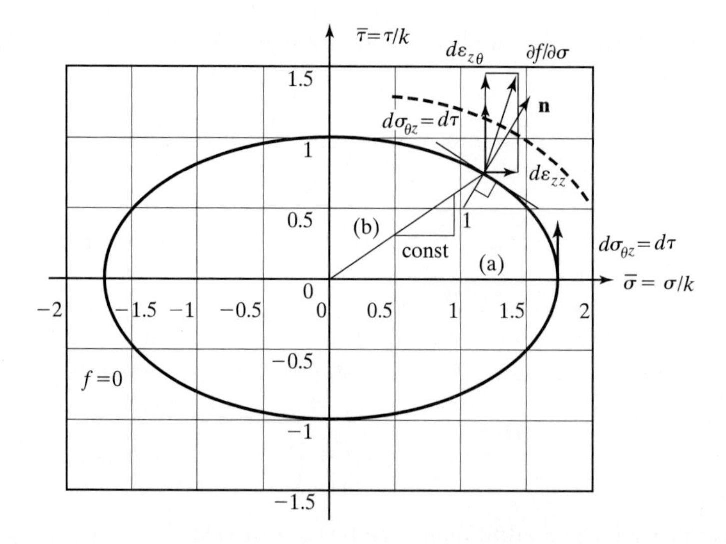

FIGURE 8.12: Combined tension–torsion: hardening in the $(\bar{\sigma}, \bar{\tau})$-plane.

- Things are different in loading case (b). The directions $\partial f/\partial\sigma_{zz}$ and $\partial f/\partial\sigma_{z\theta}$ read

$$\frac{\partial f}{\partial \sigma_{zz}} = \frac{\partial}{\partial \sigma}\sqrt{\sigma^2/3+\tau^2} = \frac{1}{3}\bar{\sigma}$$

$$\frac{\partial f}{\partial \sigma_{z\theta}} = \frac{\partial}{\partial \tau}\sqrt{\sigma^2/3+\tau^2} = \bar{\tau}$$

Thus, the plastic multiplier is

$$d\lambda = \frac{1}{H}\left(\bar{\sigma}/3, \bar{\tau}\right)\begin{pmatrix} 0 \\ d\tau \end{pmatrix} = \frac{\bar{\tau}d\tau}{H}$$

The directions $(\partial f/\partial\sigma, \partial f/\partial\tau) = (\bar{\sigma}/3, \bar{\tau})$ taken by the plastic increments $d\varepsilon^p_{zz}$ and $d\varepsilon^p_{z\theta}$ are displayed in Figure 8.12. We should note however that $(\partial f/\partial\sigma, \partial f/\partial\tau)$ is not normal to the displayed loading function (see Figure 8.12).

8.6.3 Elastic Ideal Plastic Loading

Ideal Plasticity.

Loading case 2(a), in which $d\sigma_{zz} = \sigma_{z\theta} = 0$, ensures the consistency condition of ideal plasticity:

$$df = \frac{\partial f}{\partial \boldsymbol{\sigma}} : d\boldsymbol{\sigma} = 0$$

Hence, in contrast to the hardening case, there is plastic flow at an undetermined rate $d\lambda$:

$$d\boldsymbol{\varepsilon}^p = \frac{d\lambda}{2\sqrt{3}}\left(-\mathbf{e}_r \otimes \mathbf{e}_r - \mathbf{e}_\theta \otimes \mathbf{e}_\theta + 2\mathbf{e}_z \otimes \mathbf{e}_z\right)$$

We note that the application of torsion stress $d\tau$ does not provoke a plastic distortion.

For the same reason (i.e., $df = \frac{\partial f}{\partial \boldsymbol{\sigma}} : d\boldsymbol{\sigma} = 0$), loading case 2(b) cannot be carried out on an ideal plastic material. The additional torsion stress leads in the ideal plastic case to $df > 0$, which is not permitted (loading point outside elasticity domain).

Elastic Loading.

For the assumed homogeneous stress state in the thin-walled hollow cylinder, $\boldsymbol{\sigma} = \sigma \mathbf{e}_z \otimes \mathbf{e}_z + \tau (\mathbf{e}_\theta \otimes \mathbf{e}_z + \mathbf{e}_z \otimes \mathbf{e}_\theta)$, the strain for an isotropic linear elastic material reads

$$(\varepsilon_{ij}) = \begin{pmatrix} -\frac{\nu\sigma}{E} & 0 & 0 \\ & -\frac{\nu\sigma}{E} & \frac{\tau}{2G} \\ \text{sym} & & \frac{\sigma}{E} \end{pmatrix}$$

where $2G = E/(1+\nu)$. The length increase during elastic loading is obtained from

$$\varepsilon_{zz} = \frac{\delta}{L} = \frac{\sigma}{E} \rightarrow \delta = \frac{L\sigma}{E}$$

The distortion is $\gamma = 2\varepsilon_{\theta z}$, and following the geometric relation displayed in Figure 8.11 (see also training set in Chapter 2, Section 2.5), we obtain the torsion angle α from

$$L\gamma = R\alpha \rightarrow \alpha = \frac{L}{R} 2\varepsilon_{\theta z} = \frac{L\tau}{RG}$$

Elastoplastic Loading.

Admissible elastoplastic range means here loading states satisfying

$$f = df = \frac{\partial f}{\partial \boldsymbol{\sigma}} : d\boldsymbol{\sigma} = 0$$

that is, here $f = 0 \Leftrightarrow \sqrt{J_2} = k$, and

$$df = 0 : \frac{1}{2k} \begin{pmatrix} -\frac{\sigma}{3} & 0 & 0 \\ 0 & -\frac{\sigma}{3} & \tau \\ 0 & \tau & \frac{2\sigma}{3} \end{pmatrix} : \begin{pmatrix} 0 & 0 & 0 \\ 0 & 0 & d\tau \\ 0 & d\tau & d\sigma \end{pmatrix} = \frac{\sigma d\sigma}{3k} + \frac{\tau d\tau}{k} = 0$$

For these loading states, $d\boldsymbol{\varepsilon} = d\boldsymbol{\varepsilon}^{\text{el}} + d\boldsymbol{\varepsilon}^{\text{p}}$, and $d\boldsymbol{\varepsilon}^{\text{p}} = d\lambda \mathbf{s}/2k$:

$$(d\varepsilon_{ij}) = \begin{pmatrix} -\frac{\nu d\sigma}{E} & 0 & 0 \\ & -\frac{\nu d\sigma}{E} & \frac{d\tau}{2G} \\ \text{sym} & & \frac{d\sigma}{E} \end{pmatrix} + \frac{d\lambda}{2k} \begin{pmatrix} -\frac{\sigma}{3} & 0 & 0 \\ 0 & -\frac{\sigma}{3} & \tau \\ 0 & \tau & \frac{2\sigma}{3} \end{pmatrix}$$

The change in radius, torsion angle, and length are obtained from the components of $d\varepsilon$ (which preserve their geometric significance also during plastic deformation)[8]:

$$d\varepsilon_{rr} = \frac{dR}{R} = -\frac{\nu d\sigma}{E} - \frac{d\lambda\sigma}{6k} \quad \rightarrow \quad dR = -R\left(\frac{\nu d\sigma}{E} + \frac{d\lambda\sigma}{6k}\right)$$

$$d\varepsilon_{\theta z} = \frac{Rd\alpha}{2L} = \frac{d\tau}{2G} + \frac{d\lambda\tau}{2k} \quad \rightarrow \quad d\alpha = \frac{L}{R}\left(\frac{d\tau}{G} + \frac{d\lambda\tau}{k}\right)$$

$$d\varepsilon_{zz} = \frac{dL}{L} = \frac{d\sigma}{E} + \frac{d\lambda\sigma}{3k} \quad \rightarrow \quad dL = L\left(\frac{d\sigma}{E} + \frac{d\lambda\sigma}{3k}\right)$$

Plastic Loading.

We need to check whether an increasing torsion angle (i.e., $d\alpha > 0$), at a constant cylinder length, $dL = 0$, corresponds to a plastic loading. To this end, we use the previous obtained results for dR, $d\alpha$, and dL. The condition $dL = 0$ delivers the plastic multiplier:

$$dL = 0 : d\lambda = -3\frac{kd\sigma}{E\sigma} \geq 0$$

We note that $d\lambda = 0$ for $d\sigma = 0$, and $d\lambda > 0$ for $d\sigma < 0$. Use of the expression for $d\lambda$ in the relation giving $d\alpha$ yields

$$d\alpha = \frac{L}{R}\left(\frac{d\tau}{G} - 3\frac{d\sigma}{E}\frac{\tau}{\sigma}\right)$$

Therefore, a positive torsion stress increment $d\tau > 0$ together with $d\sigma < 0$ ensures the positivity of $d\alpha$. On the other hand, for $d\tau > 0$, $d\sigma$ cannot be zero during plastic loading, without violating the consistency condition:

$$df = 0 : \sigma d\sigma + 3\tau\, d\tau = 0$$

Hence, $d\alpha > 0$ corresponds necessarily to plastic loading.

Next, the change of radius reads

$$dR = -R\left(\frac{\nu}{E} - \frac{1}{2E}\right) d\sigma$$

[8]Note that the change in radius is obtained from

$$d\varepsilon_{rr} = \frac{\partial\left(d\xi_r\right)}{\partial r} \rightarrow d\xi_r = \int d\varepsilon_{rr} dr$$

where $d\xi_r = d\xi_r(r)$ is the radial displacement increment. Given the homogeneous stress and strain state in the thin cylinder, $d\xi_r = d\varepsilon_{rr} r$. Then,the change of radius is

$$dR = d\xi_r(r = R)$$

Therefore, for the thin cylinder, with constant strain over r,

$$d\varepsilon_{rr} = \frac{d\xi_r(r = R)}{R} = \frac{dR}{R}$$

and, after integration,

$$\frac{\Delta R}{R} = \frac{\Delta\sigma}{E}\left(\frac{1}{2} - \nu\right)$$

where $\Delta\sigma = \sigma - \sigma_0$. We note that the radius increases during plastic loading, in contrast to elastic loading.

We are left with expressing the stress σ as a function of the torsion angle α. From Figure 8.13, we note that the stresses situated on the loading surface satisfy

$$f = 0 : \sigma = k\sqrt{3}\cos\phi; \quad \tau = k\sin\phi$$

The stress increments read

$$d\sigma = -d\phi k\sqrt{3}\sin\phi; \quad d\tau = d\phi k\cos\phi$$

Use in the expression for $d\alpha$ gives

$$d\alpha = \frac{Lk}{R}\left(\frac{\cos\phi}{G} + 3\frac{\sin\phi}{E}\frac{\sin\phi}{\cos\phi}\right)d\phi = \frac{Lk}{R}\left[\left(\frac{1}{G} - \frac{3}{E}\right)\cos\phi + \frac{3}{E\cos\phi}\right]d\phi$$

and, after integration, with the initial condition $\alpha(\tau = k\sin\phi = 0) = 0$,

$$\alpha(\phi) = \frac{Lk}{R}\left[\left(\frac{1}{G} - \frac{3}{E}\right)\sin\phi + \frac{3}{E}\ln\tan\left(\frac{\phi}{2} + \frac{\pi}{4}\right)\right]$$

The stress solution is then obtained in an implicit form as a function of $\phi = \phi(\alpha)$. At the onset of plastic deformation, $\phi = \alpha = 0$; and for $\phi = \pi/2$ the torsion angle becomes infinite, $\alpha \to \infty$. This plastic loading is sketched in Figure 8.13.

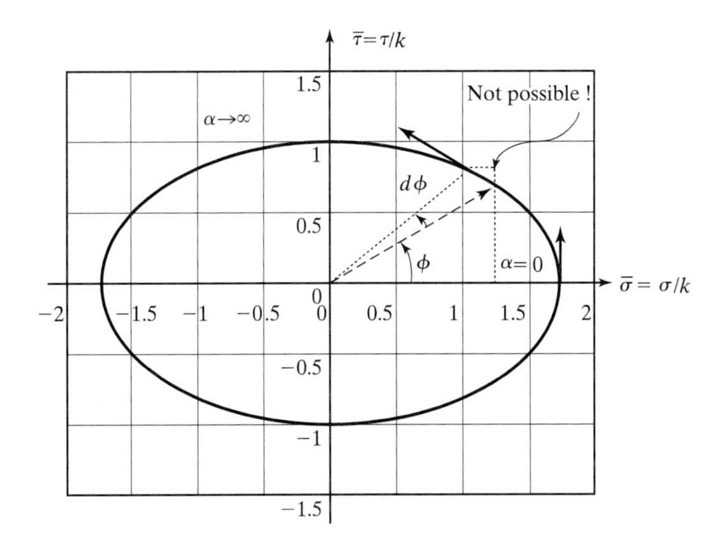

FIGURE 8.13: Ideal plastic loading in the $(\overline{\sigma}, \overline{\tau})$-plane.

8.7 PROBLEM SET: CHAMPAGNE METHOD

We consider a rigid infinite half-space with an empty hole of diameter $2a$ and length L (see Figure 8.14). In this hole we want to force a deformable cylinder sample of same length L but of a greater diameter $2R > 2a$ than the bore hole. The cylinder is initially stress free, and it is entirely forced into the hole. During this process, the cylinder preserves its length. The contact between the cylinder walls and the bore hole walls is frictionless. We want to study the stress fields and the required force F to maintain the cylinder in the bore hole. Throughout this problem set we assume isothermal and quasi-static evolutions, and body forces are disregarded.

1. **Deformation and Strain:** We consider a displacement field of the form

$$\boldsymbol{\xi} = u(r)\mathbf{e}_r + \mathbf{D}$$

with $u(r)$ the displacement in the radial direction $\mathbf{e}_r$: It is a pure radial displacement field. $\mathbf{D}$ is a rigid body displacement field, which ensures during the process that the axis of the cylinder sample coincides with the axis of the bore hole.

   (a) Determine the condition which the displacement field needs to satisfy to be kinematically admissible.

   (b) Once the cylinder sample is in the hole, we want to deal with the problem within the hypothesis of small perturbations. For the problem at hand, specify the restriction on the considered displacement field.

   (c) Determine the linearized strain tensor.

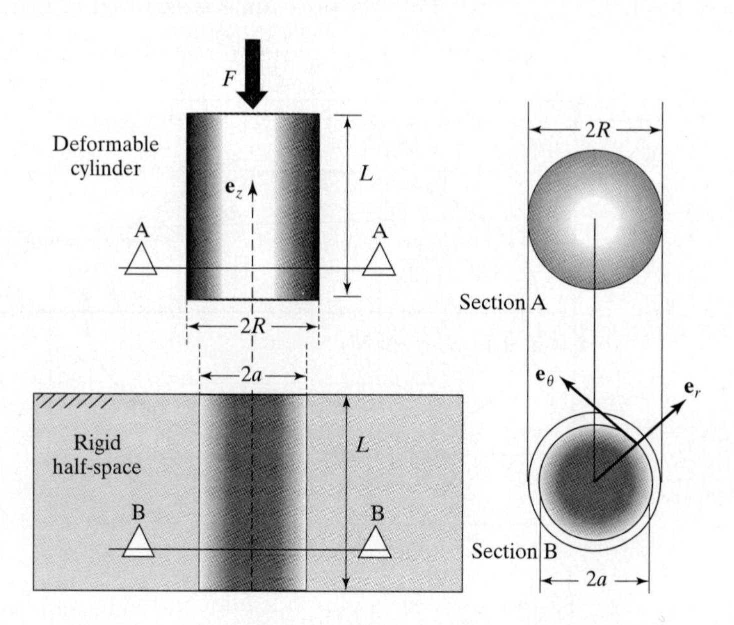

FIGURE 8.14: Problem set: Champagne method.

2. **Elasticity**: We want to determine the stress field in the sample at the end of the process, and the required force to enforce the sample in the hole. In a first approximation, we suppose a linear elastic isotropic behavior of the cylinder sample (Lamé constants λ, G).

   (a) Determine the elastic stress field $\boldsymbol{\sigma}$ in the cylinder sample.

   (b) Which equation needs to satisfy the radial displacement $u(r)$ for the elastic stress field $\boldsymbol{\sigma}$ to be statically admissible? By solving this equation, determine function $u(r)$ and the linear elastic stress solution.

   (c) Determine the force intensity F that is required to enforce the sample in the bore hole. Determine the pressure p that the surrounding rigid medium exerts on the sample.

   (d) Determine the elastic energy that is stored in the cylinder once the sample is in the bore hole.

3. **Elastic Strength Limit**: We want to determine the maximum difference in radius $R - a$ between the sample and the bore hole, when the stress reaches the elastic limit.

   (a) Plot the Mohr circle of the elastic solution determined previously.

   (b) The sample is assumed to be composed of a Tresca material, characterized by a uniaxial compressive strength f'_c. Determine the admissible radius difference $R - a$. Where and on which surfaces does the maximum shear occur?

   (c) What is the maximum admissible radius difference $R - a$ for a Mohr–Coulomb material, defined by the cohesion c and the friction angle φ?

4. **Plasticity**: More realistically, we assume now that the cylinder is composed of an isotropic ideal elastoplastic Von–Mises material (shear strength k). We want to determine the required force F to enforce the sample in the hole for a given radius difference $R - a_1 > R - a_0$. $R - a_0$ corresponds to the limit radius difference when the sample starts yielding.

   (a) Determine the radius difference $R - a_0$ that corresponds to the elastic limit state of the Von–Mises material sample.

   (b) We want to determine the elastoplastic stress field $\boldsymbol{\sigma}$ prevailing in the cylinder once the cylinder is in the hole.

      i. For the problem at hand, write the constitutive equations required to solve the problem.

      ii. Show that the consistency condition $df = 0$ leads to

      $$dp - \frac{dF}{\pi R^2} = 0$$

      where dp is the pressure increment applied from the surrounding medium on the sample; and dF the force applied on the cylinder head (see Figure 8.14). From the previous result, determine the plastic multiplier.

iii. Determine the stress tensor $\boldsymbol{\sigma}$ at the end of the elastoplastic loading process.

(c) Illustrate the elastoplastic process in the Mohr stress plane. How is the stress vector on the critical shear plane affected by the elastoplastic process?

(d) Determine the force intensity F during the elastoplastic process. Compare the result with the elastic case (i.e., Question 2c). Comment.

(e) Plot the pressure p exerted by the surrounding rigid half-space versus the radius variation for this loading process.

(f) Determine the amount of energy that is dissipated into heat during the process.

(g) Last, we want to "uncork" the cylinder sample from the rigid half-space by means of a "corkscrew" that preserves the length of the cylinder sample during uncorking. Determine the required force F to do so.

8.7.1 Deformation and Strain

Kinematically Admissable Displacement Field.

The displacement field must satisfy the displacement boundary condition on $\partial\Omega_{\xi^d}$. This is the prescribed radial displacement:

$$r = R : u^d = a - R$$

This displacement field is such that any point at the cylinder wall, defined by position vector $\mathbf{X}$ in the undeformed configuration, becomes after deformation $\mathbf{x} = \mathbf{X} + u^d\mathbf{e}_r + \mathbf{D}$ (i.e., in the r-direction, $x_r = a$, $X_r = R$; therefore, $u^d = x_r - X_r$).

Small Displacement Hypothesis.

$$|a - R| \ll R$$

The consequence of the small perturbation hypothesis is

$$u(r = R) = u(r = a)$$

Linearized Strain Tensor.

Rigid body motion does not provoke strain; $\varepsilon_{zz} = 0$ since the length in the bore hole is the same as the original length. Due to radial symmetry, the shear strains are zero. It follows that

$$\varepsilon_{rr} = \frac{\partial u(r)}{\partial r}; \quad \varepsilon_{\theta\theta} = \frac{u(r)}{r}; \quad \varepsilon_{zz} = \frac{\partial u_z}{\partial z} = 0; \quad \text{other } \varepsilon_{ij} = 0$$

8.7.2 Elasticity

Elastic Stress Field in Cylinder.

$$\boldsymbol{\sigma} = 2G\boldsymbol{\varepsilon} + \lambda\,(\mathrm{tr}\,\boldsymbol{\varepsilon})\,\mathbf{1};$$

Thus,

$$\begin{aligned}
\sigma_{rr} &= \left(2G+\lambda\right)\frac{\partial u(r)}{\partial r} + \lambda\frac{u(r)}{r} \\
\sigma_{\theta\theta} &= \left(2G+\lambda\right)\frac{u(r)}{r} + \lambda\frac{\partial u(r)}{\partial r} \\
\sigma_{zz} &= \lambda\left(\frac{\partial u(r)}{\partial r} + \frac{u(r)}{r}\right)
\end{aligned}$$

Statically Admissible Stress Field.

Equilibrium equation:

$$\frac{\partial \sigma_{rr}}{\partial r} + \frac{1}{r}\left(\sigma_{rr} - \sigma_{\theta\theta}\right) = 0$$

Thus, with the stress components previously defined (see Section 5.5),

$$u'' + u'/r - u/r^2 = 0 \rightarrow u(r) = Ar + \frac{B}{r}$$

since $u(r=0)=0 \rightarrow B=0$. The displacement boundary condition at $r=R$ gives

$$u(r=R) = a - R = AR \rightarrow A = \frac{a-R}{R}$$

Whence the elastic solution:

$$\sigma_{rr} = \sigma_{\theta\theta} = 2(G+\lambda)\left[\frac{a-R}{R}\right]$$

$$\sigma_{zz} = 2\lambda\left[\frac{a-R}{R}\right] = \nu(\sigma_{rr} + \sigma_{\theta\theta})$$

Force Intensity.

Application of the reduction formula delivers

$$-F\mathbf{e}_z = \int_{A=\pi a^2 \simeq \pi R^2} \boldsymbol{\sigma}\cdot\mathbf{e}_z da \simeq 2\lambda\left[\frac{a-R}{R}\right]\pi R^2 \mathbf{e}_z \rightarrow F = 2\pi\lambda\Big[R-a\Big]R$$

The pressure p exerted from the surrounding rigid medium is

$$\text{on } r=R: \mathbf{T}(\mathbf{n}=\mathbf{e}_r) = \boldsymbol{\sigma}\cdot\mathbf{e}_r \equiv -p\mathbf{e}_r \rightarrow p = -\sigma_{rr} = 2(G+\lambda)\left[\frac{R-a}{R}\right]$$

or, equivalently,

$$p = \frac{F}{2\nu\pi R^2}$$

Stored Energy.

Since $\boldsymbol{\xi}$ and $\boldsymbol{\sigma}$ are solutions of the linear elastic problem, Clapeyron's formula applies:

$$W(\varepsilon) = \int_{\Omega} \psi(\varepsilon)\, d\Omega = \frac{1}{2}\Big[\Phi(\boldsymbol{\xi}) + \Phi^*(\boldsymbol{\sigma})\Big]$$

Noting that

$$\Phi(\boldsymbol{\xi}) = \int_{\partial\Omega_{\mathbf{T}^d}} \mathbf{T}^d \cdot (\boldsymbol{\xi} - D)\, da = 0$$

we obtain

$$\begin{aligned} W(\varepsilon) &= \frac{1}{2}\Phi^*(\boldsymbol{\sigma}) = \frac{1}{2}\int_{\partial\Omega_{\boldsymbol{\xi}^d}} \mathbf{T}(\mathbf{e}_r)\cdot(\boldsymbol{\xi}^d - \mathbf{D}^d)\, da = \frac{1}{2}\int_{\partial\Omega_{\boldsymbol{\xi}^d} = 2\pi RL} \Big[-pu(r=R)\Big]\, da \\ &= p(R-a)L\pi R = 2\pi(G+\lambda)(R-a)^2 L \end{aligned}$$

8.7.3 Elastic Strength Limit

Mohr Circle.

Here, $\sigma_I = \sigma_{zz} \geq \sigma_{II} = \sigma_{III} = \sigma_{rr} = \sigma_{\theta\theta}$. The radius and center of the Mohr circle are

$$|\tau| = \frac{1}{2}(\sigma_I - \sigma_{III}) = G\left[\frac{R-a}{R}\right]$$

$$\sigma_c = \frac{1}{2}(\sigma_I + \sigma_{III}) = -(2\lambda + G)\left[\frac{R-a}{R}\right]$$

The Mohr circle is displayed in Figure 8.15.

Tresca Criterion.

Note that $\sigma_0 = f_c'$. The Tresca criterion for the problem delivers

$$f(\boldsymbol{\sigma}) = |\tau_{\text{crit}}| - f_c'/2 = 0 \to \max(R-a) = R\frac{f_c'}{2G}$$

Maximum shear occurs on a cone of which the surfaces are inclined with regard to $\mathbf{e}_z$-direction by (see Figure 8.15)

$$\vartheta(\mathbf{u}_I = \mathbf{e}_z, \mathbf{n}) = -\pi/4$$

Mohr–Coulomb Criterion.

$$f(\boldsymbol{\sigma}) = 0 : |\tau_{\text{crit}}| + \sigma_{\text{crit}} \tan\varphi - c = 0$$

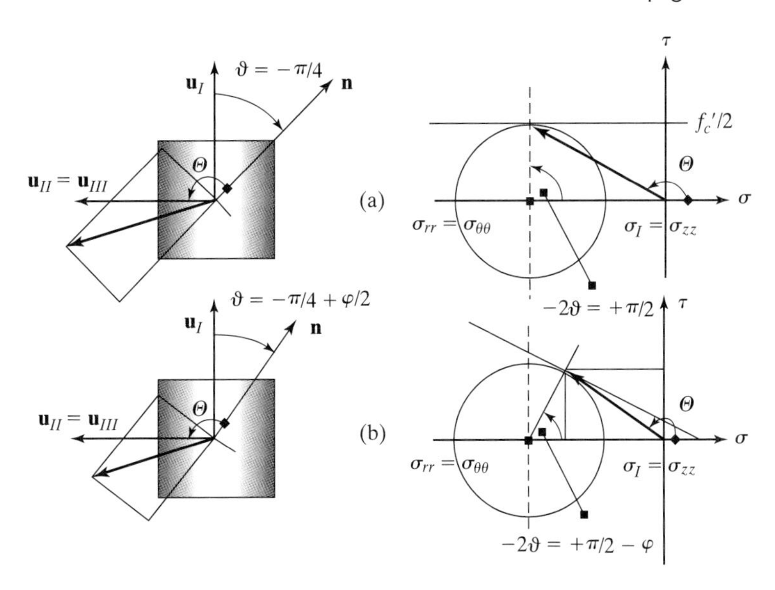

FIGURE 8.15: Elastic limits in the physical space and the Mohr plane: (a) Tresca criterion; (b) Mohr–Coulomb criterion.

If we note $|\tau_{\text{crit}}| = |\tau|\cos\varphi$, $\sigma_{\text{crit}} = \sigma_c + |\tau|\sin\varphi$, we obtain for the problem at hand

$$\max(R - a) = R\frac{c\cos\varphi}{G - \sin\varphi[G + 2\lambda]}$$

Maximum shear occurs on a cone inclined with regard to $\mathbf{e}_z$-direction by (see Figure 8.15)

$$\vartheta(\mathbf{u}_I = \mathbf{e}_z, \mathbf{n}) = \varphi/2 - \pi/4$$

8.7.4 Plasticity

Von–Mises Criterion.

$$f(\boldsymbol{\sigma}) = \sqrt{J_2} - k \leq 0$$

where $J_2 = \sqrt{\frac{1}{2}s_{ij}s_{ij}}$. The stress state is a triaxial stress state:

$$(s_{ij}) = \frac{1}{3}\left(p + \sigma_{zz}\right)\begin{pmatrix} -1 & & \\ & -1 & \\ & & 2 \end{pmatrix} \to J_2 = \frac{1}{3}\left(p + \sigma_{zz}\right)^2$$

Thus

$$f(\boldsymbol{\sigma}) = 0 : \frac{1}{\sqrt{3}}|p + \sigma_{zz}| - k = 0$$

With elastic solution, we obtain the elastic limit:

$$p + \sigma_{zz} = 2G\left[\frac{R - a(t)}{R}\right] \to f\left[a(t) = a_0\right] = 0 \to R - a_0 = R\frac{\sqrt{3}k}{2G}$$

Elastoplastic Stress Field for Von–Mises Plasticity.

We first note for Von–Mises plasticity:

$$\mathrm{tr}\,\boldsymbol{\varepsilon}^p = 0$$

Therefore, with $\varepsilon_{zz} = 0$ and other strain quantities still as previously obtained,

$$\sigma_{rr} = 2G(\varepsilon_{rr} - \varepsilon^p_{rr}) + \lambda \mathrm{tr}\,\boldsymbol{\varepsilon}$$

$$\sigma_{\theta\theta} = 2G(\varepsilon_{\theta\theta} - \varepsilon^p_{\theta\theta}) + \lambda \mathrm{tr}\,\boldsymbol{\varepsilon}$$

$$\sigma_{zz} = -2G\varepsilon^p_{zz} + \lambda \mathrm{tr}\,\boldsymbol{\varepsilon}$$

where $\mathrm{tr}\,\boldsymbol{\varepsilon} = \varepsilon_{rr} + \varepsilon_{\theta\theta}$. In addition, since the stress field in the sample at the yield limit is homogeneous, the plastic deformation in the sample is also homogeneous. The displacement, therefore, is still of the form

$$u(r) = (a(t)/R - 1)r \rightarrow \varepsilon_{rr} = \varepsilon_{\theta\theta} = \frac{a(t) - R}{R}$$

Furthermore, $\sigma_{rr} = \sigma_{\theta\theta} = -p$ during plastic loading. In addition, the plastic flow rule reads here

$$d\boldsymbol{\varepsilon}^p = d\lambda \frac{\mathbf{s}}{2\sqrt{J_2}} = \frac{d\lambda}{2\sqrt{3}} \mathrm{sign}\,(p + \sigma_{zz}) \begin{pmatrix} -1 & & \\ & -1 & \\ & & 2 \end{pmatrix}$$

where $d\lambda$ = plastic multiplier. Therefore,

$$d\varepsilon^p_{rr} = d\varepsilon^p_{\theta\theta}$$

We first assume that $\mathrm{sign}(p + \sigma_{zz}) = +1$, as this is the case when the stress reaches the yield limit (see Mohr circle), and rewrite the stress equations of state in an incremental form[9]:

$$\begin{pmatrix} d\sigma_{rr} \\ d\sigma_{\theta\theta} \\ d\sigma_{zz} \end{pmatrix} = 2\frac{da}{R} \begin{pmatrix} G + \lambda \\ G + \lambda \\ \lambda \end{pmatrix} + G\frac{d\lambda}{\sqrt{3}} \begin{pmatrix} 1 \\ 1 \\ -2 \end{pmatrix}$$

[9]Note that the prescribed displacement on the cylinder wall reads:

$$u^d(r = R) = a(t) - R$$

Thus,

$$du^d = da$$

The loading is elastic for $u^d \in [0, a_0 - R_1]$ and becomes elastoplastic for $u^d \in [a_0 - R, a_1 - R]$. The total displacement is

$$u(r) = u^d \frac{r}{R}$$

Therefore, the strains

$$\varepsilon_{rr} = \varepsilon_{\theta\theta} = u^d/R$$

and in an incremental form

$$d\varepsilon_{rr} = d\varepsilon_{\theta\theta} = du^d/R = da/R$$

with $da \leq 0$.

Next, we determine the plastic multiplier from the consistency condition $df = 0$. With $\sigma_{rr} = \sigma_{\theta\theta} = -p$ during plastic loading, the consistency condition reduces to

$$df = \frac{\partial f}{\partial \boldsymbol{\sigma}} : d\boldsymbol{\sigma} = 0 \rightarrow dp + d\sigma_{zz} = 0$$

Thus, noting $dp = -d\sigma_{rr} = -d\sigma_{\theta\theta}$:

$$-2G\frac{da}{R} - \sqrt{3}Gd\lambda = 0 \rightarrow d\lambda = -\frac{2}{\sqrt{3}}\frac{da}{R}$$

Integration gives

$$\Lambda = \int d\lambda = -\frac{2}{\sqrt{3}}\int_{\alpha=a_0}^{\alpha=a_1}\frac{d\alpha}{R} = \frac{2}{\sqrt{3}}\frac{a_0 - a_1}{R}$$

It follows that

$$\varepsilon_{rr}^p = -\frac{a_0 - a_1}{3R} = \varepsilon_{\theta\theta}^p$$

$$\varepsilon_{zz}^p = 2\frac{a_0 - a_1}{3R} = -(\varepsilon_{rr}^p + \varepsilon_{\theta\theta}^p)$$

With the plastic solution in hand, we determine the elastoplastic stress field:

$$\begin{aligned}
\sigma_{rr} &= \sigma_{\theta\theta} = 2(G+\lambda)\left[\frac{a_1 - R}{R}\right] + 2G\frac{a_0 - a_1}{3R} \\
&= 2(G+\lambda)\left[\frac{a_0 - R}{R}\right] - \left(\frac{4}{3}G + 2\lambda\right)\frac{a_0 - a_1}{R}
\end{aligned}$$

$$\begin{aligned}
\sigma_{zz} &= 2\lambda\left[\frac{a_1 - R}{R}\right] - 4G\frac{a_0 - a_1}{3R} \\
&= 2\lambda\left[\frac{a_0 - R}{R}\right] - \left(\frac{4}{3}G + 2\lambda\right)\frac{a_0 - a_1}{R}
\end{aligned}$$

We verify that $p + \sigma_{zz} = \sqrt{3}k > 0$ provided that $a_1 < a_0 < R$. The found stress solution satisfies the plasticity criterion $f(\boldsymbol{\sigma}) = 0$, the stress state is statically admissible, and the associated displacement field is kinematically admissible. In short, it is the solution of the problem.

Mohr Plane.

While the radius of the Mohr circle remains constant during plastic loading from $R_0 \rightarrow R_1$,

$$|\tau| = \frac{1}{2}(\sigma_I - \sigma_{III}) = \frac{1}{2}(\sigma_{zz} + p) = \frac{\sqrt{3}k}{2}$$

the center shifts by

$$\sigma_c(R_1) = \frac{1}{2}(\sigma_I + \sigma_{III}) = \frac{1}{2}(\sigma_{zz} - p) = \sigma_c(R_0) - \left(\frac{4}{3}G + 2\lambda\right)\frac{a_0 - a_1}{R}$$

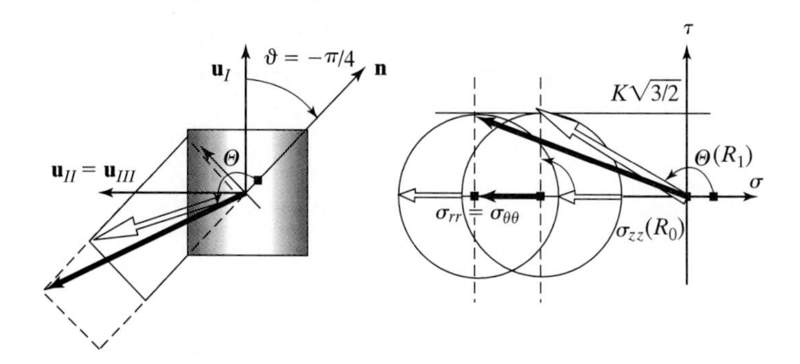

FIGURE 8.16: Stretch and rotation of stress vector during plastic deformation.

This is displayed in Figure 8.16. The stress vector on the shear plane is stretched and rotated. While the shear component of the stress vector remains constant, the normal stress component changes its value.

Force Intensity.

$$-F\mathbf{e}_z = \int_{A=\pi a^2 \simeq \pi R^2} \boldsymbol{\sigma} \cdot \mathbf{e}_z da \rightarrow F = \left[2\lambda[R - a(t)] + \frac{4G}{3} \langle a_0 - a(t) \rangle \right] \pi R$$

With yielding a greater force per unit of radius change $da < 0$ needs to be applied. This is due to the zero-plastic dilatation of the Von–Mises plasticity criterion.

Pressure p.

The loading process involves two stages:

- Elastic loading:

$$u^d \in \left[0, a_0 - R\right] \leftrightarrow p = 2(G + \lambda) \left[\frac{-u^d(t)}{R}\right]$$

- Plastic loading:

$$u^d \in \left[a_0 - R, a_1 - R\right] \leftrightarrow p = 2(G + \lambda) \left[\frac{-u^d(t)}{R}\right] - \frac{2G}{3} \frac{\langle a_0 - R - u^d(t) \rangle}{R}$$

The pressure–radial displacement curve, $p = p(u^d)$, is shown in Figure 8.17.

Energy Dissipation.

The local dissipation (i.e., per unit of volume) of the Von–Mises plasticity model is

$$\varphi dt = \boldsymbol{\sigma} : d\boldsymbol{\varepsilon}^p = d\lambda \boldsymbol{\sigma} : \frac{\partial f}{\partial \boldsymbol{\sigma}} = d\lambda \frac{\boldsymbol{\sigma} : \mathbf{s}}{2\sqrt{J_2}} = d\lambda \sqrt{J_2} \equiv d\lambda k$$

Thus, with the previous results,

$$\Delta = k \int d\lambda = \Lambda k = \frac{2k}{\sqrt{3}} \frac{a_0 - a_1}{R}$$

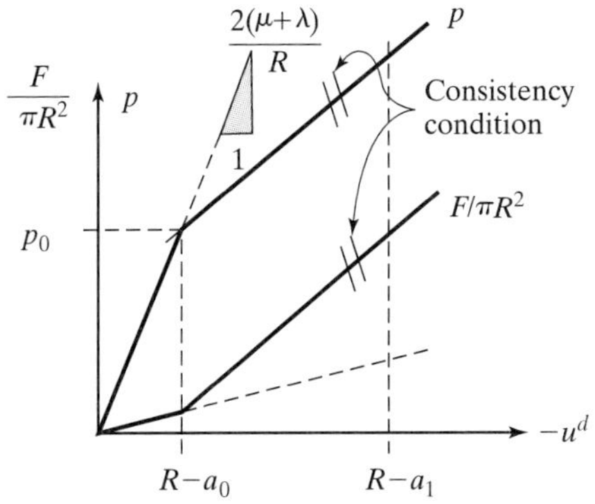

FIGURE 8.17: Load displacement curves.

Therefore, in the sample,

$$\mathcal{D} = \int_{\Omega=\pi R^2 L} \Lambda k d\Omega = \frac{2k}{\sqrt{3}}(a_0 - a_1)\pi R L$$

Energy stored in the system

$$W = \int_{\Omega=\pi R^2 L} \psi(\boldsymbol{\varepsilon}, \boldsymbol{\varepsilon}^p) d\Omega = \left[\frac{\lambda}{2}[\mathrm{tr}\,(\boldsymbol{\varepsilon} - \boldsymbol{\varepsilon}^p)]^2 + G\mathrm{tr}\,[(\boldsymbol{\varepsilon} - \boldsymbol{\varepsilon}^p)^2]\right] \pi R^2 L$$

With

$$\mathrm{tr}\,(\boldsymbol{\varepsilon} - \boldsymbol{\varepsilon}^p) = \mathrm{tr}\,\boldsymbol{\varepsilon} = \varepsilon_{rr} + \varepsilon_{\theta\theta} = 2\varepsilon_{rr}$$

and

$$\begin{aligned} \mathrm{tr}\,[(\boldsymbol{\varepsilon} - \boldsymbol{\varepsilon}^p)^2] &= (\varepsilon_{rr} - \varepsilon^p_{rr})^2 + (\varepsilon_{\theta\theta} - \varepsilon^p_{\theta\theta})^2 + (-\varepsilon^p_{zz})^2 \\ &= 2\varepsilon^2_{rr} - 4\varepsilon_{rr}\varepsilon^p_{rr} + 6\varepsilon^{p\;2}_{rr} \end{aligned}$$

we obtain

$$\begin{aligned} W &= \left[(2\lambda + 2G)\varepsilon^2_{rr} + 2G(3\varepsilon^{p\;2}_{rr} - 2\varepsilon_{rr}\varepsilon^p_{rr})\right]\pi R^2 L \\ &= \pi L \left[(2\lambda + 2G)[R - a(t)]^2 + \frac{2G}{3}\langle a_0 - a(t)\rangle^2 - \frac{4G}{3}[R - a(t)]\langle a_0 - a(t)\rangle\right] \\ &= \pi L \left[(2\lambda + 2G)[R - a(t)]^2 - \frac{2G}{3}\langle a_0 - a(t)\rangle [2R - a(t) - a_0]\right] \end{aligned}$$

Uncorking.

Uncorking corresponds to an elastic unloading, which relaxes the radial pressure exerted by the rigid half-space on the sample. Since the length of the sample is preserved (i.e., $\varepsilon_{zz} = 0$), we have after uncorking:

$$\sigma_{rr} = 0 : \sigma_{rr} = 2(G + \lambda)\varepsilon_{rr} - 2G\varepsilon^p_{rr} = 0 \rightarrow \varepsilon_{rr} = \frac{G\varepsilon^p_{rr}}{G + \lambda}$$

The vertical stress becomes

$$
\begin{aligned}
\sigma_{zz} &= -2G\varepsilon_{zz}^{p} + 2\lambda\varepsilon_{rr} = -2G\varepsilon_{zz}^{p} + \frac{2G\lambda\varepsilon_{rr}^{p}}{G+\lambda} \\
&= 4G(1+\nu)\,\varepsilon_{rr}^{p} = 2E\varepsilon_{rr}^{p} = -\frac{2E}{3R}(a_0 - a_1)
\end{aligned}
$$

Therefore, the force required to uncork the cylinder is

$$
F_{\mathrm{cr}} = \frac{2E}{3}(a_0 - a_1)\pi R
$$

Note that $F_{\mathrm{cr}} < F$, where F is the force required to enforce the cylinder in the hole.

CHAPTER 9

Limit Analysis and Yield Design

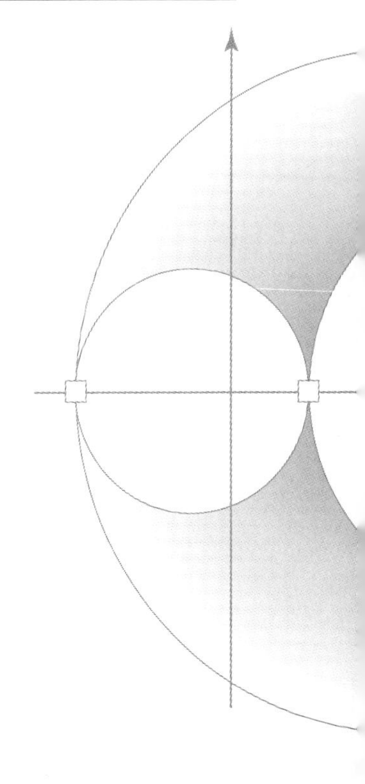

Material systems and structures collapse or fail when the work supplied from the outside cannot be stored as recoverable energy into the system. The capacity is exhausted to sustain any additional load by means of stresses in the structure, which satisfy equilibrium and which do not exceed the local material strength. If the limit load is associated with plastic failure, any additional supplied work is dissipated through plastic yielding in the material bulk and/or along narrow bands of surfaces of discontinuity. Without performing a detailed analysis of stresses and deformation prior to failure, the aim of limit analysis is to study whether or not a given structure has a sufficient capacity to dissipate the additional external work. Limit theorems provide estimates of the dissipation at plastic collapse: The lower limit theorem, based on statically and plastically compatible stress fields, underestimates the actual dissipation capacity; and the upper limit theorem, associated with kinematically admissible plastic failure mechanisms, overestimates it. These bounds are at the core of yield design and are widely used in engineering design of materials and structures.

9.1 ELEMENTS OF LIMIT ANALYSIS

The objective of limit analysis is to determine the load-bearing capacity of plastic material systems and structures. At plastic collapse, the structure has exhausted its capacity to store any additional external work $d\mathcal{W}_{\text{ext}}$ in recoverable free energy,

dW, including hardening energy. This is formally expressed by the Clausius–Duhem inequality, which at the structural level and for isothermal evolutions reads

$$d\mathcal{D} = d\mathcal{W}_{\text{ext}} - dW \geq 0 \tag{9.1}$$

where $d\mathcal{D} = \int_\Omega \varphi dt \, d\Omega$ denotes the dissipation. At plastic collapse, $dW = 0$. It follows that the additional external work $d\mathcal{W}_{\text{ext}}$ is entirely dissipated, through plastic yielding, into heat form $d\mathcal{D}$:

$$dW \equiv 0 \Leftrightarrow d\mathcal{D} = d\mathcal{W}_{\text{ext}} \geq 0 \tag{9.2}$$

Equation (9.2) defines the ultimate plastic dissipation capacity of an elastoplastic material system, without referring to the different energy states W the structure had witnessed on its way to this limit state. Limit analysis, therefore, does not require to perform the complete elastoplastic analysis of the stress and deformation prior to failure. For a given geometry and prescribed forces, limit analysis concentrates on the critical additional work increment leading to failure.

9.1.1 Plastic Collapse Load

The two underlying ideas of plastic yield analysis are as follows:

1. Plastic collapse occurs once the structure has exhausted its capacity to develop, in response to a prescribed loading, stress fields that are (a) statically compatible with the external loading, and (b) plastically admissible with the strength of the constitutive material. This suggests a *stress–strength approach* to the load-bearing capacity of material systems and structures, to evaluate *when* plastic failure occurs: Among all possible stress fields $\boldsymbol{\sigma}(\mathbf{x})$, limit analysis explores the ones that are statically compatible with prescribed body forces $\rho\mathbf{f}$ and surface forces $\mathbf{T}^d$ and that, at the same time, are compatible with the strength domain $D_k(\mathbf{x})$ of the constitutive material at any point $\mathbf{x}$ of the structure Ω:

$$\text{in } \Omega : \rho\mathbf{f} = \operatorname{div} \boldsymbol{\sigma}(\mathbf{x}); \quad \text{on } \partial\Omega_{\mathbf{T}^d} : \mathbf{T}^d = \boldsymbol{\sigma}(\mathbf{x}) \cdot \mathbf{n}(\mathbf{x}) \tag{9.3}$$

$$\forall \mathbf{x}; \quad \boldsymbol{\sigma}(\mathbf{x}) \in D_k(\mathbf{x}) \leftrightarrow f(\mathbf{x}; \boldsymbol{\sigma}(\mathbf{x})) \leq 0 \tag{9.4}$$

where $\partial\Omega_{\mathbf{T}^d}$ is the boundary of Ω where surface forces are prescribed and $f(\mathbf{x}; \boldsymbol{\sigma}(\mathbf{x}))$ denotes the scalar loading function that defines the local strength domain D_k of the material composing the structure.

2. Plastic collapse occurs at an undetermined rate. This has two implications: (a) The failure mechanism described by velocity field $\mathbf{V}(\mathbf{x})$ cannot be controlled from the outside; but (b) is set forth by the kinematics of the plastic flow which develops in the structure. The first condition means that failure cannot be prescribed by a controlled velocity field:

$$\text{on } \partial\Omega_{\mathbf{V}^d} : \mathbf{V}^d \equiv 0 \tag{9.5}$$

Therefore, by application of the theorem of virtual work (6.14) with the velocity field $\mathbf{V}(\mathbf{x})$, the critical additional work rate at failure is (see Section 6.1)[1]

$$d\mathcal{W}_{\text{ext}} = dt \int_{\Omega} \rho \mathbf{f} \cdot \mathbf{V} d\Omega + \int_{\partial\Omega_{\mathbf{T}^d}} \mathbf{T}^d \cdot \mathbf{V} da \tag{9.6}$$

The second condition means that the failure mechanism defined by $\mathbf{V}(\mathbf{x})$ must be compatible with the plastic flow in the structure bulk and/or along plastic slippage planes in Ω, which dissipates the externally supplied energy $d\mathcal{W}_{\text{ext}}$ into heat form. This suggests a *kinematic approach* to the load-bearing capacity of material systems and structures, to evaluate *how* plastic failure occurs. Among all possible velocity fields $\mathbf{V}(\mathbf{x})$, limit analysis explores whether the structure can actually afford the plastic dissipation (9.2) along velocity fields $\mathbf{V}(\mathbf{x})$ that are kinematically compatible with the boundary condition (9.5) and that, at the same time, are compatible with the kinematics of the plastic flow in the structure bulk and/or along plastic slippage planes.

A combination of the stress–strength approach and the kinematic approach defines the plastic collapse load. It is the ultimate load-bearing capacity of material systems and structures, for which statically and plastically admissible stress fields are dissipated into heat form within Ω.

Finally, the external load in most practical applications can be defined by n independent load cases. Given the additive nature of the dissipation, relations (9.2) and (9.6) can be recast in the form

$$\frac{d\mathcal{D}}{dt} = \frac{d\mathcal{W}_{\text{ext}}}{dt} = \mathbf{Q}^{\text{lim}} \cdot \mathbf{q} \geq 0 \tag{9.7}$$

where $\mathbf{Q}^{\text{lim}}$ is the load vector $(1 \times n)$, and $\mathbf{q}$ is the associated velocity vector $(n \times 1)$. The aim of limit analysis is then to determine, in the $\mathbf{Q}$-space, the load-bearing capacity of the structure, including all loads $\mathbf{Q}^{\text{lim}}$ the structure can sustain according to the stress–strength approach and the kinematic approach.

Exercise 78. We consider the excavation pit of Chapter 4 (have a quick look back at Section 4.5). In addition to the gravity body forces $\rho g \mathbf{e}_x$, the excavation pit is subjected to a uniform pressure p at $x = 0$, $y \geq 0$, as sketched in Figure 9.1. Determine the load vector $\mathbf{Q}$ and its associated velocity $\mathbf{q}$ for any velocity field $\mathbf{V}$ within the excavation pit.

The external body forces reduce to $\mathbf{f} = g\mathbf{e}_x$. The prescribed stress vector has nonzero values $\mathbf{T}^d = p\mathbf{e}_x$ only on $x = 0$, $y \geq 0$ of the boundary $\partial\Omega_{\mathbf{T}^d}$. Let

[1] That is, for

$$\int_{\partial\Omega_{\mathbf{V}^d}} \mathbf{T} \cdot \mathbf{V}^d da = 0$$

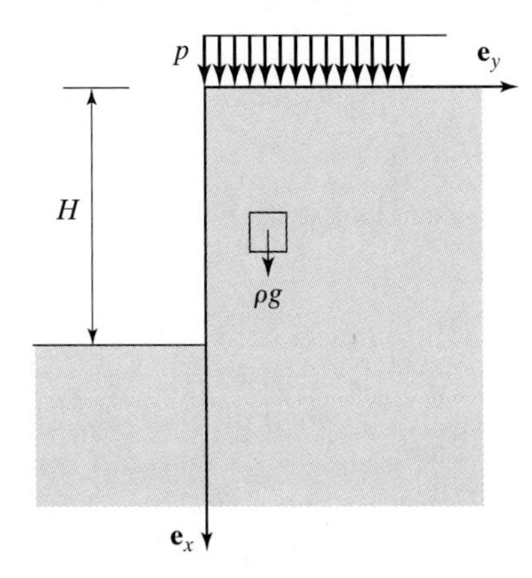

FIGURE 9.1: The excavation pit: body forces and surface forces.

$\mathbf{V} = V_x\mathbf{e}_x + V_y\mathbf{e}_y$ be any velocity field within the domain Ω. Application of (9.6) delivers the work rate of external forces in the form

$$\frac{d\mathcal{W}_{\text{ext}}}{dt} = \int_\Omega \rho g V_x dx\, dy + \int_{y=0,x=0}^{y=\infty,x=0} pV_x dy$$

A convenient choice for the load vector $\mathbf{Q}$ and its associated velocity $\mathbf{q}$ in the form of Eq. (9.7) is

$$\mathbf{Q} = \left(\begin{array}{cc} Q_1 & Q_2 \end{array} \right) = \left(\begin{array}{cc} \rho g H & p \end{array} \right); \quad \mathbf{q} = \left(\begin{array}{c} q_1 \\ q_2 \end{array} \right) = \left(\begin{array}{c} \frac{1}{H}\int_\Omega V_x dx\, dy \\ \int_{y=0,x=0}^{y=\infty,x=0} V_x dy \end{array} \right)$$

The chosen form of q_1 shows that the height H scales adequately the structural response in the x-direction due to gravity forces. The aim of limit analysis is then to determine the domain of safe loads $\mathbf{Q}$ in the $(\rho g H \times p)$-space. ■

9.1.2 Plastic Collapse Kinematics

The collapse of structures occurs in the material bulk, but also in narrow zones of intense deformation. In the framework of ideal plasticity, these narrow zones can be lumped into surfaces of discontinuity Σ across which the velocity field is discontinuous. The kinematics of plastic collapse then requires the compatibility of the velocity field $\mathbf{V}(\mathbf{x})$ with the plastic flow that occurs within the structure and along slippage planes. This involves two steps: (1) to link the dissipation associated with external load application (9.6) to the local dissipation that occurs respectively in the bulk material and on surfaces of discontinuity; and (2) to link these local kinematics to the one permitted by the plastic flow of the material.

The first task is achieved by application of the generalized divergence theorem. This divergence theorem, which generalizes the standard divergence theorem to any

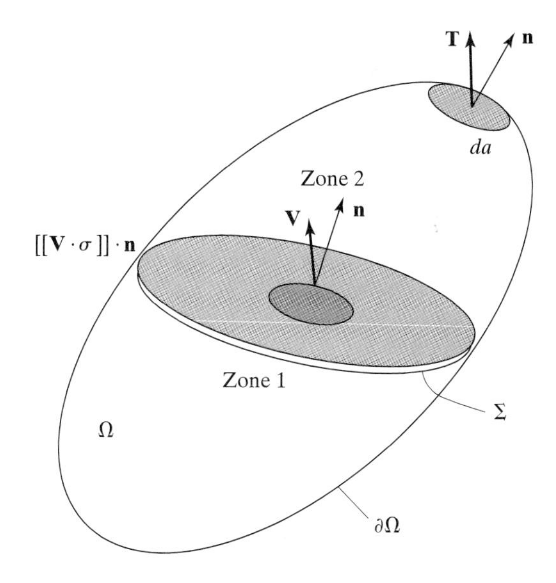

FIGURE 9.2: The generalized divergence theorem: domain Ω separated by a surface of discontinuity Σ.

domain Ω of boundary $\partial\Omega$ with one (or several) surface(s) of discontinuity Σ, reads (see Figure 9.2)[2]

$$\int_{\partial\Omega} \mathbf{f} \cdot \mathbf{n} da = \int_{\Omega} \operatorname{div} \mathbf{f} d\Omega + \int_{\Sigma} [[\mathbf{f}]] \cdot \mathbf{n} d\Sigma \tag{9.8}$$

where $[[\mathbf{f}]] = \mathbf{f}^+ - \mathbf{f}^-$ denotes the jump of physical quantity $\mathbf{f}$ over the surface of discontinuity Σ. Application of (9.8) to the surface work rate term in (9.6), $\mathbf{f} \cdot \mathbf{n} = \mathbf{T} \cdot \mathbf{V} = (\boldsymbol{\sigma} \cdot \mathbf{n}) \cdot \mathbf{V} = (\boldsymbol{\sigma} \cdot \mathbf{V}) \cdot \mathbf{n}$, yields

$$\int_{\partial\Omega} (\mathbf{V} \cdot \boldsymbol{\sigma}) \cdot \mathbf{n} da = \int_{\Omega} (\boldsymbol{\sigma} : \operatorname{grad} \mathbf{V} + (\operatorname{div} \boldsymbol{\sigma}) \cdot \mathbf{V}) \, d\Omega + \int_{\Sigma} [[\mathbf{V} \cdot \boldsymbol{\sigma}]] \cdot \mathbf{n} d\Sigma \tag{9.9}$$

The volume integral on the right-hand side of (9.9) corresponds to the standard divergence theorem applied to the surface work rate of a continuous material system (see Section 6.1.1). In return, the second term is an additional internal work

[2]The generalized divergence theorem is obtained by applying the standard divergence theorem to a domain Ω composed of two subdomains Ω_1 and Ω_2 separated by the discontinuity surface Σ (see Figure 9.2). In this derivation, the boundary is the regular boundary $\partial\Omega$ and the surface of discontinuity Σ; thus

$$\int_{\partial\Omega+\Sigma} \mathbf{f} \cdot \mathbf{n} da = \int_{\partial\Omega} \mathbf{f} \cdot \mathbf{n} da + \int_{\Sigma} \mathbf{f} \cdot \mathbf{n}_i da = \int_{\Omega} \operatorname{div} \mathbf{f} d\Omega$$

Noting that the outward normal unit vector on Σ have opposite directions (i.e., $\mathbf{n}_1 = -\mathbf{n}_2$), the sum of the individual equations leads to the generalized divergence theorem (9.8), which applies to a domain Ω containing one (or several) surface(s) of discontinuity.

rate associated with the surface of discontinuity Σ. The stress in Ω is statically admissible; meaning that the stress tensor satisfies (have a quick look back at Section 4.1)

$$\text{in } \Omega : \left\{ \begin{array}{l} \mathbf{T} = \boldsymbol{\sigma} \cdot \mathbf{n} \\ \operatorname{div} \boldsymbol{\sigma} + \rho \mathbf{f} = 0 \\ \boldsymbol{\sigma} = {}^t\boldsymbol{\sigma} \end{array} \right\} \tag{9.10}$$

$$\text{on } \partial\Omega_{\mathbf{T}^d} : \mathbf{T}^d = \mathbf{T}(\mathbf{n}) = \boldsymbol{\sigma} \cdot \mathbf{n}$$

$$\text{on } \Sigma : \mathbf{T}(\mathbf{n}) + \mathbf{T}(-\mathbf{n}) = 0$$

If we use (9.10) in (9.9) and substitute the result in (9.6), the external work rate, which is dissipated at plastic failure into heat form, becomes

$$\frac{d\mathcal{W}_{\text{ext}}}{dt} = \int_\Omega (\boldsymbol{\sigma} : \mathbf{d})\, d\Omega + \int_\Sigma [[\mathbf{V}]] \cdot \mathbf{T}(\mathbf{n})\, d\Sigma = \frac{d\mathcal{D}}{dt} \tag{9.11}$$

where $\mathbf{d}$ is the strain rate at failure:

$$\mathbf{d} = \frac{1}{2}\left(\operatorname{grad} \mathbf{V} + {}^t\operatorname{grad} \mathbf{V}\right) \tag{9.12}$$

and $[[\mathbf{V}]] = \mathbf{V}^+ - \mathbf{V}^-$ the velocity jump over the surface of discontinuity:

$$[[\mathbf{V}(\mathbf{x})]] = [[V_n]]\mathbf{n} + [[V_t]]\mathbf{t} + [[V_b]]\mathbf{b} \tag{9.13}$$

Relations (9.12) and (9.13) define the local kinematics of failure in the continuous domain Ω and on the surface of discontinuity Σ. These kinematic quantities must be compatible with the kinematics of plastic flow, defined by the plastic flow rule. This is the second condition the velocity field must satisfy, in addition to the boundary condition (9.5).

Consider first the strain rate $\mathbf{d}$ in Ω. We first note that relation (9.11) formally identifies stress $\boldsymbol{\sigma}$ as the driving force of the strain rate $\mathbf{d}$: It is the force that provokes locally the plastic dissipation. Since no energy is stored in the structure, $dW = 0$, it follows $d\boldsymbol{\sigma} = 0$, which corresponds to the case of ideal plasticity, for which the material has reached everywhere in Ω the ultimate strength:

$$\text{in } \Omega : f(\boldsymbol{\sigma}(\mathbf{x})) = 0 \tag{9.14}$$

Hence, the material undergoes a free plastic flow; meaning that the strain rate is purely plastic, $\mathbf{d} \equiv \mathbf{d}^p$. The intensity of the plastic flow remains undetermined, but the orientation is specified by the normality rule of ideal plasticity:

$$\text{in } \Omega : \mathbf{d} \equiv \mathbf{d}^p = \dot{\lambda}\frac{\partial f}{\partial \boldsymbol{\sigma}}; \quad \dot{\lambda} \geq 0; \quad f \leq 0; \quad \dot{\lambda} f = 0 \tag{9.15}$$

where $\dot{\lambda}$ is the plastic multiplier, which expresses the intensity of plastic flow. The application of an associated plastic flow rule (9.15) refers to the principle of maximum plastic work (see Section 8.1.5): The material, during plastic collapse, is assumed to dissipate locally the externally applied work at the highest possible rate.

We proceed in a similar way with the surface dissipation term in (9.11), which formally identifies the stress vector $\mathbf{T}(\mathbf{n})$ on surface Σ as the driving force of velocity jump $[[\mathbf{V}]]$:

$$\hat{\varphi} = [[\mathbf{V}]] \cdot \mathbf{T}(\mathbf{n}) = \sigma_{nn}[[V_n]] + \sigma_{nt}[[V_t]] + \sigma_{nb}[[V_b]] \geq 0 \tag{9.16}$$

where σ_{nn}, σ_{nt}, and σ_{nb} are the components of the stress vector $\mathbf{T}(\mathbf{n})$, which is continuous over Σ and which satisfies a surface strength criterion:

$$\text{on } \Sigma : f(\mathbf{T}(\mathbf{n})) = 0 \tag{9.17}$$

Furthermore, since $d\mathcal{W} = 0$ in the adjacent continuous medium, it is the same in Σ, and it follows that $d\mathbf{T}(\mathbf{n}) = 0 \Leftrightarrow [[\mathbf{V}]] \equiv \mathbf{V}^p$; meaning that the velocity jump over Σ is purely plastic. Then, evoking the principle of maximum plastic work for these surface systems, the orientation of $\mathbf{V}^p$ is specified by a surface flow rule that obeys the normality rule,

$$\text{on } \Sigma : [[\mathbf{V}]] \equiv \mathbf{V}^p = \dot{\lambda}\frac{\partial f}{\partial \mathbf{T}}; \quad \dot{\lambda} \geq 0; \quad f \leq 0; \quad \dot{\lambda} f = 0 \tag{9.18}$$

The flow rule has three components:

$$[[V_n]] = V_n^p = \dot{\lambda}\frac{\partial f}{\partial \sigma_{nn}}; \quad [[V_t]] = V_t^p = \dot{\lambda}\frac{\partial f}{\partial \sigma_{nt}}; \quad [[V_b]] = V_b^p = \dot{\lambda}\frac{\partial f}{\partial \sigma_{nb}} \tag{9.19}$$

The normal component $[[V_n]] = [[\mathbf{V}]] \cdot \mathbf{n}$ corresponds to the normal opening of the surface of discontinuity, and $[[V_t]] = [[\mathbf{V}]] \cdot \mathbf{t}$ and $[[V_b]] = [[\mathbf{V}]] \cdot \mathbf{b}$ are tangential velocities related to the plastic sliding of the surface.

In summary, Eqs. (9.15) and (9.18) specify the local kinematics of plastic flow in Ω and on Σ, according to the normality rule of plastic dissipative mechanisms, leading to the maximum dissipated plastic work $d\mathcal{W}^p$ we associate with the plastic collapse:

$$\frac{d\mathcal{D}}{dt} = \frac{d\mathcal{W}_{\text{ext}}}{dt} = \frac{d\mathcal{W}^p}{dt} = \mathbf{Q}^{\text{lim}} \cdot \mathbf{q} \tag{9.20}$$

The yield functions (9.15) and (9.17) need to be convex functions of their arguments, $\boldsymbol{\sigma}(\mathbf{x})$ and $\mathbf{T}(\mathbf{n})$, respectively. We will explore this convexity condition further in the two coming sections for the derivation of lower and upper bounds of the actual dissipation rate (9.20).

9.2 LOWER LIMIT THEOREM

The two elements of yield design are a statically and plastically admissible stress state, and a kinematically compatible velocity field satisfying the normality rule of plastic flow. Limit theorems explore the case when one or the other condition is relaxed. The first case we consider focuses on the compatibility of statically and plastically admissible stress states only, relaxing the condition of plastic flow compatibility.

9.2.1 The Theorem

The static equilibrium of a structure subjected to prescribed body forces $\rho\mathbf{f}d\Omega$ and surface force $\mathbf{T}^d da$ requires at least the existence of one statically admissible stress field $\boldsymbol{\sigma}'(\mathbf{x})$ satisfying

$$\begin{gathered}\text{in } \Omega: \left\{\begin{array}{l}\mathbf{T}' = \boldsymbol{\sigma}' \cdot \mathbf{n} \\ \operatorname{div} \boldsymbol{\sigma}' + \rho\mathbf{f} = 0 \\ \boldsymbol{\sigma}' = {}^t\boldsymbol{\sigma}'\end{array}\right\} \\ \text{on } \partial\Omega_{\mathbf{T}^d}: \mathbf{T}^d \equiv \mathbf{T}'(\mathbf{n}) = \boldsymbol{\sigma}' \cdot \mathbf{n} \\ \text{on } \Sigma: \mathbf{T}'(\mathbf{n}) + \mathbf{T}'(-\mathbf{n}) = 0\end{gathered} \tag{9.21}$$

A necessary condition for a structure to sustain the external loading is that the material strength sustains everywhere in Ω the stress fields induced by external loading:

$$\forall \mathbf{x} \text{ in } \Omega; \quad \boldsymbol{\sigma}'(\mathbf{x}) \in D_k(\mathbf{x}) \Leftrightarrow f(\mathbf{x}; \boldsymbol{\sigma}'(\mathbf{x})) \leq 0 \tag{9.22}$$

$D_k(\mathbf{x})$ represents the ultimate domain of plastically admissible stress states at any point $\mathbf{x}$ in Ω. The term *ultimate* indicates that no further hardening effects are available, since $dW = 0$ (which includes $d\mathcal{U} = 0$).

Conditions (9.21) and (9.22) are strictly the same stated for the plastic collapse solution $\boldsymbol{\sigma}(\mathbf{x})$ [i.e., (9.10) and (9.14)]. They focus on the compatibility between the quasi-static equilibrium of the system (9.21) and the local strength capacity of the material; but they do not evoke the compatibility between stress $\boldsymbol{\sigma}'(\mathbf{x})$ and the kinematics of the free plastic flow, $\mathbf{d} = \mathbf{d}^p$ associated with the collapse stress solution $\boldsymbol{\sigma}(\mathbf{x})$ through (9.15). This is the underlying idea of the lower bound theorem. According to the principle of maximum plastic work (8.37), the plastic work rate $\boldsymbol{\sigma}(\mathbf{x}) : \mathbf{d}^p$ at failure is always greater than or equal to the one realized by a stress field $\boldsymbol{\sigma}'(\mathbf{x})$ that satisfies only the stress–strength compatibility:

$$\forall \mathbf{x} \text{ in } \Omega; \quad (\boldsymbol{\sigma}(\mathbf{x}) - \boldsymbol{\sigma}'(\mathbf{x})) : \mathbf{d}^p(\mathbf{x}) \geq 0 \tag{9.23}$$

Integration of (9.23) over Ω yields

$$\frac{d\mathcal{D}}{dt} = \int_\Omega (\boldsymbol{\sigma} : \mathbf{d}^p)\, d\Omega \geq \int_\Omega (\boldsymbol{\sigma}' : \mathbf{d}^p)\, d\Omega \tag{9.24}$$

The left-hand side represents the dissipation rate at failure, while the right-hand side appears as a lower estimate of this dissipation rate. If we apply the divergence theorem to both sides, we obtain

$$\int_\Omega \rho\mathbf{f} \cdot \mathbf{V} d\Omega + \int_{\partial\Omega_{\mathbf{T}^d}} \left(\mathbf{T}^d\right) \cdot \mathbf{V} da \geq \int_\Omega \rho\mathbf{f}' \cdot \mathbf{V} d\Omega + \int_{\partial\Omega_{\mathbf{T}^d}} \left(\mathbf{T}^d\right)' \cdot \mathbf{V} da \tag{9.25}$$

or, equivalently, in the form of (9.7),

$$\int_\Omega \Big(\boldsymbol{\sigma}(\mathbf{x}) - \boldsymbol{\sigma}'(\mathbf{x})\Big) : \mathbf{d}^p(\mathbf{x})\, d\Omega = \left(\mathbf{Q}^{\lim} - \mathbf{Q}'\right) \cdot \mathbf{q} \geq 0 \tag{9.26}$$

where $\mathbf{Q}^{\lim}$ is the load vector associated with stress solution $\boldsymbol{\sigma}(\mathbf{x})$, and $\mathbf{Q}'$ the one associated with the statically and plastically admissible stress field $\boldsymbol{\sigma}'(\mathbf{x})$. Relation (9.26) translates the principle of maximum plastic work from the material level to the structural level: In the $\mathbf{Q}$-space, the load $\mathbf{Q}^{\lim}$ provokes a free plastic yielding of the structure in the $\mathbf{q}$-direction. It is the maximum overall plastic work rate among all candidates $\mathbf{Q}'$ that can be sustained by statically admissible and plastically admissible stress fields $\boldsymbol{\sigma}'(\mathbf{x})$. In addition, the principle of maximum work rate (9.26) reveals the convexity of the domain of safe loads in the $\mathbf{Q}$-space, which lends itself to a similar interpretation as developed in Section 7.1.3: In the 1D case, the convex domain of safe loads reduces to the segment $Q^- \leq Q \leq Q^+$, with $Q^- \leq 0$ and $Q^+ \geq 0$. For $Q^- < Q < Q^+$, $Q - Q'$ may be positive or negative. Therefore, the only solution that satisfies (9.26) is $q = 0$: Plastic yielding cannot occur, and the load, therefore, is safe. For $Q = Q^+$ (respectively, $Q = Q^-$), q can take any positive value (respectively, negative value), and plastic yielding may well occur. Finally, since $Q' = 0$ is a safe load [it is in static equilibrium with the plastically admissible stress field $\boldsymbol{\sigma}'(\mathbf{x}) = 0$], use in (9.26) yields $Q^{\lim} q \geq 0$: The velocity field q at yielding has the same sign as the limit load $Q^{\lim}$ that provokes it.

The results can be summarized in form of the lower limit theorem:

THEOREM 6 of the Lower Limit: Any stress field $\boldsymbol{\sigma}'(\mathbf{x})$ that is statically admissible with the loading $\mathbf{Q}'$ and that is everywhere below or at yield, $\boldsymbol{\sigma}'(\mathbf{x}) \in D_k(\mathbf{x})$, delivers a lower bound $\mathbf{Q}' \cdot \mathbf{q}$ to the actual dissipation rate $\mathbf{Q}^{\lim} \cdot \mathbf{q}$ of the ultimate limit load $\mathbf{Q}^{\lim}$ along velocity field $\mathbf{q}$:

$$\mathbf{Q}' \cdot \mathbf{q} \leq \mathbf{Q}^{\lim} \cdot \mathbf{q} = \int_{\Omega} \max_{\substack{\boldsymbol{\sigma}'(\mathbf{x}) SA \\ \boldsymbol{\sigma}'(\mathbf{x}) \in D_k(\mathbf{x})}} \left[\boldsymbol{\sigma}'(\mathbf{x}) : \mathbf{d}^p(\mathbf{x})\right] d\Omega \tag{9.27}$$

In the $\mathbf{Q}$-space the loading $\mathbf{Q}'$ belongs to a convex domain of safe loads.

Exercise 79. We consider a hollow sphere (see Figure 9.3) of inner and outer radius a and b, respectively. A pressure p is exerted on the inner surface $r = a$. The external

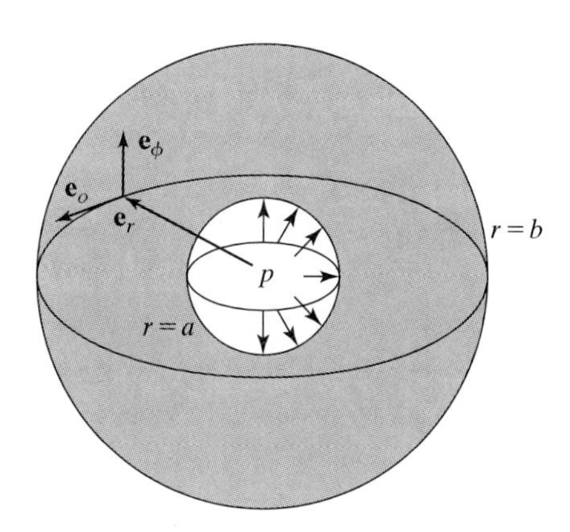

FIGURE 9.3: Hollow sphere with internal pressure.

surface $r = b$ is stress free. The material is a homogeneous plastic material that obeys a Von–Mises strength criterion, with σ_0 the tensile strength. Determine lower bounds of the limit pressure $p^{\lim}$ the hollow sphere can support.

Given the spherical symmetry of the problem, we look for statically admissible stress fields of the form

$$(\sigma_{ij}(r)) = \begin{pmatrix} \sigma_{rr}(r) & 0 & 0 \\ 0 & \sigma_{\theta\theta}(r) & 0 \\ 0 & 0 & \sigma_{\phi\phi} = \sigma_{\theta\theta} \end{pmatrix}$$

The Von–Mises plastic criterion is independent of the mean stress (have a quick look back at Section 8.1.3). This means that the relevant stresses can be obtained by subtracting from the actual tensor $\boldsymbol{\sigma}$ any spherical tensor $a\mathbf{1}$. If we choose $\sigma_{\theta\theta}\mathbf{1}$, we obtain a uniaxial stress tensor, having $\sigma_{rr} - \sigma_{\theta\theta}$ as the only nonzero stress. Therefore, any plastically admissible stress field must satisfy

$$\text{in } r \in [a, b]; \quad |\sigma_{rr} - \sigma_{\theta\theta}| \leq \sigma_0$$

Due to spherical symmetry, and $\sigma_{\phi\phi} = \sigma_{\theta\theta}$, the only equilibrium field equation to satisfy is (have a quick look back at the definition of the spherical div operator in spherical coordinates in Section 2.6.3)

$$\text{in } r \in [a, b]; \quad \frac{d\sigma_{rr}}{dr} + 2\frac{\sigma_{rr} - \sigma_{\theta\theta}}{r} = 0$$

In addition, the stress boundary conditions are

$$\text{on } \partial\Omega_{\mathbf{T}^d} : \sigma_{rr}(r = a) = -p; \quad \sigma_{rr}(r = b) = 0$$

Any stress field satisfying the three previous relations provides a lower bound p to the limit pressure $p^{\lim}$ that the hollow sphere can sustain.

Choice 1: We consider a stress field for which

$$\sigma'_{\theta\theta} = \sigma = \text{const}$$

The equilibrium equation reduces to

$$\frac{d\sigma'_{rr}}{dr} + 2\frac{\sigma'_{rr}}{r} = \frac{\sigma}{r}$$

and, after integration,

$$\sigma'_{rr} = \sigma + \frac{C}{r^2}$$

where C is an integration constant. The boundary conditions yield

$$|\sigma'_{rr} - \sigma'_{\theta\theta}| = \frac{pa^2}{b^2 - a^2}\frac{b^2}{r^2}$$

The maximum of $|\sigma'_{rr} - \sigma'_{\theta\theta}|$ is obtained at $r = a$. The stress field is plastically admissible everywhere provided that

$$p \leq p'_1 = \sigma_0\left(1 - \frac{a^2}{b^2}\right) \leq p^{\lim}$$

The value p'_1 is a lower bound of the limit pressure $p^{\lim}$.

Choice 2: The first stress field we considered restricts plastic yielding to $r = a$. A better approximation of $p^{\lim}$ can be obtained for a stress field that provokes yielding everywhere:

$$\forall \mathbf{x} \text{ in } r \in [a, b]; \quad \sigma'_{rr} - \sigma'_{\theta\theta} = -\sigma_0$$

In this case, the equilibrium equation reduces to

$$\frac{d\sigma'_{rr}}{dr} = \frac{2\sigma_0}{r}$$

The solution is

$$\sigma'_{rr} = -\sigma_0 \ln \frac{C}{r^2}$$

where C is an integration constant. The boundary condition $\sigma'_{rr}(r = b) = 0$ yields

$$\sigma'_{rr} = -2\sigma_0 \ln \frac{b}{r}$$

Finally, the second boundary condition $\sigma'_{rr}(r = a) = -p$ delivers a second lower bound p'_2 of the limit pressure $p^{\lim}$:

$$p'_2 = 2\sigma_0 \ln \frac{b}{a} \leq p^{\lim}$$

As expected, $p'_1 \leq p'_2 \leq p^{\lim}$. ■

Remark The lower limit theorem expressed by (9.27) is restricted to continuous material systems, without surfaces of discontinuity. Considering such surfaces of discontinuity, however, is not a problem, given the following three elements of the lower bound theorem:

1. Stress equilibrium for a surface of discontinuity means foremost the continuity of the stress vector over Σ (i.e., $[[\mathbf{T}'(\mathbf{n})]] = 0$), which is part of the very definition (9.21) of a statically admissible stress field.
2. Strength compatibility means that the stress vector $\mathbf{T}'$ must satisfy a surface strength criterion of the form (9.17)

$$\forall \mathbf{x} \text{ on } \Sigma; \quad \mathbf{T}'(\mathbf{x}) \in D_k(\mathbf{x}) \Leftrightarrow f\left(\mathbf{T}'(\mathbf{n})\right) = 0 \tag{9.28}$$

3. The principle of maximum plastic work applied to the surface system reads on a material level

$$\forall \mathbf{x} \text{ on } \Sigma; \quad \left(\mathbf{T}(\mathbf{x}) - \mathbf{T}'(\mathbf{x})\right) \cdot \mathbf{V}^p(\mathbf{x}) \geq 0 \tag{9.29}$$

and on a structural level

$$\int_\Sigma \mathbf{T}(\mathbf{x}) \cdot \mathbf{V}^p(\mathbf{x})\, da \geq \int_\Sigma \mathbf{T}'(\mathbf{x}) \cdot \mathbf{V}^p(\mathbf{x})\, da \tag{9.30}$$

The dissipation contribution of the surfaces of discontinuity need to be added to the volumetric dissipations in (9.24).

Finally, application of the generalized divergence theorem (9.11) to the sum of (9.24) and (9.30) delivers the lower bound theorem in the form

$$\begin{aligned}\mathbf{Q}^{\lim} \cdot \mathbf{q} &= \int_{\Omega} \max_{\boldsymbol{\sigma}'(\mathbf{x}) SA \boldsymbol{\sigma}'(\mathbf{x}) \in D_k(\mathbf{x})} [\boldsymbol{\sigma}'(\mathbf{x}) : \mathbf{d}(\mathbf{x})] \, d\Omega \\ &+ \int_{\Sigma} \max_{\mathbf{T}'(\mathbf{x}) SA \mathbf{T}'(\mathbf{x}) \in D_k(\mathbf{x})} [\mathbf{T}'(\mathbf{x}) : [[\mathbf{V}(\mathbf{x})]]] \, da \qquad (9.31)\end{aligned}$$

9.2.2 Static Approach from Inside: Constructing the Domain of Safe Loads

The convexity of the domain of safe loads furnishes a systematic way to construct it. In most cases the external loading can be reduced to an n-parameter loading of the form

$$\mathbf{Q} = \begin{pmatrix} Q_1 & Q_2 & \cdots & Q_n \end{pmatrix} \qquad (9.32)$$

For purpose of argument, consider, in the $\mathbf{Q}$-space, the positive loading path along the Q_1-axis of coordinates, as sketched in Figure 9.4a:

$$\mathbf{Q} = \begin{pmatrix} Q_1 > 0 & 0 & \cdots & 0 \end{pmatrix} \qquad (9.33)$$

If we increase monotonically $Q_1 > 0$, we obtain according to the lower limit theorem a lower bound Q_1^+ for the limit load. This limit load is associated with a stress field $\boldsymbol{\sigma}_1^+(\mathbf{x})$ that is statically compatible with Q_1^+, and plastically admissible everywhere:

$$\forall \mathbf{x}; \quad \boldsymbol{\sigma}_1^+(\mathbf{x}) \in D_k(\mathbf{x}) \Leftrightarrow f(\mathbf{x}; \boldsymbol{\sigma}_1^+(\mathbf{x})) \leq 0 \qquad (9.34)$$

Similarly, the lower limit theorem applied to any Q_i-axis delivers a lower bound Q_i^+ associated with the statically and plastically admissible stress field $\boldsymbol{\sigma}_i^+(\mathbf{x})$ satisfying

$$\forall \mathbf{x}; \quad \boldsymbol{\sigma}_i^+(\mathbf{x}) \in D_k(\mathbf{x}) \Leftrightarrow f(\mathbf{x}; \boldsymbol{\sigma}_i^+(\mathbf{x})) \leq 0 \qquad (9.35)$$

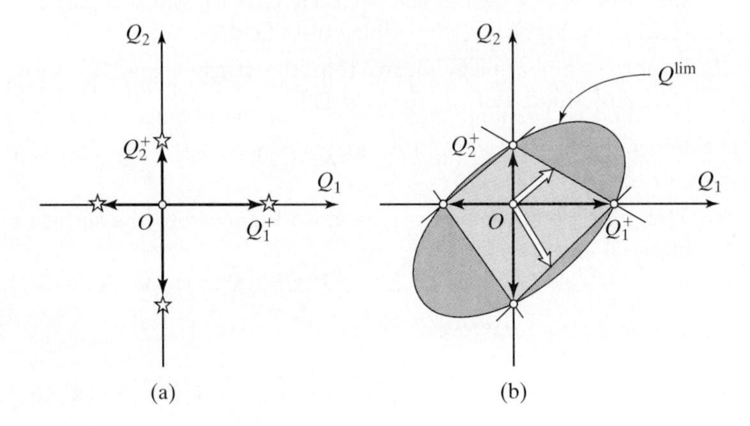

FIGURE 9.4: Construction of the domain of safe loads from the inside: (a) loading along the Q_i-axis; (b) convexity of the domain of safe loads.

Consider now a combination of these individual limit loads along Q_i-axis:

$$\mathbf{Q}' = \left(\alpha_1 Q_1^+ \quad \alpha_2 Q_2^+ \quad ... \quad \alpha_n Q_n^+ \right); \quad \sum_{i=1,n} \alpha_i = 1 \tag{9.36}$$

Due to the linearity of the set of conditions for statically admissible stress fields, the stress field $\sum_{i=1,n} \alpha_i \boldsymbol{\sigma}_i^+(\mathbf{x})$ is statically admissible with the loading $\mathbf{Q}'$ defined by (9.36). Furthermore, the convexity of the loading function implies

$$f\left(\mathbf{x}; \sum_{i=1,n} \alpha_i \boldsymbol{\sigma}_i^+(\mathbf{x})\right) \leq \sum_{i=1,n} \alpha_i \, f\left(\mathbf{x}; \boldsymbol{\sigma}_i^+(\mathbf{x})\right) \tag{9.37}$$

Finally, from a combination of (9.35) and (9.37) the stress field $\sum_{i=1,n} \alpha_i \boldsymbol{\sigma}_i^+(\mathbf{x})$ turns out to be also plastically admissible everywhere:

$$\forall \mathbf{x}; \; f\left(\mathbf{x}; \sum_{i=1,n} \alpha_i \boldsymbol{\sigma}_i^+(\mathbf{x})\right) \leq 0 \tag{9.38}$$

As a consequence, all the loads $\mathbf{Q}'$ defined by (9.36) are safe loads. The approach equally applies to negative loading paths $Q_i < 0$. The approach is displayed in Figure 9.4b. It illustrates the construction of a lower limit load domain, based on only the compatibility of the quasi-static equilibrium of the system (9.21) and the local convex strength domain of the material (9.37). The convexity of the local strength domain, which is a property associated with the maximum plastic work hypothesis, implies the convexity of the domain of safe loads $\mathbf{Q}'$. Since $\mathbf{Q}' \equiv 0$ is a safe load, all loading paths emerging from the origin $\mathbf{Q}' \equiv 0$ to any load $\mathbf{Q}'$ defined by (9.36) are safe.

Exercise 80. In the training set of Chapter 4, we developed one statically admissible and plastically admissible solution for the critical height of an excavation pit subjected to gravity forces, $\rho g \mathbf{e}_x$, in a Tresca-type material (have a quick look back at Section 4.5). We now know that the determined critical height $H_{\text{crit}} = \sigma_0 / \rho g$ is only a lower bound $H_{\text{crit}} = H^{\text{inf}}$ of the maximum limit height H^{lim}. In addition to gravity forces, consider a uniform pressure p at $x = 0$, $y \geq 0$ (see Figure 9.5a). In a first step, determine a lower bound of the limit pressure p^{lim}, in the absence of gravity forces. Then, by considering the two separate load cases, deduce, in the $(\rho g H \times p)$-space, a safe lower bound of the critical height.

The boundary conditions of the excavation pit subjected to a vertical uniform pressure p are (see Figure 9.5a)

$$\begin{aligned} &x = 0, \; y \geq 0; \quad \sigma'_{xx} = -p; \quad \sigma'_{xy} = \sigma_{xz} = 0 \\ &x = H, \; y \leq 0; \quad \sigma'_{yy} = \sigma'_{xy} = \sigma'_{xz} = 0 \\ &0 \leq x \leq H, \; y = 0; \quad \sigma'_{xx} = \sigma'_{xy} = \sigma'_{xz} = 0 \end{aligned}$$

In the absence of gravity forces, the field equilibrium equation reduces to $\operatorname{div} \boldsymbol{\sigma}' = 0$. It is satisfied by a stress field that is constant within separated zones. A statically

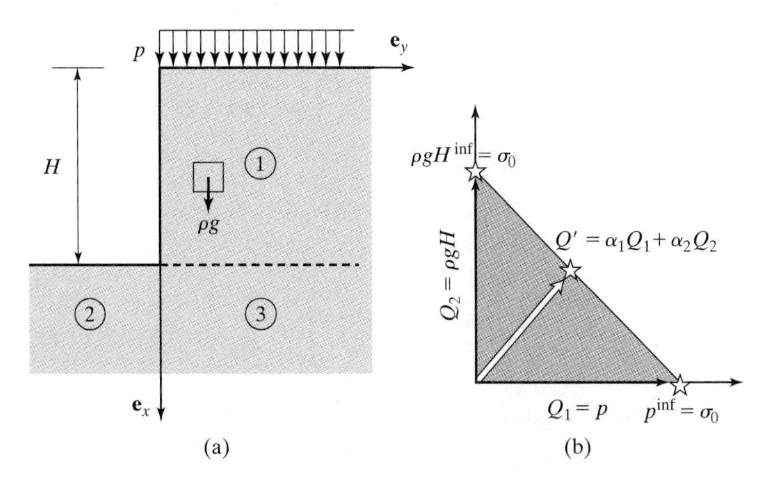

FIGURE 9.5: Excavation pit revisited from inside: (a) two-load case system; (b) convex domain of safe lower bound loads in the $(\rho g H \times p)$-space.

admissible stress field that satisfies the equilibrium equation, the boundary conditions, and the continuity of the stress vector is

$$\begin{array}{l}0 \leq x \leq H,\ y \geq 0; \quad \sigma'_{xx} = -p; \quad \sigma'_{yy} = \sigma'_{xy} = \sigma'_{zz} = \sigma'_{xz} = 0 \\ H \leq x,\ y \geq 0; \quad \sigma'_{xx} = -p; \quad \sigma'_{yy} = \sigma'_{zz} = -p; \quad \sigma'_{xy} = \sigma'_{xz} = 0 \\ H \leq x, \quad y \leq 0; \quad \sigma'_{xx} = \sigma'_{xy} = \sigma'_{xz} = 0; \quad \sigma'_{yy} = \sigma'_{zz} = -p\end{array}$$

The Tresca criterion reads $\max|\sigma_I - \sigma_J| \leq \sigma_0$, where σ_I and σ_J stand for principal stresses. For the stress field in hand, $\max|\sigma'_I - \sigma'_J| = p$. The stress field is plastically admissible provided that $p \leq \sigma_0$. Hence, for the surface pressure load case, a lower bound of the limit pressure is $p^{\inf} = \sigma_0 \leq p^{\lim}$.
Consider now, in addition, the gravity force load case, for which the limit height was found to be $H^{\inf} = \sigma_0/\rho g \leq H^{\lim}$. Applying the convexity property of the domain of safe loads to the $(\rho g H \times p)$-space, the domain of safe height and pressure is included in the triangle delimited by the three straight lines:

$$p = 0; \quad H = 0; \quad p + \rho g H \leq \sigma_0$$

Thus,

$$0 \leq H^{\inf} = \frac{\sigma_0 - p}{\rho g} \leq \frac{\sigma_0}{\rho g} \leq H^{\lim}$$

The $(\rho g H \times p)$-space is sketched in Figure 9.5b. ■

9.3 UPPER LIMIT THEOREM

The lower limit theorem explores the compatibility between the static equilibrium conditions and the strength of the constitutive material. It delivers a lower bound of limit loads. The upper limit theorem explores kinematically admissible velocity fields and the associated dissipation capacity of material systems and structures. At failure, the externally supplied work can no more be stored in the system in the form of elastic energy but must be dissipated through plastic yielding. Whether or not the structure has this capacity to dissipate the externally supplied work rate at failure into heat is the focus of the upper limit theorem.

9.3.1 The Theorem

For clarity, we first restrict ourselves to continuous material systems without surfaces of discontinuity. Let $\mathbf{V}'(\mathbf{x})$ be a possible velocity field at failure, related to the plastic strain rate $\mathbf{d}^{p\prime}(\mathbf{x})$ through (9.12), and to stress $\boldsymbol{\sigma}'(\mathbf{x})$ through a plastic flow rule (9.15). According to the principle of maximum plastic work, the plastic work realized by $\boldsymbol{\sigma}'(\mathbf{x}) : \mathbf{d}^{p\prime}(\mathbf{x})$ is always greater than or equal to the plastic work of any stress that is not associated with $\mathbf{d}^{p\prime}(\mathbf{x})$ through a plastic work rate:

$$\forall \mathbf{x};\quad \forall \boldsymbol{\sigma}(\mathbf{x}) \in D_k(\mathbf{x});\quad (\boldsymbol{\sigma}'(\mathbf{x}) - \boldsymbol{\sigma}(\mathbf{x})) : \mathbf{d}^{p\prime}(\mathbf{x}) \geq 0 \tag{9.39}$$

Note that $\boldsymbol{\sigma}(\mathbf{x})$ is the stress solution of the problem. But since we evoke only the principle of maximum plastic work, it is the stress $\boldsymbol{\sigma}'(\mathbf{x})$ related to $\mathbf{d}^{p\prime}(\mathbf{x})$ by the normality rule that realizes the maximum of plastic work among all plastically admissible stress states, including the stress state of the plastic collapse solution $\boldsymbol{\sigma}(\mathbf{x})$. In addition, since $\boldsymbol{\sigma}'(\mathbf{x})$ is associated with $\mathbf{d}^{p\prime}$ through a flow rule, which establishes a unique relation between these two quantities, the maximum available plastic work rate the material can locally dissipate depends only on $\mathbf{d}^{p\prime}(\mathbf{x})$; thus

$$\begin{aligned} \forall \mathbf{x};\quad \forall \boldsymbol{\sigma}(\mathbf{x}) \in D_k(\mathbf{x});\quad \boldsymbol{\sigma}(\mathbf{x}) : \mathbf{d}^{p\prime}(\mathbf{x}) &\leq \varphi'\left(\mathbf{x}; \mathbf{d}^{p\prime}(\mathbf{x})\right) \\ &= \max_{\boldsymbol{\sigma}'(\mathbf{X}) \in D_k(\mathbf{X})} \left[\boldsymbol{\sigma}'(\mathbf{x}) : \mathbf{d}^{p\prime}(\mathbf{x})\right] \end{aligned} \tag{9.40}$$

where $\varphi'\left(\mathbf{x}; \mathbf{d}^{p\prime}(\mathbf{x})\right)$ is the maximum local dissipation rate associated with $\mathbf{V}'(\mathbf{x})$, $\mathbf{d}^{p\prime}(\mathbf{x})$ and $\boldsymbol{\sigma}'(\mathbf{x})$.

Exercise 81. We consider a Von–Mises material with σ_0 as tensile strength. Derive the maximum local dissipation rate $\varphi' = \varphi'\left(\mathbf{x}; \mathbf{d}^{p\prime}(\mathbf{x})\right)$ as a function of the plastic strain rate $\mathbf{d}^{p\prime}(\mathbf{x})$.

The Von–Mises yield surface was introduced in Section 8.1.3. We consider it here in the form

$$f(\boldsymbol{\sigma}) = \frac{1}{2}\mathbf{s} : \mathbf{s} - k^2$$

where $\mathbf{s} = \boldsymbol{\sigma} + \frac{1}{3}(\operatorname{tr}\boldsymbol{\sigma})\mathbf{1}$ is the stress deviator, and $k = \sigma_0/\sqrt{3}$ the shear strength. In the principal stress space $\sigma_I \times \sigma_{II} \times \sigma_{III}$, the loading function $f(\boldsymbol{\sigma}) = 0$ is a cylinder of radius $\sqrt{2}k$ generated around the hydrostatic axis $\sigma_I = \sigma_{II} = \sigma_{III}$ (see Figure 8.2). Given this geometry, it is readily understood that the plastic strain rate $\mathbf{d}^{p\prime}$ that satisfies the normality condition (9.15), that is,

$$\frac{\partial f}{\partial \boldsymbol{\sigma}} = \mathbf{s}$$

satisfies the incompressibility condition

$$d'_I + d'_{II} + d'_{III} = \operatorname{tr} \mathbf{d}^{p\prime} = 0$$

Note that if $\mathbf{d}^{p\prime}$ was not normal to the loading surface, $\operatorname{tr} \mathbf{d}^{p\prime} \neq 0$, and the dissipation $\varphi'\left(\mathbf{d}^{p\prime}\right) = \max_{f(\boldsymbol{\sigma})\leq 0} \boldsymbol{\sigma} : \mathbf{d}^{p\prime} = ... + \sigma_m \operatorname{tr} \mathbf{d}^{p\prime}$ would be infinite for $\sigma_I = \sigma_{II} = \sigma_{III} = \sigma_m \to \infty$. By contrast, for a plastic strain rate normal to the yield surface, the maximum dissipation the material can afford is

$$\operatorname{tr} \mathbf{d}^{p\prime} = 0;\ \varphi'\left(\mathbf{d}^{p\prime}\right) = \max_{f(\boldsymbol{\sigma}')\leq 0} \boldsymbol{\sigma}' : \mathbf{d}^{p\prime} = \max_{f(\boldsymbol{\sigma}')\leq 0} \mathbf{s}' : \mathbf{d}^{p\prime}$$

The normality rule, therefore, defines a one-to-one relation between stress $\mathbf{s}$ and the plastic strain rate $\mathbf{d}^{p\prime}$ from

$$f(\boldsymbol{\sigma}') = \frac{1}{2}\mathbf{s}' : \mathbf{s}' - \sigma_0^2/3 = 0;\ \ \mathbf{d}^{p\prime} = \dot{\lambda}\frac{\partial f}{\partial \boldsymbol{\sigma}'} = \dot{\lambda}\mathbf{s}';\ \ \dot{\lambda} \geq 0$$

In detail, a combination of the loading function and the flow rule delivers the plastic multiplier:

$$f(\boldsymbol{\sigma}') = 0 \Leftrightarrow \frac{1}{2\dot{\lambda}^2}\mathbf{d}^{p\prime} : \mathbf{d}^{p\prime} - \sigma_0^2/3 = 0 \to \dot{\lambda} = \frac{1}{\sigma_0}\sqrt{\frac{3}{2}\mathbf{d}^{p\prime} : \mathbf{d}^{p\prime}}$$

A development of the maximum plastic work yields

$$\begin{aligned}\operatorname{tr} \mathbf{d}^{p\prime} = 0;\ \varphi'\left(\mathbf{d}^{p\prime}\right) &= \max_{f(\boldsymbol{\sigma}')\leq 0} \mathbf{s}' : \mathbf{d}^{p\prime} = \max_{f(\boldsymbol{\sigma}')\leq 0} \frac{1}{\dot{\lambda}}\mathbf{d}^{p\prime} : \mathbf{d}^{p\prime} \\ &= \sigma_0\sqrt{\frac{2}{3}} \max_{f(\boldsymbol{\sigma}')\leq 0} \sqrt{\mathbf{d}^{p\prime} : \mathbf{d}^{p\prime}}\end{aligned}$$

The maximum local dissipation rate turns out to be a unique function of the plastic strain rate, $\varphi' = \varphi'\left(\mathbf{d}^{p\prime}\right)$. Note that we only evoked the yield function and the normality rule. ■

Equation (9.40) defines the maximum dissipation capacity at a material point $\mathbf{x}$. An integration of (9.40) over the whole domain Ω occupied by the structure yields

$$\forall \mathbf{x};\ \ \forall \boldsymbol{\sigma}(\mathbf{x}) \in D_k(\mathbf{x});\ \ \int_\Omega \boldsymbol{\sigma}(\mathbf{x}) : \mathbf{d}^{p\prime}(\mathbf{x})d\Omega \leq \int_\Omega \varphi'\left(\mathbf{x}; \mathbf{d}^{p\prime}(\mathbf{x})\right) d\Omega = \frac{d\mathcal{D}}{dt}\left(\mathbf{V}'\right) \tag{9.41}$$

where $\frac{d\mathcal{D}}{dt}\left(\mathbf{V}'\right)$ is the maximum dissipation rate the structure can afford, given a specific failure mechanism defined by $\mathbf{V}'$. In addition, we note now that $\mathbf{d}^{p\prime}$ derives, at failure, from a kinematically admissible velocity field, that is,

$$\mathbf{d}^{p\prime} = \mathbf{d}' = \frac{1}{2}\left[\operatorname{grad} \mathbf{V}' + {}^t\operatorname{grad} \mathbf{V}'\right] \tag{9.42}$$

The left-hand side of (9.41) can be developed, by inverse application of the divergence theorem, in the form

$$\int_\Omega \boldsymbol{\sigma}(\mathbf{x}) : \mathbf{d}' d\Omega = \int_\Omega \rho\mathbf{f} \cdot \mathbf{V}' d\Omega + \int_{\partial\Omega_{\mathbf{T}^d}} \mathbf{T}^d \cdot \mathbf{V}' da = \mathbf{Q}^{\lim} \cdot \mathbf{q}' \tag{9.43}$$

Thus,

$$\forall \mathbf{V}': \quad \mathbf{Q}^{\lim} \cdot \mathbf{q}' \leq \frac{d\mathcal{D}}{dt}(\mathbf{V}') \tag{9.44}$$

It is instructive to compare inequality (9.44) with relation (9.7), which defines the dissipation associated with the plastic collapse solution, for which $\mathbf{V}$ is kinematically admissible, and $\boldsymbol{\sigma}$ statically and plastically admissible. This comparison shows that inequality (9.44) defines an upper bound of the capacity of a material system or structure to dissipate the external work rate of volume forces $\rho \mathbf{f} d\Omega$ and prescribed surface forces $\mathbf{T}^d da$ into heat form:

$$\mathbf{Q}^{\lim} \cdot \mathbf{q} \leq \mathbf{Q}^{\lim} \cdot \mathbf{q}' \leq \frac{d\mathcal{D}}{dt}(\mathbf{V}') \tag{9.45}$$

Therefore, given this inequality, the structure will certainly collapse under the external load $\mathbf{Q}^{\lim}$, if there exists at least one kinematically admissible velocity field $\mathbf{V}'$, for which the externally supplied work rate $\mathbf{Q}^{\lim} \cdot \mathbf{q}'$ cannot entirely be converted into plastic dissipation $\frac{d\mathcal{D}}{dt}(\mathbf{V}')$:

$$\begin{aligned} &\mathbf{Q}^{\lim} \textit{ not } \text{sustained by the structure} \\ &\Rightarrow \exists \mathbf{V}'(\mathbf{x}): \quad \mathbf{Q}^{\lim} \cdot \mathbf{q}' \geq \frac{d\mathcal{D}}{dt}(\mathbf{V}') \end{aligned} \tag{9.46}$$

Equation (9.46) sets out a formidable optimization problem: to search for the lowest upper bound by means of kinematically admissible velocity fields at failure. The results can be summarized in form of the upper limit theorem.

THEOREM 7 of the Upper Limit : Any kinematically admissible velocity field $\mathbf{V}'$ delivers an upper bound $\frac{d\mathcal{D}}{dt}(\mathbf{V}')$ to the actual dissipation rate the limit load $\mathbf{Q}^{\lim}$ realizes along the actual velocity field $\mathbf{q}$:

$$\mathbf{Q}^{\lim} \cdot \mathbf{q} = \min_{\text{on } \partial\Omega_{\mathbf{V}^d}: \mathbf{V}' = \mathbf{V}^d \equiv 0} \left[\frac{d\mathcal{D}}{dt}(\mathbf{V}') \right] \tag{9.47}$$

The dissipation rate $\frac{d\mathcal{D}}{dt}(\mathbf{V}')$ is the maximum dissipation the material system can afford.

Exercise 82. We consider again the hollow sphere, made of a Von–Mises material (tensile strength σ_0), subjected at the inner surface $r = a$ to a pressure p (see Figure 9.3; Exercise 79 in Section 9.2.1). The outer surface $r = b$ is stress free. Using the upper limit theorem, determine an upper bound of the limit pressure $p^{\lim}$ that the hollow sphere can support.

Given the spherical symmetry of the problem, we choose a radial velocity field:

$$\mathbf{V}' = u(r)\, \mathbf{e}_r$$

The first condition we explore for a Von–Mises material is the incompressibility condition, which ensures that the dissipation associated with $\mathbf{V}'$ is not infinite (see Exercise 81, Section 9.3.1):

$$\operatorname{div} \mathbf{V}' = \operatorname{tr} \mathbf{d}^{p\prime} = 0 \Rightarrow \frac{d\mathcal{D}}{dt}(\mathbf{V}') \neq \infty$$

For the chosen radial velocity field, the incompressibility condition reads

$$d'_{rr} + d_{\theta\theta} + d_{\phi\phi} = \frac{du}{dr} + 2\frac{u}{r} = 0$$

which integrates into

$$u(r) = V'\frac{a^2}{r^2}$$

where V' stands for the radial velocity of the inner surface where the pressure is applied. The associated plastic strain rate reads

$$\mathbf{d}^p(r) = V'\frac{a^2}{r^3}\begin{pmatrix} -2 & 0 & 0 \\ 0 & 1 & 0 \\ 0 & 0 & 1 \end{pmatrix}$$

The local maximum dissipation for a Von–Mises material was determined in Exercise 81

$$\operatorname{tr}\mathbf{d}^{p\prime} = 0; \quad \varphi'\left(\mathbf{d}^{p\prime}\right) = \sigma_0\sqrt{\frac{2}{3}}\sqrt{\mathbf{d}^{p\prime} : \mathbf{d}^{p\prime}} = 2V'\sigma_0\frac{a^2}{r^3}$$

The maximum plastic dissipation the hollow cylinder can afford is obtained from the volume integral of $\varphi'\left(\mathbf{d}^{p\prime}\right)$:

$$\frac{d\mathcal{D}}{dt}(u(r)) = \int_{r=a}^{r=b}\int_{\theta=0}^{\theta=\pi}\int_{\phi=0}^{\phi=2\pi}\left(2V'\sigma_0\frac{a^2}{r^3}\right)r^2\sin\theta\, d\theta\, d\phi = 8\pi a^2 V'\sigma_0\ln\frac{b}{a}$$

According to (9.44), this quantity is to be compared to the external work rate realized by pressure p. It reads

$$Q^{\lim}q' = 4\pi a^2 V' p^{\lim}$$

Finally, according to the upper limit theorem, the pressure $p^{\lim}$ will not be sustained by the cylinder if the external work rate is greater than the plastic work rate the cylinder can afford. It follows from (9.44) that

$$4\pi a^2 V' p^{\lim} \leq 8\pi a^2 V'\sigma_0\ln\frac{b}{a} \rightarrow p^{\lim} \leq p_2 = 2\sigma_0\ln\frac{b}{a}$$

Therefore, $p_2 = 2\sigma_0\ln\frac{b}{a}$ constitutes an upper bound of the limit pressure $p^{\lim}$ the cylinder can sustain.
Last, in Exercise 79 of Section 9.2.1, we showed that $p'_2 = 2\sigma_0\ln\frac{b}{a}$ is also a lower bound of the limit pressure,

$$p^{\lim} \geq p'_2 = 2\sigma_0\ln\frac{b}{a}$$

Hence, from the combination of the lower and the upper bound theorem, we conclude that $p_2 = p'_2$ is actually the limit load:

$$p^{\lim} = 2\sigma_0\ln\frac{b}{a}$$

Note, this exercise illustrates how the combination of the two limit theorems provides a powerful tool to determine the domain of safe loads without analyzing the behavior of the structure prior to collapse. The limit load is totally independent of the elastic properties of the structure, since it relies only on the maximum plastic work rate the structure can afford in response to the external work rate supply. ■

9.3.2 Kinematic Fields Involving Surfaces of Discontinuity

For clarity, we restricted ourselves to continuous material systems in which the velocity field $\mathbf{V}'$ is related by a gradient operator to the plastic strain rate at collapse. The theory, however, extends to material systems with surfaces of discontinuity Σ, across which the velocity vector is not continuous, if we evoke the principle of maximum plastic dissipation (9.39) for the dissipation of the surface system (9.16):

$$\forall \mathbf{x};\quad \forall \mathbf{T}(\mathbf{x}) \in D_k(\mathbf{x});\quad \left(\mathbf{T}'(\mathbf{x}) - \mathbf{T}(\mathbf{x})\right) \cdot \mathbf{V}^{p\prime}(\mathbf{x}) \geq 0 \tag{9.48}$$

$\mathbf{V}^{p\prime} = [[\mathbf{V}']]$ is the plastic velocity jump over the surface of discontinuity, which is related through a flow rule of the form (9.18) to the stress vector $\mathbf{T}'(\mathbf{x})$. Then, analogously to (9.40), the maximum available plastic work rate the material can locally dissipate along $\mathbf{V}^{p\prime}$ is

$$\begin{aligned} \forall \mathbf{x} \text{ on } \Sigma;\quad \forall \mathbf{T}(\mathbf{x}) \in D_k(\mathbf{x});\quad \mathbf{T}(\mathbf{x}) \cdot \mathbf{V}^{p\prime} \leq \ & \hat{\varphi}' \left(\mathbf{x}; \mathbf{V}^{p\prime}(\mathbf{x})\right) \\ = \ & \max_{\mathbf{T}'(\mathbf{x}) \in D_k(\mathbf{x})} \left[\mathbf{T}'(\mathbf{x}) \cdot \mathbf{V}^{p\prime}(\mathbf{x})\right] \end{aligned} \tag{9.49}$$

The maximum local surface dissipation capacity $\hat{\varphi}'\left(\mathbf{x}; \mathbf{V}^{p\prime}\right)$ depends only on $\mathbf{V}^{p\prime}$. The contribution of all surfaces of discontinuities Σ needs to be added to the volumetric one, expressed by (9.41); thus

$$\begin{aligned} \int_\Omega \boldsymbol{\sigma}(\mathbf{x}) \ & : \ \mathbf{d}^{p\prime}(\mathbf{x})\, d\Omega + \int_\Sigma \mathbf{T}(\mathbf{x}) \cdot \mathbf{V}^{p\prime}(\mathbf{x})\, da \\ & \leq \int_\Omega \varphi'\left(\mathbf{x}; \mathbf{d}^{p\prime}(\mathbf{x})\right) d\Omega + \int_\Sigma \hat{\varphi}'\left(\mathbf{x}; \mathbf{V}^{p\prime}(\mathbf{x})\right) da \end{aligned} \tag{9.50}$$

Finally, with the help of (9.42) and the generalized divergence theorem (9.8), we obtain the upper limit theorem in the form

$$\mathbf{Q}^{\lim} \cdot \mathbf{q} \leq \mathbf{Q}^{\lim} \cdot \mathbf{q}' \leq \frac{d\mathcal{D}}{dt}\left(\mathbf{V}'\right) \tag{9.51}$$

where

$$\frac{d\mathcal{D}}{dt}\left(\mathbf{V}'\right) = \int_\Omega \varphi'\left(\mathbf{x}; \mathbf{d}^{p\prime}(\mathbf{x}) \equiv \mathbf{d}'(\mathbf{x})\right) d\Omega + \int_\Sigma \hat{\varphi}'\left(\mathbf{x}; \mathbf{V}^{p\prime}(\mathbf{x}) \equiv [[\mathbf{V}'(\mathbf{x})]]\right) da \tag{9.52}$$

The maximum available plastic work $\frac{d\mathcal{D}}{dt}\left(\mathbf{V}'\right)$ that can be dissipated into heat form accounts for two terms: one related to the dissipation in the structural bulk, the other to the dissipation capacity of all surfaces of discontinuity.

9.3.3 Kinematic Approach from Outside versus Static Approach from Inside

The lower limit theorem is referred to as the static approach from inside, because it is based on statically and plastically admissible stress fields, which underestimate the admissible dissipation at failure. The upper limit theorem, because of its focus on kinematically admissible velocity fields, is also referred to as the kinematic approach from outside, since it overestimates the admissible dissipation at failure. Figure 9.6 illustrates this approach from the outside for a one-loading parameter and a two-loading parameter system.

The upper limit theorem is generally easier to work out than the lower limit theorem, because it is easier to find kinematically admissible velocity fields that satisfy condition (9.5) than statically admissible stress fields that satisfy the set of equations (9.21). But the focus of an optimized yield design will focus on framing the real plastic solution; that is,

$$\mathbf{Q}' \cdot \mathbf{q} \leq \mathbf{Q}^{\text{lim}} \cdot \mathbf{q} \leq \mathbf{Q}^{\text{lim}} \cdot \mathbf{q}' \tag{9.53}$$

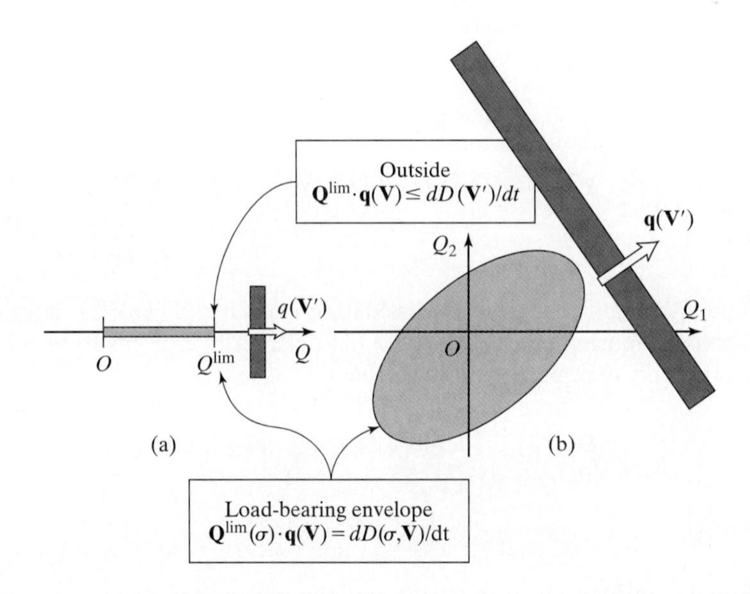

FIGURE 9.6: Kinematic approach from outside: (a) one-parameter loading system; (b) two-parameter loading system (after Salençon, 1990).[3]

9.3.4 A Classical Exercise: The Excavation Pit Revisited from Outside

The critical height of the excavation pit studied in the training set of Section 4.5 is a lower or safe bound of the limit height H^{lim}:

$$\boldsymbol{\sigma}' \text{ SA}; \quad \boldsymbol{\sigma}' \in D_k : H_{\text{crit}} = \frac{\sigma_0}{\rho g} \leq H^{\text{lim}}$$

[3]Salençon, F. (1990). "An Introduction to the Yield Design Theory and its Application to Soil Mechanics." *Eur. J. Mech., A/Solids*, Vol. 9, No. 5, 477–500.

where σ_0 is the compressive strength of the pure cohesive soil, and ρg the volume force density. Focus of this classical exercise is the determination of an upper bound of $H^{\lim}$, using the kinematic approach from outside with kinematically admissible failure mechanisms involving surfaces of velocity discontinuity.

Dissipation Rate along a Surface of Discontinuity.

For the problem at hand, we need to determine the maximum available plastic work that can be dissipated by the material in the surface of discontinuity Σ, by a stress vector $\mathbf{T} = \sigma\mathbf{n} + \tau\mathbf{t}$ along the plastic velocity jump $\mathbf{V}^p = V_n^p\mathbf{n} + V_t^p\mathbf{t}$, where $V_n^p = [[V_n^p]] = [[\mathbf{V}]] \cdot \mathbf{n}$ is the normal velocity jump, and $V_t^p = [[V_t]] = [[\mathbf{V}]] \cdot \mathbf{t}$ is the tangential velocity jump over Σ. In the $\mathbf{T}$-space [i.e., in the Mohr plane $(\sigma \times \tau)$], the Tresca criterion $f(\mathbf{T}) = 0$ restricts the stress vector $\mathbf{T}$ to the band defined by two straight lines $|\tau| \leq \sigma_0/2$, where $\sigma_0/2$ is the shear strength (see, for instance, Figure 4.8). In the Mohr plane, the normality rule requires that the velocity discontinuity is oriented in the direction of the shear stress axis; that is,

$$\frac{\partial f}{\partial \mathbf{T}} = \text{sign}\,(\tau)\,\mathbf{t} \Leftrightarrow V_n^p = [[V_n]] = 0$$

Note that the dissipation would be infinite if $\mathbf{V}^p = [[\mathbf{V}]]$ were not normal to the loading surface $f = |\tau| - \frac{\sigma_0}{2} = 0$, and thus $[[V_n]] \neq 0$. In fact, since there is no strength limit on the hydrostatic axis, the normal stress may go to infinity $\sigma = \mathbf{T} \cdot \mathbf{n} \to \infty$, and the dissipation as well, $\sigma[[V_n]] \to \infty$. $[[V_n]] \neq 0$ means that the two material surfaces of Σ would separate from each other, and the material incompressibility would not be preserved. By contrast, for $[[V_n]] = 0$, the maximum dissipation has a finite value, defined by the Tresca criterion, and the surface flow rule:

$$f(\mathbf{T}) = |\tau| - \frac{\sigma_0}{2} = 0; \quad [[\mathbf{V}]] = \mathbf{V}^p = \dot{\lambda}\frac{\partial f}{\partial \mathbf{T}} = \dot{\lambda}\text{sign}\,(\tau)\,\mathbf{t} \to \dot{\lambda} = |V_t^p|$$

In this case, dissipation (9.16) reads

$$[[V_n]] = 0 : \hat{\varphi} = \mathbf{T} \cdot [[\mathbf{V}]] = \tau\dot{\lambda}\text{sign}\,(\tau) = \frac{\sigma_0}{2}\dot{\lambda} = \frac{\sigma_0}{2}|V_t^p|$$

In summary,

$$\begin{aligned}
\mathbf{T} &= \sigma\mathbf{n} + \tau\mathbf{t}; \quad f(\mathbf{T}) = |\tau| - \frac{\sigma_0}{2} \\
[[V_n]] &= 0; \quad \hat{\varphi}(\mathbf{x}; [[\mathbf{V}(\mathbf{x})]]) = \frac{\sigma_0}{2}|V_t^p| \\
[[V_n]] &\neq 0; \quad \hat{\varphi}(\mathbf{x}; [[\mathbf{V}(\mathbf{x})]]) = \infty
\end{aligned}$$

A Possible Block Failure Mechanism.

The first failure mechanism we consider is sketched in Figure 9.7a. The failure is assumed to occur through the sliding of a rigid triangular block along a straight

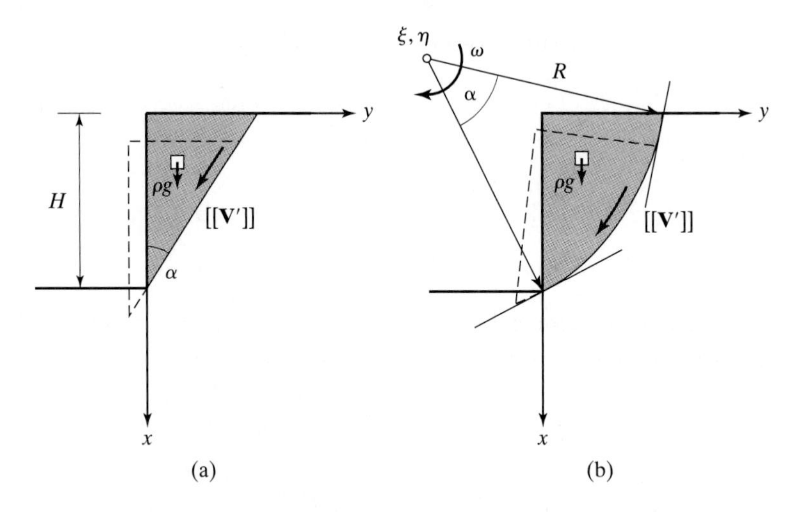

FIGURE 9.7: Excavation pit revisited from outside: (a) block failure mechanism; (b) failure mechanism along a circle.

line, inclined with regard to the x-axis by an angle α.[4] In order for $\hat{\varphi}'\left(\mathbf{x};[[\mathbf{V}'(\mathbf{x})]]\right)$ to remain finite, the velocity discontinuity across the straight line must be zero (i.e., $[[V_n]]=0$). This is the case for a constant velocity field $\mathbf{V}'$ in the sliding triangle of the form

$$\mathbf{V}' = q_x'\mathbf{e}_x + q_y'\mathbf{e}_y$$

for which

$$q_x' = V'\cos\alpha; \quad q_y' = -V'\sin\alpha$$

Dissipation Rate and External Work Rate.

The displacement jump at the interface lies along the straight line and reads $[[\mathbf{V}'(\mathbf{x})]] = V'\mathbf{t}$ so that

$$\hat{\varphi}'\left(\mathbf{x};[[\mathbf{V}'(\mathbf{x})]]\right) = \frac{\sigma_0}{2}\left|V'\right|$$

The dissipation rate only results from the energy dissipation in the slip plane Σ of surface $1_z H/\cos\alpha$:

$$\frac{d\mathcal{D}}{dt}(\mathbf{V}') = \int_\Sigma \hat{\varphi}'\left(\mathbf{x};[[\mathbf{V}'(\mathbf{x})]]\right) da = 1_z\frac{\sigma_0 H}{2\cos\alpha}\left|V'\right|$$

The external work rate that results from the body forces $\rho\mathbf{f} = \rho g\mathbf{e}_x$ reads here

$$\mathbf{Q}^{\text{lim}}\cdot\mathbf{q}' = \int_\Omega \rho\mathbf{f}\cdot\mathbf{V}'d\Omega = \int_\Omega \rho g q_x' d\Omega = 1_z\frac{\rho g H^2\sin\alpha}{2}V'$$

[4]The question may arise whether the considered block failure mechanism is kinematically admissible. The block with the designated velocity field would necessarily enter the solid at $y=H$, as displayed in Figure 9.7a. However, we should note that the surface of discontinuity is not a velocity boundary, and that there is no restriction *a priori* imposed on the velocity jump $[[\mathbf{V}'(\mathbf{x})]] = \mathbf{V}'^{+} - \mathbf{V}'^{-}$ over the surface. In turn, the velocity fields in the adjacent continua must be kinematically admissible and differentiable within the continuous domains, where they are defined.

Minimization of the Upper Bound Solution.

The failure will occur if the external work rate $\mathbf{Q}^{\lim} \cdot \mathbf{q}'$ is greater than the plastic work rate $\frac{d\mathcal{D}}{dt}(\mathbf{V}')$ that the soil can afford. Therefore, with $V' > 0$ (the external work rate must be positive to provoke the collapse), the failure will occur if the height of the excavation pit is greater than

$$H \geq \frac{\sigma_0}{\rho g} \times \frac{1}{\cos\alpha \sin\alpha}$$

The right-hand member has a minimum for $\alpha = \pi/4$. Therefore, the upper bound approach delivers an upper bound to the limit height:

$$H^{\lim} \frac{\rho g}{\sigma_0} \leq 2$$

Improvement of the Upper Bound.

We consider a second block failure mechanism along a circle of radius R that goes through the bottom of the excavation pit. The block is subjected to a rotation of angular velocity ω around a center ξ, η, as displayed in Figure 9.7b. The velocity field is of the form

$$\mathbf{V}' = \omega r \mathbf{e}_\theta$$

where r is the radial coordinate. The displacement jump lies along the circle of radius R with $|[[\mathbf{V}(\mathbf{x})]] \cdot \mathbf{t}| = \omega R$ ($\omega > 0$!) so that

$$\frac{d\mathcal{D}}{dt}(\mathbf{V}') = 1_z \frac{\sigma_0}{2} \omega \alpha R^2$$

The determination of the external work rate requires some algebra, leading to

$$\mathbf{Q}^{\lim} \cdot \mathbf{q}' = \int_\Omega \left[\rho g \mathbf{e}_x \cdot \omega r \mathbf{e}_\theta \right] d\Omega = 1_z \int_0^H \rho g \omega x (x - \xi) dx$$

The height H scales the problem in the x-direction, and a more convenient form for the external work rate is

$$\mathbf{Q}^{\lim} \cdot \mathbf{q}' = 1_z \omega H^3 \int_0^1 \bar{x}(\bar{x} - \bar{\xi})\, d\bar{x}$$

where we let $\bar{x} = x/H$ and $\bar{\xi} = \xi/H$. The failure will occur if $\mathbf{Q}^{\lim} \cdot \mathbf{q}' > \frac{d\mathcal{D}}{dt}(\mathbf{V}')$. An upper bound of the limit height $H^{\lim}$ is therefore given by

$$\frac{\rho g H^{\lim}}{\sigma_0} \leq \frac{(R/H)^2 \alpha}{2 \int_0^1 \bar{x}(\bar{x} - \bar{\xi})\, d\bar{x}}$$

For instance, if we let $\bar{\xi} = \bar{\eta} = 0$, for which $R/H = 1$, and $\alpha = \pi/2$, we obtain

$$\frac{\rho g H^{\lim}}{\sigma_0} \leq \frac{\pi}{4 \int_0^1 \bar{x}^2 d\bar{x}} = \frac{3\pi}{4} = 2.36$$

This is not an improvement with regard to the previously found upper bound for which $\rho g H^{\lim}/\sigma_0 \leq 2$. Further minimization consists of finding the minimum with regard to the kinematic yield design parameters R, α, and $\bar{\xi}$. This gives the best upper bound for the considered failure mechanism. The solution is attributed to Taylor (1937)[5]:

$$\frac{\rho g H^{\lim}}{\sigma_0} \leq 1.9; \quad \bar{\xi} = -1.2; \quad \bar{\eta} = -1.4$$

This optimized solution delivers an improved (i.e., smaller) upper bound for the critical height of the excavation pit. The failure mechanism along a circle comes closer to the real dissipation mechanism at collapse.

Finally, a combination of this result with the conservative stress result found in Section 4.5 allows us to frame the unknown collapse solution:

$$1 \leq H^{\lim}\frac{\rho g}{\sigma_0} \leq 1.9$$

Excavation Pit with Tension Cutoff Criterion.

In general, soil cannot sustain significant tensile stresses, which are permitted by a Tresca criterion. To improve the analysis of the collapse of the excavation pit, we consider a Tresca-type material with a tension cutoff:

$$f(\boldsymbol{\sigma}) = \max\{|\sigma_I - \sigma_J| - \sigma_0, \sigma_I\} \leq 0$$

In the Mohr plane, shown in Figure 9.8a, the Tresca criterion with tension cutoff restricts the stress vector $\mathbf{T}$ to the domain limited by the two straight lines $|\tau| \leq \frac{\sigma_0}{2}$, and a half-circle of radius $\sigma_0/2$ centered around $\sigma = -\sigma_0/2$ and closed at $\sigma = 0$. The figure also indicates the orientation of the velocity jumps $[[V_t]]$ and $[[V_n]]$, according to the normality rule. For any stress vector situated on the half-circle, $[[V_n]] \neq 0$, which indicates that the material surfaces of Σ can actually separate at a finite dissipation rate. For a normal separation, for which $[[V_t]] = 0, [[V_n]] \neq 0$, the dissipation rate $\hat{\varphi}(\mathbf{x}; [[\mathbf{V}(\mathbf{x})]]) = \max_{f(\mathbf{T})\leq 0}(\mathbf{T} \cdot [[\mathbf{V}(\mathbf{x})]])$ is even zero.

The block failure mechanism we consider is sketched in Figure 9.8b:

- The block $ODBC$ rotates around B with velocity ω:

$$u'_x = -\omega y; \quad u'_y = \omega x$$

- The block $ODBC$ separates normally from the remainder of the excavation pit:

$$\text{on } x = 0; \quad y \in [0, H - l] : [[V'_t]] = u'_x = 0, [[V'_n]] = u'_y = \omega x$$

[5]Taylor, D. W. (1937). "Stability of Earth Slopes", *J. Boston Soc. Civil Engineers*, Vol. 24, No. 3, 337–386.

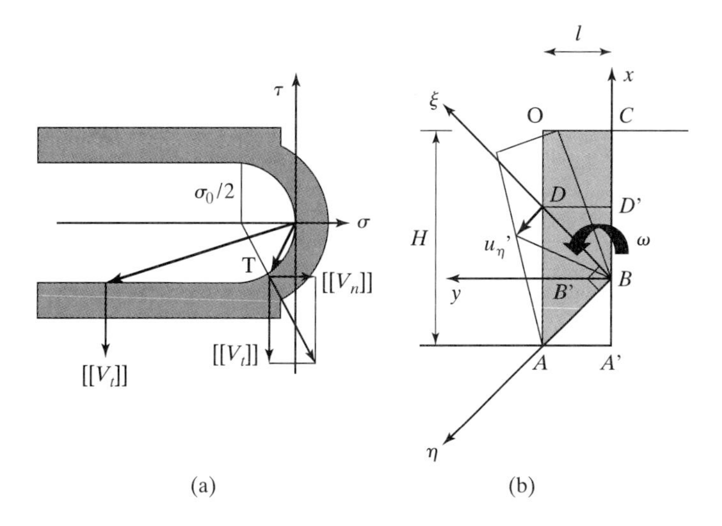

FIGURE 9.8: Excavation pit with tension cutoff criterion: (a) tension cutoff criterion in the Mohr stress plane; (b) considered failure mechanism.

As a consequence of the rotation of the rigid block, the soil continuum in the triangle ABD with $BD = BA$ and $BD \perp BA$ deforms in a simple shear following a velocity field in the local coordinate system $\xi \times \eta$ of the form (see Figure 9.8b)

$$u'_{\xi} = 0; \quad u'_{\eta} = \omega\xi$$

The strain rate associated with this velocity field in ABD is

$$d'_{\xi\xi} = d'_{\eta\eta} = 0; \quad d'_{\xi\eta} = d'_{\eta\xi} = \frac{1}{2}\omega$$

Along BC, $[[V'_t]] = 0$ and $[[V'_n]] = \omega x \neq 0$. For this pure normal separation, $\hat{\varphi}'\left(\mathbf{x}; [[\mathbf{V}'(\mathbf{x})]]\right) = 0$, and since this is the only velocity jump in which block $ODBC$ is involved (see Figure 9.5b), the contribution of the block mechanism to the overall dissipation capacity is zero. The only contribution results from the dissipation rate in the triangle ABD (i.e., from the uniform simple shear deformation), for which the principal strains are

$$d'_I = \frac{1}{2}\omega; \quad d'_{II} = 0; \quad d'_{III} = -\frac{1}{2}\omega$$

and the maximum dissipation rate is

$$\varphi'\left(\mathbf{d}^{p\prime}\right) = \max_{f(\boldsymbol{\sigma})\leq 0} \left(\sigma'_I d'_I + \sigma'_{III} d'_{III}\right)$$

where σ'_I and σ'_{III} are principal stresses, related to the principal strain rates by the normality rule:

$$d'_I = \dot{\lambda}\frac{\partial f}{\partial \sigma'_I}; \quad d'_{III} = \dot{\lambda}\frac{\partial f}{\partial \sigma'_{III}}; \quad \dot{\lambda} \geq 0$$

Given the pure shear deformation, the stresses are restricted by the Tresca criterion, that is,

$$f(\boldsymbol{\sigma}) = |\sigma'_I - \sigma'_J| - \sigma_0 = 0; \quad d'_I = \dot{\lambda}\mathrm{sign}\,(\sigma'_I - \sigma'_{III}) = -d'_{III}$$

We then obtain the maximum dissipation rate:

$$\varphi'\left(\mathbf{d}^{p\prime}\right) = \max_{f(\boldsymbol{\sigma})\leq 0}\left[\left(\sigma'_I - \sigma'_{III}\right)d'_I\right] = \sigma_0\dot{\lambda} = \frac{1}{2}\sigma_0\left(|d'_I| + |d_{III}|'\right) = \frac{1}{2}\sigma_0\omega$$

Integration over triangle ABD (of volume $\Omega = 1_z l^2$) yields the overall dissipation rate:

$$\frac{d\mathcal{D}}{dt}(\mathbf{V}') = \int_\Omega \varphi'\left(\mathbf{x}; \mathbf{d}^{p\prime}(\mathbf{x})\right)d\Omega = \frac{1}{2}\sigma_0\omega l^2 1_z$$

where l stands for length OC.

The external work rate has two terms, one related to the block $ODBC$, the other related to triangle ABD:

$$\mathbf{Q}^{\mathrm{lim}}\cdot\mathbf{q}' = -1_z\rho g\int_{ODBC} u'_x\,dx\,dy + 1_z\rho g\omega O\left(l^3\right)$$

where $1_z\rho g\omega O\left(l^3\right)$ represents the order of magnitude of the contribution of triangle ABD, and $u'_x = -\omega y$. The contribution of the block $ODBC$ is situated between the contribution of only the upper rectangular block $ODD'C$ (with $\int_{ODD'C} y dx\,dy = (H-2l)\,l^2/2$), and the whole rectangular block $OAA'C$ (with $\int_{OA'AC} y dx\,dy = Hl^2/2$). Therefore, the external work rate $\mathbf{Q}^{\mathrm{lim}}\cdot\mathbf{q}'$ has the following two limits:

$$1_z\rho g\omega\,(H-2l)\,\frac{l^2}{2} + 1_z\rho g\omega O\left(l^3\right) \leq \mathbf{Q}^{\mathrm{lim}}\cdot\mathbf{q}' \leq 1_z\rho g\omega H\frac{l^2}{2} + 1_z\rho g\omega O\left(l^3\right)$$

According to the upper limit theorem, $\mathbf{Q}^{\mathrm{lim}}\cdot\mathbf{q}' \leq \frac{d\mathcal{D}}{dt}(\mathbf{V}')$ delivers an upper bound, say H^{sup}, of the limit height H^{lim}. Using the two limits of the external work rate, this upper bound is situated between

$$\frac{1}{1+O\left(l/H^{\mathrm{sup}}\right)} \leq \frac{\rho g H^{\mathrm{sup}}}{\sigma_0} \leq \frac{1+2\left(l/H^{\mathrm{sup}}\right)\rho g H^{\mathrm{sup}}/\sigma_0}{1+O\left(l/H^{\mathrm{sup}}\right)}$$

Thus for $l/H^{\mathrm{sup}} \to 0$, for which the two limits tend to unity, $\rho g H^{\mathrm{sup}}/\sigma_0 = 1$:

$$H^{\mathrm{lim}} \leq H^{\mathrm{sup}} = \frac{\sigma_0}{\rho g}$$

Finally, a comparison of this upper bound with the lower bound found in Section 4.5, for which $H_{\mathrm{crit}} = \sigma_0/\rho g \leq H^{\mathrm{lim}}$, leads to the conclusion that the limit height of the excavation pit actually is

$$H_{\mathrm{crit}} = H^{\mathrm{sup}} \Leftrightarrow H^{\mathrm{lim}} = \frac{\sigma_0}{\rho g}$$

This result is due to Drucker (1953) and is reported in Salençon (1983).[6]

[6]Drucker, D. C. (1953). "Limit Analysis of Two and Three Dimensional Soil Mechanics Problems", *J. Mech. Phys. Solids*, Vol. 1, No. 4, 217–226.

Salençon, J. (1983). *Calcul à la Rupture et Analyse Limite*, Presse de l'Ecole Nationale des Ponts et Chaussées, Paris, France.

Excavation Pit Subjected to an Additional Uniform Pressure.

As a last case study, consider that the excavation pit is subjected to an additional pressure on surface $x = 0$, $y \geq 0$ (see Figure 9.3). In Exercise 80 of Section 9.2.2, the domain of safe height and pressure was found to be situated within the domain

$$p = 0; \quad H = 0; \quad p + \rho g H \leq \sigma_0$$

The aim of this exercise is to estimate the domain of critical height and critical pressure from the outside, by means of a kinematically admissible velocity field. The failure mechanism we consider is the block mechanism sketched in Figure 9.9a, for which

$$\mathbf{V}' = V' \left(\cos\alpha \mathbf{e}_x - \sin\alpha \mathbf{e}_y\right)$$

The external work rate is obtained by adding to the volume force work rate the additional surface force work rate developed by the prescribed stress vector $\mathbf{T}^d = +p\mathbf{e}_x$ along $\mathbf{V}'$:

$$\begin{aligned}\mathbf{Q}^{\lim} \cdot \mathbf{q}' &= \int_\Omega \rho \mathbf{f} \cdot \mathbf{V}' d\Omega + \int_{\partial\Omega_{\mathbf{T}^d} = 1_z H \tan\alpha} \mathbf{T}^d \cdot \mathbf{V}' da \\ &= 1_z V' \left[\frac{\rho g H^2 \sin\alpha}{2} + pH \sin\alpha\right]\end{aligned}$$

The maximum dissipation capacity is unchanged:

$$\frac{d\mathcal{D}}{dt}(\mathbf{V}') = \int_\Sigma \hat{\varphi}' \left(\mathbf{x}; [[\mathbf{V}'(\mathbf{x})]]\right) da = 1_z \frac{\sigma_0 H}{2\cos\alpha} |V'|$$

The excavation pit will certainly collapse for a height H and a pressure p, for which $\mathbf{Q}^{\lim} \cdot \mathbf{q}' \geq \frac{d\mathcal{D}}{dt}(\mathbf{V}')$, that is, for

$$p + \frac{\rho g H}{2} \geq \frac{\sigma_0}{2\cos\alpha \sin\alpha}$$

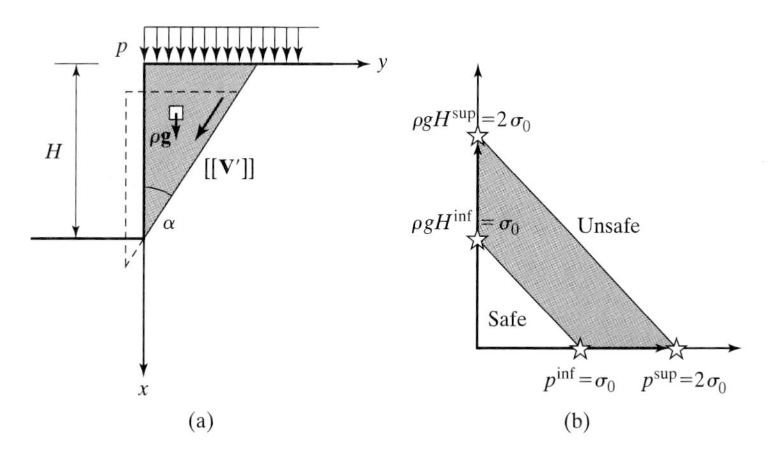

FIGURE 9.9: Excavation pit revisited from outside: (a) considered failure mechanism; (b) convex domain of unsafe upper and safe lower bound loads in the $(\rho g H \times p)$-space.

The right-hand side has a minimum for $\alpha = \pi/4$. Therefore, an upper bound of the domain of critical height and a critical pressure is included in the triangle delimited by the three straight lines:

$$p = 0; \quad H = 0; \quad p + \frac{\rho g H}{2} \leq \sigma_0$$

Thus, in combination with the lower bound solution,

$$0 \leq \frac{\sigma_0 - p}{\rho g} = H^{\text{inf}} \leq H^{\text{lim}} \leq H^{\text{sup}} = \frac{2\,(\sigma_0 - p)}{\rho g}$$

The safe lower and the unsafe upper bound of the actual critical height and critical pressure are shown in Figure 9.9b.

9.4 APPLICATION TO STRUCTURAL ELEMENTS

We have derived the limit theorems for continuum material systems. They equally apply to structural elements, trusses, beams, plates, and shells. It suffices to replace the equilibrium conditions and velocity fields of the continuum theory by the appropriate equilibrium conditions and velocity fields of structural elements; and to replace material strength criteria by section strength criteria. The static approach (from inside) is considered first.

9.4.1 Static Approach from Inside Applied to Structural Elements

Exercise 83. Figure 9.10a shows a beam of length $2l$ loaded in bending at $x = l$ by a normal force of intensity P. The beam is clamped at $x = 0$, and simply supported at $x = 2l$. The strength criterion of the beam's section is expressed by a section criterion:

$$\forall x; \; f(M(x)) = |M(x)| - M_0 \leq 0$$

where $M(x)$ is the bending moment, and M_0 the maximum moment the section can support. Determine the limit load P^{lim} of the structure.

The beam is statically indetermined. This means that there are more force and moment unknowns of the problem than equilibrium conditions. We choose the

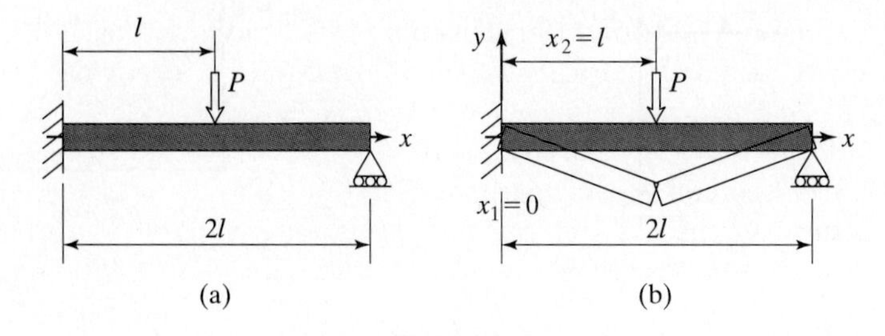

FIGURE 9.10: Application of the limit theorems to structural elements: (a) studied static system; (b) considered failure mechanism.

vertical reaction, say R, at $x = 2l$ as the hyperstatic force. The bending moment along x is

$$\begin{aligned} l &\leq x \leq 2l : M'(x) = R(2l - x) \\ 0 &\leq x \leq l : M'(x) = R(2l - x) - P(l - x) \end{aligned}$$

These expressions ensure that the equilibrium is satisfied everywhere in the structure. They play the role of the equilibrium field equations of continuum material systems. Hence, $M'(x)$ is one statically admissible field of bending moments. In addition, $M'(x)$ is said to be plastically admissible if it satisfies the section strength criterion, $\forall x;\ f(M(x)) \leq 0$. For concentrated point load cases, the moment field is piecewise linear. The moment field, therefore, is plastically admissible as soon as the moments at the end sides of each segment are compatible with the strength criterion. For the problem in hand, this delivers the following conditions:

$$M(x) \in D_k : |M\,(x = l)| = |R|\,l \leq M_0; \quad |M\,(x = 0)| = |2R - P|\,l \leq M_0$$

The two conditions are satisfied for

$$-\frac{M_0}{l} \leq R \leq \frac{M_0}{l}; \quad -\frac{M_0}{l} \leq 2R - P \leq \frac{M_0}{l}$$

or, equivalently,

$$-\frac{M_0}{l} \leq R \leq \frac{M_0}{l}; \quad -\frac{M_0}{2l} - \frac{P}{2} \leq R \leq \frac{M_0}{2l} + \frac{P}{2}$$

These inequalities are satisfied provided that

$$-\frac{M_0}{l} \leq \frac{M_0}{2l} + \frac{P}{2}; \quad -\frac{M_0}{2l} - \frac{P}{2} \leq \frac{M_0}{l}$$

This delivers the maximum admissible force intensity:

$$\left|P^{\lim}\right| = \frac{3M_0}{l}$$

It is the exact value, since we have explored all possible statically admissible bending moment fields. ■

Exercise 83 shows the elements of the static approach applied to structural elements, based on statically admissible section force/moment fields, and a section-specific strength criterion. The force/moment fields are defined by the equilibrium equations of beam, truss, shell, and plate theories, which replace the field equations (9.21) of the continuum mechanics theory. They can be found in textbooks on structural mechanics and will not be developed here. The section-type strength criteria restrict the section forces and moments, defined by the reduction formulas $\mathcal{N} = \int_{A_i} (\boldsymbol{\sigma} \cdot \mathbf{e}_x)\, da$ and $\mathcal{M} = \int_{A_i} \mathbf{OM} \times (\boldsymbol{\sigma} \cdot \mathbf{e}_x)\, da$ (see Section 3.3.3), to section-specific strength properties, that is,

$$(\mathcal{N}(x_i), \mathcal{M}(x_i)) \in D_A \Leftrightarrow f(x_i, \mathcal{M}, \mathcal{N}) \leq 0 \tag{9.54}$$

For instance, in Exercise 83 we considered a Tresca-type moment strength criterion, which is of the general form[7]

$$f(x_i, \mathcal{M}) = \sqrt{\mathcal{M}^2 - M_x^2} - M_0 = \sqrt{M_y^2 + M_z^2} - M_0 \leq 0 \qquad (9.55)$$

This strength criterion restricts the admissible values of $M_y = \int_a z\sigma_{xx}(y,z)\,da$ and $M_z = -\int_a y\sigma_{xx}(y,z)\,da$, to a maximum value M_0, which is called the plastic moment. It does not restrict the torsion moment $M_x = \int_a (y\sigma_{zx} - z\sigma_{yx})\,da$ (i.e., the shear stresses σ_{zx} and σ_{yx}). Furthermore, strength criterion (9.55) is clearly restricted to double symmetric sections, for which $\max |M_y|_{M_z=0} = \max |M_z|_{M_y=0} = M_0$. This underlines that section type strength criteria needed to account for both the local material behavior and the section geometry. We have developed such a section-specific beam strength criterion in Problem Section 4.7 in Chapter 4 (see also Problem Section 9.7). Such force–moment interaction curves are at the core of the yield design of structures.

Finally, with these elements in hand, the lower limit theorem (9.31) for structural systems subjected to concentrated forces and moments can formally be stated in the form

$$\sum_{i=1}^{N} \left[\mathcal{N}'(x_i) \cdot \mathbf{u}(\mathbf{V}(x_i)) + \mathcal{M}'(x_i) \cdot \dot{\boldsymbol{\theta}}(\mathbf{V}(x_i))\right] \leq \mathbf{Q}^{\lim} \cdot \mathbf{q} \qquad (9.56)$$

The left-hand side represents the external work rate realized by statically and plastically admissible forces and moments, $\mathcal{N}'$ and $\mathcal{M}'$, along, respectively, the velocity $\mathbf{u}(\mathbf{V}(\mathbf{x}_i))$ and the rotation rate $\dot{\boldsymbol{\theta}}(\mathbf{V}(\mathbf{x}_i))$ at point x_i.

9.4.2 Kinematic Approach from Outside: Plastic Hinge Design

The most common application of the upper limit theorem in structural mechanics is in the form of structural discontinuities in beams, plates, and shells, introducing the notion of plastic hinges and yield lines, respectively. Consider a velocity jump in a beam of the form

$$\mathbf{V}^p(x_i) = [[\mathbf{V}(x_i)]] = \mathbf{OM}\,(x_i) \times [[\dot{\boldsymbol{\theta}}\,(x_i)]] = \begin{pmatrix} y[[\dot{\theta}_z]] - z[[\dot{\theta}_y]] \\ z[[\dot{\theta}_x]] \\ -y[[\dot{\theta}_x]] \end{pmatrix} \qquad (9.57)$$

where $\mathbf{OM} = y\mathbf{e}_y + z\mathbf{e}_z$ is the position vector in the beam's section at point x_i; and $[[\dot{\boldsymbol{\theta}}]] = \dot{\boldsymbol{\theta}}^+ - \dot{\boldsymbol{\theta}}^-$ denotes the rotation rate discontinuity in the section (see Figure 9.11). This velocity jump corresponds to a plastic hinge, over which—at

[7] By Tresca-type criterion, we mean a criterion that restricts the inplane components of a vector quantity on a surface. For instance, the Tresca criterion on a surface reads

$$f(\mathbf{T}(\mathbf{n})) = \sqrt{\mathbf{T}^2 - \sigma^2} - \sigma_0/2 \leq 0$$

where $\sigma = \mathbf{n} \cdot \mathbf{T}(\mathbf{n})$ is the normal stress.

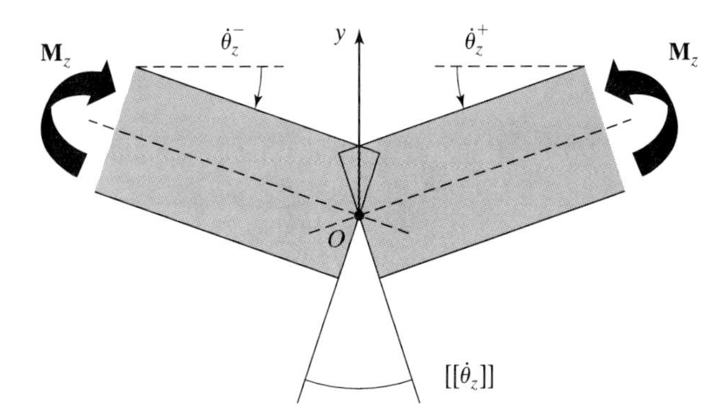

FIGURE 9.11: Plastic hinge: example of a rotational discontinuity in a beam.

collapse—externally supplied energy is dissipated. Use in (9.11) yields the dissipation rate associated with the plastic hinge discontinuity:

$$\begin{aligned} \frac{d\mathcal{D}}{dt}|_{x_i} &= \int_{\Sigma(x_i)} \mathbf{T}(\mathbf{e}_x) \cdot [[\mathbf{V}(x_i)]]\, da \\ &= \int_{\Sigma(x_i)} \left(\sigma_{xx}\left(y[[\dot{\theta}_z]] - z[[\dot{\theta}_y]]\right) + [(\sigma_{xy}z - \sigma_{xz}y)\,[[\dot{\theta}_x]]\right) da \end{aligned} \tag{9.58}$$

or, equivalently, using the reduction formulas,

$$\frac{d\mathcal{D}}{dt}|_{x_i} = \mathcal{M}(x_i) \cdot [[\dot{\boldsymbol{\theta}}(x_i)]] = M_x[[\dot{\theta}_x]] + M_y[[\dot{\theta}_y]] + M_z[[\dot{\theta}_z]] \tag{9.59}$$

where $\mathcal{M}(x_i)$ and $[[\dot{\boldsymbol{\theta}}(x_i)]]$, by analogy with (9.18), are associated by a section type flow rule satisfying the normality rule:

$$\text{on } x_i : [[\dot{\boldsymbol{\theta}}(x_i)]] = \dot{\lambda}\frac{\partial f}{\partial \mathcal{M}} = \dot{\lambda}\frac{\partial f}{\partial M_x}\mathbf{e}_x + \dot{\lambda}\frac{\partial f}{\partial M_y}\mathbf{e}_y + \dot{\lambda}\frac{\partial f}{\partial M_z}\mathbf{e}_z \tag{9.60}$$

The normality rule (9.60) establishes a relation between the moment $\mathcal{M}$ and the rotation rate discontinuity $[[\dot{\boldsymbol{\theta}}(x_i)]]$, such that the associated dissipation $\mathring{\varphi}_i$ is a function of $[[\dot{\boldsymbol{\theta}}(x_i)]]$ only.

Finally, if we evoke (for a last time) the principle of maximum plastic work, here applied to the moment–rotation rate couple, such that

$$\begin{aligned} \forall x_i;\ \ \forall \mathcal{M}(x_i) \in D_A(x_i);\ \ \mathcal{M}(x_i)[[\dot{\boldsymbol{\theta}}'(x_i)]] &\leq \mathring{\varphi}_i\left(x_i, [[\dot{\boldsymbol{\theta}}'(x_i)]]\right) \\ &= \max_{\mathcal{M}(x_i)\in D_A} \left[\mathcal{M}'(x_i) \cdot [[\dot{\boldsymbol{\theta}}'(x_i)]]\right] \end{aligned} \tag{9.61}$$

the upper bound theorem for a structural system with plastic hinges is obtained:

$$\mathbf{Q}^{\lim} \cdot \mathbf{q} \leq \mathbf{Q}^{\lim} \cdot \mathbf{q}' \leq \frac{d\mathcal{D}}{dt}\left([[\dot{\boldsymbol{\theta}}']]\right) = \sum_i \mathring{\varphi}_i\left(x_i, [[\dot{\boldsymbol{\theta}}'(x_i)]]\right) \tag{9.62}$$

where $\frac{d\mathcal{D}}{dt}\left([[\dot{\boldsymbol{\theta}}']]\right)$ represents the maximum dissipation capacity the structure can afford by creation of kinematically admissible plastic hinges.

Exercise 84. We consider the hyperstatic beam shown in Figure 9.10. Using the upper bound theorem, show that the solution $\left|P^{\lim}\right| = 3M_0/l$ found in Exercise 83 in Section 9.4.1 is actually the exact solution of the yield design problem.

In a first step, we need to determine the maximum dissipation capacity of the plastic hinge, that is,

$$\mathring{\varphi}_i\left(x_i, [[\dot{\boldsymbol{\theta}}'(x_i)]]\right) = \max_{f(\mathcal{M}(x_i))\leq 0}\left[\mathcal{M}(x_i)\cdot[[\dot{\boldsymbol{\theta}}(x_i)]]\right]$$

For the Tresca-type section strength criterion (9.55), the flow rule (9.60) reads

$$\begin{aligned}
f(\mathcal{M}) &= \sqrt{M_y^2+M_z^2}-M_0=0;\quad [[\dot{\theta}_x]]=0;\quad [[\dot{\theta}_y]]=\dot{\lambda}\frac{M_y}{M_0};\quad [[\dot{\theta}_z]]=\dot{\lambda}\frac{M_z}{M_0}\\
&\Rightarrow \dot{\lambda}=\sqrt{[[\dot{\theta}_y]]^2+[[\dot{\theta}_z]]^2}\\
M_y &= M_0\frac{[[\dot{\theta}_y]]}{\sqrt{[[\dot{\theta}_y]]^2+[[\dot{\theta}_z]]^2}};\quad M_z=M_0\frac{[[\dot{\theta}_z]]}{\sqrt{[[\dot{\theta}_y]]^2+[[\dot{\theta}_z]]^2}}
\end{aligned}$$

The maximum dissipation capacity then reads

$$\mathring{\varphi}_i\left(x_i, [[\dot{\boldsymbol{\theta}}'(x_i)]]\right) = \max_{f(\mathcal{M}(x_i))\leq 0}\left(M_y[[\dot{\theta}_y]]+M_z[[\dot{\theta}_z]]\right) = M_0\sqrt{[[\dot{\theta}_y]]^2+[[\dot{\theta}_z]]^2}$$

In a second step, we design a specific failure mechanism, which ensures that the structure remains kinematically admissible. The structure displayed in Figure 9.10b can afford two plastic hinges. The corresponding velocity vector reads

$$\begin{aligned}
0 &\leq x\leq l:\mathbf{V}'(x)=-V'\frac{x}{l}\mathbf{e}_y\\
l &\leq x\leq 2l:\mathbf{V}'(x)=-V'\left(\frac{2l-x}{l}\right)\mathbf{e}_y
\end{aligned}$$

and the rotation rate is

$$\begin{aligned}
0 &< x<l:\dot{\boldsymbol{\theta}}'(x)=-\frac{V'}{l}\,\mathbf{e}_z\\
l &< x<2l:\dot{\boldsymbol{\theta}}'(x)=+\frac{V'}{l}\,\mathbf{e}_z
\end{aligned}$$

The rotation rate has two discontinuities, one at $x=0$, for which

$$[[\dot{\theta}_z(x=0)]]=-\frac{V'}{l}\rightarrow\mathring{\varphi}_1\left(x_1=0,[[\dot{\boldsymbol{\theta}}'(0)]]\right)=\frac{M_0}{l}\left|V'\right|$$

the other at $x = l$, for which

$$[[\dot{\theta}_z(x = l)]] = 2\frac{V'}{l} \rightarrow \mathring{\varphi}_2\left(x_2 = l, [[\dot{\boldsymbol{\theta}}'(l)]]\right) = \frac{M_0}{l}2\left|V'\right|$$

Hence, the maximum dissipation capacity of the structure is

$$\frac{d\mathcal{D}}{dt}\left([[\dot{\boldsymbol{\theta}}']]\right) = \mathring{\varphi}_1 + \mathring{\varphi}_2 = \frac{M_0}{l}3\left|V'\right|$$

In a third step, we determine the external work rate realized along $\mathbf{V}'(x)$. For the concentrated load, it is

$$\mathbf{Q}^{\text{lim}} \cdot \mathbf{q}' = (-P\mathbf{e}_y) \cdot (-V'\mathbf{e}_y)$$

Finally, according to (9.62), the structure certainly fails if $\mathbf{Q}^{\text{lim}} \cdot \mathbf{q}' \geq \frac{d\mathcal{D}}{dt}\left([[\dot{\boldsymbol{\theta}}']]\right)$, that is, for

$$PV' \geq 3\frac{M_0}{l}\left|V'\right|$$

It follows that an upper bound to the limit load intensity is:

$$\left|P^{\text{lim}}\right| \leq |P| = \frac{3M_0}{l}$$

From a comparison of this upper bound with the lower bound, we verify that the found limit load $\left|P^{\text{lim}}\right| = 3M_0/l$ is actually the load-bearing capacity of the structure. ■

Remark We restricted ourselves to plastic hinges, characterized by the velocity jump (9.57). The model can be extended to include discontinuities associated with tension and shear hinges in the beam section. In this case, the upper limit theorem reads

$$\mathbf{Q}^{\text{lim}} \cdot \mathbf{q} \leq \mathbf{Q}^{\text{lim}} \cdot \mathbf{q}' \leq \frac{d\mathcal{D}}{dt}\left([[\dot{\boldsymbol{\theta}}']]; [[\mathbf{u}']]\right) \tag{9.63}$$

where $\frac{d\mathcal{D}}{dt}\left([[\dot{\boldsymbol{\theta}}']]; [[\mathbf{u}'(x)]]\right)$ is the maximum dissipation the structure can afford by the creation of rotational discontinuities and tension and shear discontinuities:

$$\frac{d\mathcal{D}}{dt}\left([[\dot{\boldsymbol{\theta}}']]; [[\mathbf{u}'(x)]]\right) = \sum_i \max_{f(\mathcal{M},\mathcal{N})\leq 0}\left(\mathcal{M}'(x_i) \cdot [[\dot{\boldsymbol{\theta}}'(x_i)]] + \mathcal{N}'(x_j) \cdot [[\mathbf{u}'(x_j)]]\right) \tag{9.64}$$

The moment and force quantities are restricted by a section strength criterion of the form (9.54) and related by flow rules to the rotation rate jump $[[\dot{\boldsymbol{\theta}}'(x_i)]]$ and the velocity jump $[[\mathbf{u}'(x_j)]]$:

$$[[\dot{\boldsymbol{\theta}}'(x_i)]] = \dot{\lambda}\frac{\partial f(\mathcal{M},\mathcal{N})}{\partial \mathcal{M}}; \quad [[\mathbf{u}'(x_j)]] = \dot{\lambda}\frac{\partial f(\mathcal{M},\mathcal{N})}{\partial \mathcal{N}(x_i)} \tag{9.65}$$

9.5 TRAINING SET: STRENGTH DOMAIN OF FIBER-REINFORCED COMPOSITE MATERIALS

Fiber-reinforced composite materials have become increasingly popular in engineering applications ranging from aerospace applications (e.g., fiber-reinforced polymer matrix composites) to geotechnical applications (reinforced earth, nailing, bolts, geotextiles, etc.). These engineered materials are fine tailored to overcome an intrinsic drawback of the matrix behavior. The aim of this training set is to develop safe estimates for the effective strength domain of composite materials, based on the lower bound theorem. For the purpose of analysis, we will restrict ourselves to a simple unidirectional reinforced composite material[8] displayed in Figure 9.12a.

9.5.1 Lower and Upper Bounds

We consider a representative unit cell, extracted from the composite material, as shown in Figure 9.5b. The unit cell is a prismatic cylinder of arbitrary length along the fiber oriented by unit normal $\mathbf{e}_f$. In this unit cell of volume taken equal to unity $V = 1$, V_f and $V_m = V - V_f$ denote the regions occupied by the matrix and the fiber. Naturally, we introduce the fiber volume fraction:

$$v = \frac{V_f}{V}; \quad 1 - v = \frac{V_m}{V} \tag{9.66}$$

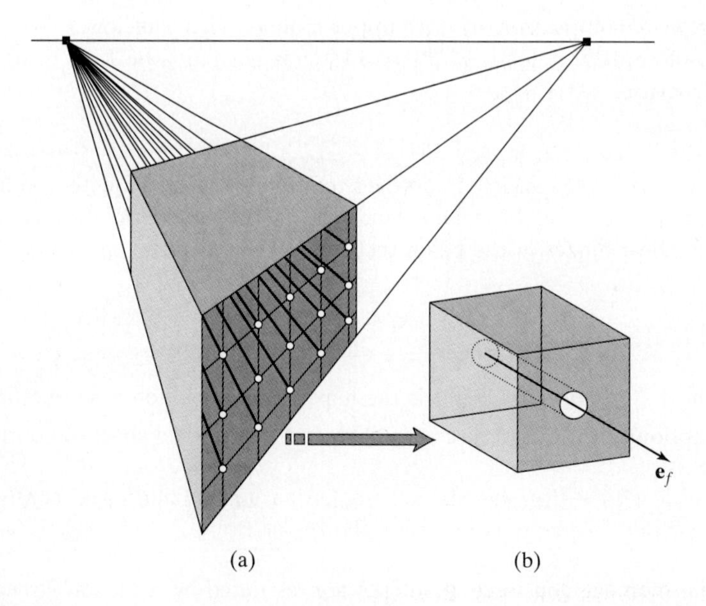

FIGURE 9.12: (a) Unidirectional fiber-reinforced composite with (b) representative unit cell.

[8]For the extension to multidirectional composite materials, see the reference paper of de Buhan and Talierco (1991), which inspired this exercise:
de Buhan, P., and Talierco, A. (1991). "A Homogenization Approach to the Yield Strength of Composite Materials." *Eur. J. Mech., A/Solids*, Vol. 10, No. 2, 129–154.

We want to determine the dissipation rate realized in the unit cell subjected to a macroscopic stress $\mathbf{\Sigma}$ on the boundary ∂V:

$$\text{on } \partial V : \mathbf{T}^d = \mathbf{\Sigma} \cdot \mathbf{N} \equiv \boldsymbol{\sigma} \cdot \mathbf{n} \tag{9.67}$$

where $\boldsymbol{\sigma}$ is a microscopic stress field. We assume that $\mathbf{T}^d$ is constant on ∂V, we can apply the Hill lemma developed in Section 6.1.2, to the external work rate to which the unit volume is subjected. This leads to the external work rate, which is dissipated at plastic failure:

$$\frac{d\mathcal{W}_{\text{ext}}}{dt} = \mathbf{\Sigma} : \mathbf{D} = \langle \boldsymbol{\sigma} \rangle : \langle \mathbf{d} \rangle = \frac{d\mathcal{D}}{dt} \tag{9.68}$$

where $\langle \boldsymbol{\sigma} \rangle$ and $\langle \mathbf{d} \rangle$ are the volume averages

$$\langle \boldsymbol{\sigma} \rangle = \int_{V=1} \boldsymbol{\sigma}(\mathbf{x})\, dV; \quad \langle \mathbf{d} \rangle = \int_{V=1} \mathbf{d}(\mathbf{x})\, dV \tag{9.69}$$

Application of the lower limit theorem delivers here

$$\mathbf{\Sigma}' : \mathbf{D} \leq \mathbf{\Sigma}^{\text{lim}} : \mathbf{D} = \frac{d\mathcal{D}}{dt} \tag{9.70}$$

where $\mathbf{\Sigma}' = \langle \boldsymbol{\sigma}' \rangle$ is a lower bound of the actual limit stress $\mathbf{\Sigma}^{\text{lim}} = \langle \boldsymbol{\sigma}^{\text{lim}} \rangle$, provided that the microscopic stress fields $\boldsymbol{\sigma}'(\mathbf{x})$ are statically and plastically admissible in the unit cell. For the two-phase composite material composed of matrix and fiber, the stress fields must satisfy

$$\text{in } V_m \;:\; \boldsymbol{\sigma}'(\mathbf{x}) = \boldsymbol{\sigma}'_m \text{ SA}, \boldsymbol{\sigma}'_m \in D_m \Leftrightarrow f_m(\boldsymbol{\sigma}'_m) \leq 0 \tag{9.71}$$

$$\text{in } V_f \;:\; \boldsymbol{\sigma}'(\mathbf{x}) = \boldsymbol{\sigma}'_f \text{ SA}, \boldsymbol{\sigma}'_f \in D_f \Leftrightarrow f_f(\boldsymbol{\sigma}'_f) \leq 0 \tag{9.72}$$

where $\boldsymbol{\sigma}'_m$ and $\boldsymbol{\sigma}'_f$ are statically admissible stress fields at the microscopic level of the heterogeneous material, and D_m and D_f are the strength domains of matrix and fiber, defined by the plasticity criterion $f_m(\boldsymbol{\sigma}_m)$ and $f_f(\boldsymbol{\sigma}_f)$, respectively.

Analogously, application of the upper limit theorem gives here

$$\mathbf{\Sigma}^{\text{lim}} : \mathbf{D} \leq \mathbf{\Sigma}^{\text{lim}} : \mathbf{D}' \leq \frac{d\mathcal{D}}{dt}(\mathbf{v}') \tag{9.73}$$

where $\frac{d\mathcal{D}}{dt}(\mathbf{v}')$ is the maximum dissipation capacity the composite material can afford in both matrix and fiber along kinematically admissible velocity fields $\mathbf{v}'(\mathbf{x})$.

9.5.2 Voigt–Reuss Strength Bounds

The most conservative choice for a macroscopic strength domain consists of choosing a constant stress field in V, similar to the Reuss bound (see Section 6.5.4) such that the average macroscopic stress, say $\mathbf{\Sigma}'$, reads

$$\mathbf{\Sigma}' = \int_{V=1} \boldsymbol{\sigma}'(\mathbf{x}) dV = \langle \boldsymbol{\sigma}'(\mathbf{x}) \rangle = \boldsymbol{\sigma}'_m = \boldsymbol{\sigma}'_f \tag{9.74}$$

Exercise 85. Reuss Strength Bound: We consider that matrix and fiber obey to a Von–Mises strength criterion. By developing the conditions that the stresses $\boldsymbol{\sigma}'_m$ and $\boldsymbol{\sigma}'_f$ need to satisfy, develop a lower bound of the macroscopic strength domain of the composite material.

To be statically admissible, $\boldsymbol{\sigma}'_m$ and $\boldsymbol{\sigma}'_f$ need to satisfy

1. The stress boundary condition (9.67);
2. The equilibrium equations:

$$\text{in } V_m : \text{div } \boldsymbol{\sigma}'_m = 0; \quad \text{in } V_f : \text{div } \boldsymbol{\sigma}'_f = 0$$

3. The continuity of the stress vector on the fiber-matrix interface:

$$\text{on } \Sigma : [[\mathbf{t}'(\mathbf{n}_f)]] = \boldsymbol{\sigma}'_m \cdot \mathbf{n}_f - \boldsymbol{\sigma}'_f \cdot \mathbf{n}_f = 0$$

where $\mathbf{n}_f$ is the orientation of the interface, which is normal to the fiber direction $\mathbf{e}_f$ (*i.e.*, $\mathbf{n}_f \cdot \mathbf{e}_f = 0$).

These conditions are met by the stress field (9.74).
To be plastically admissible, $\boldsymbol{\sigma}'_m$ and $\boldsymbol{\sigma}'_f$ need to satisfy the local strength criteria of the matrix (9.71) and the fiber (9.72). A combination of the strength criteria with (9.74) indicates that a safe lower bound for the macroscopic strength domain is the minimum of both strength domains, respectively, the maximum of both yield functions

$$\text{in } V : F(\boldsymbol{\Sigma}') = \max[f_m(\boldsymbol{\sigma}'_m = \boldsymbol{\Sigma}'); \quad f_f(\boldsymbol{\sigma}'_f = \boldsymbol{\Sigma}')] \leq 0$$

For a composite material composed of two Von–Mises materials, having $\sigma_0^m = \sqrt{3}k_m$ and $\sigma_0^f = \sqrt{3}k_f$ as uniaxial strengths (have a quick look back at Section 8.1.3), the macroscopic strength criterion is obtained:

$$F(\boldsymbol{\Sigma}' = \boldsymbol{\sigma}'_m = \boldsymbol{\sigma}'_f) = \frac{1}{2}\mathbf{S}' : \mathbf{S}' - \min\left(k_m^2, k_f^2\right) \leq 0 \Rightarrow \min(\sigma_0^m; \sigma_0^f) = \Sigma'_0 \leq \Sigma_0$$

where $\Sigma'_0 = \mathbf{N} \cdot \boldsymbol{\Sigma}' \cdot \mathbf{N}$ is a safe lower bound to the uniaxial macroscopic strength of the fiber-reinforced composite material. ■

A Voigt strength bound is obtained if we assume a microscopic strain rate field that is constant in V:

$$\mathbf{D}' = \langle \mathbf{d}' \rangle = \mathbf{d}'_m = \mathbf{d}'_f \tag{9.75}$$

where $\mathbf{d}'_m$ and $\mathbf{d}'_f$ are the strain rates in the matrix and fiber, associated with a velocity field:

$$\mathbf{v}' = \mathbf{d}'_m \cdot \mathbf{x} = \mathbf{d}'_f \cdot \mathbf{x} \tag{9.76}$$

Exercise 86. Voigt Strength Bound: By developing the conditions that the strain rate fields (9.75) must satisfy, develop an upper bound of the strength domain of a fiber-reinforced composite composed of a Von–Mises matrix and a Von–Mises fiber phase.

To be kinematically admissible, the velocity field must satisfy (9.5). Given that the boundary ∂V is a stress boundary, $\mathbf{v}'$ is kinematically admissible.
The strain rate fields must be compatible with the kinematics of plastic flow, at failure, defined by the flow rule of the Von–Mises materials:

$$\mathbf{d}'_m \equiv \mathbf{d}^{p\prime}_m = \dot{\lambda}_m \frac{\partial f_m}{\partial \boldsymbol{\sigma}'_m} = \dot{\lambda}_m \mathbf{s}'_m; \quad \mathbf{d}'_f \equiv \mathbf{d}^{p\prime}_f = \dot{\lambda}_f \frac{\partial f_f}{\partial \boldsymbol{\sigma}'_f} = \dot{\lambda}_f \mathbf{s}'_f$$

where $\dot{\lambda}_m$ and $\dot{\lambda}_f$ denote plastic multipliers. The maximum dissipation capacity for a Von–Mises material was developed in Section 9.3.1, and reads here for the matrix and the fiber phase

$$\varphi'\left(\mathbf{d}^{p\prime}_m\right) = \sigma_0^m \sqrt{\frac{2}{3}} \max_{f(\boldsymbol{\sigma}'_m)\leq 0} \sqrt{\mathbf{d}^{p\prime}_m : \mathbf{d}^{p\prime}_m}; \quad \varphi'\left(\mathbf{d}^{p\prime}_f\right) = \sigma_0^f \sqrt{\frac{2}{3}} \max_{f(\boldsymbol{\sigma}'_f)\leq 0} \sqrt{\mathbf{d}^{p\prime}_f : \mathbf{d}^{p\prime}_f}$$

and, after integration, using (9.75), over the unit cell

$$\frac{d\mathcal{D}}{dt}\left(\mathbf{v}'\right) = \sqrt{\frac{2}{3}} \left[(1-v)\,\sigma_0^m + v\sigma_0^f\right] \sqrt{\mathbf{D}' : \mathbf{D}'}$$

Furthermore, the strain rate field $\mathbf{D}'$ needs to satisfy the incompressibility condition of the Von–Mises flow rule (i.e., $\operatorname{tr}\mathbf{d}'_m \propto \operatorname{tr}\mathbf{s}'_m = 0$ and $\operatorname{tr}\mathbf{d}'_f \propto \operatorname{tr}\mathbf{s}'_f = 0$). Expressed in principal strain rates, $D'_I \geq D'_{II} \geq D'_{III}$, the incompressibility is satified by a strain rate field of the form

$$\mathbf{D}' = \begin{pmatrix} D'_I & 0 & 0 \\ 0 & D'_{II} = -\left(D'_I + D'_{III}\right) & 0 \\ 0 & 0 & D'_{III} \end{pmatrix}$$

The maximum dissipation capacity the composite can afford is

$$\frac{d\mathcal{D}}{dt}\left(\mathbf{v}'\right) = 2\sqrt{\frac{1}{3}} \left[(1-v)\,\sigma_0^m + v\sigma_0^f\right] \sqrt{D_1'^2 + D_1' D_3' + D_3'^2}$$

If this dissipation capacity is greater than the external work rate $\boldsymbol{\Sigma}^{\text{lim}} : \mathbf{D}'$, the composite will surely fail, that is, for

$$\begin{aligned} \boldsymbol{\Sigma}^{\text{lim}} : \mathbf{D}' &= \left(\Sigma_I^{\text{lim}} - \Sigma_{II}^{\text{lim}}\right) D_1' + \left(\Sigma_{III}^{\text{lim}} - \Sigma_{II}^{\text{lim}}\right) D_3' \\ &\geq 2\sqrt{\frac{1}{3}} \left[(1-v)\,\sigma_0^m + v\sigma_0^f\right] \sqrt{D_1'^2 + D_1' D_3' + D_3'^2} \end{aligned}$$

where $\Sigma_1^{\text{lim}} \geq \Sigma_2^{\text{lim}} \geq \Sigma_3^{\text{lim}}$ are principal macroscopic limit stresses. The right-hand side defines an upper bound to the actual dissipation the material can develop. Looking for the minimum of this upper bound, we need to find the minimum of the maximum dissipation capacity $\frac{d\mathcal{D}}{dt}\left(D'_I, D'_{III}\right)$,

$$\frac{\partial}{\partial D'_I}\left[\frac{d\mathcal{D}}{dt}\left(D'_I, D'_{III}\right)\right] = \frac{\partial}{\partial D'_{III}}\left[\frac{d\mathcal{D}}{dt}\left(D'_I, D'_{III}\right)\right] = 0$$

The minimum of the maximum dissipation capacity is obtained for

$$D'_{III} = -\frac{1}{2}D'_I \rightarrow \mathbf{D}' = D'_I \begin{pmatrix} 1 & 0 & 0 \\ 0 & -1/2 & 0 \\ 0 & 0 & -1/2 \end{pmatrix}$$

Use in the upper bound theorem (9.73) gives

$$\mathbf{\Sigma}^{\lim} : \mathbf{D}' = \left(\Sigma_I^{\lim} - \frac{1}{2}\left(\Sigma_I^{\lim} + \Sigma_{III}^{\lim}\right)\right) D'_I \leq \left[(1-v)\,\sigma_0^m + v\sigma_0^f\right] |D'_I|$$

This inequality delivers an upper bound of the triaxial strength:

$$\left|\Sigma_I^{\lim} - \frac{1}{2}\left(\Sigma_{II}^{\lim} + \Sigma_{III}^{\lim}\right)\right| \leq (1-v)\,\sigma_0^m + v\sigma_0^f$$

Note that the strength domain is independent of the direction of the fiber reinforcement. ■

In summary, a combination of the strength bounds based, respectively, on a constant stress field (9.74) and a constant strain rate field (9.75) frames the uniaxial strength of the composite material:

$$\min(\sigma_0^m; \sigma_0^f) \leq \left|\Sigma_0^{\lim}\right| \leq (1-v)\,\sigma_0^m + v\sigma_0^f \tag{9.77}$$

Clearly, the Reuss strength lower bound is too conservative to account for one of the very reasons for reinforcing a matrix by fibers, which is to overcome the intrinsic strength weakness of the matrix. On the other hand, the Voigt strength upper bound certainly overestimates the actual strength of the composite, since it does not account for the directional reinforcement of the composite. A refined analysis is required to provide a more realistic estimate of the strength domain of composite materials.

9.5.3 An Improved Lower Bound: Enriched Statically Admissible Stress Field

To improve the lower bound, let us consider a stress field that differs in fiber and matrix:

$$\text{in } V_f : \boldsymbol{\sigma}'_f = \boldsymbol{\sigma}'_m + \delta\sigma \mathbf{e}_f \otimes \mathbf{e}_f; \quad \text{in } V_m : \boldsymbol{\sigma}'_m = \boldsymbol{\sigma}'_m \tag{9.78}$$

where $\delta\sigma = \sigma'_f - \sigma'_m$ is the difference in axial stress in the fiber direction $\mathbf{e}_f$ between the axial fiber stress, $\sigma'_f = \mathbf{e}_f \cdot \boldsymbol{\sigma}'_f \cdot \mathbf{e}_f$, and the axial matrix stress, $\sigma'_m = \mathbf{e}_f \cdot \boldsymbol{\sigma}'_m \cdot \mathbf{e}_f$. This stress field is statically admissible, since it ensures the continuity of the stress vector over the interface Σ, $[[\mathbf{t}'(\mathbf{n}_f \perp \mathbf{e}_f)]] = 0$. For the stress field (9.78), the average stress $\mathbf{\Sigma}'$ in the unit cell is

$$\mathbf{\Sigma}' = \frac{1}{V}\int_{V=1} \boldsymbol{\sigma}'(\mathbf{x})\, d\Omega = v\boldsymbol{\sigma}'_f + (1-v)\boldsymbol{\sigma}'_m = \boldsymbol{\sigma}'_m + v\delta\sigma \mathbf{e}_f \otimes \mathbf{e}_f \tag{9.79}$$

Note that $\boldsymbol{\sigma}'_m$ is not weighted by its volume fraction, but only the term $\delta\sigma \mathbf{e}_f \otimes \mathbf{e}_f$, which can be seen as the excess stress in the fiber with regard to the overall prevailing matrix stress.

Exercise 87. We consider that the matrix strength behavior is governed by a Von–Mises criterion:

$$\text{in } V_m : \ f_m\left(\boldsymbol{\sigma}_m\right) = \frac{3}{2}\mathbf{s}_m : \mathbf{s}_m - \left(\sigma_0^m\right)^2 \leq 0$$

The fibers are assumed to be governed by a uniaxial strength criterion:

$$\text{in } V_f : f_f\left(\boldsymbol{\sigma}_f\right) = \left|\sigma_{ff}\right| - \sigma_0^f \leq 0$$

where $\sigma_{ff} = \mathbf{e}_f \cdot \boldsymbol{\sigma}_f \cdot \mathbf{e}_f$ is the fiber stress component in the fiber direction.

1. Consider a uniaxial macroscopic stress state prescribed in the direction of the fiber orientation $\alpha = 0$. By using the stress field defined by (9.78), determine a lower bound $\Sigma'_{\alpha=0}$ of the actual uniaxial strength $\Sigma_0^{\lim}$ of the composite in the direction of the fiber orientation.
2. In a similar way, develop a lower bound $\Sigma'_{\alpha=\pi/2}$ for the strength $\Sigma_{\pi/2}^{\lim}$ in the direction normal to the fiber orientation.

1. Composite Strength in the Fiber Direction: The prescribed macroscopic stress in the direction of the fiber is

$$\boldsymbol{\Sigma}'\left(\alpha = 0\right) = \Sigma'\mathbf{e}_f \otimes \mathbf{e}_f$$

Using (9.78) and (9.79) yields the corresponding matrix and fiber stresses:

$$\begin{aligned}\boldsymbol{\sigma}'_m\left(\alpha = 0\right) &= \left(\Sigma' - v\,\delta\sigma\right)\mathbf{e}_f \otimes \mathbf{e}_f \\ \boldsymbol{\sigma}'_f\left(\alpha = 0\right) &= \left(\Sigma' + (1 - v)\,\delta\sigma\right)\mathbf{e}_f \otimes \mathbf{e}_f\end{aligned}$$

For a uniaxial stress field, the Von–Mises criterion reduces to

$$f_m\left(\boldsymbol{\sigma}'_m = \sigma'_m\mathbf{e}_f \otimes \mathbf{e}_f\right) = \left(\sigma'_m\right)^2 - \left(\sigma_0^m\right)^2 \leq 0$$

Thus,

$$f_m\left(\sigma'_m, \alpha = 0\right) = \Sigma'^2 - 2v\,\delta\sigma\Sigma' + \left(v\,\delta\sigma\right)^2 - \left(\sigma_0^m\right)^2 \leq 0$$

Analogously, the fiber strength criterion delivers

$$f_f\left(\sigma'_f, \alpha = 0\right) = \left|\Sigma' + (1 - v)\,\delta\sigma\right| - \sigma_0^f \leq 0$$

Both strength criteria depend in addition to the applied stress Σ' also on the stress excess $\delta\sigma$:

$$F(\Sigma', \alpha = 0) = \max_{\delta\sigma}\left(\begin{array}{c} f_m\left(\sigma'_m, \alpha = 0\right) = \Sigma'^2 - 2v\,\delta\sigma\Sigma' + \left(v\,\delta\sigma\right)^2 - \left(\sigma_0^m\right)^2 \\ f_f\left(\sigma'_f, \alpha = 0\right) = \left|\Sigma' + (1 - v)\,\delta\sigma\right| - \sigma_0^f \end{array}\right) \leq 0$$

Assuming, at failure, that matrix and fibers reach their strength limit, deliver a value for the unknown $\delta\sigma$ of the problem:

$$f_m = f_m = 0 \Rightarrow \delta\sigma = \frac{1}{(1 - 2v)}\left(\pm\sqrt{\Sigma_0'^2 + (1 - 2v)\left[\left(\sigma_0^f\right)^2 - \left(\sigma_0^m\right)^2\right]} - \Sigma'_0\right)$$

The previous result must hold for $v \to 0 \Leftrightarrow |\Sigma'_0| = \sigma_0^m$, for which $|\delta\sigma|$ realizes a minimum:

$$\lim_{v \to 0} |\delta\sigma| = \sigma_0^f - \sigma_0^m$$

Then, given this minimum, any stress excess $|\delta\sigma|$ that is smaller than $\lim_{v\to 0} |\delta\sigma|$ is a safe lower bound of the admissible stress excess:

$$|\delta\sigma| \le \lim_{v \to 0} |\delta\sigma| = \sigma_0^f - \sigma_0^m$$

Finally, use in the fiber strength criterion delivers a lower bound of the macroscopic strength criterion in the fiber direction:

$$f_f\left(\sigma'_f, \alpha = 0\right) = 0 \Leftrightarrow |\Sigma'| \le |\Sigma'_0| = v\sigma_0^f + (1 - v)\,\sigma_0^m \le \left|\Sigma_0^{\lim}\right|$$

The found lower bound, for $\alpha = 0$, coincides with the Voigt strength bound (9.77), which is an upper bound for any value of α; thus

$$\left|\Sigma_{\alpha=0}^{\lim}\right| = v\sigma_0^f + (1 - v)\,\sigma_0^m$$

2. Composite Strength in the Direction Normal to the Fiber Orientation: For stress $\boldsymbol{\Sigma}'(\alpha = \pi/2) = \Sigma'\mathbf{n}_f \otimes \mathbf{n}_f$, the matrix and fiber stresses have two components:

$$\begin{aligned} \boldsymbol{\sigma}'_m(\alpha = \pi/2) &= \Sigma'\mathbf{n}_f \otimes \mathbf{n}_f - v\delta\sigma\mathbf{e}_f \otimes \mathbf{e}_f \\ \boldsymbol{\sigma}'_f(\alpha = \pi/2) &= \Sigma'\mathbf{n}_f \otimes \mathbf{n}_f + (1 - v)\,\delta\sigma\mathbf{e}_f \otimes \mathbf{e}_f \end{aligned}$$

It is useful to note that the case $\delta\sigma = 0$ corresponds to the Reuss strength bound, for which

$$\delta\sigma = 0 \to \left|\Sigma'_{\pi/2}\right| = \min\left(\sigma_0^m, \sigma_0^f\right) = \sigma_0^m \le \left|\Sigma_{\alpha=\pi/2}^{\lim}\right|$$

Therefore, an improved lower bound in the $\mathbf{n}_f$-direction is obtained for a nonzero value of $\delta\sigma$, which gives the maximum value of the two strength criteria:

$$F(\Sigma', \alpha = \pi/2) = \max_{\delta\sigma}\left[\begin{array}{c} f_m\left(\boldsymbol{\sigma}'_m = \Sigma'\mathbf{n}_f \otimes \mathbf{n}_f - v\delta\sigma\mathbf{e}_f \otimes \mathbf{e}_f\right) \\ f_f\left(\sigma_{ff} = (1 - v)\,\delta\sigma\right) \end{array}\right] \le 0$$

The deviator stress in the matrix reads

$$\mathbf{s}'_m = \boldsymbol{\sigma}'_m - \frac{1}{3}\mathrm{tr}\,\boldsymbol{\sigma}'_m = \Sigma'\left(\mathbf{n}_f \otimes \mathbf{n}_f - \frac{1}{3}\mathbf{1}\right) - v\delta\sigma\left(\mathbf{e}_f \otimes \mathbf{e}_f - \frac{1}{3}\mathbf{1}\right)$$

The corresponding stress deviator invariant is

$$j'_m = \frac{1}{2}\mathbf{s}'_m : \mathbf{s}'_m = \frac{1}{3}\left[\Sigma'^2 + v\delta\sigma\Sigma' + (v\delta\sigma)^2\right]$$

Use in the matrix strength criterion gives

$$f_m(\Sigma', \delta\sigma, \alpha = \pi/2) = 3j'_m - (\sigma_0^m)^2 = \Sigma'^2 + v\delta\sigma\Sigma' + (v\delta\sigma)^2 - (\sigma_0^m)^2 \le 0$$

which has a maximum for

$$\frac{\partial f}{\partial(\delta\sigma)} = v\Sigma' + 2v^2\delta\sigma = 0 \Rightarrow \delta\sigma = -\frac{\Sigma'}{2v}$$

for which

$$f_m\left(\Sigma', \delta\sigma = -\frac{\Sigma'}{2v}, \alpha = \pi/2\right) = \frac{3}{4}\Sigma'^2 - (\sigma_0^m)^2 \leq 0$$

The fiber strength criterion reads

$$f_f\left(\Sigma', \delta\sigma = -\frac{\Sigma'}{2v}, \alpha = \pi/2\right) = \frac{1-v}{2v}|\Sigma'| - \sigma_0^f \leq 0$$

From a combination of the two criteria, we obtain the following lower bound:

$$|\Sigma'| \leq \left|\Sigma'_{\pi/2}\right| = \sigma_0^m \min\left(\frac{2}{\sqrt{3}}; \frac{2v}{1-v}\rho\right) \leq \left|\Sigma_{\pi/2}^{\lim}\right|$$

$\rho = \sigma_0^f/\sigma_0^m$ is the fiber-to-matrix strength ratio. This lower bound must be greater than the Reuss bound, $\left|\Sigma'_{\pi/2}\right| = \sigma_0^m$, which is the smallest lower bound. Hence, a necessary requirement for this improved lower bound is

$$\rho \geq \frac{1-v}{2v}; \quad v \geq \frac{1}{2\rho+1}$$

Finally, it is instructive to note that the fiber stress excess criterion $|\delta\sigma| \leq \sigma_0^f - \sigma_0^m$ is more severe than the fiber strength criterion, which therefore constitutes also a safe lower bound:

$$|\Sigma'| \leq \left|\Sigma'_{\pi/2}\right| = \sigma_0^m \min\left(\frac{2}{\sqrt{3}}; \quad 2v(\rho-1)\right) \leq \left|\Sigma_{\pi/2}^{\lim}\right|$$

for

$$\rho \geq \frac{1+2v}{2v}; \; v \geq \frac{1}{2(\rho-1)}$$

■

The found estimates for $|\Sigma'_0|$ and $\left|\Sigma'_{\pi/2}\right|$ show that the statically admissible stress field (9.78) gives a lower bound provided it satisfies the matrix strength criterion and the stress excess criterion. It follows that for any direction $\alpha = \alpha(\mathbf{e}_f, \mathbf{n})$,

$$\forall\alpha;\; F(\mathbf{\Sigma}', \alpha) = \max_{\delta\sigma}\begin{pmatrix} f_m\left(\boldsymbol{\sigma}'_m = \mathbf{\Sigma}' - v\delta\sigma\mathbf{e}_f \otimes \mathbf{e}_f\right) \\ |\delta\sigma| - \left(\sigma_0^f - \sigma_0^m\right) \end{pmatrix} \leq 0 \tag{9.80}$$

This result is due to de Buhan and Salençon (1987).[9]

Exercise 88. By generalizing the previous results, determine the uniaxial strength criterion for any direction $\alpha = \alpha(\mathbf{e}_f, \mathbf{n})$ the fibers enclose with the direction of load application $\mathbf{n}$. Display the result for a fiber volume fraction of $v = 10\%$, and a matrix-to-fiber strength ratio of $\rho = \sigma_0^f/\sigma_0^m = 6$.

[9] de Buhan, P., and Salençon, J. (1987). "Yield Strength of Reinforced Soils as Anisotropic Media." *IUTAM Symposium on Yielding, Damage and Fracture of Anisotropic Solids*, ed. by. J. P. Boehler, Mech. Eng. Publ., London, 791–803.

The deviator stress in the matrix is expressed as a function of the macroscopic deviator stress $\mathbf{S}'$ and the fiber stress excess $\delta\sigma$:

$$\mathbf{s}'_m = \mathbf{S}' - v\delta\sigma\left(\mathbf{e}_f \otimes \mathbf{e}_f - \frac{1}{3}\mathbf{1}\right)$$

The matrix stress deviator invariant reads

$$j_2^{m\prime} = \frac{1}{2}\mathbf{s}'_m : \mathbf{s}'_m = J'_2 - v\,\delta\sigma\,S'_{ff} + \frac{1}{3}(v\delta\sigma)^2$$

where $J'_2 = \frac{1}{2}\mathbf{S}' : \mathbf{S}'$, and $S'_{ff} = \mathbf{e}_f \cdot \mathbf{S}' \cdot \mathbf{e}_f$. Use in the matrix strength criterion gives

$$f_m\left(\boldsymbol{\sigma}'_m\right) = 3j_2^{m\prime} - (\sigma_0^m)^2 = \frac{3}{2}\mathbf{S}' : \mathbf{S}' - 3v\,\delta\sigma\,S'_{ff} + (v\delta\sigma)^2 - (\sigma_0^m)^2 \leq 0$$

The matrix strength criterion has a maximum for

$$\frac{\partial f_m}{\partial\left(\delta\sigma\right)} = -3vS'_{ff} + 2v^2\delta\sigma = 0; \quad \frac{\partial^2 f}{\partial\left(\delta\sigma\right)^2} = 2v^2 \geq 0 \Rightarrow \delta\sigma = \frac{3}{2v}S'_{ff}$$

for which

$$f_m\left(\boldsymbol{\sigma}'_m = \boldsymbol{\Sigma}' - \frac{3}{2}S'_{ff}\mathbf{e}_f \otimes \mathbf{e}_f\right) = 3J'_2 - \frac{9}{4}S'^2_{ff} - (\sigma_0^m)^2 \leq 0$$

The lower bound of the macroscopic strength domain is then defined by the following criterion:

$$F(\boldsymbol{\Sigma}', \alpha) = \max\left(\begin{array}{c} f_m\left(\boldsymbol{\Sigma}' - \frac{3}{2}S'_{ff}\mathbf{e}_f \otimes \mathbf{e}_f\right) \\ f_m\left(\boldsymbol{\Sigma}' - v\left(\sigma_0^f - \sigma_0^m\right)\mathbf{e}_f \otimes \mathbf{e}_f\right) \end{array}\right) \leq 0$$

Consider now a uniaxial tensile macroscopic stress state $\Sigma'\mathbf{n} > 0$ prescribed in the $\mathbf{n}$-direction onto the unit cell. The corresponding macroscopic stress tensor and deviator stress read

$$\boldsymbol{\Sigma}' = \Sigma'\mathbf{n} \otimes \mathbf{n} \rightarrow \mathbf{S}' = \frac{1}{3}\Sigma'\left(2\mathbf{n} \otimes \mathbf{n} - \mathbf{t} \otimes \mathbf{t} - \mathbf{b} \otimes \mathbf{b}\right)$$

where $\mathbf{t} \cdot \mathbf{n} = 0$ and $\mathbf{n} \times \mathbf{t} = \mathbf{b}$ define the directions normal to the directions of the macroscopic stress application. The macroscopic stress deviator invariant reads

$$J'_2 = \frac{1}{2}\mathbf{S}' : \mathbf{S}' = \frac{1}{3}\Sigma'^2$$

We consider fibers in the $(\mathbf{n} \times \mathbf{t})$-plane, for which $\mathbf{b} \cdot \mathbf{e}_f \equiv 0$, $\mathbf{e}_f \cdot \mathbf{n} \otimes \mathbf{n} \cdot \mathbf{e}_f = \cos^2 \alpha(\mathbf{e}_f, \mathbf{n})$ and $\mathbf{e}_f \cdot \mathbf{t} \otimes \mathbf{t} \cdot \mathbf{e}_f = \sin^2 \alpha(\mathbf{e}_f, \mathbf{n})$.[10] The deviator stress component in the $\mathbf{e}_f$-direction reads

$$S'_{ff} = \mathbf{e}_f \cdot \left(\frac{1}{3} \Sigma' \left(2\mathbf{n} \otimes \mathbf{n} - \mathbf{t} \otimes \mathbf{t} - \mathbf{b} \otimes \mathbf{b} \right) \right) \cdot \mathbf{e}_f = \frac{1}{3} \Sigma' \left(2 - 3 \sin^2 \alpha \right)$$

Use of the expressions of J'_2 and S'_{ff} in the strength criteria gives

$$F\left(\Sigma', \alpha\right) = \max \begin{pmatrix} \Sigma'^2 \left(1 - \frac{1}{4}\left(2 - 3\sin^2\alpha\right)^2\right) - \left(\sigma_0^m\right)^2 \\ \Sigma'^2 - v(\sigma_0^f - \sigma_0^m)\, \Sigma' \left|2 - 3\sin^2\alpha\right| + v^2(\sigma_0^f - \sigma_0^m)^2 - \left(\sigma_0^m\right)^2 \end{pmatrix} \leq 0$$

or, in a dimensionless form,

$$F\left(\bar{\Sigma}' = \Sigma'/\sigma_0^m, \alpha\right) = \max \begin{pmatrix} \bar{\Sigma}'^2 \left(1 - \frac{1}{4}\left(2 - 3\sin^2\alpha\right)^2\right) - 1 \\ \bar{\Sigma}'^2 - v\bar{\rho}\, \bar{\Sigma}' \left|2 - 3\sin^2\alpha\right| + \left(v\bar{\rho}\right)^2 - 1 \end{pmatrix} \leq 0$$

where $\bar{\rho} = \rho - 1$. This yields the following lower bound:

$$\left|\bar{\Sigma}'\right| \leq \left|\bar{\Sigma}'_\alpha\right| \leq \left|\Sigma_\alpha^{\lim}\right|$$

Figure 9.13 displays this lower bound of the uniaxial strength domain for the strength ratio of $\rho = 6$ and $v = 10\%$. The angle α^* indicates the point where the fiber stress excess reaches its critical value $\delta\sigma/\sigma_0^m = \bar{\rho}$:

$$\alpha^* = \arcsin \sqrt{\frac{2}{3} - \frac{v\bar{\rho}}{3} \sqrt{\frac{4}{(v\bar{\rho})^2 + 1}}}$$

For $\alpha \leq \alpha^* \Leftrightarrow \delta\sigma/\sigma_0^m = \bar{\rho}$, the lower bound solution predicts that the fiber stress excess is at its maximum admissible value; and for $\alpha > \alpha^* \Leftrightarrow \delta\sigma/\sigma_0^m < \bar{\rho}$, it is the inverse. For the given data set, $\alpha^* = 37.377$.

$$\left|\bar{\Sigma}'_\alpha\right| = \max \left[\min \begin{pmatrix} \left(1 - \left(1 - \frac{3}{2}\sin^2\alpha\right)^2\right)^{-1/2} \\ \hline v\bar{\rho}\left|1 - \frac{3}{2}\sin^2\alpha\right| + \sqrt{1 + (v\bar{\rho})^2 \left[\left(1 - \frac{3}{2}\sin^2\alpha\right)^2 - 1\right]} \end{pmatrix}, 1 \right]$$

[10]Note that

$$\begin{aligned} \mathbf{e}_f \cdot \mathbf{t} \otimes \mathbf{t} \cdot \mathbf{e}_f &= (\mathbf{t} \cdot \mathbf{e}_f)(\mathbf{e}_f \cdot \mathbf{t}) \\ &= \cos\alpha'(\mathbf{t} \cdot \mathbf{e}_f) \cos\alpha''(\mathbf{e}_f \cdot \mathbf{t}) \\ &= \cos(\pi/2 - \alpha(\mathbf{e}_f \cdot \mathbf{n})) \cos(\alpha(\mathbf{e}_f \cdot \mathbf{n}) - \pi/2) \\ &= \sin^2 \alpha(\mathbf{e}_f, \mathbf{n}) \end{aligned}$$

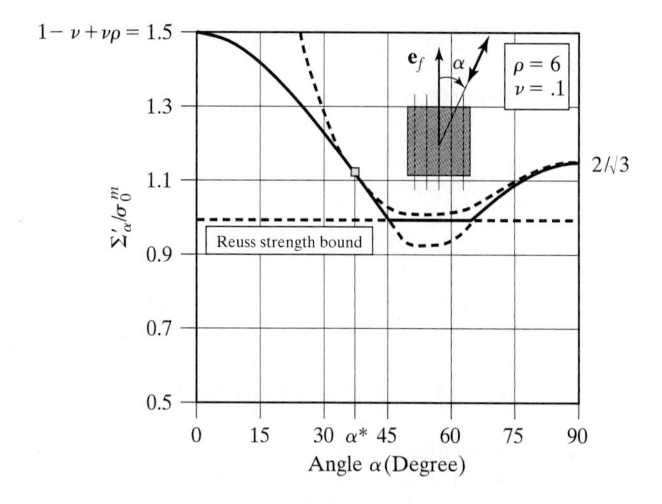

FIGURE 9.13: Lower bound of the macroscopic uniaxial tensile–compressive strength domain of a unidirectionally reinforced Von–Mises matrix.

■

9.5.4 Refinement with Interface Strength Criterion

The improved lower bound is based on the assumption of a perfect interface between matrix and fibers. Often, however, it is precisely in this interface transition zone where defects and flaws resulting from the manufacturing process restrict the strength, leading to debonding and delamination failure of the composite material. In the case of pure debonding, the shear stress $\tau = \mathbf{e}_f \cdot \mathbf{t}(\mathbf{n}_f)$ at the interface is limited to a maximum shear strength value k_{int}:

$$\mathbf{t}(\mathbf{n}_f) \in D_{\text{int}} \Leftrightarrow f_{\text{int}}(\mathbf{t}(\mathbf{n}_f)) = |\tau| - k_{\text{int}} \leq 0 \tag{9.81}$$

In return, delamination can be associated with a tension cutoff criterion, which limits the normal component $\sigma = \mathbf{n}_f \cdot \mathbf{t}(\mathbf{n}_f)$ of stress vector $\mathbf{t}(\mathbf{n}_f)$ acting upon the interface to a prescribed value σ_0^{int}:

$$\mathbf{t}(\mathbf{n}_f) \in D_{\text{int}} \Leftrightarrow f(\mathbf{t}(\mathbf{n}_f)) = \sigma - \sigma_0^{\text{int}} \leq 0 \tag{9.82}$$

Finally, in order to be statically admissible, the stress vector $\mathbf{t}(\mathbf{n}_f)$ must be continuous over the interface:

$$[[\mathbf{t}(\mathbf{n}_f)]] = (\boldsymbol{\sigma}_m - \boldsymbol{\sigma}_f) \cdot \mathbf{n}_f = 0 \tag{9.83}$$

A lower bound of the strength domain refers to the weakest link of the material system composed of matrix, fiber, and interface:

$$F\left(\boldsymbol{\Sigma}'\right) = \max\left[f_m\left(\boldsymbol{\sigma}'_m\right); f_f\left(\boldsymbol{\sigma}'_f\right), f_{\text{int}}(\mathbf{t}'(\mathbf{n}_f))\right] \leq 0 \tag{9.84}$$

Exercise 89. We consider again the composite composed of a Von–Mises matrix reinforced by unidirectionally oriented fibers. The stress fields are defined by (9.78). Show the effect of, respectively, an interface shear strength criterion and a delamination strength criterion on the uniaxial macroscopic strength domain of the composite.

For the stress field (9.78), and a uniaxial macroscopic stress state, the stress components $\tau' = \mathbf{e}_f \cdot \mathbf{t}(\mathbf{n}_f)$ and $\sigma = \mathbf{n}_f \cdot \mathbf{t}(\mathbf{n}_f)$ read

$$\tau' = \mathbf{e}_f \cdot \left(\mathbf{\Sigma}' - v\delta\sigma \mathbf{e}_f \otimes \mathbf{e}_f\right) \cdot \mathbf{n}_f = \mathbf{e}_f \cdot \mathbf{\Sigma}' \cdot \mathbf{n}_f = \frac{1}{2}\Sigma' \sin 2\alpha$$

$$\sigma' = \mathbf{n}_f \cdot \left(\mathbf{\Sigma}' - v\delta\sigma \mathbf{e}_f \otimes \mathbf{e}_f\right) \cdot \mathbf{n}_f = \mathbf{n}_f \cdot \mathbf{\Sigma}' \cdot \mathbf{n}_f = \Sigma' \sin^2 \alpha$$

where $\alpha = \alpha(\mathbf{e}_f, \mathbf{n})$ is still the angle between the fiber orientation and the direction of load application. Remarkably, but not surprising, we find that the fiber stress excess $\delta\sigma$ does not enter the interface criteria (9.81) and (9.82). Therefore, the interface strength criteria are additional restrictions on the uniaxial strength domain of fiber-reinforced composite materials:

$$\mathbf{t}'(\mathbf{n}_f) \in D_{\text{int}} \Leftrightarrow \left\{ \begin{array}{c} f(\mathbf{t}'(\mathbf{n}_f)) = |\Sigma' \sin 2\alpha| - 2k_{\text{int}} \leq 0 \\ f(\mathbf{t}'(\mathbf{n}_f)) = \Sigma' \sin^2 \alpha - \sigma_0^{\text{int}} \leq 0 \end{array} \right\}$$

Figure 9.14 illustrates the effect of the shear strength interface criterion on the uniaxial strength domain of the previously considered fiber-reinforced composite material ($\rho = 6$, $v = 10\%$), for $2k_{\text{int}}/\sigma_0^m = 1$. Figure 9.15 shows the same for the tension cutoff criterion, for $\sigma_0^{\text{int}}/\sigma_0^m = 1$. As suggested by physical evidence, and displayed in Figure 9.15, the delamination strength criterion only affects the uniaxial tensile strength domain.

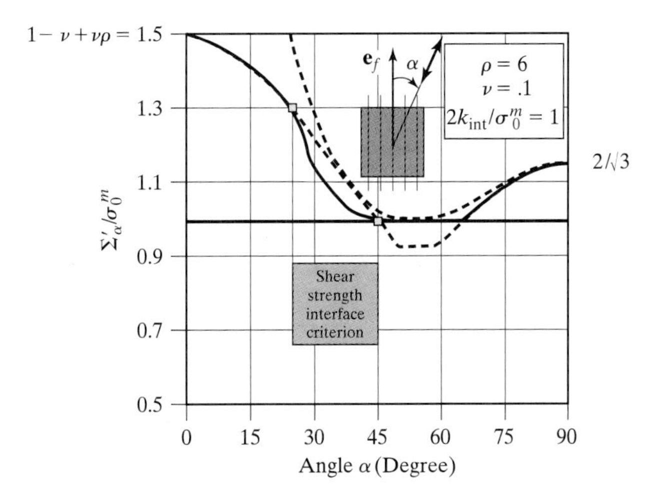

FIGURE 9.14: Effect of a shear strength interface criterion on the macroscopic uniaxial strength domain of a unidirectionally reinforced Von–Mises composite.

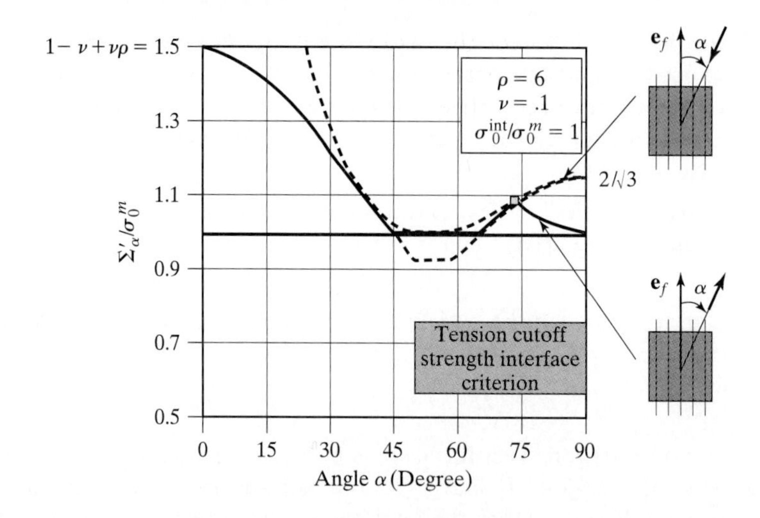

FIGURE 9.15: Effect of a tension cutoff interface criterion on the uniaxial tensile strength domain of a unidirectionally fiber reinforced Von–Mises composite.

■

Remark The strength domain of composite materials is restricted by the strength of the composite components and their interface. Albeit a lower bound, the developed estimates are useful tools to investigate the failure mechanisms of a specific composite material. For instance, in the case of a Von–Mises matrix reinforced by fibers, an experimentally determined tensile–compressive strength in the direction of the fiber orientation $\alpha = 0$, which is below the composite strength, that is,

$$\alpha = 0 : \max |\Sigma'_{\text{exp}}| < |\Sigma_0^{\text{lim}}| = \sigma_0^m(1-v) + v\sigma_0^f$$

indicates that a shear debonding mechanism at the interface between matrix and fiber affects the overall uniaxial strength behavior. In return, if the tensile strength normal to the fiber direction is lower than the composite strength, in combination with a pronounced asymmetric behavior in compression (superscript $-$) and tension (superscript $+$), that is,

$$\alpha = \pi/2 : \max \Sigma'^{+}_{\text{exp}} < \Sigma'^{+}_{\pi/2}; \quad |\max \Sigma'^{-}_{\text{exp}}| > \max \Sigma'^{+}_{\text{exp}}$$

a delamination mechanism at the interface is likely to affect the uniaxial strength domain. For a specific matrix behavior, the developed estimates give some indications where to search for critical failure mechanisms. This illustrates the beneficial use of limit analysis for engineering design and fine tailoring of fiber reinforced composite materials.

9.6 APPENDIX CHAPTER 9: DISSIPATION FUNCTIONS

The dissipation function for a Tresca material and a Von–Mises material were derived in this chapter from a combination of the strength criterion and the associated flow rule. Analogously, the dissipation functions of other plasticity models can be derived and found in standard textbooks devoted to limit analysis and yield design.[11] Some are presented in this appendix.

9.6.1 Tresca Criterion

See Section 4.4.1:

$$f(\boldsymbol{\sigma}) = \sigma_I - \sigma_{III} - \sigma_0 \leq 0$$

where $\sigma_I \geq \sigma_{II} \geq \sigma_{III}$ are principal stresses, and σ_0 = uniaxial strength. The corresponding dissipation functions read

$$\text{in } \Omega : \varphi(\mathbf{d}^p) = \left\{ \begin{array}{c} +\infty \text{ if } \operatorname{tr} \mathbf{d}^p \neq 0 \\ \frac{1}{2}\sigma_0 \left(|d_I^p| + |d_{II}^p| + |d_{III}^p|\right) \text{ if } \operatorname{tr} \mathbf{d}^p = 0 \end{array} \right\}$$

$$\text{on } \Sigma : \hat{\varphi}(\mathbf{V}^p) = \left\{ \begin{array}{c} +\infty \text{ if } |\mathbf{V}^p \cdot \mathbf{n}| \neq 0 \\ \frac{1}{2}\sigma_0 |\mathbf{V}^p| \text{ if } |\mathbf{V}^p \cdot \mathbf{n}| = 0 \end{array} \right\}$$

where $d_I^p \geq d_{II}^p \geq d_{III}^p$ are principal strain rates.

9.6.2 Von–Mises Criterion

See Section 8.1.3:

$$f(\boldsymbol{\sigma}) = \frac{3}{2}\mathbf{s} : \mathbf{s} - \sigma_0^2 \leq 0$$

where $\mathbf{s} = \boldsymbol{\sigma} - \sigma_m \mathbf{1}$, and σ_0 = uniaxial tensile strength. The corresponding dissipation functions read

$$\text{in } \Omega : \varphi(\mathbf{d}^p) = \left\{ \begin{array}{c} +\infty \text{ if } \operatorname{tr} \mathbf{d}^p \neq 0 \\ \sqrt{\frac{2}{3}}\sigma_0 \sqrt{\mathbf{d}^p : \mathbf{d}^p} \text{ if } \operatorname{tr} \mathbf{d}^p = 0 \end{array} \right\}$$

$$\text{on } \Sigma : \hat{\varphi}(\mathbf{V}^p) = \left\{ \begin{array}{c} +\infty \text{ if } |\mathbf{V}^p \cdot \mathbf{n}| \neq 0 \\ \sqrt{\frac{1}{3}}\sigma_0 |\mathbf{V}^p| \text{ if } |\mathbf{V}^p \cdot \mathbf{n}| = 0 \end{array} \right\}$$

9.6.3 Mohr–Coulomb Criterion

See Section 4.4.2:

$$f(\boldsymbol{\sigma}) = \sigma_I (1 + \sin\varphi) - \sigma_{III}(1 - \sin\varphi) - 2c\cos\varphi \leq 0$$

where c is the cohesion, and φ the friction angle, related to the uniaxial tensile strength and compression strength, f_t' and f_c', by

$$f_t' = \frac{2c\cos\varphi}{1 + \sin\varphi}; \quad f_c' = \frac{2c\cos\varphi}{1 - \sin\varphi}$$

[11] For instance, see

Salençon, J. (1983). *Calcul à la Rupture et Analyse Limite*, Presse de l'Ecole Nationale des Ponts et Chaussées, Paris, France.

Chen, W. F., and Han, D. J. (1988). *Plasticity for Structural Engineers*, Springer-Verlag, New York.

The corresponding dissipation functions read

$$\text{in } \Omega : \varphi(\mathbf{d}^p) = \left\{ \begin{array}{c} +\infty \text{ if } \operatorname{tr} \mathbf{d}^p < (|d_I^p| + |d_{II}^p| + |d_{III}^p|) \sin\varphi \\ \rho \operatorname{tr} \mathbf{d}^p \text{ if } \operatorname{tr} \mathbf{d}^p \geq (|d_I^p| + |d_{II}^p| + |d_{III}^p|) \sin\varphi \end{array} \right\}$$

$$\text{on } \Sigma : \hat{\varphi}(\mathbf{V}^p) = \left\{ \begin{array}{c} +\infty \text{ if } \mathbf{V}^p \cdot \mathbf{n} < |\mathbf{V}^p| \sin\varphi \\ \rho(\mathbf{V}^p \cdot \mathbf{n}) \text{ if } \mathbf{V}^p \cdot \mathbf{n} \geq |\mathbf{V}^p| \sin\varphi \end{array} \right\}$$

where $\rho = c \cot\varphi$ is the cohesive pressure.

9.6.4 Drucker–Prager Criterion

See Section 8.1.4:

$$f(\boldsymbol{\sigma}) = \sqrt{J_2} + \alpha(\sigma_m - \rho) \leq 0$$

where $\sqrt{J_2} = \sqrt{\frac{1}{2}\mathbf{s} : \mathbf{s}}$; α is the friction coefficient, and $\rho = c/\alpha$ is the cohesion pressure, related to the uniaxial tensile strength and compression strength, f_t' and f_c', by

$$f_c' = \rho \frac{3\alpha}{\sqrt{3} - \alpha}; \quad f_t' = \rho \frac{3\alpha}{\sqrt{3} + \alpha}$$

The corresponding dissipation functions read

$$\text{in } \Omega : \varphi(\mathbf{d}^p) = \left\{ \begin{array}{c} +\infty \text{ if } \operatorname{tr} \mathbf{d}^p < \sqrt{2}\alpha\sqrt{\mathbf{d}^p : \mathbf{d}^p - \frac{1}{3}(\operatorname{tr} \mathbf{d}^p)^2} \\ \rho \operatorname{tr} \mathbf{d}^p \text{ if } \operatorname{tr} \mathbf{d}^p \geq \sqrt{2}\alpha\sqrt{\mathbf{d}^p : \mathbf{d}^p - \frac{1}{3}(\operatorname{tr} \mathbf{d}^p)^2} \end{array} \right\}$$

$$\text{on } \Sigma : \hat{\varphi}(\mathbf{V}^p) = \left\{ \begin{array}{c} +\infty \text{ if } \mathbf{V}^p \cdot \mathbf{n} < |\mathbf{V}^p| \sin\varphi \\ \rho(\mathbf{V}^p \cdot \mathbf{n}) \text{ if } \mathbf{V}^p \cdot \mathbf{n} \geq |\mathbf{V}^p| \sin\varphi \end{array} \right\}$$

where φ is the friction angle related to the friction coefficient α by

$$\alpha = \frac{3 \sin\varphi}{\sqrt{3(3 + \sin^2\varphi)}}; \quad \sin\varphi = \alpha\sqrt{\left(\frac{3}{3 - \alpha^2}\right)}$$

9.7 PROBLEM SET: SECTION STRENGTH FOR COMBINED BENDING AND AXIAL FORCE

In Problem Section 4.7, we developed a section strength criterion for a rectangular section $A = 4a^2$, composed of a Tresca material (tensile strength σ_0) subjected to combined bending and axial load. The approach we employed was based on statically admissible stresses that satisfy locally the Tresca criterion. Hence, in the context of yield design, the found solution constitutes a lower bound. The aim of this exercise is twofold: (1) to improve the lower bound, and (2) to complete the analysis by an upper bound. Consider the beam displayed in Figure 9.16a, clamped at $x = 0$, and subjected at $x = L$ to a normal force $\mathcal{N}^d(L) = N\mathbf{e}_x$ and a bending moment $\mathcal{M}^d = M\mathbf{e}_z$.

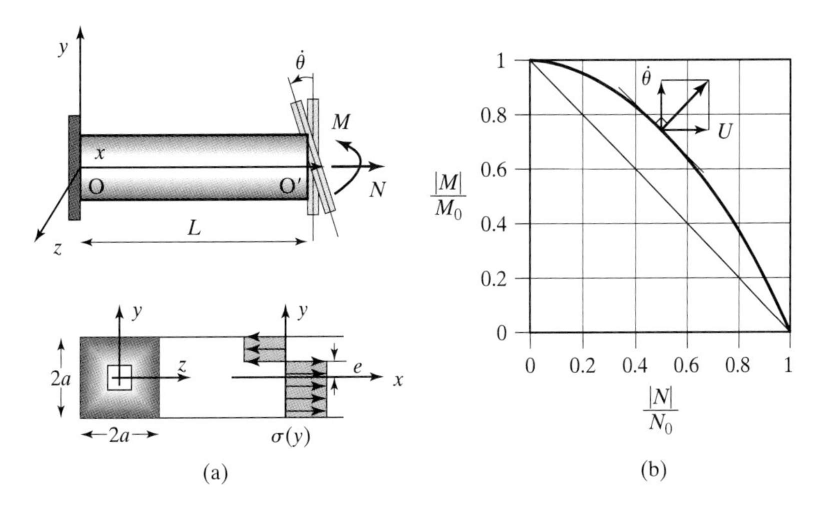

FIGURE 9.16: Combined bending and axial force strength domain of a rectangular beam: (a) structure with statically and plastically admissible stress field; (b) improved section strength domain.

1. **Lower Bound**: Consider a block stress field $\boldsymbol{\sigma}'(x)$, which is everywhere in the section at the strength limit under combined bending and axial force. For this stress field, develop an improved lower bound of the section strength criterion developed in Problem Section 4.7.
2. **Upper Bound**:

   (a) **Pure Bending**: Consider first the pure bending case. The velocity field we consider is of the form[12]

   $$\text{in } \Omega : \mathbf{V}' = -\dot{\theta} y \frac{x}{L} \mathbf{e}_x + \frac{\dot{\theta}}{2L} \left(x^2 + \alpha (y^2 - z^2) \right) \mathbf{e}_y + \alpha \frac{\dot{\theta}}{L} yz \mathbf{e}_z$$

   where $\dot{\theta}$ denotes the rotation rate around the z-axis. Develop an upper bound of the maximum plastic moment the section can sustain.

   (b) **Combined Bending and Axial Force**: By an appropriate modification of the previous velocity field, develop an upper bound of the strength domain for combined bending and axial force. Compare the result with the lower bound. Conclude.

9.7.1 Improved Lower Bound

The stress field we consider is of the form

$$\boldsymbol{\sigma}'(\mathbf{x}) = |\sigma| \operatorname{sign}(e - y)\, \mathbf{e}_x \otimes \mathbf{e}_x$$

[12]Note that the velocity field is of a similar form as the displacement field of the beam in elastic bending we studied in Problem Section 5.7. Indeed, it suffices to replace coefficient α by the Poisson ratio ν, to obtain the elastic velocity field.

$|e| \leq a$ is the section coordinate, where the uniaxial stress changes its sign. With the same arguments developed in Problem Section 4.7, the block stress state is statically admissible. It is plastically admissible provided $|\sigma| \leq \sigma_0$. The normal force and moment (around z-axis) read

$$N' = \int_{A=4a^2} |\sigma| \operatorname{sign}(e-y)\, da = 4\,|\sigma|\, ae$$

$$M' = -\int_{A=4a^2} y\,|\sigma| \operatorname{sign}(e-y)\, da = 2a^3\,|\sigma| \left(1-(e/a)^2\right)$$

In order to be plastically admissible, the generalized forces must separately satisfy

$$|N'| = 4\sigma_0 a^2\,|e/a| \leq N_0 = 4\sigma_0 a^2$$

$$|M'| = 2a^3\sigma_0\left(1-(e/a)^2\right) \leq M_0 = 2a^3\sigma_0$$

Finally, a combination of the two expressions delivers the following lower bound strength criterion of a rectangular section:

$$f(N', M') = \frac{|M'|}{M_0} + \left(\frac{N'}{N_0}\right)^2 - 1 \leq 0$$

A comparison with the strength domain developed in Problem Section 4.7 is displayed in Figure 9.16b. It shows that the found solution is an improved lower bound.

9.7.2 Upper Bound

Pure Bending.

The velocity field $\mathbf{V}'$ is kinematically admissible, since it satisfies the zero velocity boundary condition at $x = 0$:

$$x = 0 : \mathbf{V}^d \cdot \mathbf{e}_x = 0$$

All other surfaces are part of the stress boundary.

The associated strain rate components read (see Problem Section 6.10):

$$d'_{xx} = -\frac{\dot{\theta} y}{L};\quad d'_{yy} = d'_{zz} = \alpha\frac{\dot{\theta} y}{L};\quad d'_{xy} = d'_{yz} = d'_{xz} = 0$$

The absence of shear components indicates that the strain rate components are principal strains.

For a Tresca criterion, we need to satisfy the incompressibility criterion, that is,

$$\operatorname{tr}\mathbf{d}' = \operatorname{tr}\mathbf{d}^{p\prime} = \frac{\dot{\theta} y}{L}(2\alpha - 1) = 0 \rightarrow \alpha = 1/2$$

Use in the maximum dissipation capacity of a Tresca material gives

$$\varphi'\left(\mathbf{d}^{p\prime} = \mathbf{d}'\right) = \sigma_0 \left|\frac{\dot{\theta} y}{L}\right|$$

Integration delivers the maximum dissipation capacity of the beam section:

$$\frac{d\mathcal{D}}{dt}\left(\dot{\theta}\right) = \frac{\sigma_0}{L}|\dot{\theta}| \int_{\Omega} |y| \, d\Omega = 2a^3\sigma_0|\dot{\theta}|$$

The beam will collapse if the externally supplied work rate, $\mathbf{Q}^{\text{lim}} \cdot \mathbf{q}' = M^{\text{lim}}\dot{\theta}$, is greater than the maximum dissipation capacity, that is, for

$$\mathbf{Q}^{\text{lim}} \cdot \mathbf{q}' = M^{\text{lim}}\dot{\theta} \geq \sigma_0 \left|\dot{\theta}\right| 2a^3$$

This leads to the following upper bound of the admissible bending moment the beam can support:

$$\left|M^{\text{lim}}\right| \leq M_0 = 2a^3\sigma_0$$

This is the same value provided by the lower bound for $N' = 0$; thus, $\left|M^{\text{lim}}\right| = 2a^3\sigma_0$.

Combined Bending: Axial Force.

The additional velocity field we consider is the field we considered throughout Chapter 1 to study the deformation of axial deformation:

$$\text{in } \Omega : \mathbf{V}' = \mathbf{V}'\left(\dot{\theta}, \alpha\right) + \frac{U}{L}\left[x\mathbf{e}_x - \alpha\left(y\mathbf{e}_y + z\mathbf{e}_z\right)\right]$$

where $U = \mathbf{V}'(L) \cdot \mathbf{e}_x$ is the axial velocity at $x = L$, in the center of the section, $y = z = 0$. The corresponding strain rate reads

$$d'_{xx} = -\frac{\dot{\theta}y}{L} + \frac{U}{L}; \quad d'_{yy} = d'_{zz} = \alpha\left(\frac{\dot{\theta}y}{L} - \frac{U}{L}\right); \quad d'_{xy} = d'_{yz} = d'_{xz} = 0$$

The incompressibility condition of the Tresca material delivers

$$\operatorname{tr}\mathbf{d}' = \operatorname{tr}\mathbf{d}^{p\prime} = \frac{1}{L}\left(\dot{\theta}y - U\right)(2\alpha - 1) = 0$$

This incompressibility condition is satisfied for

$$\forall\alpha; \; y = e = U/\dot{\theta}$$

$$\forall\,|y| \leq a; \; \alpha = 1/2$$

Given these conditions, the dissipation rate reads

$$\varphi'\left(\mathbf{d}^{p\prime} = \mathbf{d}'\right) = \frac{\sigma_0}{L}|\dot{\theta}||e - y|$$

Integration gives

$$\frac{d\mathcal{D}}{dt}\left(\dot{\theta}, U\right) = 2a\sigma_0|\dot{\theta}| \int_{-a}^{+a} |e - y| dy = 2a\sigma_0\left(a^2 + e^2\right)|\dot{\theta}| = M_0|\dot{\theta}| + N_0\frac{U^2}{2a|\dot{\theta}|}$$

where $M_0 = 2a^3\sigma_0$, and $N_0 = 4a^2\sigma_0$. The structure certainly collapses if the external work rate $\mathbf{Q}^{\lim} \cdot \mathbf{q}' = M^{\lim}\dot{\theta} + N^{\lim}U$ is greater than the maximum affordable dissipation $\frac{d\mathcal{D}}{dt}\left(\dot{\theta}, U\right)$, that is, for

$$\mathbf{Q}^{\lim} \cdot \mathbf{q}' - \frac{d\mathcal{D}}{dt}\left(\dot{\theta}, U\right) = M^{\lim}\dot{\theta} + N^{\lim}U - M_0|\dot{\theta}| - N_0\frac{U^2}{2a|\dot{\theta}|} \geq 0$$

Hence, the minimum of all upper bounds for the given velocity field is obtained from minimizing the previous expression with regard to the two unknowns, $\dot{\theta}$ and U:

$$\begin{aligned}
\frac{\partial}{\partial\dot{\theta}}\left(\mathbf{Q}^{\lim} \cdot \mathbf{q}' - \frac{d\mathcal{D}}{dt}\left(\dot{\theta}, U\right)\right) &= M^{\lim} - M_0\left[1 - \frac{N_0 a}{2M_0}\left(\frac{e}{a}\right)^2\right]\operatorname{sign}\left(\dot{\theta}\right) = 0 \\
\frac{\partial}{\partial U}\left(\mathbf{Q}^{\lim} \cdot \mathbf{q}' - \frac{d\mathcal{D}}{dt}\left(\dot{\theta}, U\right)\right) &= N^{\lim} - N_0\frac{|e|}{a} = 0
\end{aligned}$$

A combination of these two relations gives

$$\frac{\left|M^{\lim}\right|}{M_0} + \left(\frac{N^{\lim}}{N_0}\right)^2 - 1 = 0$$

for which $\mathbf{Q}^{\lim} \cdot \mathbf{q}' - \frac{d\mathcal{D}}{dt}\left(\dot{\theta}, U\right) = 0$. This criterion coincides with the lower bound section criterion found in Section 9.7.1. Therefore, the strength criterion for a rectangular section subjected to combined bending and axial forces reads

$$f(N, M) = \frac{|M|}{M_0} + \left(\frac{N}{N_0}\right)^2 - 1 \leq 0$$

9.8 PROBLEM SET: DESIGN THICKNESS OF A PRESSURE VESSEL

The focus of this problem set is the engineering design of a pressure vessel composed of a thin wall cylinder and a calotte of internal radius R, as sketched in Figure 9.17. The vessel is subjected to a known maximum service pressure p. The aim of the design is the determination of the smallest thickness $\min e$ of the vessel wall to sustain this pressure.

1. **Lower Bound**:

   (a) Design Stresses: We assume that the wall thickness e is small in comparison with the tube radius R, such that the stress components $\sigma_{\theta\theta}$ and σ_{zz} in a section can be assumed constant. Under this assumption, show that

   $$\sigma_{\theta\theta} = \frac{pR}{e}; \quad \sigma_{zz} = \frac{pR}{2e}$$

   (b) The pressure vessel is made of a material of the Von–Mises type, for which the stresses are restricted by the yield criterion:

   $$f(\boldsymbol{\sigma}) = \sqrt{\frac{3}{2}\mathbf{s} : \mathbf{s}} - \sigma_0 \leq 0$$

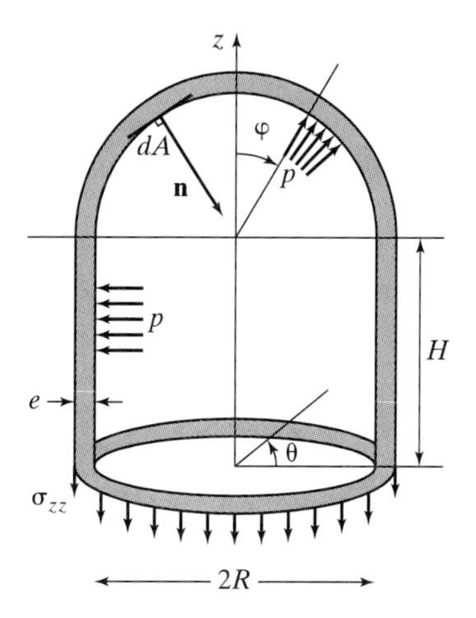

FIGURE 9.17: Design problem: thickness of a pressure vessel.

where $\mathbf{s}$ is the stress deviator, and σ_0 the uniaxial yield strength. By assuming that the wall thickness e was very small compared to the radius, show that a 'safe' lower bound of the design thickness is:

$$\frac{\sqrt{3}}{2}\frac{p}{\sigma_0} = \frac{e^+}{R} \geq \frac{e}{R}$$

2. **Upper Bound**: We consider a radial velocity field in both the cylinder wall and the (half-sphere) calotte:

$$\begin{array}{rl} \text{in the cylinder wall:} & \mathbf{V}' = v(r)\mathbf{e}_r \\ \text{in the calotte:} & \mathbf{V}' = u(r)\mathbf{e}_r \end{array}$$

   (a) For a Von–Mises material at plastic collapse, determine functions $v(r)$ and $u(r)$ close to a multiplying constant.

   (b) Determine the maximum dissipation capacity the entire structure can afford. Distinguish the volumetric dissipation rate, and any surface dissipation rate the structure may develop.

   (c) By assuming that the wall thickness e was very small compared to the radius, show that the upper limit theorem delivers a second design thickness of the form:

$$\frac{e}{R} \geq \frac{e^-}{R} = \frac{e^+}{R} \times \mathcal{F}(...)$$

   where $\mathcal{F}(...)$ is a function of some relevant parameters to be determined. By means of a detailed yield design reasoning, determine the relevant extremum of this function.

   (d) By comparing the lower and the upper design thickness, conclude on the relevant dissipative mechanism in the structure.

Remark The infinitesimal surface and volume of a sphere is:

$$dA = r^2 \sin\varphi d\varphi d\theta; \quad dV = r^2 dr \sin\varphi d\varphi d\theta$$

9.8.1 Lower Bound

Design Stresses.

The hoop stress for a thin wall cylinder was derived in Chapter 3, Section 3.5.3., from the projection of all stresses acting on a cylinder in the $\mathbf{e}_y$-direction:

$$-2e\langle\sigma_{\theta\theta}\rangle + \int_A -p\mathbf{n}\cdot\mathbf{e}_y da = 0$$

where $\langle\sigma_{\theta\theta}\rangle$ is the average hoop stress in the cylinder wall, and $\mathbf{n} = -\mathbf{e}_r = -[\cos\theta\mathbf{e}_x + \sin\theta\mathbf{e}_y]$ is the unit normal to the inner surface of the cylinder. Thus, for an assumed constant stress distribution, for which $\langle\sigma_{\theta\theta}\rangle = \sigma_{\theta\theta}$:

$$2e\sigma_{\theta\theta} = p\int_0^{\pi} \sin\theta R d\theta = 2pR \rightarrow \sigma_{\theta\theta} = \frac{pR}{e}$$

Analogously, projection of the stresses in the $\mathbf{e}_z$-direction (see Figure 9.17) reads:

$$-2\pi e R\sigma_{zz} + \int_\Sigma -p\mathbf{n}\cdot\mathbf{e}_z dA = 0$$

where $dA = R^2\sin\varphi d\varphi d\theta$ is the infinitesimal surface dA in spherical coordinates oriented by unit normal $\mathbf{n}$, for which $-\mathbf{n}\cdot\mathbf{e}_z = \mathbf{e}_r\cdot\mathbf{e}_z = \cos\varphi$. Thus,

$$2\pi e R\sigma_{zz} = p\int_0^{2\pi} d\theta \int_0^{\pi/2} \cos\varphi\sin\varphi R^2 d\varphi = \pi R^2 p \rightarrow \sigma_{zz} = \frac{pR}{2e}$$

Lower Limit Theorem.

We first note that the statically admissible stresses $\sigma_{\theta\theta} = pR/e$ and $\sigma_{zz} = pR/2e = \sigma_{\theta\theta}/2$ are much greater than the maximum radial stress $|\max\sigma_{rr}| = p$ for a thin wall vessel $R/e \gg 1$. Therefore, neglecting the radial stress, the Von–Mises criterion reads here:

$$\sqrt{\frac{3}{2}\mathbf{s}':\mathbf{s}'} = \sqrt{3J_2'} = \sqrt{3\times\frac{1}{6}\left[(\sigma'_{\theta\theta} - \sigma'_{zz})^2 + \sigma'^2_{\theta\theta} + \sigma'^2_{zz}\right]} = \frac{\sqrt{3}}{2}\frac{pR}{e} \le \sigma_0$$

The lower limit theorem implies:

$$p^+ = \frac{2\sigma_0 e}{\sqrt{3}R} \le p^{\lim}$$

Thus, a safe lower bound to the design thickness is:

$$\frac{e^+}{R} \le \frac{\sqrt{3}}{2}\frac{p^{\lim}}{\sigma_0}$$

9.8.2 Upper Bound

Zero Volume Dilatation Condition.

We consider a radial velocity field in both the cylinder wall:

$$\mathbf{V}' = v\,(r)\,\mathbf{e}_r$$

and the calotte:

$$\mathbf{V}' = u(r)\mathbf{e}_r$$

The strain in the cylinder wall is:

$$d'_{rr} = \frac{\partial v}{\partial r}; \quad d'_{\theta\theta} = \frac{v}{r}$$

and in the calotte:

$$d'_{rr} = \frac{\partial u}{\partial r}; \quad d'_{\theta\theta} = d_{\varphi\varphi} = \frac{u(r)}{r}$$

For the Von–Mises material, we need to ensure the zero dilatation condition, so that the dissipation rate remains finite. This means for the cylinder:

$$\operatorname{tr}\mathbf{d}' = \frac{\partial v}{\partial r} + \frac{v}{r} = 0 \Rightarrow v\,(r) = \frac{C_1}{r}$$

and for the calotte:

$$\operatorname{tr}\mathbf{d}' = \frac{\partial u}{\partial r} + 2\frac{u(r)}{r} = 0 \Rightarrow u\,(r) = \frac{C_2}{r^2}$$

where $C_1 = v(R)R = V'R > 0$ and $C_2 = u(R)R^2 = U'R^2 > 0$ are two integration constants, related to the radial velocity of the inner cylinder wall V' and of the inner calotte wall U'. Note that there is no reason that V' and U' coincide.

The associated strain rates read, in the cylinder:

$$\mathbf{d}' = \mathbf{d}^{p\prime} = -\frac{V'R}{r^2}\begin{pmatrix} 1 & & \\ & -1 & \\ & & 0 \end{pmatrix}$$

and in the calotte:

$$\mathbf{d}' = \mathbf{d}^{p\prime} = -\frac{U'R^2}{r^3}\begin{pmatrix} 2 & & \\ & -1 & \\ & & -1 \end{pmatrix}$$

Maximum Dissipation Capacity.

Volume dissipation capacity: The maximum local volumetric dissipation capacity of the Von–Mises material is in the cylinder:

$$\varphi\left(\mathbf{d}^{p\prime}\right) = \sigma_0\sqrt{\frac{2}{3}}\sqrt{\mathbf{d}^{p\prime}:\mathbf{d}^{p\prime}} = \frac{2V'R}{\sqrt{3}r^2}\sigma_0$$

and in the calotte:

$$\varphi\left(\mathbf{d}^{p\prime}\right) = \sigma_0\sqrt{\frac{2}{3}}\sqrt{\mathbf{d}^{p\prime}:\mathbf{d}^{p\prime}} = \frac{2U'R^2}{r^3}\sigma_0$$

Integration (cylinder: $dV = rdrd\theta dz$; sphere: $dV = r^2 dr\sin\varphi d\varphi d\theta$) delivers:

$$\begin{aligned} \text{in } \Omega: \frac{d\mathcal{D}}{dt}\left(\mathbf{V}'\right) &= \frac{4\pi V'}{\sqrt{3}}\sigma_0 HR\int_R^{R+e}\frac{dr}{r} + 4\pi U'R^2\sigma_0\int_0^{\pi}\int_R^{R+e}\frac{dr}{r}\sin\varphi d\varphi \\ &= \left(\frac{V'}{\sqrt{3}}\frac{H}{R} + 2U'\right)4\pi\sigma_0 R^2\ln\left(1+\frac{e}{R}\right) \simeq \left(\frac{V'}{\sqrt{3}}\frac{H}{R} + 2U'\right)4\pi\sigma_0 R^2\frac{e}{R} \end{aligned}$$

where we developed the 'ln' in a power function.[13]

Surface Dissipation Capacity: In addition, there is a surface dissipation because of the velocity jump in the joint between the cylinder and the calotte:

$$\text{on } d\Sigma = rdrd\theta;\ |\mathbf{V}^p\cdot\mathbf{e}_z| = 0 : \hat{\varphi}\left(\mathbf{V}^{p\prime} = [[\mathbf{V}']]\right) = \sqrt{\frac{1}{3}}\sigma_0\left|\mathbf{V}^{p\prime}\right| = \sqrt{\frac{1}{3}}\sigma_0\left|V'\frac{R}{r} - U'\frac{R^2}{r^2}\right|$$

Integration gives:

$$\begin{aligned} \frac{d\mathcal{D}}{dt}\left([[\mathbf{V}']] = \mathbf{V}^p\right) &= \int_{\theta=0}^{\theta=2\pi}\int_{r=R}^{r=R+e}\sqrt{\frac{1}{3}}\sigma_0\left|V'\frac{R}{r} - U'\frac{R^2}{r^2}\right| rdrd\theta \\ &= \sqrt{\frac{1}{3}}\sigma_0 2\pi\int_{r=R}^{r=R+e}\sqrt{\frac{1}{3}}\sigma_0 R\left|V' - U'\frac{R}{r}\right| dr \\ &= \sqrt{\frac{1}{3}}\sigma_0 2\pi R^2 V'\int_1^{1+\varepsilon}\left|1 - \frac{\upsilon}{\rho}\right| d\rho \geq 0 \end{aligned}$$

where $\rho = r/R$, $\varepsilon = e/R$, and $\upsilon = U'/V'$.

---

[13] Note that

$$\ln(1+\varepsilon) = \varepsilon - \frac{1}{2}\varepsilon^2 + \frac{1}{3}\varepsilon^3 - \frac{1}{4}\varepsilon^4 + O\left(\varepsilon^5\right)$$

We have a look on the integral:

$$\int_1^{1+\varepsilon} \left|1 - \frac{\upsilon}{\rho}\right| d\rho = \text{sig}\left(1 - \frac{\upsilon}{1+\varepsilon}\right)(1 + \varepsilon + (\ln \upsilon - \ln(1+\varepsilon))\upsilon)$$
$$+\text{sig}(-1+\upsilon)(1 + (\ln \upsilon)\upsilon)$$
$$= \left[\text{sig}\left(1 - \frac{\upsilon}{1+\varepsilon}\right) - \text{sig}(1-\upsilon)\right](1 + (\ln \upsilon)\upsilon)$$
$$+\text{sig}\left(1 - \frac{\upsilon}{1+\varepsilon}\right)(\varepsilon - \ln(1+\varepsilon)\upsilon)$$

which simplifies for $\varepsilon \ll 1$ to:

$$\int_1^{1+\varepsilon} \left|1 - \frac{\upsilon}{\rho}\right| d\rho \simeq \varepsilon |1 - \upsilon|$$

The surface dissipation, therefore, reads:

$$\text{on } \Sigma : \frac{d\mathcal{D}}{dt}\Big([[\mathbf{V}']] = \mathbf{V}^p \Big) \simeq \sqrt{\frac{1}{3}}\sigma_0 2\pi R^2 V' |1 - \upsilon| \frac{e}{R}$$

Upper Bound Theorem.

The total dissipation capacity, therefore, reads:

$$\frac{d\mathcal{D}}{dt}(\mathbf{V}', [[\mathbf{V}']]) = \sigma_0 2\pi R^2 V' \left[\left(\frac{2}{\sqrt{3}}\frac{H}{R} + 4\upsilon\right) + \sqrt{\frac{1}{3}}|1-\upsilon|\right]\frac{e}{R}$$

The structure certainly fails if this dissipation capacity is greater than the external work rate, which reads here:

in the cylinder ($\mathbf{T}^d = p\mathbf{e}_r$; $\mathbf{V}'(r = R) = \frac{C_1}{R}\mathbf{e}_r$; $da = Rd\theta dz$):

$$\frac{dW^{ext}}{dt} = \int_{\partial\Omega = 2\pi RH} \mathbf{T}^d \cdot \mathbf{V}' da = 2\pi p V' RH$$

in the calotte ($\mathbf{T}^d = p\mathbf{e}_r$; $\mathbf{V}'(r = R) = \frac{C_2}{R^2}\mathbf{e}_r$; $dA = R^2 \sin\varphi d\varphi d\theta$):

$$\frac{dW^{ext}}{dt} = \int_{\partial\Omega} \mathbf{T}^d \cdot \mathbf{V}' dA = 4\pi p U' R^2$$

Thus, the structure certainly fails if:

$$\mathbf{Q}^{\text{lim}} \cdot \mathbf{q}' = 2\pi R^2 p V' \left(\frac{H}{R} + 2\upsilon\right) \geq \sigma_0 2\pi R^2 V' \left[\frac{2}{\sqrt{3}}\frac{H}{R} + 4\upsilon + \sqrt{\frac{1}{3}}|1-\upsilon|\right]\frac{e}{R}$$

and after rearrangement:

$$\text{Collapse if} : \frac{e}{R} \leq \frac{e^-}{R} = \frac{\sqrt{3}p^{\text{lim}}}{2\sigma_0}\mathcal{F}\left(\upsilon, \frac{R}{H}\right)$$

where:

$$\mathcal{F}\left(v, \frac{R}{H}\right) = \frac{1 + 2v\frac{R}{H}}{1 + 2\sqrt{3}v\frac{R}{H} + \frac{1}{2}\left|1 - v\right|\frac{R}{H}}$$

Given that $v \geq 0$, the maximum of function $\mathcal{F}(v, R/H)$ is 1, which gives the highest value of e^-, for which the structure may eventually fail. Hence, the design thickness obtained from the upper limit theorem is:

$$\frac{e}{R} \geq \frac{e^-}{R} = \frac{\sqrt{3}p^{\text{lim}}}{2\sigma_0}$$

Conclusion.

This upper bound design thickness e^- coincides with the lower bound design thickness e^+, which therefore is the actual solution of the design problem:

$$\frac{e^+}{R} = \frac{e}{R} = \frac{\sqrt{3}p^{\text{lim}}}{2\sigma_0} = \frac{e^-}{R}$$

Further Reading Volume I

SELECTED TEXTBOOKS

Ashby, M. F., Jones, D. R. H. (1994). *Engineering Materials 1 & 2*, Pergamon Press, Oxford, UK.

Barenblatt, G. I. (1996). *Scaling, Self-Similarity, and Intermediate Asymptotics*, Cambridge Texts in Applied Mathematics, Cambridge University Press, Cambridge, UK.

Bathe, K.-J. (1996). *Finite Element Procedures*, Prentice Hall, Upper Saddle River, NJ.

Bažant, Z. P., Cedolin, L. (1991). *Stability of Structures*, Oxford University Press, Oxford, UK.

Chen, W. F., Han, D. J. (1988). *Plasticity for Structural Engineers*, Springer-Verlag, New York.

Coussy, O. (1995). *Mechanics of Porous Continua*, John Wiley & Sons, Chichester, UK.

de Groot, S. R., Mazur, P. (1983). *Non-Equilibrium Thermodynamics*, Dover Publications, New York.

Francois, D., Pineau, A., Zaoui, A. (1995). *Comportemenent Mécanique des Matériaux. Elasticité et Plasticité*, Hermes, Paris, France.

Germain, P. (1973). *Cours de Mécanique des Milieux Continus. Tome 1—Théorie Générale*, Masson et Cie, Paris, France.

Halphen, B., Salençon, J. (1987). *Elasto-plasticité*, Presse de l'Ecole Nationale des Ponts et Chaussées, Paris, France.

Huerre, P. (1997). *Mécanique des Fluides. Tome I. Cours*, Département de Mécanique, Ecole Polytechnique, France.

Jirasek, M, Bažant, Z.P. (2001). *Inelastic Analysis of Structures*, John Wiley & Sons, Chichester, UK.

Lemaitre, J., Chaboche, J.-L. (1998). *Mechanics of Solid Materials*, Cambridge University Press, Cambridge, UK.

Mandel, J. (1978). *Propriétés Mécaniques des Matériaux*, Eyrolles, Paris, France.

Mang, H., Hofstetter, G. (2000). *Festigkeitslehre*, Springer-Verlag, Wien, Austria.

Salençon, J. (1983). *Calcul à la Rupture et Analyse Limite*, Presse de l'Ecole Nationale des Ponts et Chaussées, Paris, France.

Salençon, J. (1988). *Mécanique des Milieux Continus.* Vols. I and II, Ellipses, Paris, France.

Sonin, A. A. (1997). *The Physical Basis of Dimensional Analysis*, Department of Mechanical Engineering, MIT, Cambridge, MA, 52 pages.

Suresh, S. (1998). *Fatigue of Materials*, 2nd Edition, Cambridge University Press, Cambridge, UK.

Truesdell, C. (Ed.) (1972). *Mechanics of Solids II*, Encyclopedia of Physics, Vol. VIa/2, Springer-Verlag, Berlin, Germany.

SELECTED ARTICLES AND REPORTS

Borodich, F. M. (1998). "Similarity Methods in Hertz Contact Problems and Their Relations with the Meyer Hardness Test." *Technical Report TR/MAT/FMB/98-98*, Glasgow Caledonian University, UK.

Buckingham, E. (1914). "On Physically Similar Systems. Illustrations of the Use of Dimensional Analysis." *Physical Review*, Vol. 4, 345–376.

de Buhan, P., Salençon, J. (1987). "Yield Strength of Reinforced Soils as Anisotropic Media." *IUTAM Symposium on Yielding, Damage and Fracture of Anisotropic Solids*, Edited. by J. P. Boehler, Mech. Eng. Publ., London, 791–803.

de Buhan, P., Taliercio, A. (1991). "A Homogenization Approach to the Yield Strength of Composite Materials." *Eur. J. Mech.*, A/Solids, Vol. 10, No. 2, 129–154.

Drucker, D.C. (1953). "Limit Analysis of Two and Three Dimensional Soil Mechanics Problems." *J. Mech. Phys. Solids*, Vol. 1, No. 4, 217–226.

Halsey, T. C., Levine, A. J. (1998). "How Sandcastles Fall." *Physical Review Letters*, Vol. 80, No. 14, 3141–3144.

Salençon, J. (1990). "An Introduction to the Yield Design Theory and its Application to Soil Mechanics." *Eur. J. Mech.*, A/Solids, Vol. 9, No. 5, 477–500.

Taylor, D. W. (1937). "Stability of Earth Slopes." *J. Boston Soc. Civil Engineers*, Vol. 24, No. 3, 337–386.

Index

1D-think model
  creep hesitancy, 258
  cyclic hardening plasticity, 237
  elasticity, 113
  friction element, 210
  frozen energy, 221
  hardening plasticity, 218
  ideal plasticity, 212
  variational method
    displacement-based, 158
    stress-based, 165
  viscoplasticity, 231

Base dimensions, 126, 141
Beam
  2-Layer, 207
  complementary energy, 171
  elastic bending, 146
  finite bending deformation, 22
  forces and moments, 63
  hypothesis
    infinitesimal deformation, 26
    navier–bernoulli, 29
  microflexural structures, 202
  section strength criterion, 104, 350, 368
  statically admissible stresses, 100
Bending modulus bounds, 202
Boundary conditions, 128
  displacement, 128
  regular displacement, 155
  regular stress, 154
  stress, 128
  velocity at failure, 322
Bounds
  modulus
    bending, 207
    bulk, 173
    shear, 200
    Young's, 179
  upper and lower
    elasticity, 171
    yield limit, 340
  Voigt–Reuss
    elasticity, 180
    strength, 355
Boussinesqu problem
  Poisson ratio, 125

refined analysis, 140
bulk modulus
energy bounds, 163, 170, 173
hydrostatic test, 121

Cam–Clay plasticity, 289
flow rule, 291
hardening rule, 292
plasticity criterion, 289
thermodynamic consistency, 294
Champaign method, 310
Clapeyron's formula, 172, 174, 192, 197, 204
Clausius–Duhem inequality
1D-hardening plasticity, 227
1D-ideal plasticity, 215
3D-hardening plasticity, 278
structural level, 322
Cohesion, 87, 272
pressure, 272, 368
complementary evolution law
1D-hardening plasticity, 228
1D-ideal plasticity, 216
3D-hardening plasticity, 281
cyclic plasticity, 240
normal dissipative mechanism, 236
Composite cylinder model, 200
Consistency condition
1D-hardening plasticity, 220
1D-ideal plasticity, 211
3D-hardening plasticity, 283
3D-ideal plasticity, 266
Contact conditions
example, 133
frictionless contact, 128
perfect adhesion, 128
Convexity
complementary energy, 168
definition function, 157
dissipation potential, 235
elasticity domain
1D-hardening plasticity, 226
1D-ideal plasticity, 214
3D-hardening plasticity, 282
3D-ideal plasticity, 266
free energy, 161
safe load domain (lower bound), 329
Creep, 231
hesitancy, 258
plastic, 232, 263
Critical states, 289–297
Cam–Clay, 289
plasticity, 294
viscoplasticity, 297

Deformation
finite, 2
gradient, 3
infinitesimal, 30
Deviator plane, 268, 269
Dilatation
linear
finite deformation theory, 14
linear theory, 33
plastic, 271, 272
coefficient, 273
thermal
coefficient, 117, 217, 281
coefficient tensor, 119
isotropic coefficient, 123
Dimension function, 141
Dimensional analysis
Boussinesq problem, 125, 140
Displacement, 16
based variational method, 160
gradient, 31
strain-displacement relation, 16
Dissipation
capacity, 322, 335
functions, 366
intrinsic, 215
maximum plastic, 226
maximum rate, 335
Von-Mises material, 336
non-negativity, 276
plastic, 276
rate at failure, 328
Distortion
equivalent plastic, 270, 271, 286
finite deformation theory, 14, 15
linear theory, 33
Divergence theorem, 64

generalized, 324
Driving force
hardening deformation, 228, 279
plastic deformation, 214, 216, 279
strain rate, 326
total deformation, 213
velocity jump, 327
viscoplastic deformation, 235
Drucker–Prager plasticity, 271
dissipation function, 368
flow rule, 272
isotropic hardening, 286
kinematic hardening, 287
yield surface, 271

Elastic stiffness
4th order tensor, 119
isothermal and adiabatic, 118
isotropic tensor, 124
Elastic yield limit, 136
Elasticity, 113
1D-thermodynamics, 115
1D-think Model, 113
direct solving methods, 127
displacement method, 129
stress method, 131
isotropic
compliance tensor, 124
material constants, 121, 124
stiffness tensor, 124
stability requirement, 124
stiffness tensor, 119
uniqueness of solution, 162
Energy
complementary, 156, 165, 279
definition, 156
free, 116, 156, 215, 278, 321
isotropic linear thermoelasticity, 121
linear thermoelasticity, 118
linear thermoelastoplasticity, 279
frozen, 222, 224, 226, 278
Helmholtz, 116, 215, 278
potential, 156
strain, 152
Excavation pit
block failure, 341, 343
external work rate, 323
lower bound, 333
strength approach, 90
tension cut-off criterion, 344
upper bound, 347
Exponent matrix, 141

Finite element method, 165
problem set, 182
Flow rule, 211, 220, 267, 281
associated, 274, 326
Cam–Clay, 290–294
plasticity, 291
viscoplasticity, 296
Drucker–Prager plasticity, 272
non-associated, 274, 282
section type specific, 351
Von-Mises plasticity, 269
Forces
body, 57
surfaces, 57
Friction
angle, 87, 105, 272, 367
coefficient, 272, 368
element, 210

Gravity dam, 182

Hardening
3-truss analogy, 245
force
1D, 219, 223, 225
3D, 279
Cam–Clay plasticity, 289
isotropic
1D, 229
3D, 284
kinematic
1D, 229
3D, 285
modulus
1D, 220
1D energy interpretation, 225
3D, 283
Cam–Clay plasticity, 294
plastic compatibility, 219

plasticity
1D, 218
plasticity models, 283
positive, 220
variables, 285–287, 292
Hardening rule, 220, 281
Cam–Clay, 292–296
plasticity, 292
viscoplasticity, 296
Drucker–Prager, 286–287
isotropic hardening, 286
kinematic hardening, 287
non-associated, 282
Heat
adiabatic conditions, 117
adiabatic dissipation, 218
dissipation, 218
latent heat of deformation, 118
latent heat of elastic deformation, 218
volume heat capacity, 117, 217, 281
Heterogeneous material system
bending modulus bounds, 202
bulk modulus bounds, 163, 170, 172
Hill lemma, 153
torsion modulus bounds, 194
Young's modulus bounds, 173
Hill lemma, 153, 155, 355
Hollow sphere
lower limit theorem, 329
upper limit theorem, 337
Homogenization, 154
Hydrostatic axis, 77, 268, 272
Hypothesis
action–reaction "law", 59
continuity, 3
infinitesimal deformation, 30, 127
local contact forces, 56
maximum plastic work, 281
small displacements, 127
small perturbations, 127

Ideal
1D-elastoplasticity, 212
1D-plasticity, 210
3D-plasticity models, 265
plasticity, 220
Invariants
strain, 120
stress, 76, 267
Isotropic
3D-plasticity criterion, 267
behavior, 77
complementary energy, 169
free energy, 162
plasticity models, 265
thermoelasticity, 120

Jacobian of deformation
finite theory, 6
linear theory, 32

Kinematic approach from outside
plastic hinge design, 350
upper limit theorem, 334
yield design, 323
Kinematically Admissible
displacement field, 45, 161
velocity field, 323
Kinematics
plastic flow, 291, 322
Kronecker delta, 32
Kuhn–Tucker conditions, 211, 219, 267

Lamé constants, 121
Law
the first, 115, 215, 278
the second, 115, 215, 278
Legendre–Fenchel transform, 166, 279
Lemma
Hill, 153, 355
tetrahedron, 60
Limit analysis, 321
Load-bearing capacity, 322
Loading function
1D-hardening plasticity, 219
1D-ideal plasticity, 211
3D-hardening plasticity, 283
3D-ideal plasticity, 266
Cam–Clay model, 289

cyclic plasticity, 237
Drucker–Prager, 272
isotropic hardening, 286
kinematic hardening, 287
Von-Mises, 268
Loading rate, 231
Lower limit theorem, 327
heterogeneous material system, 355
structural elements, 350

Mandel's model, 258
Mass conservation, 58
Material laws
necessity of, 112
Material surface
oriented, 8, 32
Material vector, 4, 32
Maxwell symmetry
3D-plasticity, 279
thermoelasticity, 117, 120
Micro-hardness, 297
Micro-indentation, 297
Micromechanics, 153
Mohr plane
strain space, 39
stress space, 79
Mohr–Coulomb criterion
deviator plane, 272
dissipation function, 367
dry friction, 108
effective friction, 110
strength criterion, 87
triaxial test, 97
Momentum balance, 56
cylinder coordinates, 70
discrete material points, 57
equation of motion, 64
symmetry of stress tensor, 67
Morphology
Composite cylinder model, 201

Normal dissipative mechanism, 235
Normality rule, 274, 326
1D-hardening plasticity, 226
1D-ideal plasticity, 214
hardening plasticity, 282
surface of discontinuity, 327

Operator
Cartesian coordinates, 50
curl, 36
cylinder coordinates, 51
differential, 50
divergence, 32
gradient, 5
gradient (linear theory), 32
spherical coordinates, 52
Orthogonality properties, 10
Overstress, 232

Physical linearization
1D-ideal thermoelastoplasticity, 217
1D-thermoelasticity, 117
3D-thermoelasticity, 118
Pi-theorem, 126, 141
Plastic
collapse, 322
collapse load, 322
creep, 232, 263
hinge, 350
hinge discontinuity, 351
multiplier, 326
1D-hardening plasticity, 220
1D-ideal plasticity, 211
3D-hardening plasticity, 283
3D-ideal plasticity, 267
Plastically admissible stresses, 266
Lower limit theorem, 328
Plasticity
1D-energy approach, 210
cyclic, 236
models, 265
strain-based formulation, 218
thermodynamics
1D-hardening plasticity, 226
1D-ideal plasticity, 215
3D-hardening plasticity, 277
Poisson ratio, 124
dimensional analysis, 125
Polar decomposition
finite deformation, 13
linear theory, 35
Potential

dissipation, 366
elasticity, 113
hardening, 293
plastic, 274
viscoplastic dissipation, 236
Pressure vessel formula
elasticity solution, 136
stress approach, 68
Principle of maximum plastic work
1D-hardening plasticity, 225
1D-ideal plasticity, 214
3D-hardening plasticity, 282
3D-ideal plasticity, 277
limit analysis, 326
structural elements, 351
structural level, 329
upper limit theorem, 335

Ratchetting, 245
Rate effects, 230, 234
Reduction formulae, 63, 72
Relaxation
hardening force, 260, 264
plastic stress, 233
time, 233, 263
Representative elementary volume, 3, 154
Rigid body motion, 9, 31

Scale
macroscopic, 3
microscopic, 153
Shear
deformation
single and double shear, 37
single shear, 17
modulus
from triaxial test, 122
torsion bounds, 194
Softening, 220
Spontaneous evolution, 216
Stéfani model, 237
State equations
1D-ideal thermoelastoplasticity, 217
1D-thermoelasticity, 117
3D-thermoelasticity, 119
incremental, 287
inversion, 279
isotropic thermoelasticity, 121
thermoelastoplasticity, 278
State variables
1D-hardening plasticity, 226
1D-thermoelastoplasticity, 216
3D-hardening plasticity, 278
choice of hardening variable, 223
cyclic plasticity, 239
external, 116, 216, 278
internal, 216, 278
Static approach from inside
lower limit theorem, 332
structural elements, 348
yield design, 322
Statically admissible stresses
definition, 74
fiber-reinforced composite, 358
lower limit theorem, 328
stress-based variational method, 168
Strain measures, 9
Cauchy dilatation, 10
Green–Lagrange, 12
linearized strain tensor, 33
strain gage rosette, 42
Strength criterion, 84
combined bending and axial forces, 103, 368
fiber-reinforced composite materials, 354
interface, 364
Mohr–Coulomb, 87
pressure dependent, 271
section-type, 349
shear, 267
tension cut-off, 84
Tresca, 85
Stress, 56
based variational method, 168
Cauchy stress tensor, 59
concentration factor, 222
deviator, 77
hydrostatic, 81
invariants, 76, 267
octrahedral, 79, 268

plane, 83
principal, 68
simple shear, 83
statically admissible, 74
uniaxial tension, 82
Stress vector
components, 75
deviator plane, 268, 272
normal contact force assumption, 57
principal stresses, 76
Structural elements
limit analysis, 348
variational methods, 171
Surface of discontinuity, 324
dissipation rate, 341
lower limit theorem, 331
upper limit theorem, 339
System of unit, 126

Tangent modulus
1D-hardening plasticity, 220
3D-hardening plasticity, 287
Tension cut-off
excavation pit, 344
strength criterion, 84
Tensor notation, 20
scalar product, 21
tensor product, 21
vector product, 21
Tetrahedron lemma, 60
Theorem
dynamic
moment, 59
resultant, 59
limit
lower, 329
upper, 337
minimum
complementary energy, 169
potential energy, 161
superposition, 129
virtual work, 151, 152
Thermodynamics
1D-hardening plasticity, 226
1D-ideal plasticity, 215
1D-thermoelasticity, 115
1D-viscoplasticity, 234
3D-hardening plasticity, 277
driving force of dissipation, 216
irreversible processes, 278
Linear 3D-thermoelasticity, 118
restriction, 282
Thermoelasticity, 112
Torsion
and extension, 48
deformation, 44
elasticity bounds, 194
thin wall cylinder in plasticity, 302
Transport formulae
elementary volume
finite deformation, 6
linear theory, 32
finite deformation, 6
linearization, 32
oriented material surface
finite deformation, 8
linear theory, 32
Tresca criterion
beam section strength criterion, 103
dissipation function, 367
excavation pit, 92
moment section strength, 350
strength criterion, 85
Triaxial test
Mohr–Coulomb criterion, 97
shear modulus, 122
stresses, 94, 95

Upper limit theorem, 334
heterogeneous material system, 355

Variational methods, 151
displacement based, 158
stress based, 165
Velocity jump, 326
Virtual work theorem, 151
plastic collapse, 323
Viscoplasticity
1D, 230
1D-thermodynamics, 234
Cam–Clay, 295

Viscosity, 231
Volume average
    bending stiffness, 205, 207
    bulk modulus, 164, 170, 173
    lame constants, 176
    stress and strain, 154
    torsion stiffness, 198, 199
    Young's modulus, 178
Von–Mises plasticity, 267
    dissipation function, 367
    flow rule, 269
    yield surface, 267

Work, 156
    external, 152, 321
    hardening, 282
    hardening–softening, 281
    internal, 151
    maximum plastic, 327
    plastic, 214, 276, 282
    rate, 326

Yield design, 321
Young's modulus, 124
    energy bounds, 173

Greek Alphabet and Transliteration

Name of Letter	Greek Alphabet		Transliteration
Alpha	Α	α	a
Beta	Β	β	b
Gamma	Γ	γ	g
Delta	Δ	δ ∂	d
Epsilon	Ε	ϵ	e
Zeta	Ζ	ζ	z
Eta	Η	η	ē
Theta	Θ	θ ϑ	th
Iota	Ι	ι	i
Kappa	Κ	κ	k
Lambda	Λ	λ	l
Mu	Μ	μ	m
Nu	Ν	ν	n
Xi	Ξ	ξ	x
Omicron	Ο	o	o
Pi	Π	π	p
Rho	Ρ	ρ	r; *initially*, rh
Sigma	Σ	σ s	s
Tau	Τ	τ	t
Upsilon	Υ	υ	u; *except after* a, e, ë, i, *often* y
Phi	Φ	ϕ φ	ph
Chi	Χ	χ	kh
Psi	Ψ	ψ	ps
Omega	Ω	ω	ō